D1155935

HISTORY OF TECHNOLOGY SERIES 1

Measuring instruments: tools of knowledge and control

Frontispiece: *Emperor Shah Jehan had this inlaid carved marble screen made, c1645, in the Red Fort, Delhi, India to show that his justice 'was weighed like a scale'. This symbolic use has been adopted throughout time.*

Measuring instruments: tools of knowledge and control

P.H.SYDENHAM, M.E., Ph.D., F.Inst.M.C., F.I.I.C., A.M.I.E. Aust.

Head and Professor
School of Electronic Engineering
South Australian Institute of Technology
Adelaide
Australia

PETER PEREGRINUS LTD.

in association with the

SCIENCE MUSEUM, LONDON

Published by: Peter Peregrinus Ltd., Stevenage, UK, and New York

British Library Cataloguing in Publication Data

Sydenham, P.H.
Measuring instruments — (History of technology).
1. Measuring instruments — History
I. Title II. Science Museum
III. Series
620'.0044 QC100.5 79-41298

ISBN 0-906048-19-2

Printed in England by A.Wheaton & Co., Ltd., Exeter

Contents

Foreword

If the rise of science and technology has been one of the major features of western civilisation, this rise has been acutely dependent on the development of scientific instruments with which to observe the physical universe. As de Broglie (who first postulated the wave properties of the electron) put it: 'Left to itself theoretical science would always tend to rest on its laurels; but experiment, by becoming continually more exact and delicate, has shown us more clearly each day that "there are more things in heaven and earth than are dreamt of in your philosophy" ' and so the scientific instrument is 'one of the essentials of intellectual progress'.

Too often scientific instruments have been taken for granted, and it has been assumed that their development and operation could be left to *Niebelungen* of an entirely lower order than the giants who formulate theory. But if de Broglie's appreciation is not enough, let us recall von Helmholtz who – for all his many contributions to physical theory – declared that he had often 'spent more intellectual effort in getting an instrument that was out of order to work properly, than he had in framing the theory which the instrument was designed to test'.

The history of scientific instruments is therefore an essential part of the history of western civilisation, and yet it has been even more neglected than the history of science itself. Historians in general are now becoming aware that they can no longer ignore the profound interaction of science and technology with all other aspects of society and civilisation: and historians of science in turn cannot afford to ignore the history of instruments and measurement. Even such a general social document as Magna Carta had as one of its clauses the establishment of a common system of measurement.

Dr. Sydenham has therefore set himself the task of filling an important gap by writing a history of measuring instruments, and for this he has two essential qualifications: enthusiasm for instrument design, and the achievement of having developed some notable instruments himself. He can therefore write with insight of the problems which have faced the instrument designers of the past, and the ways in which they have arrived at their solutions. It is a testimony both to him and to the University at Armidale that enthusiasm and understanding support, in the preparation of this work have so signally triumphed over geographical remoteness from the historical centres of instrument development.

R.V. Jones
June, 1979

Preface

Measuring instruments and related apparatus are the tools used by man to obtain knowledge and to assist control the quality of human existence. Since the earliest times measuring apparatus has played a dominant role in the growth of civilisation. This book traces the development of suitable ideas, and their practical implementation as measuring instruments, from classical antiquity through to the first half of the twentieth century.

It is my intention to show that the use of measuring instruments is a prime characteristic of man and that they, although appearing to be part of a vastly complex, incoherent and disordered array of ideas and implementations, are properly worthy of recognition as the hardware manifestations of more fundamental ideas and concepts that form a powerful and vital discipline called measurement science.

Measuring instruments have been seen for too long as mere by-products of the application needs of disciplines. The time has come at which man recognises that progress, and control of the status quo, depend on measurements to such an extent that study of the design and application of measuring apparatus should now be given greater emphasis than has been the case in the past.

The account shows how workers build on the experience of those who went beforehand, and how instruments steadily evolve rather than occur through revolutionary discovery. It is the applications, not the principles, of new instruments that are associated with such great changes as were, for instance, brought about by the introduction of telegraphy. Design skills and the availability of ideas progress steadily and relentlessly as each worker adds a little more to the accomplishments of man. Man's measuring accomplishments, the instruments, embody that which he has actually been able to achieve.

These artefacts of instrumentation portray's man's progress as powerfully and surely as do the architectural structures that have been interpreted as evidence of the growth of civilisation. Instruments are not the dreams but the realisations of dreams; they embody man's understanding of a scientific principle in as near perfect a form as is necessary or, in many cases, as was possible to implement in the given circumstances.

A main thread of the account is concerned with electrical method but, as is shown, these techniques form only a subset of the whole set of measurement ability.

Many books have been published on various historical aspects of measurement but none, as yet, has attempted to put all measurements into a single historical perspective. It has been my aim to provide a readable, scholarly work in which the presented material is accurate and reliable. It was not feasible in a single volume to tie all statements to the primary literature so a compromise has been struck in which, hopefully, reliable books and information about existing artefacts form the prime sources used. Much, however, must be left to the reader to interpret using the citations given.

As the first ever attempt to provide an authoritative prime account of the history of measurement instrumentation it will undoubtedly be weak in parts, and, perhaps, on occasions in error. Information taken from secondary sources has mostly been crosschecked using several accounts to provide accuracy; a process, however, that made it clear that many apparently clarified circumstances are not as well researched as one might believe. Naturally, I would be most pleased to learn of errors and to receive additional information of relevance.

This study developed out of a curiosity about the fundamental issues involved in measurement, such factors as how the instrument industry emerged; why is it that man has taken so long to realise that measurement can be achieved through a structured path of design and development and why is it that the general view is that transducers typify this field of endeavour rather than transducers being seen more correctly as the products of much more orderly, fundamental issues. I desired to know how the current situation developed and what were the trends and lessons to be learned from the vast experience amassed by those before us. This account does not answer all of these questions but does provide an ordered introduction into the literature and collections that would be needed for efficient study of such issues.

An understanding of the historical development of a subject provides the current generation with added confidence to proceed. Many times it is shown that major steps in confirmation of scientific knowledge were made with the crudest of apparatus. The important factor is that the measuring equipment leads the user to formulating useful knowledge, not just accumulating data lacking properly reasoned meaningful content.

This study developed from a small photographic exhibition assembled at the University of New England, Australia as part of a residential teaching course. From there it was enlarged into two papers published in the journal *Measurement & Control*, from which developed the structure of this considerably expanded account.

The book begins with two chapters that provide background for interpretation of the historical material that forms the bulk of the book. Chapter 1 is concerned with the intangible, nonmaterial aspects of measuring instruments, placing the more familiar hardware into perspective with applications and the body of knowledge. Chapter 2 deals with the tangible factors that restrict the growth of hardware creation and application, these being such factors as the nature of science and technology, design methods, inventiveness, availability of materials, processes, tools of manufacture and basic physical principles. Institutional mechanisms that encouraged the development, adoption and proper use of measuring instruments and the economic factors are also discussed.

With the background established, the development of hardware is then discussed beginning from the earliest times and covering until around 1950 AD. This period is covered in four chapters, each being progressively shorter in length as seemed appropriate, to cover progress of the various stages. Chapter 3 deals with the time from around 300 BC to 1500 AD. For that period it was possible to discuss the few variables in use. Electricity is briefly discussed providing background for its application in later times. Chapter 4 is about the first three centuries in which the new natural philosophy became the form of science to be practised until today. In this period optical and mechanical methods predominated. Chapter 5 covers the 19th century, the age wherein electrical method found great application after the voltaic cell, and electromagnetic methods became available to find extensive application in communication. To finish the chronological sequence the last chapter, Chapter 6, reviews the instrumentation advances and developments of the first 50 years of the twentieth century. The first of two appendixes provides detail of published biographical works relevant to persons concerned with instrument development and application. The second appendix provides a guide to the collections of instrument artefacts existing around the world.

Much of the pioneering work contributed by non-English-speaking nations is already well recorded in the English language literature and, therefore, becomes part of this account. It is, however, regrettable that this study is mainly concerned with the information that can be located in English-language sources. Perhaps this contribution will provide a better standpoint for a wider study to be made than would exist without its publication.

Acknowledgments

Owing to the need to present illustrations from early sources there were some instances where copyright ownership could not be traced. I would be pleased to hear from any person whose work has not been acknowledged accordingly. My sincere apologies are extended in such cases. The sources of all illustrations are given, first, in recognition of any copyright ownership but also as the sources can often provide the reader with additional detail.

A work of this nature has naturally involved the help of numerous persons over many years. Primarily, I must express appreciation to Professor R.V. Jones F.R.S., University of Aberdeen, for the interest that he has instilled in me since I first became interested in measuring instruments as a career many years ago. Furthermore, I am most grateful for the Foreword he has kindly contributed.

Likewise, my thanks go to Professor L. Finkelstein, The City University, London, for continually stimulating my awareness of the need 'to tidy up' the current situation regarding measurement science by ordering several of its important aspects, history being one.

Many museum staff have offered valuable aid, especially when I have needed a person on the spot to locate illustrations and offer advice about literature and dates. In the Science Museum, London, Mrs. A. McConnell and the staff of the Publications Department were my main sources of aid. At the National Museum of History and Science, Washington, DC it was Dr. B. Finn and his staff who provided the most assistance. I received considerable help, of one kind or another, in each of the Museums listed in Appendix 2.

Nearing the completion of the text I became aware of the biased slant toward British and Continental-European events. To balance this I was fortunate to obtain a short-term Visitor Smithsonian Institution Award that allowed me to spend several days with the staff of the National Museum of History and Technology in Washington DC. During that time I was able to make good use of the specialised library to expand the information adding more about the history of early American measurement science.

The Science Museum Library, London also proved to be invaluable when compiling the Appendix on biographical works.

I wish to acknowledge the prompt and generous assistance given me by Mr. W. Weinstein of the National Bureau of Standards in the USA.

Closer to home I am grateful for the skills of Mr. W. Webster and his staff at the Central Photographic Laboratory, University of New England for the high quality photographic copying that they have contributed using, in the main, books from my personal collection on the history of measurement.

Mr. M. Bone drew most of the line diagrams converting, very poor quality 'icons' into usable diagrams.

The work was typed by Ms. D. McGrady and Mrs. C. Drummond, whose skill at putting in the right word whenever my script was unreadable is to be highly commended.

I also wish to thank Mr. R. Luxton and staff of the University of New England Library for assistance with locations of material and checking of bibliographic references.

Mr. D. Rickard and Mr. A. Paine helped with the supply of early documents of George Kent and Leeds & Northrup. I am also grateful to all those organisations who supplied photographs and their permission to use them, these being given in the caption of the figures.

Providing the publisher with an acceptable typescript and illustrations is but a part of the whole task of publishing a book. I wish to thank Mr. T.D. Hills and the production staff of Peter Peregrinus Ltd. for their devotion and interest in this work. It was not possible to always provide first-class illustrative material, this being the reason for a few, less than perfect, illustrations.

This book is not the book to end all books on the topic but one to begin them. I hope other authors will find it useful and that it will encourage others to take up this subject.

Finally, I wish to acknowledge a debt to the many who have made contributions that I have used in preparing this work. I hope I have done justice to their endeavours and have helped the readers of this book to appreciate their work.

Peter H. Sydenham
University of New England
Armidale, N.S.W.
Australia
January, 1979

Explanation of source and acknowledgment for illustrations

Illustrations are of contemporary etchings, existent artefacts or reconstructions, where possible, to provide as factual an account as practicable. Illustrations provided may, however, be subject to artist's licence, lack of skill or misinterpretation.

Every attempt has been made to ascertain the state of copyright ownership for figures used from published sources; owing to the age of some texts extracted from this was not always established. The source used is referenced in each case to assist further studies. References refer to those given at rear of this work where complete detail is to be found of the work and its publisher. The Author would welcome notification of errors or omissions relating to this information.

Original numbering and lettering has, in several instances, been retained to preserve the original nature of the information. Such letters and captioning are generally not referred to in the text. Every reasonable attempt has been made to supply the publisher with high-quality photographic copies. In some cases it was necessary to make use of material of a lower standard as it was the only available.

Chapter 1

Measurements for knowledge and control

1.1 What is a measurement?

Man seeks to know everything there is to be known about every topic possible yet paradoxically the more a subject is investigated the greater is the realisation that our understanding of it is often based on phenomena we cannot explain. Explanation of radio communication, for example, relies on the use of what we term electromagnetic fields. Much is known about the effects of these fields but no appreciation exists of what is actually there. As another example scientists up to about 100 years ago were still debating the true nature of heat; was it really a substance that flowed or was it, as it was finally established, a property of energy of rapid motion?

Measurements, although one of man's oldest abilities, possess an equally elusive basic entity. Today man conducts gigantic numbers of measurements with machines called measuring instruments. They range in complexity from the simplest device, such as the eyeglass having a ruled scale, to the sophistication of the electron microscope like that shown in Fig. 1.1. The eyeglass provides visual gain of about x10, the electron microscope over x 1 000 000.

Time has enabled man to produce more sophisticated measuring tools but it has not yet allowed him to explain very much about the process itself.

In all measuring situations a similar basic mechanism appears to be implemented, one that defies complete understanding yet can be used to obtain knowledge about the whole variety of processes chosen for study.

This first Chapter investigates the philosophical aspect of measurements; the things that can be identified as occurring in all measurements regardless of the hardware form of implementation. Followers of the 'general systems theory' talk of the existence of something additional to the sum of the parts forming a system. Similarly in measurement it must be recognised that there is more to a measuring instrument, when it is in use, than the existence of what is physically observed as the apparatus itself.

Man relates in a quantitative way to the physical environment through connecting links set-up via measurements. Our senses couple us to whatever system interests us. An example situation is shown in Fig. 1.2. Once the link is established, either directly with unaided naturally given senses, or with the aid of measuring instruments that extended these capabilities, the observers' understanding of the subject of the link begins

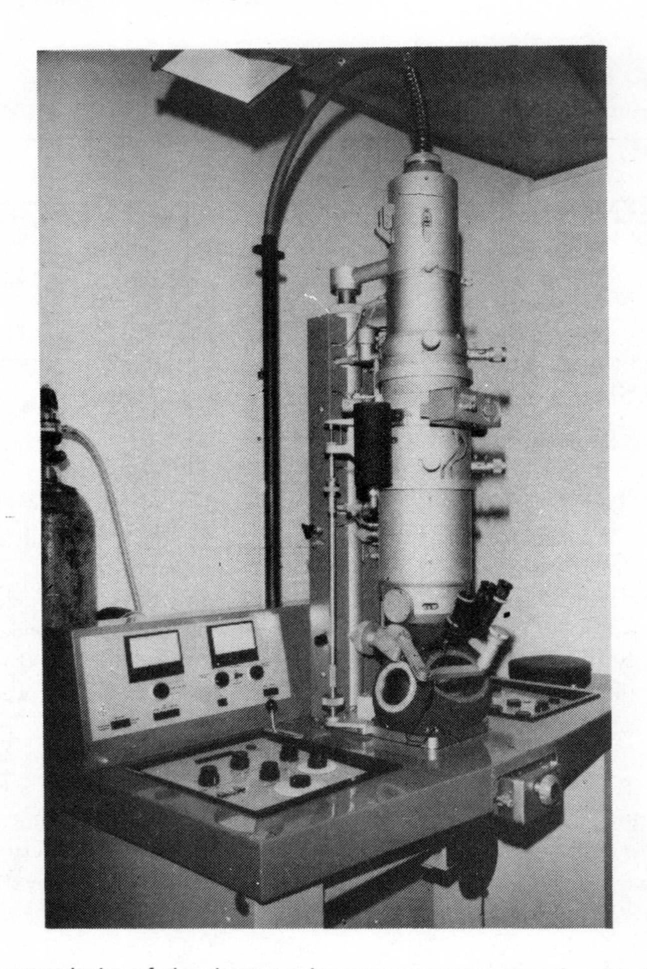

Fig. 1.1 *The complexity of the electron microscope contrasts strongly with that of a simple eyeglass as a means to increase visual sensitivity*
[Courtesy University of New England, N.S.W. Copyright Knatos Ltd., AEI Scientific Instruments, England]

to change in an irreversible manner. At some stage he may choose to replace himself with hardware that replaces his sensing function so that he is released. In this case the link is set to extract fixed information, usually for the purposes of gaining control of selected variables through engineered feedback control. This link, as well as acting as a communication channel, converts the basic information available within the system under study, and appearing at the input of the measuring stage, into knowledge sought by the observer connected to the output of that measuring stage. We can, therefore, justifiably regard measuring instruments as tools for converting information into knowledge; they are information-converting and communicating machines.

Although man has been creating measuring links in huge abundance for thousands of years it is only in the comparatively recent times of this 20th century that people have begun to quantitatively question what happens in a measurement, this being in the hope of discovering a scientific law that will enable machines to design the links. The general, and most common, attitude has been to define measurement in terms of the procedure used in practice and the physical result of that procedure—in a similar way, therefore, to how we handle the electromagnetic field situation.

A considerable part of the effort expended by philosophers has been on the study of the theory of knowledge, a subject entitled epistemology. It deals with the generation of the general conditions whereby knowledge is acquired. It is concerned with perception, the degree of certainty and the relationship between the measuring process taking place and the senses used to convey some entity to the observer. As yet no laws have emerged that would enable an engineer or scientist to construct a machine that would gain knowledge to the same degree of efficiency as physiological man. Indeed there is great difference of opinion on what knowledge is, as there is also about such related terms as information, cognition and recognition.

Evidence exists that early civilisations did give consideration to philosophical questions of measurement. The following extract from *The Dialogues of Plato* (Jowett, 1953) shows this:

> What measure is there of the relations of pleasure to pain other than excess and defect, which means that they become greater and smaller, and more and fewer, and differ in degree? For if any one says: 'Yes, Socrates, but immediate pleasure differs widely from future pleasure and pain' - to that I should reply: And do

Fig. 1.2 *Using ultrasonic radiation, this observer is able to use his eyes to see the cross-sectional composition of live meat. The measuring system provides the interface between observer and subject*

[Courtesy University of New England, N.S.W.]

> they differ in anything but pleasure and pain? There can be no other measure of them. And do you, like a skillful weigher, put into the balance the pleasures and the pains, and their nearness and distance, and weigh them, and then say which outweighs the other. If you weigh pleasures against pleasures, you of course take the more and greater; or if you weight pains against pains, you take the fewer and the less; or if pleasures against pains, then you choose that course of action in which the painful is exceeded by the pleasant, whether the distance by the near or the near by the distant; and you avoid that course of action in which the pleasant is exceeded by the painful. Would you not admit, my friends, that this is true?

A translated passage, from Aristotle's work (Africa, 1968), clearly brings out the, still common, situation showing that philosophical thought about measurement does not necessarily link up with the skills of the practising craftsman who finds out how to measure, and often measure well, without having the capability to explain why:

> We notice that the geometricians are quite unable to apply their scientific proofs in practice. When it comes to dividing a piece of land, or to any other operation on magnitudes and spaces, the surveyors can do it because of their experience, but those who are concerned with mathematics and with the reasons for these things, while they may know how it is to be done, cannot do it.

While in Padua, just before moving to Florence in 1610, Galileo Galilei (1564-1642) is said to have expressed his attitudes on the subject:

> Count what is countable, measure what is measurable, and what is not measurable, make measurable.

Much later in time this concept was dealt with in more detail in Lord Kelvin's famous statement, made to the Institution of Civil Engineers on 3 May, 1883. Lord Kelvin is shown with some of his famous instrument products in Fig. 1.3. Quoted in more length than we usually see reported he said:

> In physical science a first essential step in the direction of learning any subject is to find principles of numerical reckoning and methods for practicably measuring some quality connected with it. I often say that when you can measure what you are speaking about, and express it in numbers, you know something about it; but when you cannot measure it, when you cannot express it in numbers, your knowledge is of a meagre and unsatisfactory kind: it may be the beginning of knowledge, but you have scarcely, in your thoughts, advanced to the stage of science, whatever the matter may be.

Thus became established a paradigm of science that appears to have existed unquestioned until recent times. Whether it is a paradigm or not depends on how the entire passage is interpreted. Lord Kelvin did not state that the purpose of measurement is to obtain numerical formats only. There has, however, certainly been a preoccupation with number gathering since then. Today there exists a movement toward providing measuring output formats that better match our visual and aural senses than does the simple numerical form.

Fig. 1.3 *Portrait of Lord Kelvin (Sir William Thomson)*
[In possession of the Institution of Electrical Engineers, London]

Feinstein (1971), an authority on clinical epidemiology at Yale University, has written critically of what he terms the 'curse of Kelvin':

> One outdated paradigmatic concept – an extension of beliefs stated by Lord Kelvin – is the idea that scientific data must be expressed objectively in the form of dimensional measurements. This concept provided major enlightenment when it first became accepted as a paradigm; it has now led to major intellectual crises that remain unsolved by various *ad hoc* modifications of the basic paradigm; and is now being used to substitute for enlightened thought or to thwart it.

A reviewer of Papanek (1975) says this on the value of producing numbers to represent situations:

> Perhaps the oldest of man's intellectual myths is the one that links measurement with truth and justice. A modern manifestation of the myth is in the craze for quantisation. Let's pin numbers to everything because that makes grading easier! Whether the values so ranked are strictly commensurable, let alone mensurable, is a question too often brushed aside. When the urgent need is for a guide to swift decision, and nowadays that is most of the time, the reflexive preference is for a quantised 'scientific' approximation, even where a subjective hunch has as much or a better chance of being right.

Finkelstein's (1975) treatment shows that both symbols and numbers are valid forms of representations with symbols being the qualitative form, numbers the quantitative form, the latter being a more advanced kind of knowledge, that more properly termed measurement.

Today numbers can, and are being, produced by measuring instruments at a gigantic rate. Magnetic-storage tapes carrying measurement data that were compiled from past space explorations fill warehouses! Only a fraction of this information becomes knowledge. Never before has it been so easy to generate numbers. Unfortunately, man still has little appreciation of how best to convert this basic information into knowledge in an efficient manner that is based on adequate philosophical understanding of measurement.

Professor Heike Onnes, the man to first liquify helium gas, desired to have the words 'Door meten tot weten' (Dutch for 'through measurement to knowledge') placed over the doors of his Physics Laboratories at Leyden University, Holland in 1882. It is an easy statement to make, one that most people would agree with, but firm engineering procedures for creating an economic and efficient design from a firm knowledge-seeking basis still eludes people making measurements.

The above linguistic explanation does not lead the reader very far in understanding what constitutes a measurement. Few rules, aide-mémoires and generalities about good practice arise from such general thinking. To look deeper it is necessary to create models of the measurement process. Three different models are now given. They are certainly different yet each describes the same process; each explains different aspects of the process. Presumably there is a single model that incorporates all aspects presented but at this time no unified theory has been suggested that accounts for all features. But first some explanation on how ideas are conveyed.

1.2 Representing ideas

There exists four principal means by which ideas are conveyed between people practising engineering and science: words, icons, mathematical models and real models. The first makes use of word descriptions, the linguistic approach. Verbal transfer provides a communication media for specific and persuasive transfer of ideas. It is often (but not always) the first stage of understanding of a subject. It can be a most expressive method but rarely serves the sciences as the sole method.

Next follows the formation of ideas to a stage where pictorial representation is possible. In the engineering and scientific disciplines these pictures, called icons, provide a symbolic representation of the subject. Icons do not usually present enough information to enable physical construction. They convey an idea in ways not possible without recourse to very lengthy word descriptions. Several forms used in science and engineering are given in Fig. 1.4. The iconic stage expresses an idea in generality without necessarily giving all of the specific quantities by which it can be realised.

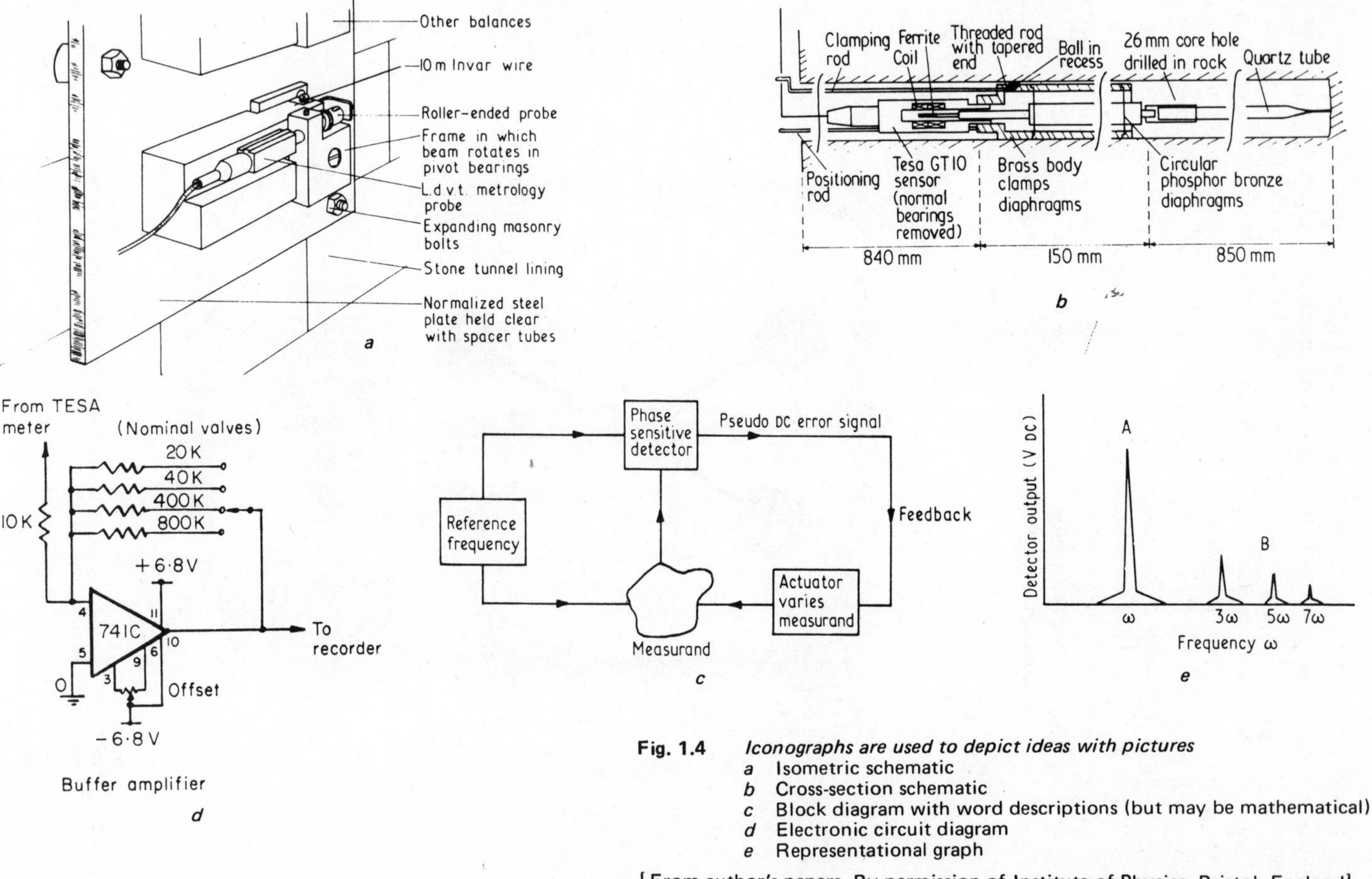

Fig. 1.4 *Iconographs are used to depict ideas with pictures*
a Isometric schematic
b Cross-section schematic
c Block diagram with word descriptions (but may be mathematical)
d Electronic circuit diagram
e Representational graph

[From author's papers. By permission of Institute of Physics, Bristol, England]

A third form of representation of physical apparatus is that in which actual physical models are made which incorporate features of operation and construction. As an example, 19th-century US patent law required scaled-down models of inventions to be lodged with the patent application. Fig. 1.5 is one of these submitted in connection with a claim of 1855. As another example, common practice when manufacturing an instrument to be marketed in large numbers, is the development of the prototype instrument which is close to, but not exactly the same as, the final product sold on the market. Ray and Ray (1974) have studied the place of models in invention.

Each of the above three methods of conveying ideas can be used to progressively move a design toward a specific physical construction. Generality must eventually give way to the concrete. At some stage, following either the linguistic, iconic or physically

Fig. 1.5 *Models can be used to represent an idea as does this patent model of Thomas Silver's marine steam-engine governor, 1855*

[Courtesy National Museum of History and Technology, Smithsonian Institution, Washington DC Photo 69390]

model description, the designer wishing to use scientific methods of investigation to best advantage will seek to create a mathematical model of the subject. This involves establishing relevant mathematical equations that describe the behaviour of the interconnected system blocks operating within given numerical boundaries of its parameters. The procedure for arriving at the mathematical model of a physical system is termed system identification. The procedure is illustrated in Fig. 1.6.

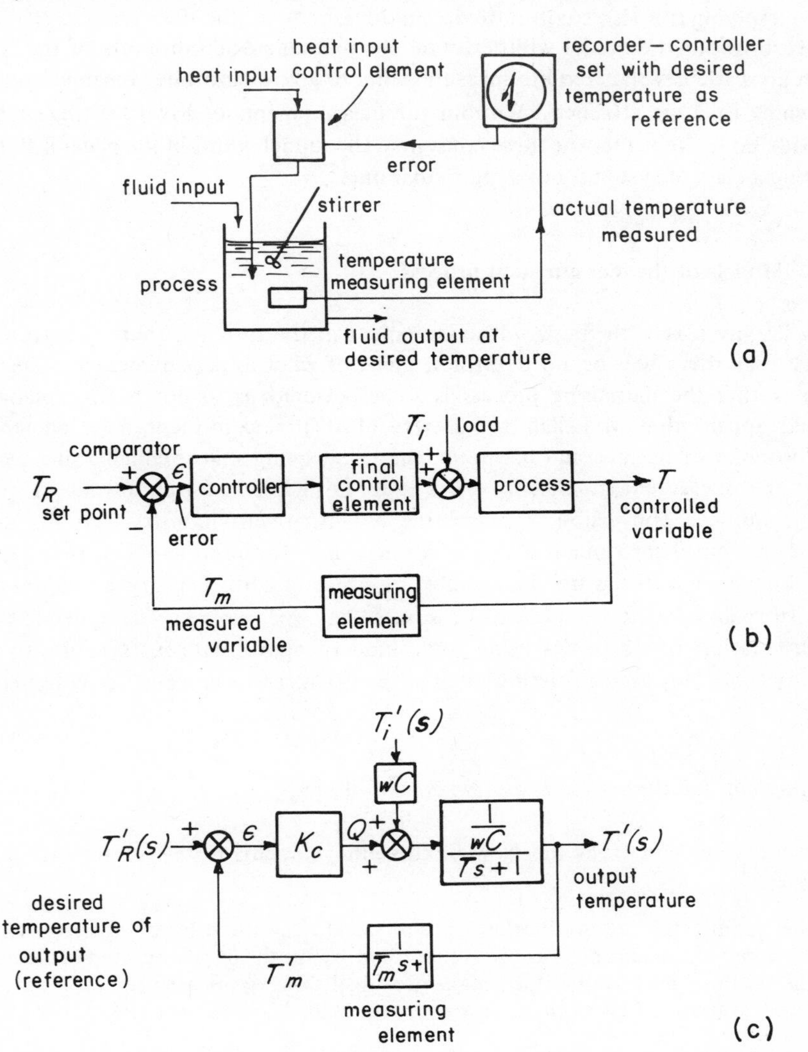

Fig. 1.6 *Stages in realising the identified mathematical model of a controlled temperature system*
a Schematic as per physical layout
b Block diagram realised in linguistic terms
c Mathematical equations entered with coefficients yet to be given values
[Drawn to author's instruction as adaptions from Coughanowr and Koppel (1965)]

A linguistic description can lead to various types of icons, including the well known (to engineers and scientists, that is) block diagram. Once the interaction between these blocks is established appropriate mathematical formulas are realised that represent the behaviour of the boxes. The next stage is to identify the numerical value of the general coefficients of the formulas. The mathematical model then becomes specific. Tests are then run to establish if indeed the model is an adequate representation of known situations implying it is also so of untried situations.

Throughout the model will be found the mathematical constants of the formulae; each are variables that require measurements to give quantitative meaning to the units to which they are attached. Without the measurements of a real existing subject that provide concrete values for these constants, the model is still in the general state representing a class of systems, not a particular one.

1.3 Models of the measurement process

Pick up any text with the word measurement in the title and there is a strong probability that there will be no definition given of what a measurement is. The implication is that the measuring process is so self-evident as to not need explanation. It would appear that McLellan and Dewey (1907) were influenced by ancient world philosophies of balance and harmony, the result being a most enlightening account of what is a measurement. Science owes it to Lord Kelvin for expressing his paradigm about number conversion, for creating an instrument industry and for designing numerous novel instruments. Yet in his account, Thomson and Tait (1883) of what was known about Natural Philosophy in the late 19th century a chapter entitled 'Measures and instruments' clearly assumes that 'measurement' is understood as the determination of quantities using some kind of appropriate instruments to make a comparison. This kind of definition is so prevalent as to be called the popular definition:

The popular definition of a measurement: model 1

A recent definition from the New Caxton Encyclopedia (1969) states that measurement is:

> The quantitative determination of a physical magnitude by comparison with a fixed magnitude adopted as the standard, or by means of a calibrated instrument. The result of measurement is thus a numerical value expressing the ratio between the magnitude under examination and a standard magnitude regarded as a unit....

This expresses the usual concept given in most texts. The terms used in the above definition can be interpreted widely without diminishing the idea portrayed. For example, the engineer will understand this as a process involving physical instruments whereas the research worker in education may see the instruments as survey forms and questionnaires.

A similar definition was given in Sloanes Dictionary (1894) but a study of various physical science works of the 1830-1900 period, such as Deschanel's (1891) *Natural Philosophy* volumes, *Cassell's Educational Course* (Wells, 1856), Pepper's (1874) *'Cyclopaedic science'* and Thomson and Tait (1883) *'Treatise on Natural Philosophy'* revealed that authors saw no need to present a definition. What a measurement is, therefore, apparently went largely unquestioned as it had for centuries. Kollrausch (1883) and Holman (1894) each specifically deal with measurements and the errors arising but similarly do not define what is a measurement.

Accounts of the historical development of common units of measurement such as volume, length and area by Klein (1975) and Ellis (1973) well document that the concept of a measurement given above was known in its practical implementation from the earliest times. The word measure itself can be traced from the Latin language, as can metrology, a synonomous term, to Ancient Greek.

The process of comparing an unknown against a standard unit having a divided scale was certainly in use by the late Stone-Age peoples. The shadow clock (Fig. 1.7) of the Egyptian civilisation clearly demonstrates application of a scale and a comparison.

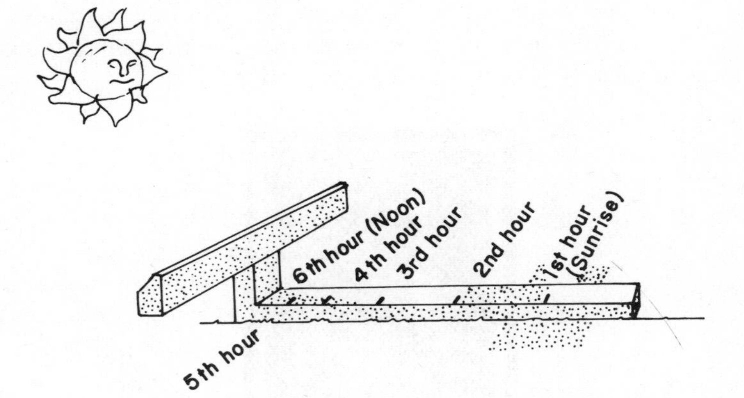

Fig. 1.7 *This Egyptian shadow clock dates from Thutmose III (about 3400 years old). It provided measurement of time by comparing the position of the bar shadow with the scale marked on the body.*

As a description of what a measurement is and entails, the popular model falls far short of complete definition as will be seen when two other models are discussed below. It does, however, begin to characterise some parameters of the measurement situation.

This model can be represented in pictorial form as is given in Fig. 1.8. Study of the definition and the iconic representation reveal that to make a measurement it is first necessary to decide the attributes of the subject. A standard for each of these is then defined enabling the magnitude of the unknowns to be expressed as a numerical value using a comparison process suited to the parameter involved.

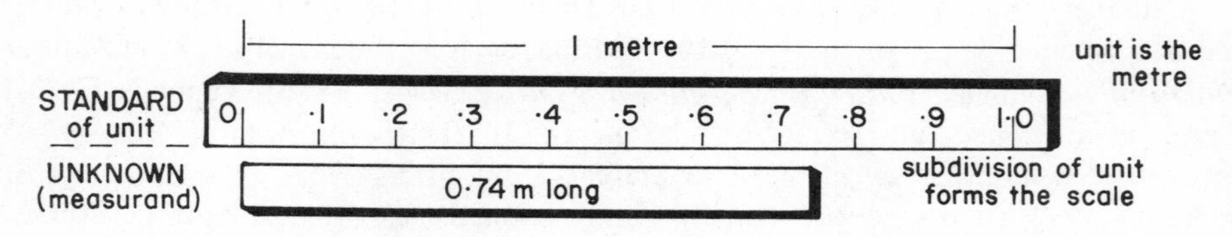

Fig. 1.8 *Pictorial representation of popular concept of a measurement in which an unknown is compared with a standard defining the unit. A scale allows differences to be expressed.*

In the physical sciences the standard—and in consequence the unit—and the comparison process have generally been easy to identify. If the standard can be realised as a physical object, or some characteristic of an apparatus, then it appears to follow that the comparison process is also realisable as a hardware form of instrument. The same is not true in the empirical sciences such as social science, geography, economics and psychology. There the major problem is usually to define the attributes that are relevant and that can be measured as unique entities. Even the simplest subject will exhibit a range of parameters with varying difficulty (Fig. 1.9).

Fig. 1.9 *A postage stamp is a simple object yet it defies total description by objective measurement. Obvious physical attributes are size, weight, shape, edge pattern: they are easy to measure. Colour, texture, pattern, beauty and value pose more difficult measurement problems.*

Some measurable variables were obviously self-evident to early man. Length, mass, time and desired units of area and volume had units and standards established very early in the story of man's civilisation. The Arcesilas Vase (*c*550 BC) shows scales in use to weigh the valuable silphium plant in Cyrene, North Africa, (Moldenke and Moldenke, 1951). Fig. 1.10 shows the scene on that vase. However, all units and standards have needed continual improvement at Klein (1975) has well shown; some of the early definitions promote mirth today (see Fig. 1.11). Standards can be good

Fig. 1.10 *Weighing the silphium herb: the Arcesilas vase from Ancient Cyrene*
[Courtesy Bibliothèque Nationale, Paris. Photo No. 72 C2866]

Fig. 1.11 *The 16th century German unit of length, the rood, was defined as the total length of 16 people's feet. A contemporary artist drew this criticism of its definition*
[Courtesy *PTB-Mitteilungen*, Vol. 79, No. 5, 1969]

enough for contemporary purposes but will always need improvement in definition as practical application catches up with the established definition.

The popular model, being self-evident, needs little explanation and is an adequate general representation of a measurement for many purposes. Other models of the same process, however, show that the popular definition is not complete in its representation of the process created.

Mathematical definition: model 2

Far less evident is the fact that the comparison process used forms a representation of the original physical variable by an alternative concept that is usually expressed in numerical form giving quantity to a defined unit. Lord Kelvin's quotation, given above in Section 1.1, expresses this.

This understanding of a measurement can be expressed in mathematical form using set theory because the action of using the magnitude of several defined units to describe a physical situation is a process of mapping from the real world set to a representational set.

This model of a measurement has traceable origins to the 1950 era. It has been extended by Finkelstein (1975) in recent times;

The following passage from his paper provides an appreciation of this kind of model of a measurement. It has little obvious similarity with the popular definition.

> Measurement is an empirical operational procedure of assignment of numbers to members of some class of aspects of characteristics of the empirical world according to a well-defined rule. The rule is so framed that the number assigned to an entity, describes it. That is, the relations between numbers assigned to different elements of the class, correspond to empirical relations between the elements to which they are assigned.
>
> Let us now express this more formally along the lines developed by Suppes and Zinnes.
>
> Take a well-defined, non-empty class of extra-mathematical entities Q. Let there exist upon that class a set of empirical relations $\mathcal{R} = \{R_1, \ldots, R_n\}$. Let us further consider a set of numbers N (in general a subset of the set of real numbers Re) and let there be defined on that set a set of numerical relations $\mathcal{P} = \{P_1, \ldots, P_n\}$. Let there exist a mapping M with domain Q and a range in N, M: $Q \to N$ which is a homomorphism of the empirical relational system $\langle Q, \mathcal{R}\rangle$ and the numerical relational system $\langle N, \mathcal{P}\rangle$. Then the triplet $\mathcal{P} \Rightarrow \langle Q, N, M\rangle$ constitutes a scale of measurement of Q.
>
> It is required that M be a well-defined operational procedure. It will be called the fundamental measurement procedure of the scale.
>
> $n_i \in N$ the image of $q_i \in Q$ under M will be denoted by $n_i = M(q_i)$. n_i will be called here the measure of q_i on scale S and q_i the measurand, and Q the measurand class.
>
> There will in general be other procedures of mapping from Q into N denoted by M' : $Q \to N$. such that $M(q_i) = M'(q_i)$ either for all $q_i \in Q'$ or for $q_i \in Q'$ where $Q' \subset Q$. Any such procedure is a measurement procedure for Q (or Q') on scale $\mathcal{P}$.

An iconograph of this model is given in Fig. 1.12. Lack of training in the set-theory symbols and procedures may mask the concepts conveyed by this modern form of mathematical symbology. As an aid to understanding the above description, the measurement process assigns a set of numbers (other symbols can be used but are less common) having a relationship with each other that represents, on a one-to-one basis, the quantities and their relationships existing in the system to be measured.

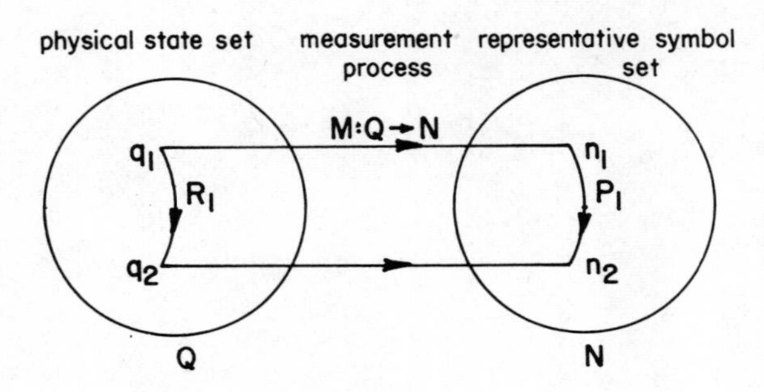

Fig. 1.12 *Pictorial representation of the set-theoretical model of a measurement* [Finkelstein (1975), courtesy Institute of Measurement and Control, London]

Considerable effort has been expended on the study of measurement theory. Krantz *et al.* (1971) and Pfanzagl (1968) have summarised the state of understanding.

This mathematical model, by examples seen elsewhere in other disciplines, would appear to be a development heading in the right direction. Mathematical models are usually more amenable to general study than are the real systems. Equations can be simulated in machines and machines can be designed around the mathematical bases. To the measurement practitioner, however, this model must be regarded, at present, as too advanced for current application in most practical design situations. It emphasises the need for equivalence between the real scale and the representative one, it reveals the nature of the mapping process and it shows why it is important to identify and isolate the various attributes forming the individual sets of a multiset measurement process.

Measurement as an information selection process: model 3

An instrument that performs the interface part of the measuring process is often termed the sensor. Fig. 1.13 shows a sensing head being used to detect cracks in shell cases in 1890, DeVries (1971). Senses are given to man naturally as sight through the eye-sensing mechanism, hearing through the ear system, touch through vibration-sensing and smell and taste through the organs in the upper breathing and eating passages. Man communicates with man through these senses (Fig. 1.14). Measuring instruments may be used to enhance the natural senses (Fig. 1.15) or to replace them (Fig. 1.16); more on this aspect later in this Chapter.

Fig. 1.13 *Schizeophone, invented by De Place in 1890*
a Sensing head
b Recording room

[De Vries (1971), p. 168 from *De Natuur,* 1895]

Fig. 1.14 *Physiological senses enable people to form links of measurement and communication between each other. Telegraphy class at the Industrial and Technological Museum, Melbourne, Australia, 1872*

[*Illustrated Australian News,* 18 June 1872. Courtesy Science Museum of Victoria, Neg. 1116]

Fig. 1.15 *Measuring instruments enable the physiological senses to be linked to systems in ways not possible directly. Nachet's 1850s multiple microscope enabled four observers to see a magnified subject simultaneously*
[Lardner (1862), Vol. IX, p.33, Fig. 46]

The observing stage, be it human or a constructed machine, is connected to the chosen attributes of the existing system under study by the sensor stage as is shown generally by Fig. 1.17. The measuring system conveys information about the system being studied through to the observer, be that a natural machine or a man-made one.

The word 'information' needs clarification because two distinctly different usages become involved in understanding this third model of a measurement presented here.

A first use is the common language, lay understanding of information as a collection of ideas, facts, identities, concepts and attributes that define a subject or object. The second use, met by engineers and scientists, is when the word is applied with respect to a body of knowledge known as information theory (which is somewhat synonomous with communication theory). Information theory is not about the theoretical considerations of the first use of the word where the meaning of the information is all important: it is about the quantity conveyed in a message passing through a communication channel. The meaning ascribed to that quantity is given no cognisance in information theory (nor is it in the Information Science body of knowledge). Shannon and Weaver (1949) stress this in their extension of information theory, which has its origins in the 1920s with Sziland and Nyquist studies.

A measuring system, in contrast to a data transfer or communication system, has the dual task of conveying correct messages in the information theory sense plus the

Fig. 1.16 ***Relative humidity sensor giving visual indication and proportional electrical signal. A measuring instrument can replace the observer in the link between the process and its controller.***
[Courtesy Foster Cambridge Ltd.]

added task of selecting meaningful information from the total information existing about the system under observation. Various writers, such as Stein (1970*a*), have denoted the total information existing about a system as its latent information. It would seem that any object possesses an infinite amount of latent information. As additional attributes are identified to attempt a total information summary, more attributes are revealed. Try characterising a simple object (such as the postage stamp of Fig. 1.9) in terms of its measurable parameters; there seems to be no end to the list of attributes that can be identified!

A key role, therefore, of the measuring instrument is that of selecting, that is, filtering, from the latent information available the specific information required. Information having defined meaning becomes knowledge. The meaning ascribed to some particular measurement data is entirely a matter of codification by the user. Without becoming deeply embroiled in philosophical debate about what knowledge is in general, it suffices here to define knowledge as information that has been given a certain meaning by the observer and subsequent users. Measuring-instrument output signals in themselves are not necessarily useful knowledge; the coding applied may be incorrect, the signal may contain extraneous information that did not enter the measuring stage from the system under study (the noise of the system that is always present to some extent).

Despite centuries of thought about a theory of knowledge, epistemological study

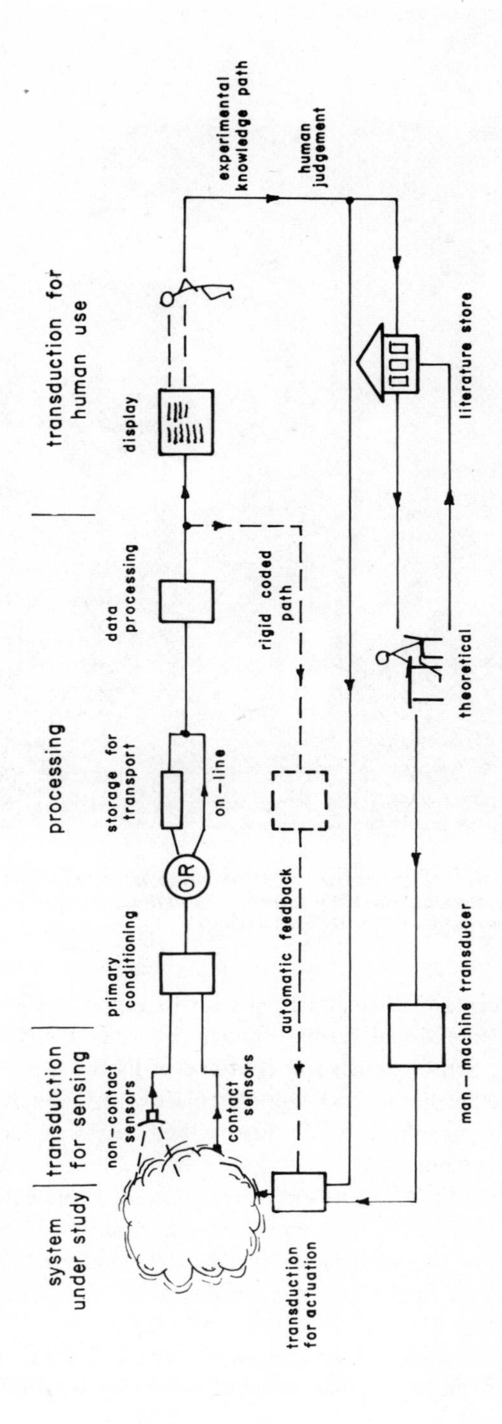

Fig. 1.17 *Generalised schematic of role of sensor in observation and control*

Fig. 1.18 *The energy level of the carrier of information is unimportant. Here ripple control signals are injected into 66/13kV substation grid lines.*
[Courtesy Zellweger Uster Ltd., Switzerland]

seems unable to produce a mathematical basis for its representation. Without this the meaning of measurements would seem to remain a property subjectively tied to the mind of the observer. Sensor design will continue to have a qualitative craft and art component that the scientific method cannot replace with objective rules. Because of this factor designers are unable to totally justify their design choice finally implemented as a measuring instrument.

The more uniquely are the parameters of value isolated, the easier the task of the measurement system. In many cases, however, practicalities of measuring variables lead to the sensing stage providing more filtered-latent information than is needed. Post-processing is used to further filter information in the conversion-to-knowledge process.

This model of the measurement process justifies the concept that measuring instruments are information machines that convey and code, with meaning, knowledge sought.

The 'information' model highlights another characteristic of measurements. Information flows from the system under study to an observing point. It is not possible to give a general physical meaning to the entity called information but it is observed that it flows on an energy or a mass-transfer carrier. Measurements are always associated with energy exchange or mass transport. The magnitude of the energy level of the carrier is inconsequential to information flow. Information signals are commonly regarded as possessing given power levels but the information part proper can be argued as a quite different entity. For example, ripple control of electricity (Fig. 1.18) uses low-level information signals conveyed over megawatt capacity energy links.

Stein (1964) has this to say about the relationship of energy and information transfer:

> Measurement consists of information transfer with accompanying energy transfer. Energy cannot be drawn from a system without altering its behaviour—hence all measurements affect the quantity to be measured. Measurement is therefore a carefully balanced combination of applied physics (energy conversion) and applied mathematics (information transfer).

The 'information' model of a measurement demonstrates the need to ensure that the measuring process does not alter the characteristics of the system under study by influencing its energy balance. Observing the stars with an optical telescope receives information by transmission of the energy of photons: the effect of the telescope on the stars is so minute as to be totally ignorable. When a situation of adequate noninteraction is met, the measurements soon prove satisfactory to their creators and the instruments are rapidly accepted in common practice having shown ease of use and giving reliable results.

This was the case with mass, length and time measurements of the past. More delicate energy balance situations, such as are found in, say, electrical-resistance temperature measurement of gas flow, act to slow-up development of the measurements as the interactions are often difficult to detect and allow for.

The form of the energy carrier need not remain the same throughout the information carrier channel. The schematic (Fig. 1.19) of the humidity meter shown in Fig. 1.16 shows this. Sensors are one type of a broader class of devices known as transducers. Transducers are defined as equipment that convert one form of energy into another. Actuators also lie in the transducer class, they are concerned with power conversion, not information.

The sensing stage has the purpose of converting the energy form, whilst retaining the coded information signal, whenever it is appropriate. For example, the temperature of a water bath is an expression of the level of thermal energy available. The output of the measuring stage might, however, need to be in electrical form so conversion from thermal to electrical energy is required to form an energy link on which to convey the measurement information.

The energy or mass-transfer link only begins to convey measurement information when its steady-state value is modulated in some way with reference to a standard condition, see Stein (1970*b*). It is the variation that carries the meaning coded into

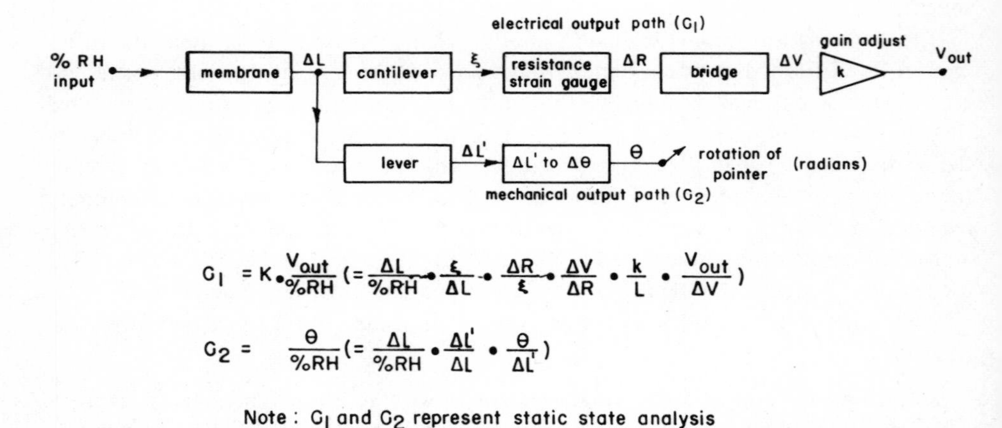

Fig. 1.19 *Energy-conversion stages through the relative humidity sensor shown in Fig. 1.16*

the modulation method. A blank sheet of paper or a steady level light beam contain no message unless the spatial or temporal steady-state condition, respectively, are varied in some way according to codes that assign meaning. Some feature, such as given level change, must be defined as the unit of measurement. Stein (1970*b*) lists the examples shown in Fig. 1.20.

Thus the information model creates awareness of modulation methods which in turn suggest clearly that measuring instruments have dynamic as well as static performance characteristics. These considerations are not so evident in the 'popular' and 'mathematical' models stated above.

The sensing and communicating stage forms a link between two systems, one, or both, of which is effected by the other. Energy or mass transfer is always involved yet the magnitude.or form of the carrier is not a factor deciding the degree of the effect. So, as was stated earlier, there exists an entity unexplained in its substance yet used prolifically to seek knowledge by knowledge of its manifestation as worldly experiences.

1.4 Units, standards and scales

The models of measurement clearly bring out the three basic entities required i.e. units, standards of those units and scales that suit the units.

A unit of measurement is the word, term or symbol representing the meaning code chosen to represent a given parameter (or entity in the Finkelstein passage above). It gives specific meaning to the numbers generated by a measurement. Without statement of the unit involved the resulting numbers remain as unconverted information and cannot be regarded as knowledge. Some examples in current use are the metre,

Waveshape	Instantaneous local	Incremental local	Instantaneous regional	Common name	Simple examples
	Individual property	Patterns of properties			
Constant level	amplitude			analog data	mercury-in-glass thermometer
		amplitude frequency		wave shape analysis signature analysis	electrocardiograph spectroscopy
			amplitude frequency		black & white photography color photography
Sine wave	amplitude			AM - amplitude modulation	AM radio
	frequency			FM - frequency modulation	doppler effect, vibrating-wires
	phase			PM - phase modulation	photoelasticity, Kerr cell
		frequency		FCM -frequency code modulation	temperature-sensitive paints
			amplitude		antenna characteristics
Pulse trains	amplitude			PAM - pulse amplitude modulation	
	frequency			PFM - pulse frequency modulation	FM radio or tape recorder
	position			PPM - pulse position modulation	
	duty cycle			PWM - pulse width modulation or PDM - pulse duration modulation	
		amplitude		PCM - pulse code modulation	digital data
			amplitude		computer card

Fig. 1.20 *Some information-carrying methods* [Reprinted from Stein, P, ***J.Metals,*** October, 1969, p.41. Courtesy of American Institute of Mining, Metallurgical, and Petroleum Engrs. (AIME)]

Fig. 1.21 *Standard mass, coded K^{20}, used to define the kilogram in the USA* [Courtesy US Government, National Bureau of Standards]

kilogram, second, Henry, Kelvin and the Pascal.

Standards of measurement are the examples realised in a physical way to define the meaning given to the specified unit. They are entirely of man's choice, nothing about the natural world defines standards, they are a contrivance of man. However, they are often based upon naturally occurring phenomena when these possess the required degree of definition. Perry (1955) and Woodward (1972) are historical accounts about standards.

When standards can be realised as physical processess that can be defined in terms of physical science they are called physical standards. The standard of mass (in the SI system) is a certain piece of metal held by one international authority. It has the unit kilogram (Fig. 1.21).

Although it does seem that progress in science is reflected by the ability to produce physical standards—because realisation of such implies ability to actually measure the subject of the unit concerned—progress can be made with less well defined standards. Plato's words, given in Section 1.1 speak of pleasure and pain. These are examples of attributes for which man has not yet, and may never be able to quantify into units having physical standards. In biological measurements a control group of subjects are maintained unchanged as the standard for comparison with those that are.

Man's struggle to produce satisfactory physical units has been investigated in depth by Klein (1974). Dresner (1971) provides information about the many units in use. Over the thousands of years of development of measurement units, thousand upon thousand of units, each with their own declared standard, have been defined. By the 18th century, havoc and chaos was the norm in trade and taxation because of the multiplicity of similar units, often with the same name yet having a different physical standard and numerical value. The use of unfamiliar number systems also helped confusion. Clarke (1866) illustrates the differences in various length standards. Out of this disorder developed the metric system. A law proposing a metric system, was prepared in the Netherlands in 1809, coming into use a decade later. In 1837 the French Government decided, after four decades of discussion, to introduce a complete metric system based on the decimal number system and with certain basic physical units. Various forms (see Fig. 1.22) of metric systems were subsequently adopted and through a long succession of international agreements the Systeme International d'Unites (called SI for short) has now almost become the sole global basis of physical units and their standards. It is the most rationalised and systematised system yet introduced but as it still contains such anomolies as using the kilogram rather than gram without a prefix of magnitude, there may yet be need for more changes in the future.

Much has been written about the metric system and its defining standards. As it is of only passing interest here readers are referred to Ellis (1973) for a basic lay intro-

Fig. 1.22 *First set of metric standards used in Holland.*
[Courtesy of Utrecht University Museum: Photographer, Jac. P. Stolp]

duction, to Klein (1975) for some history leading to its adoption and to Feirer (1977) for a full treatment of its structure and current-day usage. Everett (1876) provides insight into the 19th century appreciation of this aspect of measurements. Sas and Piddock (2947) explain the SI system of electrical units.

Another kind of standard exists; the standard specification is a form of standard often confused with physical standards. Conceptually, these are linguistic realisations of concepts. A standard specification explains what the defined product or service should be like. They provide the standard against which conformity is judged by subjective or objective methods of measurement. Standards have become recognised as a subject in their own right in recent times, Verman (1973).

When the unknown is compared with the standard the difference expressed is the magnitude in terms of the unit. When two objects are identical in every desired respect they are declared equal. But how is difference expressed when it occurs? The ranking process leads to the idea of scales that express the relationship of the magnitude resulting.

Many kinds of scales have been defined; Ellis (1966), Finkelstein (1975) and Krantz *et al.* (1971). The study of scales is now a well developed mathematical aspect of philosophy. Full description is beyond the needs of this work. To assist understanding of the development of measuring-tools it suffices to state that lack of understanding of the nature of the scale relevant to a measuring task has been and still is, a major difficulty in the development of many measuring tasks.

The degree of difference can be exemplified with some examples. Increments engraved in a simple ruler indicate subdivision into proportionate divisions of a standard representing the unit of length. Each increment is of the same distance interval as the next. In contrast the Beaufort scale of wind activity (Fig. 1.23) represents worsening conditions over a scale of grades which have no obvious interval relationship. A clock keeps time in equal increments of hour units yet the geological time scale lacks the same degree of increment equality.

1.5 Measurement and the scientific method

Science is the study of gaining knowledge and ordering it into its most general form so as to reduce the number of individual facts needed to be known. Approach to knowledge search by early civilisations before the Dark Ages, the period roughly agreed upon as the 1st to 14th centuries AD, was generally on the basis that mental reasoning alone was enough to learn about the world. The role of observation was minimal by today's standards. Exceptions were people like Hippocrates (*c*460-*c*377 BC) who used observations to decide the medical treatment that he administered to his patients. Knowledge was, however, gradually obtained in the fields of astronomy, zoology, hydraulics, pneumatics, botany, mineralogy and physics.

By the start of the Dark Ages, when the advanced civilisations succumbed to less stable and less civilised tribes, it was felt that the fundamentals of science were all known. Advance then moved over to the Arabian regions, where much of the know-

Beaufort Number	Discription of wind	Effects of wind	Speed in m.p.h.	Standard symbol
0	Calm	Smoke rises vertically: water mirror-smooth; sea unruffled	Less than 1	
1	Light air	Direction shown by smoke drift, but not by weather-vanes	1-3	
2	Light breeze	Wind felt on face; leaves rustle; weather-vanes move	4-7	
3	Gentle breeze	Leaves and twigs in motion; light flag extended	8-12	
4	Moderate breeze	Dust and loose paper raised; small branches moved	13-18	
5	Fresh breeze	Small trees in leaf begin to sway; crests form on waves	19-24	
6	Strong breeze	Large branches in motion; telegraph wires hum; large waves begin to form	25-31	
7	Moderate gale	Whole trees in motion; walking against wind difficult	32-38	
8	Fresh gale	Twigs break off trees; foam blown off waves in dense streaks	39-46	
9	Strong gale	Slight structural damage occurs; chimney-pots and slates fall	47-54	
10	Whole gale	Trees uprooted; considerable structural damage; high, overhanging waves	55-63	
11	Storm	Sea completely covered with foam; wide-spread damage; rare inland	64-75	
12	Hurricane	Countryside devastated; experienced only in tropical cyclones	Above 75	

Fig. 1.23 *Abridged Beaufort scale*

ledge of antiquity was preserved. They did have an interest in observation, especially in optics, medicine and astronomy. Through them the known body of knowledge eventually returned to Europe in the period following about 1200 AD. The attitudes inherited through the Dark Ages gave way to a new enlightened viewpoint that observation and reason were partners in the practice and philosophy of science. Roger Bacon (*c*1214-1294) was among the first to stress the use of experimental method. Galileo Galilei (1564-1642) was another scientist, among many, who followed with this attitude.

In 1605 Sir Francis Bacon, a politician, philosopher and writer, wrote his famous *Advancement of Learning* in which he set out how to totally reconstruct human knowledge and obtain it by a new technique in which observation was vital. He sought to replace the deductive system of antiquity with the inductive method of scientific investigation. He later published, in 1620, *Instauratio Magna* which contained the famous iconic representation of his idea that is given in Fig. 1.24. Carter and Muir

Fig. 1.24 *Title page to 'Instauratio Magna' by Sir Francis Bacon, 1620*
[Courtesy Dawson Rare Books, London]

(1967) had this to say of it:

> The accumulation of facts was, to Bacon, the basis of a new philosophy epitomised by the famous engraved title page showing a ship boldly sailing beyond the Pillars of Hercules (the limits of the old world). He conceived a massive plan for the reorganization of scientific method and gave purposeful thought to the relation of science to public and social life . . . Bacon made no contribution to science itself, but his insistence on making science experimental, and factual, rather than speculative and philosophical, had powerful consequences.

Many historians regard the work of Francis Bacon as a prime factor in the establishment of the Royal Society, which gained its Royal charter in 1662 from King Charles II.

This period saw the acceptance and great expansion of the 'Natural philosophy', a term used by Newton to denote the investigation of the laws of the natural world and the deduction of results that cannot be directly observed.

The new trend recognised the importance of observation. Although debate continues strongly on the nature and correctness of the scientific procedures used today, and definitions of the method abound, it is sufficient to quote Bertrand Russell (1931):

> In arriving at a scientific law there are three stages: the first consists in observing the significant facts; the second in arriving at a hypothesis, which, if it is true, would account for these facts; the third deducting from this hypothesis consequences which can be tested by observation.

This definition shows the first step is to use experimental and other observational procedures to supply data that is regarded as relevant. Subsequent reappraisal of this data, in light of the learning process advancing the understanding, assists its more effectual evaluation.

Then follows a stage in which mental processes operate to sort the data. To quote Charles Darwin:

> my mind seems to have become a machine for grinding general laws out of a large collection of facts.

Thus evolves the hypothesis which then must be tested with as many test situations as is reasonable to assume test the breadth of validity sought from the hypothetical law realised. This third stage is one of more observation and experiment.

Observations, measurements and experiments are closely related concepts that form the important experimental part in the current scientific method. Other writers' views of this theme are to be found in Dellow (1970), Herschel (1851), Jones (1967*a*), Lilley (1951), Ross (1941) and Whipple (1931, 1972).

The importance of correct, reliable measurements in the scientific procedure has been stressed by many writers of this century. Westaway (1937) had this to say:

> The more that exact measurement enters into any branch of Science, the more highly is that branch developed. It is for this reason that Chemistry and Physics are so far in advance of Botany and Geology. And the reason why we can obtain so much clearer notions of, for instance, an area or a weight, than of, say, wisdom or chivalry, is because the former are *measurable*, the latter not. It is

> of the first importance in Science that we should, whenever possible, obtain precise quantitative statements of phenomena, and thus we see why it is that the introduction of a new scientific instrument so often leads to a marked advance in our knowledge.

Clerk-Maxwell's words delivered in his Inaugural Lecture at Cambridge University in 1871, were quoted in recent times to express the importance of measurement to the advancement of science by Jones (1970).

> This characteristic of modern experiments—that they consist principally of measurements—is so prominent, that the opinion seems to have got abroad that, in a few years, all the great physical constants will have been approximately estimated, and that the only occupation which will then be left to men of science will be to carry on these measurements to another place of decimals . . But the history of science shows that even during that phase of her progress in which she devotes herself to improving the accuracy of the numerical measurement of quantities with which she has long been familiar, she is preparing the materials for the subjugation of new regions, which would have remained unknown if she had been contented with the rough methods of her early pioneers.

Knowledge is the result of scientific endeavour. That endeavour rests heavily upon the ability to make the measurements to which logical reasoning has assisted in the assignment of codes of meaning to the data produced by the measuring systems utilised.

1.6 Measurements as parameters of control

In the above Section the role of measurements in enquiring about new knowledge was explained. New knowledge, however, is not man's only need for informational measurement equipment.

Even before new knowledge was purposefully sought, man had realised that many parameters of daily life required regulation and standardisation and that this was possible through measurement.

Trade, commerce, religion, agriculture, taxation, time-keeping and travel on land and at sea each generated the need to make manual measurements using acceptable standardised units. One theme of what enables the rise of civilisation is that settled communities were able to conduct the essential tasks of survival with some time to spare. This time was used to improve the lot of early man, increasingly releasing him to search for yet more ways to make life easier and fuller. Measurements clearly enabled routine to be established, they clearly facilitated satisfactory trade, they clearly improved farming success and assisted construction (Fig. 1.25). It is reasonable, therefore, to claim that the 'story of weights and measures is indeed largely the story of civilisation' Ellis (1973).

Today society still requires numerous routine measurements to regulate its functions. Trade still requires accepted standards. Time-keeping is even more a factor in scheduling our availability. A great deal of modern instrumentation is used to control,

Fig. 1.25 *Surveyors at work in Ancient Egypt*
[Photo. Science Museum, London]

rather than gain, new knowledge in the scientific sense. Where simple devices suffice they are still used today e.g. the wooden school ruler, the plumb bob, spring balance and beam-balance weighing machines.

The measuring system used in a control loop devoid of a human link has its meaning codes assigned for its working life by the human designer. Its purpose is to main-

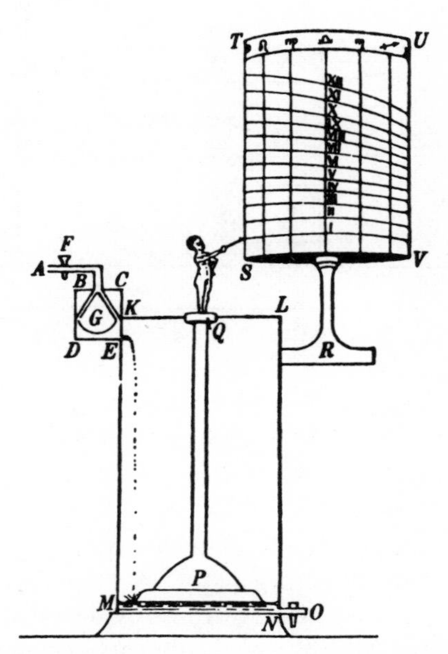

Fig. 1.26 *Diels reconstructed this water-clock of Ktesibios (3rd century BC) from a written description*
[H. Diels 'Antike Technik', Teubner, Leipzig, 1924]

tain a given fixed or programmed state of order. Knowledge about the process under control is generated by the measurement link but it is used solely to adjust the process accordingly when things deviate from the standard performance set.

The use of measurements in closed-loop control can be traced to around 300 BC when historical records strongly suggest a float valve was incorporated into a water clock devised by Ktesibios. A reconstruction by Diels is given in Fig. 1.26. Mayr (1970, 1971) has made an extensive study of the origins of feedback controls. In his book many examples show that sensing devices were used in many ways to regulate energy flows. Many of these early devices, however, leave doubt about the actual details.

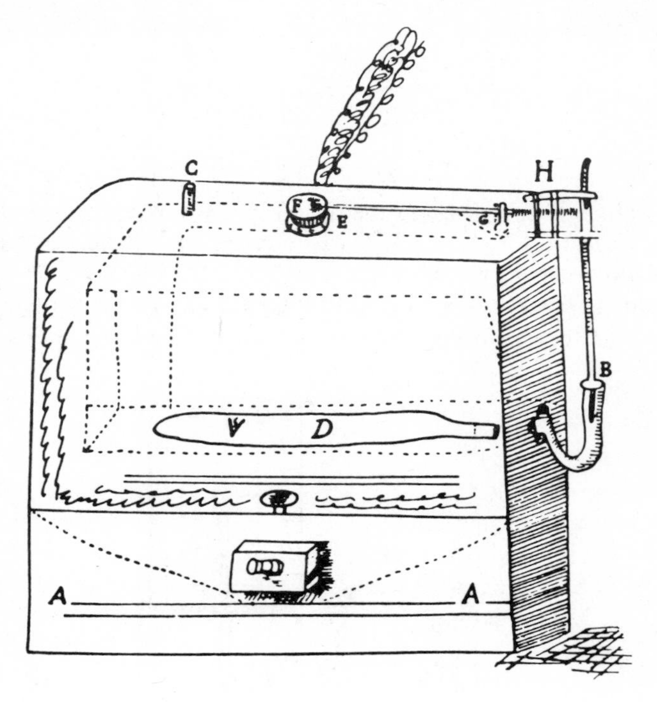

Fig. 1.27 *Drebbel's chicken incubator — first known temperature regulator c1620*
[Courtesy, Syndics of Cambridge University Library, MS Ll.5.8, page 218]

Devices of European origin that were reported from the 16th century onward leave little doubt that sensing devices were used for control purposes. In a purposeful manner Drebbel's incubator of the early 1600s is recorded as sketches (Fig. 1.27) made of it by others. From these it is easy to identify a temperature-sensing element that operates a heat vent. This device seems to have influenced many later designs.

Study of these temperature regulators and other feedback systems such as float valves, pressure regulators, windmill wind-seeking and shaft-speed control in mills, and

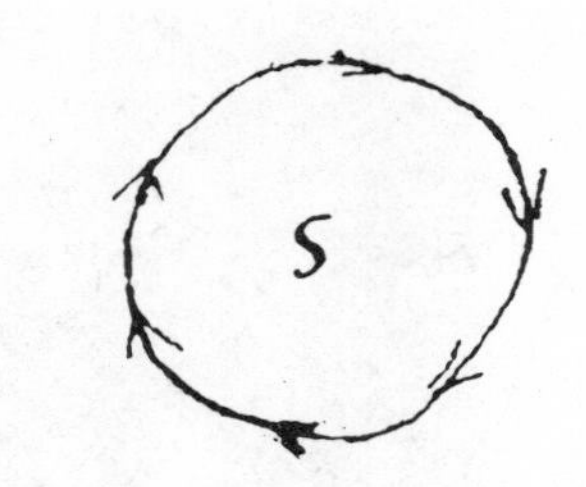

a

Maxwell's equations

magnetic intensity vector **H**

$$\nabla^2 \mathrm{H} = \mu\epsilon \frac{\partial^2 \mathbf{H}}{\partial t^2} + \mu\sigma \frac{\partial \mathbf{H}}{\partial t}$$

electric intensity vector **E**

$$\nabla^2 \mathrm{E} = \mu\epsilon \frac{\partial^2 \mathbf{E}}{\partial t^2} + \mu\sigma \frac{\partial \mathbf{E}}{\partial t}$$

in which symbols are

Scalar quantities

ϵ = permittivity
μ = permeability
σ = conductivity
Q = charge density
$q = \iiint_V Q\, dV$ = total charge within V
$\phi = \iint_S \mathbf{N} \cdot \mathbf{B}\, dS$ = total magnetic flux passing through S
$i = \iint_S \mathbf{N} \cdot \mathbf{J}\, dS$ = total current flowing through S

Vector quantities

$\mathbf{E}$ = electric intensity
$\mathbf{H}$ = magnetic intensity
$\mathbf{D} = \epsilon\mathbf{E}$ = electric flux density
$\mathbf{B} = \mu\mathbf{H}$ = magnetic flux density
$\mathbf{J}$ = current density

b

Fig. 1.28 *Stages in development from idea to commercial use: evolution of radio as an example*
a Linguistic or iconic description
b Mathematical modelling
[Thompson (1898), Fig. 2, p.88]

Fig. 1.28 *Stages in development from idea to commercial use: evolution of radio as an example*
(c) Verification of laws
[Courtesy Science Museum, London]

later in steam engines, implies from their construction that their builders did recognise that the measurement function was separable from the power function.

Although distinction has been made between those measurements used to obtain new knowledge and those used to control closed-loop processes, the distinction is only one of time-scale. It would seem that man seeks to extend the body of knowledge to make eventual use of it to subjugate his environment to suit man's needs. In other words knowledge is sought to eventually bring about control of resources. The feedback loop is closed in the long-term by man recovering the knowledge from storage depositories -libraries- applying it to bring about some form of control of the system about which the knowledge is relevant.

As an example refer to Fig. 1.28. Clerk Maxwell, just before 1855, became interested in Faraday's iconic representations of electricity and magnetism. When Clerk Maxwell extended Faraday's 'lines of force' into mathematical form, he could not have realised that he had, as had Faraday, contributed knowledge (that must have appeared quite esoteric to the layman of mid-19th-century Britain) that would lead to the creation of wireless communications. This, of course, is justification for pursuing pure science: it ultimately may lead to significant application.

Fig. 1.28 *Stages in development from idea to commercial use: evolution of radio as example (d)* Commercial application
[Courtesy GEC-Marconi Electronics Ltd.]

1.7 Measuring instrument sophistication as an expression of degree of science applied

The decision to use existing, or to create new, measuring instruments in the study of a subject comes after measurable variables have been identified. This process is, in the physical sciences, usually considerably easier to realise than in many areas of the empirical sciences. Many stages of prior reasoning precede such a decision: this is, especially in engineering, not always recognised. The process can be depicted by the chart given in Fig. 1.29.

Knowledge seeking begins presumably because of certain inquisitive features of mans makeup that put his interests in directions which seem to have more relevance than others. The processes involved are complex and, as yet, not adequately known. Paradoxically, it seems a great deal of knowledge is used in a very general way from the onset to choose candidate paths of action to follow to gain the knowledge sought.

This process, which involves the cognititive elements of sensation, perception, apperception, advises the knowledge seeker that certain information is more relevant for study than other data. It appears that the biological senses involved provide data

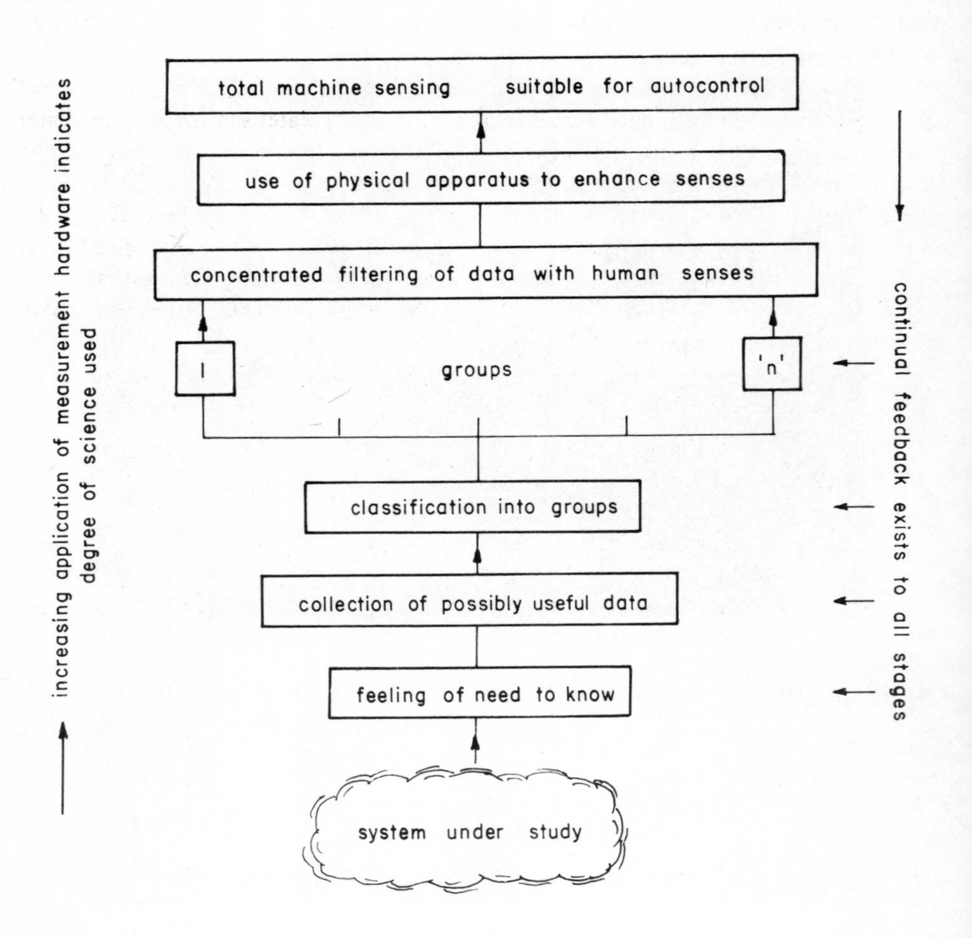

Fig. 1.29 *Simplified hierarchy of application of measuring instruments in the study of a problem*

input to the brain coding it with meaning to suit the task required. Two people viewing a plant leaf, for example, see the same object with similar senses yet both could 'see' quite different attributes. Latent information available has begun to be filtered at this stage.

The assembled data is then sorted and classified according to various kinds of similarities to detect differences. Each group forms a crude measurement standard of comparison for the others. This process can often be continued until advanced knowledge is established without using measuring instruments. Linnaeus (1707-1778) was able to make a major contribution to botany by introducing his binominial classification system, (see Fig. 1.30). Darwin's *On Origin of the Species by Means of Natural Selection* of 1859 has been recognised as probably the greatest generalisation yet.

It was made from vast quantities of data that were all assembled with little use of measuring instruments to enhance mans natural senses.

At some stage this qualitative form of science can be subjected to increasingly more quantitative methods. Attributes of the various classes became apparent in a way that allows instruments to be applied that give natural sensing greater sensitivity and greater power to move from a qualitative mode into the quantitative measurement mode. More detailed knowledge becomes available as measurements produce data that is referred against more adequate, precise and accurate standards. Thus it is, that physical measuring instruments are applied as the degree of science—which is reflected by the degree of quantification used—is improved. This line of reasoning also makes it vital that appreciation of the qualitative stages preceding measurement proper are understood as well as the instruments that evolve.

Fig. 1.30 *Linneaus resting after a botanical ramble. He devised the binomial classification system now used, reporting it in 'Systema naturae', 1758*
[Grolier Society (1946), p.3789. Courtesy Grolier Society Inc., New York]

This sentiment is not new as Lord Kelvin's quote given in Section 1.1 tells us. The words of Westaway (1937), given previously, are in sympathy with this. The concept expressed in Fig. 1.29 is simplified: in practice the stages at which hardware forms of measuring instruments are used varies widely. In some studies they are needed at the very beginning.

Finkelstein (1975) sums up the situation in this way:

> Measurement presupposes something to be measured, and measures have to be conceived and sought before they can be found in experience. Both in the historical development and logical structure of scientific knowledge the formulation of a theoretical concept, or construct, which defines a class of entities and

the relations among its members, providing a conceptual interpretation of the sensed world, precedes the development of measurement procedures and scales. It is necessary for instance to have some concept of 'degree of hotness' as a theoretical construct, interpreting the multitude of phenomena involving warmth, before one can conceive and construct a thermometer.

As measurement procedures are developed, and knowledge resulting from measurement accumulates, the concept of the measurand class becomes clearer and may to a substantial extent become identified with the operational procedures underlying the measurement process.

In some cases the concept of an entity arises from the discovery of mathematical invariances in laws arrived at by measurement, and the entity is best thought of in such mathematical terms, but in general one attempts to arrive at some qualitative conceptual framework for it, if possible.

As the subject matter becomes better known and enables unattended sensing by an observer, as needed for control or monitoring purposes, the use of measuring instruments to enhance human senses may not be appropriate. Hardware sensors then totally replace man's senses.

It is logical, therefore, to expect all of mans endeavours that require measurements to be made (most of them!) to trend steadily toward greater use of measuring instruments. Certainly, time has proven this to be so. But this is also a consequence of mans method of survival on earth. Unlike lower animals man has the ability to modify his environment to suit his biological structure. He does this usually by the use of technological developments which rarely operate in the same way as natural equivalents or are made with same materials. Comparison of natural and man-made vision sensors is given in Fig. 1.31. The knowledge man possesses is being built up of a component about the natural world plus a component about the structures that man has created. Measuring instruments are the means by which man's creations operate and they too are creations of man.

1.8 Measurements and the body of knowledge

The sum total of knowledge is termed the body of knowledge. As knowledge is a characteristic of man, not of existence, it began at zero magnitude and grew with time. No way has yet been devised to measure its magnitude in objective ways but it clearly is enlarged continuously with the passage of time.

It is formed of two groups: that about the natural world and that about the unnatural systems created by man. Man's creations grew, the natural world changes; the extent of the latent information available for conversion into knowledge therefore grows continuously.

As the body of knowledge grew various workers of the past tried to summarise all that was known. Today that must be recognised as an almost hopeless task. Collectively all knowledge must be stored in a manner whereby it is retrievable. The danger of converting latent information back into another form of latent storage via the knowledge conversion state is real: what lies in the literature is not all recoverable in an easier manner than that by which it was first generated!

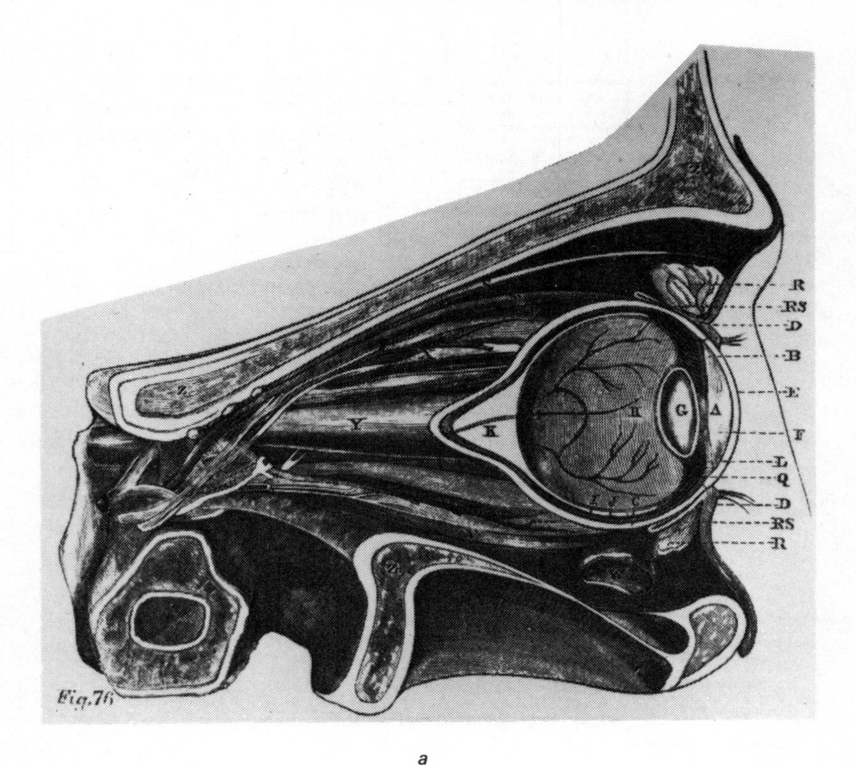

a

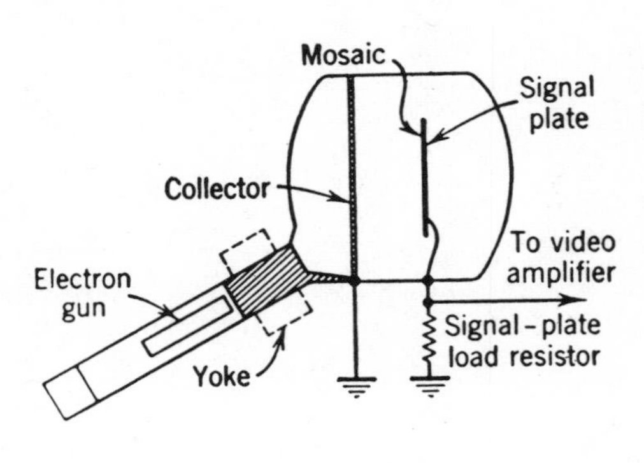

b

Fig. 1.31 *Man's creations generally use different materials and techniques as do natural systems. Here imaging sensors are contrasted*

a Longitudinal section of eye
Reprinted from PEPPER, J.B.:*Cyclopaedic Science'* [Frederick Warne, (1874)]

b RCA Iconoscope – early form of television camera tube from Kloeffler (1949) [Courtesy RCA Ltd., USA]

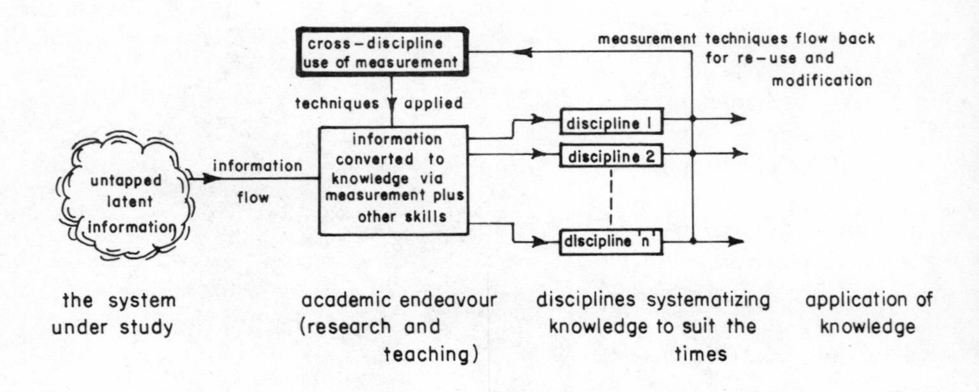

Fig. 1.32 *Relationship of measurement principles in ordering the body of knowledge*

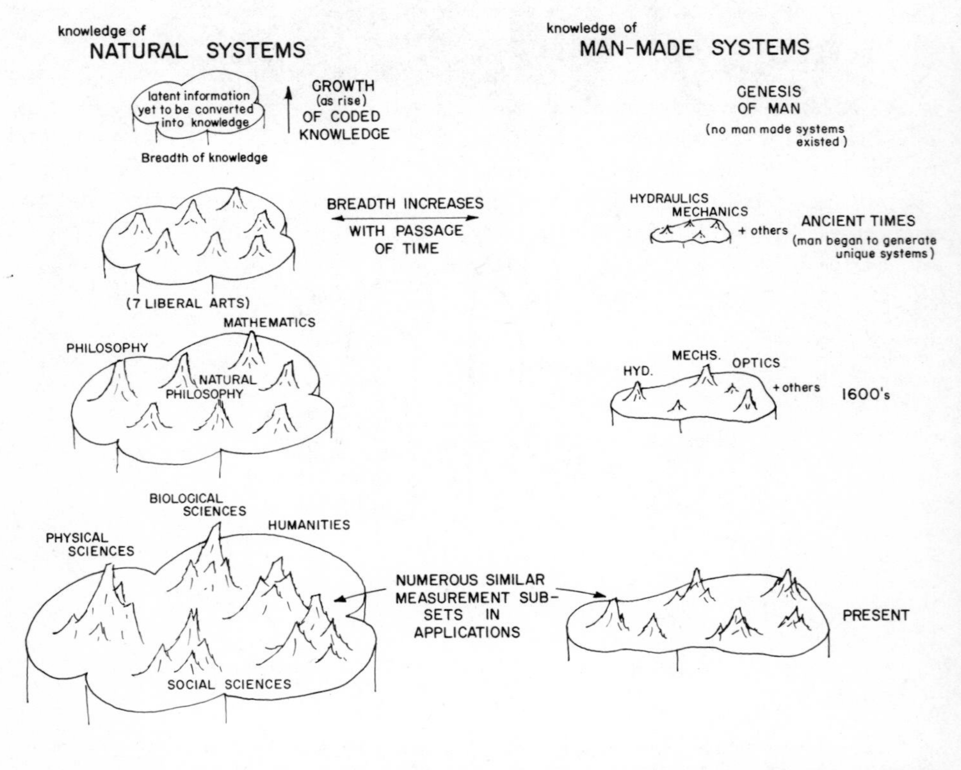

Fig. 1.33 *Epistemological mountains in the two plains of human knowledge. Measurement techniques are now duplicated on most contemporary mountains*

To retrieve knowledge it is grouped into convenient classifications. Convenience is a term in which time of action is most important. The memory span of man, especially short-term, is very limited so it has been suggested, Harman (1973), that major groupings usually total around seven. These in turn are subdivided giving the various epistemological groups.

Measurements assist in gaining knowledge and knowledge, in turn, assists new forms of measurement to be conducted. A closed-loop mechanism can be observed in the development of measurements; Fig. 1.32 depicts this.

Over the past few decades the trend toward recognition of the interdisciplinary studies that replaced the specialisms that came to us previously has highlighted the fact that not only does such a feedback process exist but that it is often duplicated (a needless waste of effort, therefore) and is often crossfertilised between epistemological groups.

The Dewey cataloguing system gives librarians a set of numerical codes each having a linguistic description of what subject matter each number represents. Of over 40 000 numerical assignments some 600 clearly relate to the measuring process. These are distributed widely over the whole body of knowledge, as classified by that system. Pictorially this means that most clusters of knowledge possess subclusters concerned with measurement method as depicted in Fig. 1.33.

At present information scientists, those people that work on the storage, coding and retrieval of knowledge, consider that the major clusters are changing to reflect the interdisciplinary attitudes. New clusterings are emerging, one which may well be that of the relatively new discipline of measurement science, the pursuit of means to convert latent information into meaningful knowledge by rigorous and objective procedures of philosophy and practice.

Chapter 2

Science and technology of measuring instruments

2.1 The role of technology

As with the term 'science', the word 'technology' is also used with an air of ease that suggests that the concept conveyed is well established and easily defined. In extensive and detailed notes issued to students at Melbourne University's Mechanical Engineering Department, P. Milner and C.J. Pengilley have listed 17 definitions of 'technology', each of which partially conveys the understanding required here. Milner's unpublished lecture on the meaning of technology, written in 1974, clearly shows that there is much that can be said and that no totally adequate definition seems possible. However, for the understanding of the development of instruments a statement by Hindle (1966) suffices:

> Science and technology have different objectives. Science seeks basic understanding . . . ideas and concepts usually expressed in linguistic or mathematical terms. Technology seeks means for making and doing things. It is a question of process, always expressible in terms of three-dimensional 'things'.

'Technology' can be regarded as synonomous with 'Engineering' for this discussion. It is the broad discipline that devises machines and structures. It uses naturally available materials and human resources to create new systems often having no parallel in Nature. Fig. 2.1 shows how a telescope is made from materials found in nature.

Technology is the sister requirement of scientific pursuit, inseparable partners in progress, each affecting the other's progress at varying degrees with time. Technological advance, however, can come about without the need for scientific method. Indeed many new machines have been the brain-child of poorly educated, quite unscientific, people who have applied an uncommonly good amount of common-sense to solving an immediate problem.

New inventions can also arise from discoveries of an effect witnessed by chance and often never being explained with scientific rigour. In the instrumentation field a good example of this is the discovery and widespread use of the barium and strontium carbonates in coating the cathodes in thermionic tubes used in electronic amplification from about 1910 onward. These dull-emitters provided greatly enhanced electron emission for a given cathode heating level and considerably increased the potential of

Fig. 2.1 *Technology produces new objects, using naturally available resources, for specific tasks. Construction of the Great Melbourne Telescope c1890.*
[Courtesy Science Museum of Victoria]

electron tubes. Little purposeful science was used to reach the decision that dull-emitters could be created.

The invention of a new device does not necessarily imply its acceptance at the same time. Social influences moderate the rate of application of new ideas. We now accept spectacles as normal technology for persons with less than ideal vision. But what must the first person to don a pair of glasses have had to contend with in the way of public comment? Imagine how noticeable a person with a nose filter would be today if that were decided to be an effective way to reduce airborne allergy defects.

History is full of examples of intolerance and nonacceptance. In 1601 AD a horse was condemned to death by burning because it could perform a number of tricks. In the 17th century Pascal came close to a similar fate for building his simple calculator. The violence of the criticism of current thinking by many of those who saw better often did not assist acceptance. Bruno was burned at the stake in 1600 AD ending a long entanglement of his ideas of astronomy with those of religious viewpoint.

Technological development relies on several factors. Materials must be available, ideas must be realised, crafts and processes of manufacture must exist to convert the ideas into objects and usually, but not necessarily so, a need (or a want) situation must exist. Such factors are the subject of this chapter. Put simply science produces new knowledge from latent information; technology uses, in part, knowledge to fashion resources into information gathering tools.

2.2 The dual nature of machines

The New Caxton Encyclopedia defines a machine as:

> Any device capable of advantageously utilizing a given form of energy or converting it to another form of energy.

The explanation given there continues at length to illustrate this definition using description of numerous mechanical and electromechanical devices classing them broadly as motive, operative, generative and transmissive machines. Throughout, the emphasis is on machines being involved with work of some kind.

The power regime of machines is concerned with emphasis on energy conversion efficiency for energy lost can represent financial loss and perhaps machine failure. Design philosophy centres on the efficiency aspect.

Largely ignored is another kind of energy converting machine, for information machine or measuring instrument sensors also are concerned with energy conversion and enhancement. The physical principles on which they are based are often the same as for work machines. The components used are often the same. What then is the distinction?

The information machine is concerned with faithful transmission of usually low energy level signals which convey the meaningful measurement information as signals modulated onto the energy carrier. The prime task here is not to minimise energy loss but to convey the information as faithfully as is necessary. Accuracy of transmission of the meaning encodement depends upon the degree of disturbance by noise sources and the extent to which the machine mechanism distorts the original signal. Energy wasted in information machines is not usually an important parameter because the energy levels are generally quite low compared with those of work machines. In contrast work machine design cares little about the accuracy of conversion: this concept, in fact, seldom arises in the work machine because the energy flow is treated more in the steady state than in a modulated form.

The components of both forms of machine will usually have the same generic form. Gearwheels, shafts, amplifying elements, transforming elements and the storage and dissipative elements of both power and information machines look alike in shape structure and have identical forms of mathematical model. The components of the information machine, however, will require tighter control over the factors that can introduce distortion and noise deviations. They can also often be made less robust compared with their power machine counterpart for the energy level is usually less. Requirements of conformity to linearity and other distorting mechanisms, however, may require robustness to restrict, say, deflection which would degrade the faithfulness of transmission.

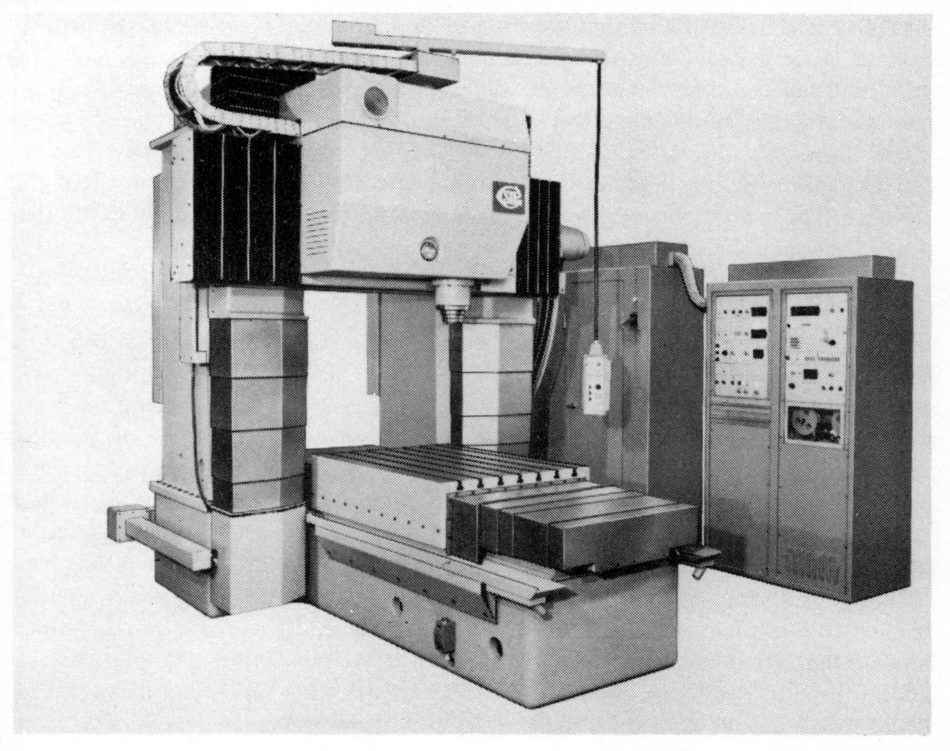

Fig. 2.2 *Subcomponents of an information machine are often of the same generic form as for a power machine. In the numerically-controlled machine tool the two functions are combined into the same components.*
[Courtesy of SIP, Société Genevoise d'Instruments de Physique]

A good example is to consider the automated machine tool used in numerically controlled manufacture of components. Such a machine is shown in Fig. 2.2. The force-loop required for metal removal often also serves as the measurement-loop for shape control. Deflection of the force-loop under the forcing load is a prime variable effecting geometrical shaping for it alters the measuring loop relationships. Such deflection has a relatively insignificant effect on the cutting action.

Design of mechanical instrument components is a task of controlling deflections and irregularities. Although the above example and statements are most easily recognised in the mechanical domain the same conceptual requirements occur in any of the energy regimes i.e. mechanical, electrical, optical, acoustic, thermal, hydraulic, to name those most commonly involved in measuring instruments.

2.3 The importance of improving instrument performance

Measurements are comparisons of the unknown with a standard. The standard is the means by which the unit of the measurement is defined and is a physical invention of man in those applications where instruments can be constructed. The comparison process is never perfect, nor is the definition of the standard. Consequently all measurements are always imperfect. They are subject to a whole host of error sources. These decide the degree of confidence that can be placed upon the measurements made. Creation and application of measuring technology is basically a case of creating a system that has optimal compromises of features so that it is just slightly better than the task requires.

Three basic measurement terms need clarification before continuing. They are very commonly used and it is clear that too few people have a proper appreciation of their meaning when used to describe the performance of measuring instruments. The terms are resolution, precision and accuracy each of which is a general descriptor only.

(i) Resolution of a measurement process is the finest interval of the measurement scale that can be discerned. Ability to resolve can be enhanced by the use of aids that increase the sensitivity. An eye-glass over a mercury-in-glass thermometer scale or electronic amplification of a displacement transducer electrical output signal each increase the resolution. Fig. 2.3 illustrates how a micrometer can be given high resolution. Merely quoting resolution of the measurement can be quite misleading. Adequate resolution is a necessary but not sufficient condition. Too often the system resolution is incorrectly quoted as the precision of the instrument. There is a vast difference between the two. In quantitative use resolution is termed the discrimination.

(ii) Precision is a statement about the closeness to a particular value that the individual measurements of a set of identically performed measurements possess. It expresses how well a measurement process repeats each time the same measurement is made - see Fig. 2.4. It is presumed that during the time taken the parameter being measured - called the measurand - remains fixed and that the scatter of values is due to the process of measurement. Alternatively, this same concept is sometimes used to describe the variation of a parameter as it changes over an interval of time or space. In this case it is assumed, too often without question, that the measuring process remains constant. The term reproducibility is generally reserved for the same concept in the situation where the measurements are not made at the same small interval of

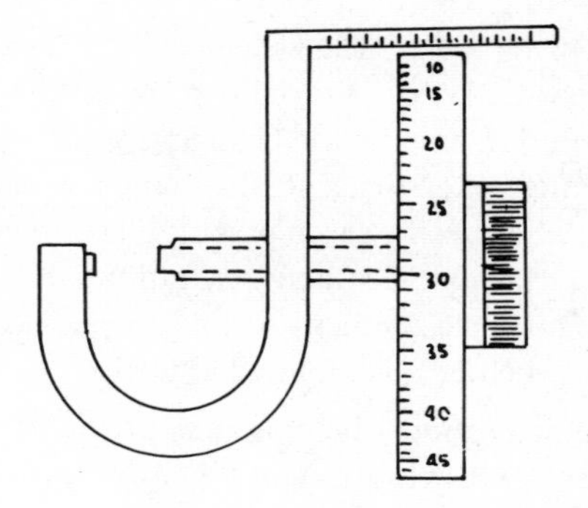

Fig. 2.3 *Adding a large drum to the dial of a screw micrometer increases its resolution but not necessarily the precision and accuracy of measurements made*

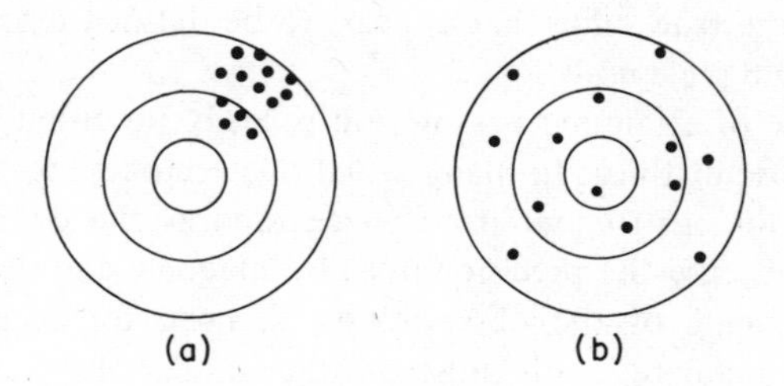

Fig. 2.4 *Rifle shooting illustrates the concepts of precision and accuracy*

(a) A precise group of shots lie close together but are not as accurate as could be; the mean is not in the centre

(b) The group is imprecise but as the mean lies in the centre the group has high accuracy

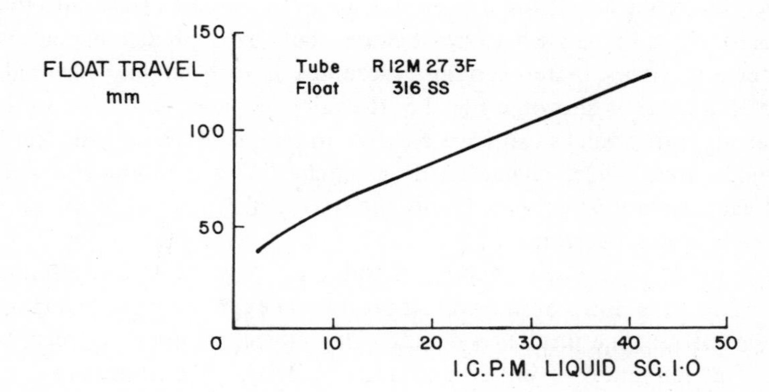

Fig. 2.5 *Systematic errors are predictable from knowledge of the state of the measurement process. Calibration chart for a flow meter.*

time but are, instead, spread over time periods of days to years. Repeated reading of the value indicated on a panel meter will yield varying values if the sensitivity of the meter is adequate to show the system variations measuring. The measuring process must have sufficient resolution to exhibit changing value of a given parameter. The smaller the scatter of values, refer Fig. 2.4, the greater is said to be the precision of the measurement process. In a specific sense precision is termed the repeatability.

Measuring instruments having adequate resolution and precision can often be useful instruments for many measurement tasks. However, there are many instances where the third parameter, accuracy, is required as well.

(iii) Accuracy is the statement about the closeness of the value (that is, for multiple measurement cases given as the best assessment taken from the scatter of values) to the value declared as the standard. As another example to that given in Fig. 2.4, a bent pointer on an indicating meter will not indicate the true value unless it were calibrated with the pointer in its bent condition. It remains precise and the resolution is the same but the value thought to be the magnitude of the parameter under observation is a little different to the true value that appears to be defined into the indicating instrument by the unit and scale used.

In order of ease of attainment resolution is easily procured, precision comes next and accuracy a difficult third. In many situations accuracy need never be considered. This arises when the instrument itself is regarded as the defining standard for the observations. In this case the precision must be adequate and the reproducibility must be within requirements of the observations. A need for accuracy arises when the results need to be compared with those of other similar observations made with other equipment . . . one is not sure if the same standards apply where the instruments themselves have been used as the defining standard.

The above definitions should not be regarded as complete but will, it is hoped, clarify useage at the level of appreciation needed for the purposes of this work.

Measurements are never perfect and errors always will occur. Errors are deviations from perfection. They arise from numerous sources but can be placed into two groups. The first is for those that arise because of causes that have a predictable nature. We call these systematic errors. After certain procedures have been implemented, such as correction to a chart as shown in Fig. 2.5, the value of these can be discovered in such a way that all future values can have a correction applied. The second kind of error comes from sources whose characteristic is stochastic or random. The value of any individual value cannot be predicted, only the statistical properties of the set of values. These are called random errors.

Error amounts are properly expressed today in terms of the uncertainty of the measured value. It is now considered bad practice to express a value giving its error, as was the general practice until this decade. The subject of errors in measurement is extensive. The reduction of error in measurement is a main incentive for the continued development of improved measurement hardware.

It would seem, from a cursory study of several tests published in the second half of the 19th century, that the concepts of resolution, precision, accuracy and the various classes of error, were not seen as topics to discuss in works on natural philosophy, engineering and science in general. It is, however, clear that scientists in all ages who were concerned with the use of observational methods recognised the importance of small measurement deviations in the pursuit of new knowledge.

Thomson and Tait (1883), when discussing the subject of 'Experience', in their discourse on natural philosophy recounted that Herschel (1738-1822) coined the term 'residual phenomena'. This described unexplained differences that might be 'excessively slight', remaining after all conceivable cause of variation had been explored in both the technique and the equipment in use to make the observation. Given that the cause is not of observer origin then it must be a true deviation between the hypothesised explanation and the real one. Thomson and Tait quoted examples of how the 'smell of electricity' resulted eventually in the discovery of ozone; of how small variations in the motion of Uranus led Adams and Le Verrier to discover a new planet; and of how work of Arago on resistance to relative motion between a magnet and a piece of copper, which began as an observation that an oscillating magnetised needle was damped when he placed a sheet of copper below the needle, led eventually to Faraday discovering the induction laws.

Residual phenomena had already led to many great scientific discoveries before the time of Herschel. Jones (1967*a*) gave account of the importance of a discrepancy of 8 minutes or arc between Kepler's predicted position of Mars and that observed by his elder Tycho Brahe. This small difference could not be explained away as instrument or observer error; a new law of planetary motion was needed. Kepler, after enormous labour, showed it was a motion of ellipses. He published the laws in 1609.

There is no shortage of similar examples to prove that improved resolution, precision and stability in instruments, combined with confident observational method, creates a situation where new knowledge may be forthcoming. This is not so surprising for such conditions place the observer in a situation where some aspect of the real world is seen for the first time. If the equipment and method is better than those that preceeded it, it must reveal new knowledge. It provides a journey into the unknown where all manner of hitherto unknown phenomena exist.

This theme can be traced over several publications of R.V. Jones (1967a, 1967b) becoming the central topic of a paper aptly titled *More and more about less and less* Jones (1968) and again in Jones (1970).

2.4 Instrument design

Design is the process whereby a quite specific concrete set of requirements are arrived at that, in the case of instruments, define a unique piece of measuring equipment. Fig. 2.6 portrays this procedure. The starting point for this process is from a position of great generality of choices coupled with an embryonic concept of what is to be achieved. The initial specifications of the measurement requirement are usually only a starting point of reference; the design process increases the designers awareness and understanding of the real need as the design proceeds.

The final hardware result is essentially the outcome of numerous tradeoffs and

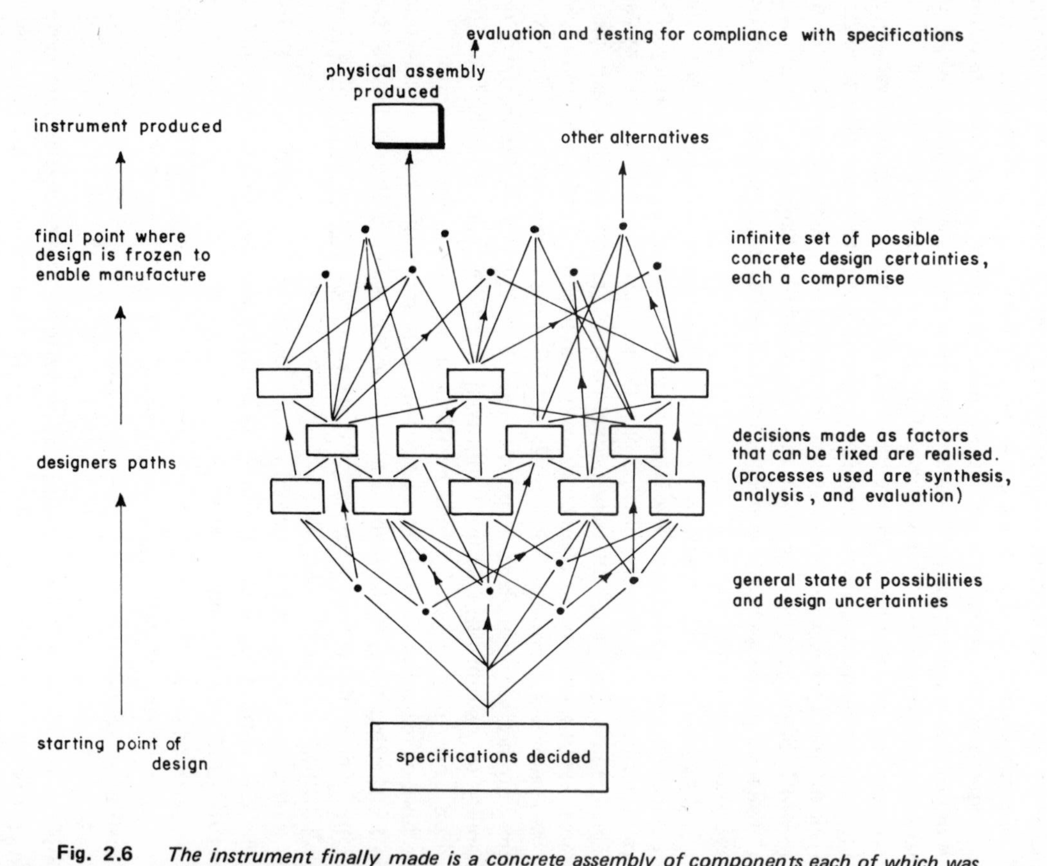

Fig. 2.6 *The instrument finally made is a concrete assembly of components each of which was selected through processes involving widespread compromise*

compromises that are made to suit the state-of-the-art existing at the time of design. Marshall (1978) discusses the process in relation to design of electrical circuits. Dominant classes of factors entering the decision are the availability of ideas, the methodology developed from the design procedure, the materials available, the processes of manufacture, the allowable financial costs, the crafts of manufacture, the designers ability, the experience of past development that is relevant and the degree of understanding of the requirement.

These remarks apply to any technological design enterprise. It is now necessary to translate them into terms of instrument design, so that the historical development of measuring instruments can be better appreciated.

Although many writers refer to the 'design of instruments' it must not be thought that all instruments are, and always were, designed in a rigorous manner using systematic procedures. Example works that illustrate the so-called instrument design method are by Horace Darwin in Glazebrook (1922) Vol. III, and Braddick (1966). Although ingenious and sophisticated instruments of many forms have been created very few developed via the use of scientific procedures even though they are based upon a scientific law. G.M.E. Williams made the following comment in the question time following a paper by Finkelstein (1963):

> When you are engaged, as we are, in trying to break down a subject in order to get the maximum understanding of principles over in the shortest possible time, we have to engage in studies of this kind. We are happy about certain fields which the author has mentioned—control theory, system design, and so forth—which we can interpret, dilute and turn over to the kind of problem which the postgraduate student can, with some difficulty, understand. But in the field of measurement and in the treatment of the technology of instruments, it is just like being in a wild wood at the present time. We know very little yet about how to unify this subject. As it goes along it will help us to shorten the path and build up the transmission of the understanding of principles. All the excellent work which has established length, mass, time and electrical quantities, and so on, is entirely subjective. No one has questioned how it comes about and what one does when trying to do it. This is what we have to do now.

Many a scientific and industrial project has been plagued with instrument problems. When the scientists of the *Challenger* were making the first global oceanographic survey in the 1870s, initial use of their instruments and apparatus under real conditions make it clear that many modifications were needed. An account of this epic voyage, prepared by Linklater (1972), however, gives little detail about the instrumentation. Futhermore the many volumes of the official record contain little more. Fig. 2.7 shows a scene from the records.

In modern times many a major project has failed for lack of adequate measurement forethought and investigation. The major problem appears always to have been that organisers give too little emphasis on establishing just what the measurement systems are supposed to do. Failure of measuring instruments can totally shut down a programme: this is not adequately recognised. It is only in very recent times that a recognisable movement has developed in academic environments centred on systematic

Fig. 2.7 *Emptying sampling water bottles during the epic HMS Challenger oceanographic sea voyage of 1872-1876*
[Photo. Courtesy Science Museum, London]

instrument design. Bosman (1978) is an account of current thinking about instrument systems design.

By contrast, organised design has evolved in the creation of electronic systems, in operations and production, in machines of work and in many other fields. For reasons yet to be studied in depth, the design of instruments is still largely an undefined procedure. Possibly a main factor is that the individual components of an instrument can usually be manufactured, altered and replaced quite readily and inexpensively (compared with those of, say, a 500 Megawatt electricity generator or a warship main-shaft gearbox). The trial and error, cut and try, procedure is workable. Furthermore the designer of an instrument very often begins from a weak specification standpoint and changes are seen to be needed as the development proceeds and the designer learns.

The creation of a measuring instrument that advances the art of information gathering is very much a creative exercise akin to that of creating a work of art. In both, physical processes and materials are assembled by the mind and craft of the maker, into a unique object.

Once a design procedure is systematised in a reasonably general way, as for example, the technique of arriving at the shape, configuration and placement of lenses in a multielement optical system, it can be recorded for others to use in similar circumstances. The person who first formulates the method may have to expend great labour and innovation to achieve this systematisation. Those that follow progressively find it easier, the grade of understanding of the basics needed being steadily reduced. From this observation it can be assumed that published works, such as books and trade pamphlets, demonstrate when a design becomes organised for general availability. The main regimes of instrument design i.e. mechanical, optical, electrical and the relevant specialist texts are now discussed in relation to design.

Burstall (1970) includes discussion of the published works on mechanical engineering. Mayr (1970) covers mechanical devices but from their control-with-feedback

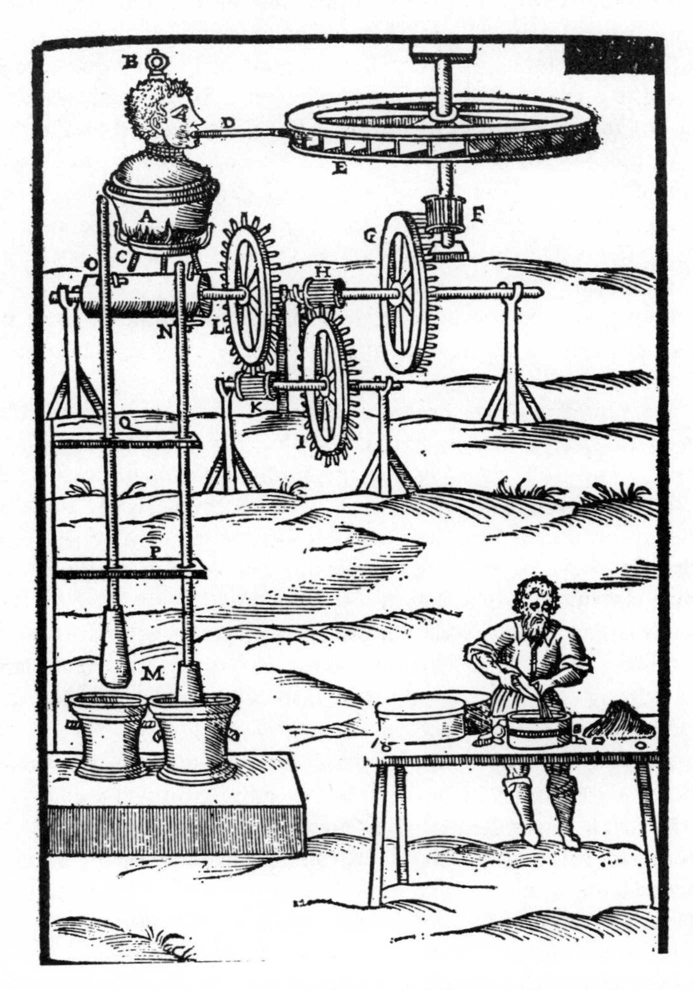

Fig. 2.8 *This figure from Branca's 'Le Machine' is typical of the many works, published by the 17th century, that catalogued mechanical and hydraulic design*
[Courtesy Dawson Rare Books, London]

aspect. Example works of the past are the *De architectura* of Vitruvius; Heron's *Pneumatica*; *Work of Archimedes on the building of locks* by an anonymous author referred to by Mayr as Pseudo-Archimedes - see Fig. 3.8; al-Jazari's *On the knowledge of ingenious geometrical mechanisms* of 1206; Leonardo da Vinci's ideas (that had less impact in his time than might be supposed); Agricola's *De re metallica* of 1556; works of Ramelli and of Haselbery of the 16th century, Giovanni Branca's *Le machine* of 1629 - a view is given in Fig. 2.8 and *Theatrium Machinarium* by an instrument maker named Leupold of 1724. These each show the well developed nature of design realisation for mechanical and fluid devices. They include a small component of measuring instruments, these apparently being seen as just another machine design exercise. Mayr (1976) expands the past played by machines in helping man think out his ideas.

Ferguson (1977) published an interesting account of the pictorial records of technology making the point that 'thinking with pictures is an essential strand in the intellectual history of technological development'. The paper has a useful bibliography on this theme. Westfall (1978) is concerned with geometric form in the development of scientific thought. Drake (1879) presents a late-19th century viewpoint.

Optical design began to find order later than mechanical apparatus but by the 18th century it, too, was explained in texts. Fig. 4.48 is an example of textbook illustrations of that period.

Electrical knowledge did not include electromagnetism until the start of the 19th century but magnetic and electrostatic principles were being applied in a reasonably regular manner by the 18th century.

Specialist measurement texts began to appear at the start of the 20th century. Examples are Darling's *Pyrometry* (1911); Glazebrook's *Dictionary of applied physics* (1922); Drysdale and Jolley's *Electrical measuring instruments* (1924) and Rolt's *Gauges and fine measurements* (1929). The prefaces to the latter three works each state that the authors understood their works to be the first comprehensive treatments made of their fields. Fig. 2.9 shows the covers of a sample of modern works concentrating on the history of various branches of measurement.

Today the situation is but a little better. There exists at present many hundreds of works on measuring devices, but few are concerned with systematic design. Most are catalogues of existing devices that the user transposes into the new requirement modifying them where necessary.

With so little basic ordering of the modern method it is not so surprising that little research has been carried out that would provide a more thorough account of design of instruments through the ages than is briefly given here.

Lack of ordered information about instruments in texts might imply that little organised knowledge is available. This is not the case. Accounts abound in the journal literature. This information, however, is recorded, not primarily as instrument information, but rather for its application to a need which is seen by the author as a more salient topic than the measurement method used. For this reason extensive overlap of ideas, multiplicity of development and lack of structured knowledge is the situation existing after hundreds of years of development.

In the English-speaking world of the 19th century the organisation of measuring

Fig. 2.9 *Specialist texts on measurement began to appear in the early 20th century. In more recent times these have begun to include many concerned with the historical aspects*

instrument knowledge had some supporters, but not many. One major study, that of the Royal Commission that began in 1970 under the chairmanship of the Duke of Devonshire, Alexander Strange, was concerned with the advancement of science in Britain. Jones (1971a) calls the findings that were issued 'the Domesday Book of British Science'. One recommendation was that a collection of scientific instruments be formed. This reflected Strange's background in which he had made and used geodetic and astronomical instruments in the course of his Indian Survey work.

The collection of instruments that did result forms much of the extensive holdings now held at the Science Museum, shown as it is today in Fig. 2.10, in London. This was the result of a Loan Exhibition of Scientific Instruments, plus other collections, that were put on show in South Kensington in 1876 - see British Association (1931) for the section on museums of London.

The South Kensington Science Museum that developed out of these displays assumed its name around 1862. Since then its presence has had a significant influence on the design of several major museums in other countries. Today it holds a place in the first three major science museums of the world. A study of instrument collections in museums has been published by Sydenham (1977, 1978 a,b). Appendix 2 is adapted from these sources.

Fig. 2.10 *Many historic measuring instruments are held at the Science Museum, London* [Photo. Science Museum, London]

Sherwood Taylor (1972) gave account of the teaching of the physical sciences in the period 1760-1800. It is interesting to see that many notable men of science that gave the world important measuring instrument principles and devices had no formal training. His list includes, for those without formal tertiary education, Ampere, Faraday and Wheatstone. Others who received education, but not in science, included Brewster and Schweigger. Those with mathematical education included Fourier, Fresnel, Gauss, Herschel, Nobili, Ohm, Talbot and Whewell. Medical background was the basic education of Henry, Oersted, Poisson, Seebeck, Wollaston and Young. Each of these people made substantial contributions to the science and practice of various aspects of measurement hardware.

Sherwood Taylor's account also discussed the popularisation of science in the 17th century. He mentioned the many scientific encyclopaedic works beginning with *Lexicon technicum* by J. Harris of 1704. He suggested that the then expanding instrument maker industries were keen to promote their products by way of popularised publications. He did not single out measuring instruments for discussion but it is evident, from his descriptions of the teaching in the traditional fields of mathematics, physics, and chemistry, that instrumentation was included, but as a topic having no integrated structure in those courses.

Journals began in earnest in the 17th century. That period has been studied by McKie (1972). Two major societies then arose; the Royal Society in London and the Royal Academy of Science in Paris. The Royal Society published the first issue of its *Philosophical transactions* on March 6, 1665. In the same year a French publication, the *Journal des Sçavans*, appeared. Both publications were concerned with the natural philosphy and its experimental aspects. The title page of the *Philosophical transactions*, given in Fig. 2.11, shows the place that instrumentation had in the reports given in such publications.

By the end of the 18th century many journals had been introduced, Sturgeon's *Annals of electricity, magnetism and chemistry* began in 1837. None, however, seems to have been devoted to measurement as a discipline in its own right. They are products of thinking of the 20th century. The place of observation, if not measurement, was, however, seen as significant by some in the 17th century. The scientific journal *Observations sur la Physique, sur l'histoire naturell, et sur les arts* began in 1771. It was significant as a specialist journal of physics which it soon became. McKie states that it was not for 'leisured amateurs' who sought entertaining reading as an apparent learning process, which, in reality, had no substance.

Because organised procedures of design did not generally exist it should not be assumed that realisation of the need did not arise. Strange, giving evidence to the Royal Commission on science mentioned above, had this to say:

> The design of scientific instruments should be systematically improved. As a rule, scientists were not good designers of instruments. A collection of scientific instruments should be made (a science museum) to aid in the development of design. I find very few persons who have really studied what I will venture to call the physiology of instruments and apparatus.

According to Jones (1967*a*), Strange well realised the need to give greater consideraion to design for most manufacturers were craftsmen seeking financial profit and were not attempting to be pioneers in a scientific pursuit. Jones, earlier in the same paper, stated relevant views of Priestley included in his 1767 *History of electricity*. Part of that quoted passage gives some insight into Priestley's attitude toward design:

> It were much to be wished, that philosophers would attend more than they do to the construction of their own machines. We might then expect to see some real and capital improvements in them; whereas little can be expected from mere mathematical instrument makers, who are seldom men of any science, and whose sole aim is to make their goods elegant and portable.

It is easier to criticise than be correct is a well known adage. It applies so aptly to this discussion for to make comment on instrument design of the past based on some positive understanding it is vital that some appreciation be gained of the controlling factors that limited a designers scope then and now.

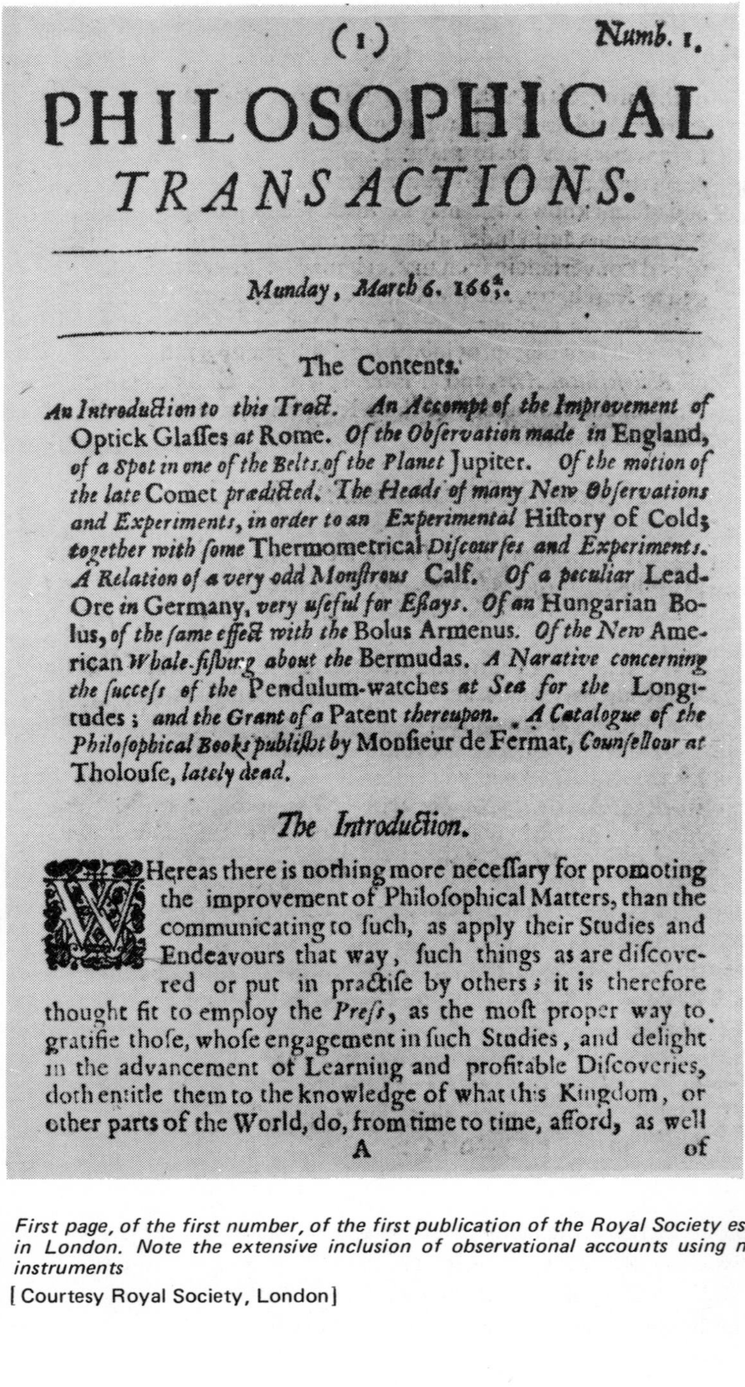

(1) Numb. 1.

PHILOSOPHICAL TRANSACTIONS.

Monday, March 6. 166$\frac{4}{5}$.

The Contents.

An Introduction to this Tract. An Accompt of the Improvement of Optick Glaſſes *at* Rome. *Of the Obſervation made in* England, *of a Spot in one of the Belts of the Planet* Jupiter. *Of the motion of the late* Comet *prædicted. The Heads of many New Obſervations and Experiments, in order to an Experimental* Hiſtory of Cold; *together with ſome* Thermometrical *Diſcourſes and Experiments. A Relation of a very odd Monſtrous* Calf. *Of a peculiar* Lead-Ore *in* Germany, *very uſeful for Eſſays. Of an* Hungarian Bolus, *of the ſame effect with the* Bolus Armenus. *Of the New* American *Whale-fiſhing about the* Bermudas. *A Narative concerning the ſucceſs of the* Pendulum-watches *at Sea for the* Longitudes; *and the Grant of a* Patent *thereupon. A Catalogue of the Philoſophical Books publiſht by* Monſieur de Fermat, *Counſellour at* Tholouſe, *lately dead.*

The Introduction.

WHereas there is nothing more neceſſary for promoting the improvement of Philoſophical Matters, than the communicating to ſuch, as apply their Studies and Endeavours that way, ſuch things as are diſcovered or put in practiſe by others; it is therefore thought fit to employ the *Preſs*, as the moſt proper way to gratifie thoſe, whoſe engagement in ſuch Studies, and delight in the advancement of Learning and profitable Diſcoveries, doth entitle them to the knowledge of what this Kingdom, or other parts of the World, do, from time to time, afford, as well

A of

Fig. 2.11 *First page, of the first number, of the first publication of the Royal Society established in London. Note the extensive inclusion of observational accounts using measuring instruments*

[Courtesy Royal Society, London]

2.5 Innovation and invention

No machine has yet been devised by man that can carry out effective original design. The power of imagination that allows physiological man to represent things and circumstances with other forms and models in his brain allows new creations to be realised that are more than copies of things observed in nature. Material resources are not enough; the human agency is needed to catalyse and arrange these into the new assembly. The key to the start of development of a new instrument, as with any new machine, are the thoughts of one or more individuals. Everyone is a potential inventor for our minds constantly create scenarios of situations that do not exist. Emerson put it this way:

> We are all inventors, each sailing out on a voyage of discovery guided each by a private chart, of which there is no duplicate.

The successful inventor is that person who has the right ideas at a time when they can be exploited and who has the ability to see that the situations are favourable for development either in the hands of others or by the inventor's own efforts. Insight into the processes involved can be gained from Hanson (1958), Jones and Thornley (1963), Glegg (1969), Leech (1972) and Holton (1978).

Despite considerable thought and research, that has attempted to define the parameters of the successful invention process, there is still no rigorous way to ensure that a good invention will result. Pouring vast sums of money into ideas teams gives no positive guarantee that benefits will accrue in proportion to expenditure. The poorly trained person on the slimmest budget may well be the source of a significant breakthrough.

The developments of measuring instruments follows this general pattern. Past events have proven this time and time again, as the following chapters of this account will show.

There exists a common misunderstanding that inventions generally appear as sudden events in history, the inventor producing them as his or her sole effort. Reality has it that most, whilst indeed being brought to the attention of the general public by an individual, usually rest on the previous work of numerous other people. Those people would not often have recognised their contribution to the specific invention that later evolved. Nevertheless, their work paved the way for others to follow.

This is well illustrated by the development of commercial radio; it is chosen here as an example for that development took place over at least 350 years and it contains numerous instrument discoveries and inventions along the way. Table 2.1 is a chronology of events seen as significant by Sydenham (1975) that each led to other necessary steps. The idea that messages could be transmitted through space can be traced back to 1558. The actual wording of Baptisa Porta's manuscript shows that the idea was conceptually correct but practically wrong. At any time in that history it is not to say that those following were necessarily influenced by the contemporary writings of others but it is clear that people of a particular time have greatest interest in the fashionable subjects and attitudes that prevail contemporaneously. Persons who differ

Table 2.1 *Chronology of events preceding commercial radio*
[From Sydenham (1975). Courtesy *Electronics Today International,* Sydney]

1558	Baptista Porta quoted use of magnetic influence
1730	S. Gray discovered the use of wires to convey static electrical charges
1745	Leyden jar devised by Kleist, Muschenbroek or Cuneus – the first capacitor
1746	Winkler discharges Leyden jar using water return circuit
1747	Watson transmitted static electricity in water and wire circuit
1788	Barthelemy suggested use of influencing magnet in a work of fiction
1795	Salva's paper presented on application of electricity to telegraphy – predicts telegraph
1800	Volta's primary battery invented
1803	Aldini used wire supported on a boat and water return path to communicate between Calais and Fort Rouge
1809	Sommering built a working 26-line galvanic telegraph using wires
1819	Oersted discovered current in a wire deflects the needle of a compass
1825	Sturgeon made first electromagnet
1827	Ohm announced his law relating current, voltage and resistance
1831	Lindsay telegraphed via ground and water circuits (probable date) Faraday discovered laws of induction
1832	Schillings five wire telegraph
1837	Anon. report appeared in *The Mechanical Magazine* on use of earth return Cooke and Wheatstone five needle telegraph patented. Immediate use on Great Western Railway Morse built first reported relay (contested claim)
1838	Steinheil of Munich worked on railway lines as 'cable' paths leading to use of earth itself Munck discovered metallic filings reduce in resistance when electrical charge flows through them
1842	Morse's celebrated cross-river wireless link using grounded plates
1843	Wheatstone's bridge circuit published
1845	Brett referred to possibility of oceanic telegraph without cables Wilkins proposed communicating between England and France using terminals dipped in earth and water at each end (1849 is more usual date quoted)

1850 Guitard noticed electrified air causes dust to cohere

1851 Cross-channel telegraph cable laid (the first underwater cable)

1852 Hightons started 20 years of work on wireless communication by conduction
Kelvin related *L, C* in resonant circuits to natural resonant frequency

1853 Kelvin published paper *Oscillatory discharge of a Leyden jar*

1854 Lindsay was granted patent of his cross-river method by conduction

1856 Varley discovered coherer principle

1858 Transatlantic cable laid
Kelvin's sensitive mirror galvanometer used on Atlantic cable

1859 Lindsay addressed British Association on ways to improve conduction method of wireless

1861 Feddersen's work experimentally proved oscillatory nature of discharge

1862 Patents issued on use of induction and leakage currents for system of telegraphy

1863 Clerk Maxwell suggested EM waves exist using theory only

1866 Usually accredited date for Varley's discovery of coherer
Sorensen (Danish) signals ship from shore by conduction

1867 Clerk Maxwell produced formulae of describing EM waves
Kelvin's syphon recorder devised and built

1870 Barbouse system of conduction wireless readied for communication with besieged Paris

1873 Clerk Maxwell published treatise *Electricity and magnetism*

1874 Proposals begin to talk of induction as well as conduction wireless

1875 Edison talked of etheric force as basis of wireless communication

1877 *Journal of telegraphy* article ridiculed proposals for a 'radiation' wireless by Loomis

1878 Hughes experiments with metallic coherer and probably detected electric waves from a spark discharge

1879 Berlin Academy of Science offered prize concerned with nature of Maxwell's theory of EM waves
Hughes observed EM waves over 600 m range
Hughes noticed coherer affect with carbon rods resting on steel plate
Brit. Patent 3132 described use of induction communication between cables at sea to a lightship

1880 Trowbridge extended Gott's idea of earth return and started on transmission without wires by a systematic study of induction effects between wires
Trowbridge induced signals in parallel wires at 2 km

1881 Smith's induction link for telephony to railcars in motion

1882 Preece completed experiments on conduction wireless between the Isle of Wight and Mainland

1883 Dolbear placed one wire of induction link in air as antenna
Lodge rediscovered coherer principle
Brit. Patent 4220 replaced one conductor with air-path requiring 'moist air to complete circuit'
Edison noticed rectifying effect in lamp (described in an 1884 patent)

1884 Dolbear demonstrated signalling with antenna
Phelps railway induction link
Preece began radio experiments including induction between wires

1885 Preece experimented with induction wires
Edison patented wire-less train telegraph using electrostatic induction

1886 Hertz obtained EM induction with close coils and Leyden jars

1888 Hertz demonstrated existence of EM waves

1890 Branly's improved coherer (based on Calzecchi-Onesti and Varley)

1891 Anon. writer in Electrician suggested 200 000 MHz waves could be used to communicate through walls
Trowbridge suggested aerials for ships

1892 Brit. patent 10161 described shore to lightship communication via induction between cables
Crookes predicted use of focused waves in *Fortnightly Review*

1894 Preece summarised his past ten years of work on wireless
Hertz died
Lodge demonstrated etheric wireless to Royal Institution and British Association using Hertzian oscillator and Branly Oscillator over 120 m
Marconi's personal claim to first recorded message through space by EM waves
Preece addressed Royal Society of Arts
Michel's ground telegraph
Gavey's Loch Ness experiments on conduction wireless telegraphy
Marconi system using Righi design of oscillator
Editorial in *The Electrician* predicted use of radio waves to detect storms

1895 Russo d'Asars, acoustic, ship wire-less telegraphy experiment achieved 100 km range (date approximate)
Popoff worked on electrical storm detection using Branly coherer and lightning conductor; incorporated decoherer device
Rutherford realised principles later used in magnetic detector
Preece made temporary link across 12 km range using induction

1897 Wilson and Evans detector (similar to Rutherford's)
Slaby began radio experiments at Charlottenburg after working with Marconi, Co-worker Count Arco
Marconi's own claim for first telegraphy between ships in motion (20 km range)

1898 Marconi devised means to electrically isolate coherer from sky rod thus introducing transformer coupling – the jigger
Lodge's system released
Braun used tuned circuits as coupled resonance
Tesla's patent for controlling route of a distant ship by radio
Zickler reported ultra-violet link over 1·3 km
Lodge and Muirhead magnetic decoherer
Preece delivered lecture to Royal Institution supporting Marconi's claims

1899 J.J. Thomson explained the Edison effect of 1883

1900 Popoff's patent of improved coherer
Dudell's patent on singing-arc continuous wave production
Car radio patented

1902 R.A. Fessenden of Pittsburgh sent voice over 2 km
Marconi patented magnetic detector

1903 Poulson's patent for using Dudell method to generate radio waves
Electrolytic detector patented

1904 Nussbaumer's musical transmission system
Hueslmeyer's radar using radio waves reflected from ships

1905 Prize offered by International Congress of Aeronautical Navigation for 480km range transmission of 1/10 HP by electrical waves
Lieben's valve work (patented in 1906 as a relay) using magnetic field control of internal current

1906 Dunwoody discovered crystal detector

1907 Fessenden patented heterodyne detection – well before its time
Lee de Forest patented triode valve

1908 Wien's quenched spark-gap

1910 Lieben built valve amplifier

1913 First USA musical transmission

1922 British Broadcasting Company came into existence

are termed radicals. Their ideas, if they do not align with popular thought, are liable, in the immediate, more to rejection and destructive criticsm than to acceptance and optimistic argument.

Measurement systems of all kind are a sub-group of the general class of all inventions. They are, therefore, subject to similar constraints.

The subject of invention has fascinated writers over all time. This could be because such works are popular with the public and, therefore, saleable. It could, also, be that the authors like to associate themselves with the creative processes of those they report. In preparing their accounts they may also have hoped to gain valuable insight about the parameters concerned.

Beckmann (1846), of the University of Göttingen, published a two volume collection of histories of inventions, discoveries and origins. He gave a little space to measuring instruments; volume 1 discusses the odometer, water clocks and mechanical clocks. A later work, Timbs (1860) gave 'stories of inventors and discoverers in science and the useful arts'. He provides accounts of the invention of the magnetic compass, the barometer, automata and speaking machines, Francis Bacon's *New Philosophy*, the watch and the chronometer, calculating devices including that of Babbage, the telescopes of many people, many optical devices, Dr. Franklin's work on electricity, photography and the electric telegraph. The current value of accounts such as these is their extensive detail of facts about the circumstances and times of the invention. Works today give far less incidental detail. Early accounts can, however, be difficult to understand for terminology was not necessarily used as it is today. The reader is, also, often under the impression that writers often stretched the truth somewhat.

An anonymously composed work, Nelson (1881), provides accounts of invention putting them in the light as triumphs, as victories of peace. It gives accounts of the telegraph, the telephone and the phonograph (which had recently been invented at that time), the spectroscope and photography. The three above-mentioned works, being sparsely illustrated, only provided verbal descriptions and were apparently meant to provide little more than pleasant reading about fascinating subjects. White (1827) published an alphabetically arranged list of inventions and discoveries.

In 1900 a far more useful account of the results of invention was published by Byrn (1900) in the USA. The author apologises in his preface for not giving much more than a cursory view of his task, that of reviewing a century of invention. It's numerous excellent illustrations, concise factual statements and patent information makes this work a milestone in the record of the outcome of invention from an eventful era, the 19th century. Chapters about measuring instruments and allied apparatus include the electric telegraph, the Atlantic cable, the telephone, electricity, medicine, the phonograph, optics, photography and X-rays.

Moving into the 20th century the work by Hall (1926) is a popular account of various technological facets of life at that time. It, however, has only a little on instruments. Larsen (1960) is well illustrated and interesting account of ideas and inventions. It does not include discussion of instruments as such but is useful for the information about the practical use of instruments at the time when electronics had just found its way into most activities. An authoritative work is a dictionary of inventions and discoveries by Carter (1974). Hanson (1958) is concerned with the conceptual bases involved.

It has been implied here that inventions are often very much a matter of chance encounter. That is the theme of the book by Batten (1968). Mary Batten there describes how Daguerre stumbled on a way to take pictures directly from nature, how Galvani's work, Fig. 2.12, by chance led to Volta's primary electric cell, and how Röentgen came upon X-rays by accident.

Fig. 2.12 *Galvani discovered that certain metals caused muscular tissue to contract. This chance discovery led to the invention of the primary electric cell.*
[James-King (1962*a*); from Galvani, L. *'De Viribus Electricitatis in Motu Musculari Commentarius'* Bologna, 1791]

The substance of invention is the combination of many things. Of these the human mind is the agency that ties them all together in the creation of the new system. Because of the great sophistication of modern technology it is easy to superficially regard ourselves as superior inventors to our forefathers. Man can only create what can be created: a fantastic new development is not so amazing once its fundamentals and operation are understood. Cleverness is hard to measure. Evidence of success given a certain background of resources suggests inventors were equally as good at invention in past times as they are today. Is the complexity of the instrumentation of an Early Warning System for defence any greater and difficult to the designers than tasks presented to those who created the Gothic cathedrals?

It would seem that inventors of the past had much the same problems to contend with as those of today. They had to work within the available materials, the available processes, the available machines of manufacture, the availability of theoretical basics and the craft skills of their times. They also often went wildly astray, Hering (1924).

These factors are the subject of the following sections.

2.6 Availability of materials

Measuring instruments, regardless of the energy form used in their operation all require physical materials to form a structure. Even electronic devices, which somehow are thought of as being quite divorced from the mechanical technologies, need materials from which to construct the microcircuits. Burstall (1963) organised his account of the history of mechanical engineering along the lines of division into themes of materials, tools and machines for each period covered. That work provides considerable indirect insight into the supply and availability situation that instrument makers had to work within. Pearsall (1974) and Glazebrook (1922) each devote chapters to materials, the former being concerned with instrument manufacture, the latter on metallurgy.

Although today we have a vast variety of materials to choose from there is still need for progress. Modern-day instrument designers still must make unwanted compromises in design to suit the materials available.

Measuring instruments, like all other products of manufacture, use an incredible variety of materials. Table 2.2 is given to assist the reader. It lists the discovery and general use of many of the materials used throughout time in instrument and scientific apparatus construction. It would be quite wrong to assume that the discovery of a material meant that it was immediately available for general use. To isolate a small quantity of a substance is a different matter to producing enough to make a viable industrial product. A substantial demand must first be established by the industry. Aluminium, for instance, was first separated as a metal by Wöhler around 1828 (dates vary a little according to which source is used), before that Oersted had already carried out important steps in the chemical extraction process. Deville, in 1854, produced the material as a commercial product but the price and purity were far from satisfactory. Electrical refining of the aluminium came later again, in the 1900 period. Pearsall (1974) reported that aluminium was £48 sterling per imperial pound in 1856 falling rapidly to £9 per pound a year later. Byrn (1900) quoted aluminium prices as $12 US per imperial pound in 1878 falling to 33¢ in 1900. Sheet aluminium was not generally available until the 1920s. Aluminium did not become a major instrument material until after that time.

Table 2.2 was assembled from various sources, the key works used being Byrn (1900), Pearsall (1974), Burstall (1970) and various encyclopaedias. The shortness of description of such a condensed table necessarily restricts the information content. The above sources do not always agree on dates. The table should, however, achieve its prime purpose, that of showing how materials gradually became available as scientific and industrial developments took place. The time lag for new materials to reach the market place varied greatly. Selenium, for example, discovered in 1818, was not in general instrument use until a century later. The same situation held for silicon and germanium which form the basic materials of the modern mammoth semiconductor industry that began in the 1950s.

High purity copper and aluminium are the results of the availability of the direct current electric generator, which became commercially useable in the 1840s.

It is often said today that simple science apparatus is a thing of the past. Necessity is the mother of invention. The materials used to produce the first telephone, by Reis in 1861, shown in the 1896 etching given in Fig. 2.13, were these: a coil of wire, a knitting needle, the skin of a German sausage, the bung of a beer barrel and a strip of platinum, Byrn (1900). Throughout time instruments have ranged from the simplest to the most complicated. Their design had to make use of the materials that were available. Seldom did the instrument need cause the development of new materials in the same way that, say, a new metal needed for a power plant component will trigger a research programme. Table 2.2 shows that many of the materials commonly used have been known to man since the earliest times. Having a raw material is, however, merely the start for it may not be in a form that suits the need nor might it be workable by the tools and machine processes of the time. It must also be remembered that materials were not made available over all geographical regions together. Materials design restrictions began to be relaxed from around the middle 18th century with the pace of new material discoveries and their marketing quickening as the industrialisation took hold in Europe in the ensuing century. Materials produced and marketed primarily to meet needs of a large industrial and consumer products market became readily procurable to the smaller consumption instrument maker.

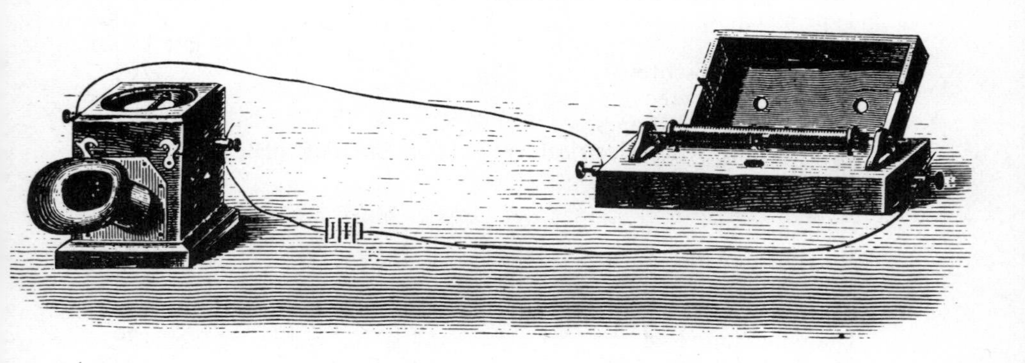

Fig. 2.13 *The Reis telephone of 1861*
[Frith and Rawson (1896), Courtesy Ward Lock Limited]

2.7 Availability of processes and machines of manufacture

Technology is the broad discipline that converts existing physical things into new forms. The conversion is carried out using tools fashioned by man. Tools make more tools and more tools enable more materials and more objects to be realised. The interdependance between these factors is obvious. Development of technological achieve-

Table 2.2 *Availability, with time, of materials used in instrument construction*

Ancient times: the BC period

Elements: Antimony, carbon, copper, gold, iron, lead, mercury, silver, sulphur, tin.

Alloys: Bronzes (3000 BC), steel (1000 BC)

Fabricated and natural: Wood, bone, flint, plywood, leather, abrasive powder, metal and wooden nails, woven cloths and ropes, concrete (Roman period) ivory, clay, glass, rock crystal.

These materials were carried through the Dark ages which ended around 1300 AD

Modern times: from 1300 AD	*AD*
Paper arrived in Germany from Far East	1320
Zinc discovered	*c*1400
Pewter in use (tin with other metals)	1400s
Type metal	*c*1500
Rolled brass	late 1500s
Phosphorous discovered	1669
Platinum discovered	*c*1735
Crucible steel for springs	*c*1740
Silver plating by nonelectrical method (Old Sheffield plate)	1743
Nickel discovered	1751
Oxygen discovered	1772
Manganese discovered	1774
Molybdenum discovered	1782
Tungsten discovered	1783
Uranium discovered	1789
Titanium discovered	1791
Beryllium discovered	1797
Chromium discovered	1798
Seamless, thin wall, brass tubing available	*c*1800
Sodium and potassium discovered (Davy)	1807
Cadmium discovered	1817
Selenium discovered (Berzelius)	*c*1817
Silicon discovered (Berzelius)	1823
German silver introduced (brass with nickel)	1824
Aluminium isolated (Wöhler)	1825
Vanadium discovered	1830
Natural rubber on cloth (Chaffee)	*c*1830

(Table 2.2 continued)

Muntz metal patent by Muntz (Brass with 40% zinc)	1832
Galvanising of iron introduced (Craufurd)	1837
Goodyear Patent gives rubber some usefulness	1837
Vulcanised rubber (Hayward)	1839
Fused quartz crystals (Gaudin)	1839
Silver plate by electro method (Elkington)	1840
Vulcanised rubber (Goodyear patent)	*c*1840
German silver obtained commercial impact	*c*1840
Gutta Percha insulation and lining material marketed in Britain	1843
Bearing metal (Babbit)	1843
Fused silica produced (Despretz)	1849
Aluminium made commercially (Wöhler)	1854
Steel by Bessemer process (Bessemer)	1856
Steel by open-hearth method	*c*1860
Materials test laboratory set up in London	1860-1880
Aluminium bronze introduced	1862
Celluloid marketed by Hyatt: first plastic material	1868
Phosphor bronze introduced	*c*1870
Nickel in common use	1880
Copper refined by electrolytic process	1885
Germanium discovered	1886
Fused quartz popularised by Boys as instrument material	1889
Artificial silks introduced	*c*1890
Carborundum manufactured (Acheson patent)	1893
Artificial diamonds attempted by Moissons (unproven)	1894
Helium discovered (Lockyer)	1895
Invar discovered (Guillaume)	1896
Neon discovered	1898
Radium discovered	1898
Free radical identified, enabling plastics era to begin	1900
Bakelite patented (Baekeland)	1907
Elinvar (Guillaume)	*c*1920
Mu-metal and Permalloy	*c*1920
Polythene developed by ICI	1935
Thermocasting plastic	1943
Plastics became reliable, general purpose, instrument material	*c*1945

ment progresses through each factor providing the next step for another factor to be realised.

The Stone-Age peoples used stone materials to fashion tools such as axes, knives and hammers. Then came bronze which could be made into additionally useful shapes by the simple process of pouring it, in its molten state, into moulds fashioned in sand. Iron followed next. It could be moulded, beaten and joined in much the same manner as bronze but it was superior in strength, in ductility and in edge keeping.

Tools made from stone, bronze and iron could be used to fashion wood into objects having purpose of utility

Fig. 2.14 *Lathes enabled cylindrical shapes to be generated with precision. They gave apparatus designers new scope. Blanchard's version of 1820 enabled patterns of non-cylindrical shape to be reproduced*
[Byrn (1900), Fig. 247, p.369 from *Scientific American*]

A considerable amount of processing of raw materials into objects was, and still is possible, using hand craft. Dramatic improvements in productivity came with the invention of machine tools that extended the hand skills. The earliest machine tool appears to be the crude lathes that date from BC times. In machine tools the material manipulation process i.e. cutting, shaping, grinding and spinning are predominantly controlled by the machine members, not the hand of the operator. The operator is given mechanical advantage over the forces involved thereby enabling faster production, better shape control and reduced operator fatigue. Evolution eventually produced machines, that, once set, produce components with little operator involvement.

Machine tools gave the manufacturer new degrees of freedom in design. The lathe enabled cylindrical shapes to be made that had geometrical precision way beyond the skill of the hand craft method. Blanchard's patented copying lathe of the 1820s is shown in Fig. 2.14. Machines and manufacturing processes developed in the main for larger consumer and commercial markets, enabled improved measuring instruments to be produced. Methods of mass production of the late 19th century lowered the prices of machine tools and materials allowing expansion of certain instrument markets.

Using similar sources as were used to prepare Table 2.2 a table has been drawn up of the times that machine tools and manufacturing processes became available. This is Table 2.3. Entries have been selected to include those certainly used at some time in the manufacture of instruments.

It would be expected that materials would have found faster acceptance into instrument products than would new processes and machine tools. The latter were substantially more costly to process and required centres of manufacture to make them before they could be applied in routine manufacture of measuring apparatus. The simpler the process of machine tool the faster it was likely to have been accepted. Early machine tools were fashioned as one-off products. It was not until the 19th century that machines mass-produced more machines of like kind. Mechanical apprentices of the 1910s often were set to the task of building a metal-working lathe from raw materials using only hand tools!

The needs of instrument manufacture had little influence on the development of the processes and machine tools needed to make them. The physical construction of instruments can be thought of as being almost totally derivative from the developments brought about by more defined and larger market needs. Reference is made later to Ramsden's dividing engine, but such special developments were rare. Even today the diversity of measuring instrument needs and the wide-spread tolerance of the use of yet another one-off design restricts the size of production runs of instruments to sizes where economy of scale of manufacture cannot be practised to much affect. Only in areas of electronic circuitry have quantity requirements resulted at the scale of market seen in consumer mechanical equipment markets.

As mass-production methods found favour the experience of that enterprise passed into instrument manufacture. A 1905 scene showing assembly of cash registers is given in Fig. 2.15. The improvements in component precision, made possible by the concepts of tight tolerance using quality machine tools, gradually put the craftsman designer-builder at a serious cost disadvantage. Today it is still necessary to manufacture prototypes by costly jobbing procedures and as many instruments in use are only made as a single unit many of the hand-crafted, build as-you-go, methods are still in use. Certain components became more easily and cheaply available. Gears, shafting, fastenings (e.g. nuts, bolts, screws, cotters, washers), sheet material, electronic subcomponents, hydraulic fittings, wheels, belts and many more items are now so common that they are invariably specified as ready-made stock lines rather than be called up

Fig. 2.15 *Mass-production methods, introduced at the end of the 19th century, enabled fine mechanism products to be made in large numbers at low price*

[Horner (1905), Vol.1 facing p.208. Courtesy NCR Corporation] as a special. It is all too easy to forget that this has not always been so and that it is only a feature of design that has come about in the past century. Unfortunately the different nature of the information machine to the work machine . . . refer to Section 2.2. . . still restricts the instrument designer in the choice of components that can be used from those available on a widespread, low-cost, basis.

2.8 Availability and use of theories and principles

It has been written earlier in this account that science is the means by which new knowledge becomes available through the use of systematic search procedures and how that knowledge is condensed into statements that are concise, yet broad in the applicability. Furthermore it has been shown that technology, as well as science, can also produce ideas of importance. From that path, however, there will not be the same degree of basic understanding of the general rules involved. In such cases science might be used to establish such basic facts after the event. In reality, the spectrum of ideas used in instrument forms ranges from devices built with no science at all to those having a clearly evident basis of science without which they could not have emerged. In this Section it is the availability of theories and enunciated principles or inventions that are investigated as these are another parameter of design that has

Table 2.3 *Availability, with time, of processes and machines of manufacture*

Ancient times: the BC period

Egyptian, Assyrian region

Bellows to produce forced air blasts
Siphoning with tubes
Wedges to move and hold objects
Twist drills using bow-string drive
Extrusion of soft metals
Cold working of bronze
Lathes of wood for wood working
Iron rasp for woodwork
Rotary corn mill
Stamping rounds
Lost-wax casting
Tempering and hardening of steels
Arrow head twist drills
Rope making by hand
Mortice and tenon joints
Dowel joints
Soldering of gold

(Table 2.3 continued)

Potters wheel
Woodcarving to an exacting degree of shape

Greek and Roman regions

Screw threads
Die sinking
Gear wheels
Water wheel as power source
Wood plane
Lathes well established

The above lists indicate availability over a span of 3000 years and as such not all developments were available immediately in all areas at all times.

Modern times: from Middle Ages	AD
Iron casting began in Middle Ages	
Mould boxes using sand	*c*1400
Printing press with moveable type	1400
Leonardo da Vinci's ideas for machines	*c*1500
Dudley's patent on iron casting	1651
Gear cutting machine (Hooke)	*c*1670
Bi-metal strips	*c*1700
Lathe of part-metal construction introduced	1700s
Differentially indexed gear cutter (Hindley)	*c*1741
Vertical boring mill	1750s
Vertical drill	1750s
Cast iron structures for machine components	1750s
Horizontal boring engine for metal	1760s
Lathe with tool holder on slide (Vaucanson)	1760
Steam engine introduced for industrial workshop use	1770s
Primitive metal gear producing machine for clock parts	1780s
Malleable wrought iron (Cort)	1784
Interchangeable parts jigged for musket locks	1785
All metal lathe (Maudesley)	1800
Circular grinding wheels	1800s
Punched-card controlled loom (Jacquard) well-known	1804
Mass-produced article (clocks of wood)	1809
Interchangeable parts for Colt revolvers	1810s
Planing machine (Roberts)	1817
Copying lathe (Blanchard)	1818
Milling machine (Whitney)	1818
Screw thread of high accuracy for measurement use	*c*1820
Lead sheaths extruded over cables (Burr)	1820
Shaping machine (Nasmyth)	1836
Solid drawn copper, brass tubes (Green) – brazed, or hot-forge welded, strips prior to this	1838
Steam hammer for large forgings (Nasmyth)	1838
Cross feed, split-nut, gear boxes on lathes	1840s
Gear generators giving epicycloid forms	1840s
Uniform threads (Whitworth)	1841

(Table 2.3 continued)

Gimlet pointed screws patented (Sloan)	1846
Sand moulds by machine manufacture	1850
Turret lathes	1850s
Compressed air tools	1850s
Photolithographic process 'perfected' (Osborne)	1859
Twist drills introduced	1860
Continuous wire mills for copper wire manufacture	1860s
Universal mill (Brown)	1861
Thread rolling (Tangye Brothers)	1863
Pocket micrometer in common use (Brown and Sharpe)	1868
Sand blasting (Tilghman)	1871
Sensitive incubator temperature control (Hearson)	1880
Oxy-hydrogen torch for cutting and welding	1880
Electric welding with carbon rod	1885
Electric welding with metal rod discovery (Elihu Thomson)	1886
Portable electric drill (Rowan)	1887
Time-motion concept enunciated (Taylor)	1890s
Resistance welding	1890s
Electric welding with metal rod in practical use	1895
Seamless metal tubing by new principle (M. and R. Mannesmann)	1895
Oxy-acetylene torch for cutting and welding	*c*1900
High-speed cutting steel (Taylor)	*c*1900
Light-weight structures developed primarily for aircraft	post1900
Hydraulic drives and controls added to machine tools	post1900
Tungsten carbide cutting tools	1920s
Die-sinking mills using pattern as source of shape (Keller)	1920s
Jig borer introduced from Swiss clock-making experience	*c*1920
Individual electric motor drives replace line shafts	1940s

influenced the development of instruments throughout history.

Table 2.4 has been prepared to provide perspective about the emergence of significant ideas or theories that have paved the way to the development and use of measuring instruments that have been, at various times, taken up into everyday practice. It is not possible to separate the development of instruments from that of science and technology in general . . . this was discussed in the previous chapter. The table, therefore, is also an overview of the historical devlopment of science, notably that part called physics.

Knowledge is grouped as seems appropriate at the time. The laws of physics have arisen out of studies that began in subjects that appeared to have no relationship to each other. For example it is most unlikely that Galileo would have connected temperature with the electric charges that he would have been familiar with. As the body of knowledge of physical things was developed it seemed most logical, to the scientists of 19th century at least, to divide it into the subject of mechanics, hydrostatics, pneumatics, heat, electricity and magnetism, light and sound – Deschanel (1891) groupings. Here they are grouped mainly as a matter of convenience into three lists. First, mechanics, hydraulics and sound which are concerned with the statics and dynamics of mass; second, electricity and magnetism for they became interdependent from the mid 19th century; and, third, optics and heat because they both pertain to radiation in the higher frequencies of the electromagnetic spectrum. The study of the entries of these divisions shows that they gradually merged at various times of history and that today the distinction is far less clear than it was in the past. Physical sciences and the technologies have now merged to such an extent that modern-day information scientists are considering new interdisciplinary groupings. Hopefully restructuring will overcome the problems of retrieval now faced as the result of the realisation that knowledge and its uses cannot be confined to such simplistic classifications.

The chronologies were prepared from many sources, the most useful for the purpose being Burstall (1970), Butler (1947), Byrn (1900), Cajori (1962), Kimball (1949), Marmery (1895), Pledge (1966) and Sydenham (1977, 1978*a*). Using those sources, entries have been extracted that relate to instruments. It must be realised that the brevity of such a perspective does not enable dates to be quoted in a way that can be taken as definite. Rarely does a new concept arise in its well understood form over a time short enough to be ascribed to a certain year. Faraday thought about and experimented with the concept that magnetic fields must be able to produce electricity for well over ten years. Dates given, therefore, only indicate when the particular idea was well enough organised and tested to be transmitted to others. It would also be wrong to assume that once clarified in the developer's mind the information was indeed always communicated. Because of many reasons some works were not published at all, some were published in coded forms and some were left to others to be published at a much later date.

Publication was no guarantee that the knowledge would be adopted immediately. Some ideas were extremely slow to find adherents; others were accepted very rapidly. Further to this must be added the complication that communication of ideas had to overcome distance, societal and language barriers.

As an example of the failure of written communication it is interesting to read the footnote added by Kelvin to the published text of his delivery of the Bakerian Lecture of 1856, where he acknowledges that his description of what is now known as the Wheatstone bridge, had formerly been published in the text of Wheatstone's delivery of the Bakerian Lecture of 1843.

Detailed use of the dates given in Table 2.4 must be confirmed by following the literature back to the primary sources or, at least, to the scholarly works that show evidence of accuracy. Many works, especially texts, are compiled from secondary and other derivatives sources, a process prone to changes of interpretation. Later sections of this book give more details of many of the events listed. History is never absolutely correct about all of its facts. History is an interpretation made by studying the facts available at the time of writing. Many developments which seemed to have established dates have been found to be incorrect as more research is made. For the purposes of this book it is sufficient to say that it is rarely the case that one person alone was responsible for what was subsequently recognised as a key step in science. The work can usually be traced to be a continuation of the contributions of others. The great scientific minds are those that have brought the contemporary state of scattered knowledge together, forming from that, a more meaningful step forward than others achieved. In reality it matters only a little who was first with an idea when evidence shows virtually simultaneous exposition. Neither does the exact date usually have much significance in the broad perspective presented here.

Finally, on the subject of dates, and their relevance to uptake of the knowledge, it must be said that the theory of the academic scientist is often inadequately expressed for the purposes of popular understanding. Indeed, many keystone publications were often unintelligible even to the contemporary scientists. Their generality often made it very difficult to envisage how the principle given could be translated into a specific unique piece of hardware. Processes of education gradually reworked the original published statement into a far more understandable and fluent concept wherein the key facts are less clouded by irrelevancies of detail of proof. There also exists the need to make the general statement apply to a specific case. This requires a mental skill that

Table 2.4 *Formulation of theories and principles of relevance to instrument design*

Ancient times

Theories and principles used in instrument construction largely are to-day classified predominantly in the discipline of physics. The initial foundations on many subgroups of physics e.g. heat, light, sound, magnetism and electricity, have roots clearly traceable to ancient times. They rarely, however, reached a maturity where formulation and proper understanding of the effect was available. The Greeks too often failed to show that their hypotheses for explaining phenomena were correct by testing them. The Romans did little more than apply Greek philosophy to the pragmatic needs of war, government and law.

Translations of Greek works reached the European world via the Arabic empires. Despite many incorrect theories these remained the key documents until their adherents became convinced that the absolute and final truth about the world had after all, not been achieved by the Greek civilisation.

Principles and theories are often best expressed in mathematical form. Mathematics was founded in ancient times but had evolved little until after the Dark Ages.

Modern times: Renaissance onward

Mechanics, hydraulics and sound

Principle of hygroscope (Alberti, Nicolaus of Cues)	*c*1450
Applied mechanics and strength of materials begun (Galileo)	*c*1600
Equilibrium of forces (Stevin)	*c*1610

(Table 2.4 continued)

Laws of orbiting bodies (Kepler)	*c*1615
Rain gauge (Castelli) – first record in modern times but existed prior to this	1639
Hydrostatic pressure exerted equally (Pascal)	1647
Interest in air pressure experiments developed	*c*1650
Vacuum as effect of reduced atmospheric pressure (von Guericke)	1654
Huygens constructed first pendulum regulated clock	1656
Pascal's law in *Traité de l'èquilibre des liquers*	1663
Laws of elasticity (Hooke)	*c*1665
Theory of compound pendulum (Huygens)	1673
Boyles law of gases (Boyle; also Marriotte)	1675
Force exerted by a jet of fluid proportional to square of velocity (Marriotte)	1686
Laws of motion, gravitation, concept of viscous shear in fluids, contraction coefficient of orifice, theory of speed of sound (Newton, the *Principia*)	1687
Researches in sound published (Sauveur)	*c*1702
Pitot effect (Pitot)	1732
Laws of fluid flow (Bernoulli), in *Hydrodynamica*	1738
Theory of columns in compression (Euler)	*c*1750
Atwood's *On the rectilinear motion and rotation of bodies*	1784
Principle of interference of waves (Young)	1800
Modulus of elasticity (Young)	1807
Vibrations of strings, rods, plates (Chladni)	*c*1809
Theory of beams for wood (Dupin)	1815
'Lissajous' figures first proposed (Bowditch)	1815
Theory of beams for metals (Duleau)	1820
Elastic limit as design basis (Navier text)	1826
Internal motion of particles in fluid (Brown)	1827
Poisson ratio (Poisson)	1829
General equations assembled for strength design (Lamé and Clapeyron)	1833
Kinematics terms introduced (Ampere)	1834
Recognition of stress raisers as cause of fatigue failure (Rankine)	1843
Bourdon discovers, by chance, the Bourdon tube	*c*1850
Autocontrol principles begin to emerge in steam-engine systems	*c*1850
Gyroscope invented (Foucault)	1852
'Lissajous' figures reinvented (Lissajous)	1855
Analysis of continuous beams (Clapeyron)	1857
Foundations laid of metallurgical science	1860s
Theory of harmony gives sound a scientific basis (Helmholtz)	1863
Combination of stresses in materials (Mohr)	1868
Phenomena of controller instability explained (Clerk-Maxwell)	1868
Servomotor concept (Farcot)	*c*1870
Strain energy used in structural analysis (Castigliano)	1873
Systematic foundation of mechanism laid (Reuleaux's *Kinematics of machines*)	1875

(Table 2.4 continued)

Stress of bodies in contact (Hertz)	*c*1880
Correct theory of velocity of sound (Laplace)	*c*1880
Two modes of fluid motion identified (Reynolds)	1883
Theory of lubrication (Reynolds)	1886
Venturi meter devised (C. Herschel)	1887
Architectural acoustics researched (Sabine; Later Jäger)	*c*1900
Specialist texts become prolific	post1900
Brownian motion researched (Perrin; Millikan; Fletcher)	*c*1905
Creep in metals under stress and elevated temperature (Andrade)	1910
Control theory advanced (Nyquist)	1932
Creep knowledge reviewed (Bailey)	1935
Metrology established as subject in Production Engineering	*c*1940
Hydraulic principles require more mathematical basis for aircraft design needs	*c*1920
Electricity and magnetism (See also Tables 2.1, 5.3 and 6.1)	
Basic magnet principles (Peregrinus)	1269
Magnetic attraction differentiated from electric attraction (Cardano)	*c*1500
Dip of earth's magnetic field discovered (Hartmann)	1544
First significant treatise on magnetism: *De magnete* (Gilbert)	1600
Secular variation of declination (Gellibrand)	*c*1620
Electric charge exists on surface (Hauksbee)	*c*1710
Conduction depends on material: use of wires (Gray)	1730
Leyden jar: first capacity (Kleist, Musschenbroek, Cunaeus)	1745
Plus-minus and negative-positive terms (Franklin)	1747
Franklin's letters on electricity (published by Collison)	*c*1750
Magnetic attraction obeys law of inverse squares (Meyer)	*c*1760
Unpublished anticipation of many laws of static electricity by Cavendish (Published later in 1879 by Clerk Maxwell)	1770s
Electric attraction obeys law of inverse squares (Coulomb)	*c*1780
Discovery of primary cell electricity (Galvani)	1780
Principle of manufacture of primary cell enunciated (Volta)	1800
Electrochemistry founded (Carlisle and Nicholson)	*c*1802
Therapeutic use of electricity (Carpue)	1803
Multiwire telegraphy demonstrated (Sommering)	1809
Electromagnetism founded as a science (Oersted)	1819
Theory of current and magnetic field relationship (Ampere)	*c*1820
Thermoelectricity principle discovered (Seebeck)	1820
Electric current conversion to mechanical motion (Faraday)	1821
Electromagnet built (Sturgeon)	1825
Law relating voltage to current (Ohm)	1826
Concept of potential used in electrics and magnetics (Green)	1828
Electrostatic induction researches (Faraday)	*c*1830
Laws of electromagnetic induction (Faraday)	1831
Electric telegraph made with electromagnetic principles (Morse)	1832
Electric motor constructed (Pixii; Saxton)	1832
Absolute units proposed for magnetism (Gauss)	1832

(Table 2.4 continued)

Anode, cathode terms introduced (Faraday)	1834
Induction effects in direction opposing them (Lenz)	1834
Transformer principle: the Ruhmkorff coil (Page)	1836
Electric lamp made (Grove)	1840
Bridge and rheostat circuits popularised (Wheatstone)	1843
Heat-pumping principle of electric junctions (Peltier)	1844
Light related to magnetism (Faraday)	1845
Diamagnetism discovered (Faraday)	1846
Concept of potential republished (Kelvin)	1846
Electric generator as converse of motor (Jacobi)	1850
Cathode rays observed	*c*1850
Absolute units of electricity (Weber)	*c*1850
Resonant circuit theory (Kelvin)	1852
Studies of discharges in vacuo (Masson)	1853
Integrating effect of cables on signals recognised (Kelvin)	*c*1860
Physical standard of resistance: mercury prism (Siemens)	1860
Principle of telephony apparatus (Riess)	1861
Discharge research moves into theoretical stage (Crookes, Maxwell, Tait, Dewar)	*c*1870
Electromagnetic theory in *Treatise on electricity and magnetism* (Clerk Maxwell)	1873
Laws of magnetic circuits shown to be like Ohms Law (Rowland)	1873
Carbon-filament lamp (Edison)	1878
Polyphase electric motor (Baily)	1879
Rectifying effect of a special lamp (Edison)	1883
Legal Ohm defined but failed to be adopted	1883
Existence of electromagnetic waves confirmed in practice (Hertz)	1888
Equation for electromotive action of ions (Nernst)	1889
Simulation experiment of magnetisation (Ewing)	1891
System of international units adopted	1893
Practical wireless communication (Marconi)	1894
X-rays discovered (Röntgen)	1895
Link made between radioactivity and other emissions (Becquerel)	1896
Tuned circuits as coupled resonance (Braun)	1898
Theory of transmission line founded (Pupin)	1899
Edison effect of rectification explained (J.J. Thomson)	1899
Theory of radioactive transformation (Rutherford and Soddy)	1902
Single atoms observed as scintillations (Crookes)	1903
Thermionic diode invented (Fleming)	1904
Radar principle stated (Hueslmeyer)	1904
Electromagnetic sensor applied to an instrument (Galitzin)	1904
Amplifying thermionic device constructed (Lieben)	1905
Crystal detector (Dunwoody; Pickard; Pierce)	1906
Heterodyne detection principle patented (Fessenden)	1907
Thermionic triode: the Audion (Lee de Forest)	1907
First USA music transmission by wireless	1913
General theory of relativity (Einstein)	1915
Fitzgerald-Lorentz electron contraction demonstrated (Bridgman)	1922

(Table 2.4 continued)

Atom split (Rutherford and Chadwick)	1922
Electron theory of metals (Sommerfeld)	1927
Integrated circuit valve receiver (Loewe)	1930
Cathode ray oscilloscope as measuring tool	*c*1930
Transistors made	1945
Integrated circuits in solid-state form	1960
Optics and heat	
Thermoscope form of thermometer (Galileo)	*c*1600
Telescope made (Lippershey)	1608
Microscope made (Joannides; Drebbel; Galileo)	*c*1610
Refraction law from experiment (Snell): unpublished	*c*1610
Theory of telescope published in *Dioptrica* (Kepler)	1611
Snell's law of sines deduced theoretically in *La Dioptrique* (Descartes)	1637
Gascoigne uses double convex lenses in telescope for astronomy	1640
Sealed, liquid-in-glass, thermometers introduced (Ferdinand II)	*c*1650
Mercury-in-glass thermometer (Boulliau)	1659
Diffraction reported in *Physico-mathesis de lumine* (Grimaldi)	1666
Reflecting optics used for telescope (Newton, or earlier, Zucchi, or Mersenne, or Gregory)	*c*1688
Divisibility of light observed (Bartholinus)	1669
Corpuscular theory of light competed with wave theory (Newton)	*c*1670
Wave theory of light (Huygens)	1687
Velocity of light shown to be finite (Römer)	*c*1678
Wave theory of light published in *Traite de la lumière* (Huygens)	1690
Mercury-in-glass thermometers popularised (Fahrenheit)	*c*1700
Finite velocity of light confirmed (Bradley)	*c*1730
Centigrade temperature scale with 0 at boiling point (Celsius)	1742
Celsius scale as used today (Strömer)	1750
Spectral lines observed (Melvill)	1752
Achromatic lens made (Dolland, perhaps earlier by Hall)	1758
Latent and specific heat concepts (Black)	*c*1760
Pyrometic fire-clay cone (Wedgewood)	*c*1782
Discovery of double diffraction . . . polarisation (Malus)	*c*1795
Heat due to motion; it is not a substance (Rumford)	1798
Solar lines investigated (Fraunhofer)	*c*1800
Infra-red rays discovered (Herschel)	*c*1800
Ultraviolet rays detected (Ritter; Wollaston)	*c*1800
Thermodynamics beginnings (Mollet, Dalton)	*c*1800
Undulatory theory of light revived (Young)	1801
Charles or Gay-Lussac law of gases published	1802
Theory of exchanges of radiation (Prevost)	*c*1810
A theory of diffraction (Fresnel)	1815
Gratings used as optical dispersive element (Fraunhofer)	*c*1820
Polarised light examined in detail (Brewster)	*c*1820
Heat flow theory began (Fourier)	1822
Thermodynamic cycle of a gas (Carnot)	1824

(Table 2.4 continued)

Thermopile devised by Nobili	1829
Photography used in researches of light (Draper)	1842
Mechanical equivalent of heat established (Joule)	1843
Plane of polarisation of light rotated by magnetic field (Faraday)	1845
Colour of hot substance depends upon its temperature, not the substance (Draper)	1847
Absolute thermodynamic temperature scale (Kelvin)	1848
Absolute measurement of speed of light (Fizeau)	1849
Velocity of light compared in different media (Foucalt)	1850
Radiant energy researched and published (Melloni)	*c*1850
Conservation of energy as greatest generalisation of century (Mayer; Joule; Colding; Helmholtz)	*c*1850
Blackbody source realised in practice (Kirchoff)	*c*1860
Emission and absorption shown to be functions of wavelength and temperature only (Kirchoff)	*c*1860
Spectrum analysis achieves significance through theoretical work (Bunsen and Kirchoff)	*c*1860
Heat established as mode of motion (Tyndall)	1862
Wavelengths of solar lines published (Ångström)	1868
Resistance temperature sensing introduced (Siemens)	1871
Selenium as a photosensitive substance (Smith)	1873
Kerr effect (Kerr)	1875
Bolometer form of total radiation detector (Langley)	1881
Optical communication link (Bell and Tainter)	1881
Ultraviolet light found to facilitate electric discharge (Hertz) Start of surface electrophysics	1887
Improved measurement of the velocity of light (Michelson and Morley)	1887
Electrical resistance of silver halides is light sensitive (Arrherrius)	1887
Spectra of elements shown to be lines series (Kayser and Runge)	*c*1890
Displacement law of radiation (Wien)	1893
Splitting of spectral line by magnetic field (Zeeman)	1896
Television concept (Dussaud and many others)	*c*1898
Photoelectric theory of vision published	1905
Third law of thermodynamics (W. Nernst)	1906
Radiation laws extended (Planck)	*c*1910
Photocells in everyday use for counting and control	*c*1930

not all people possess, especially those who devise and construct apparatus.

Chronologies appear to be uncommon in science and engineering texts. Byrn (1900) presents an extensive list of inventions through the 19th century. Glasser *et al.* (1947) have listed their selection of the milestones in radiology covering from 1600-1893 AD as one group and, later in their work, from 1895-1940. Table 2.1, given previously in this chapter, is a list of events that led to commercial radio broadcasting.

Certain books that have been published about the history of science and technology are arranged in chronological order, some with the people concerned featured in some systematic way. Examples are Marmery (1895), Hart (1924), Hogben (1938), Butler (1947), Kimball (1949), Gartmann (1960), Larsen (1960), Cajori (1962) and Pledge (1966). The name index of Pledge (1966) includes birth-death dates of the hundreds of people listed.

Biographies can also provide valuable information about the development and uptake of ideas and theories. Scholarly works will generally include a list of the works of the person concerned. Biographies can, however, be prone to loose statements because many of those about great men and women of the past are derivatives from earlier publications. There is a strong risk of inaccuracy of detail where these have been prepared by authors who did not conduct their research right back into the original papers and archives of their person.

Biographies of contributors living in the electric-sensing era (*c*1820 to present) are important because much of the information they can yield is still not available from textual sources due to the paucity of research into the history of sensing in recent historical times. Appendix 1 is a list of biographical publications that has been selected in accordance with the theme of this book.

The collected biographies of astronomers from Copernicus to Herschel - Lodge (1893) - provides information about the measurement processes and equipment involved in the so-called classical era of astronomy.

Electrical sensing had its practical beginnings in the mid 19th century through the works of Clerk Maxwell, Faraday, Tyndall and later from the engineering-scientists such as Hopkinson, Crompton, Kelvin, Edison and Bell (see Appendix 1).

Many libraries hold a special index of biographies but a difficulty that sometimes arises is how to extract them on a specific person when some titles convey no statement about the person whose life is reported and biographies about scientists and engineers form but a small part of biographical literature. Specialist libraries, such as that of the Science Museum, London have more useful biography collections due to the constraints on collection. A card system that relates subject person to the title used in print is often maintained in such cases.

Many libraries can now offer microfilm catalogues for personal use away from the library. A useful publication was the subject and author catalogue of books, published on engineering after 1930, that are held in the Science Museum, London, HMSO (1957). Ferguson (1968) is a prime source of bibliography on subjects in the history of technology.

Assuming that a new instrument principle is discovered and that it has been made available through reasonably well read primary publication channels, what is then necessary for it to become a manufactured piece of equipment?

Section 1.6 of Chapter 1 briefly showed how ideas and principles are gradually developed from the unexpressable concept to a specific exploited commercial use. Radio was chosen as the example. Commercial development of instruments rarely followed such a well supported commercial path for the demand for measuring apparatus fell way below the quantity production experienced in the dominant product areas. The rise of the instrument industry is explored later in this Chapter. Here it is appropriate to consider how the principle achieves embodiment into a universally used measuring instrument.

Initially the process can be seen to start when the discovery of a principle is made known to other people engaged in research. They, and perhaps the discoverer, will recognise that the principle has relevance to the operation of an experiment that they are about to create. For example, mid-19th century researchers on heat would have recognised that the work of Seebeck on the generating effect of electric junctions held at different temperatures enabled them to build a heat-sensing device in which the electrical signal produced indicated temperature. The method suited the, by then available, galvanometer detector and it gave a measuring scale to the measurements. Thus was born the thermoelectric, thermopile, as a crude measuring device having many different forms. Each user was able to obtain valuable results from his own apparatus but soon would realise that it was necessary to introduce some kind of standardisation amongst all users of the principle. Intertwined in this process of development would be the development of electrical variable standards, such as the definition of the volt. Encouraged by the successful use of the new instrument other scientists, and perhaps industrial users, would begin to take up the idea. At some stage the demand for similar style apparatus on this principle would reach a quantity where it became a commercial proposition for someone to make it for sale. With the thermopile, it took about 80 years to reach this stage. This may arise because an instrument maker had already made a small number of instruments under commission to a scientist and others wished to have their own set. Alternatively, as became more noticeable in the current century, the marketing body may be interested for reasons of commercial interest in profit seeking on behalf of their principals. Fig. 2.16 illustrates this procedure.

Once established as a useful scientific tool the instrument passes into the everyday routine of the appropriate use. The commercial maker, however, has only just begun to exploit his interest for there also exists the industrial market that has not yet been tapped. For this market the maker investigates the most likely demanded variations of the theme in order that the production quantity can be maximised to keep costs down. An intensive sales campaign is mounted to encourage industry to adopt the instrument as part of its routine control and monitoring operations.

Once released as readily available marketed products the developed instrument becomes available to research users at reasonable prices. Adoption widens into other areas of investigation from which new principles may be discovered.

The cycle then begins again with the new principles following this general pattern.

Of this process, passing from the laboratory model into commercial manufacture is, by far, the most complex and difficult step. It is relatively easy to build a prototype, to use it in one or two different uses and to advise others that it is suitable for them.

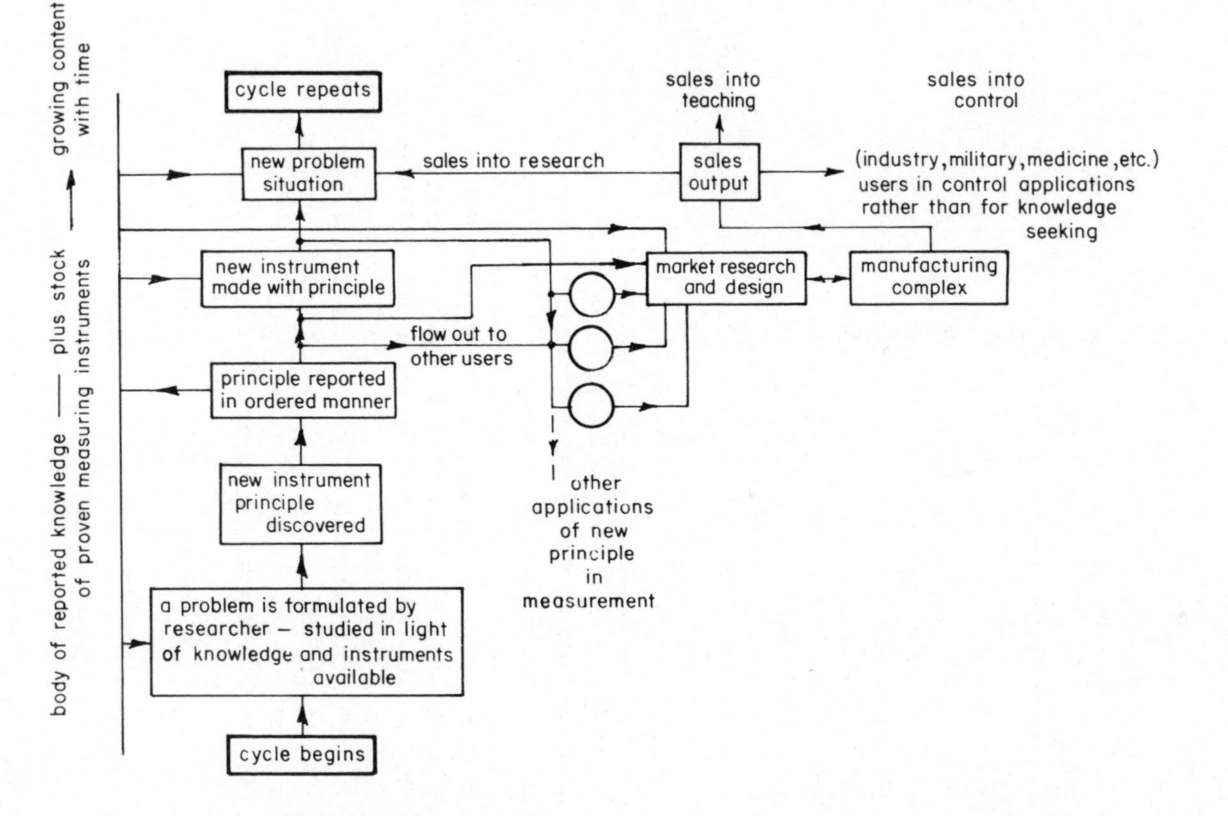

Fig. 2.16 *Simplified representation of how a new physical principle is transformed into a marketed instrument*

BWD ELECTRONICS P/L.

PROJECT Nº.....................

PROVISIONAL MODEL Nº.........

DESCRIPTION.....................

DATE COMMENCED...............

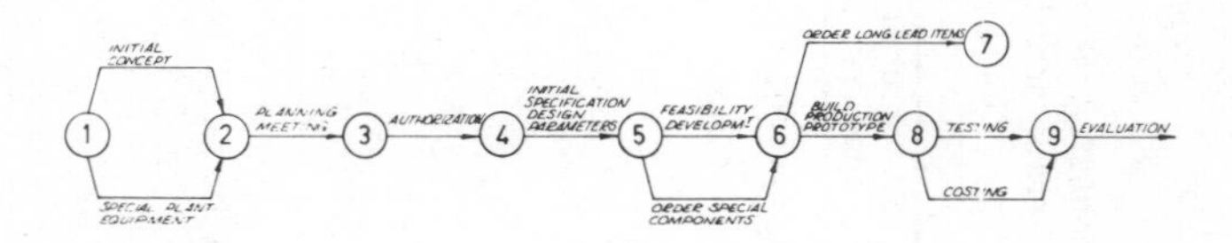

PROJECT DEVELOPMENT

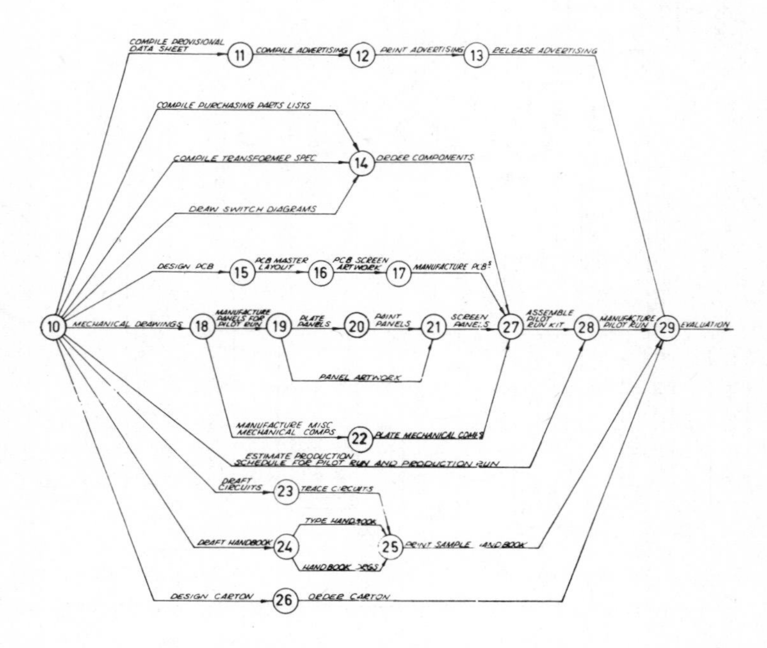

Fig. 2.17 *Stages needed to manufacture and market an electronic instrument. It can take only six weeks to reach state 10 . . . prototype completion, but then another two years may be required before the product can be released.*
[Courtesy BWD Electronics Pty. Ltd.]

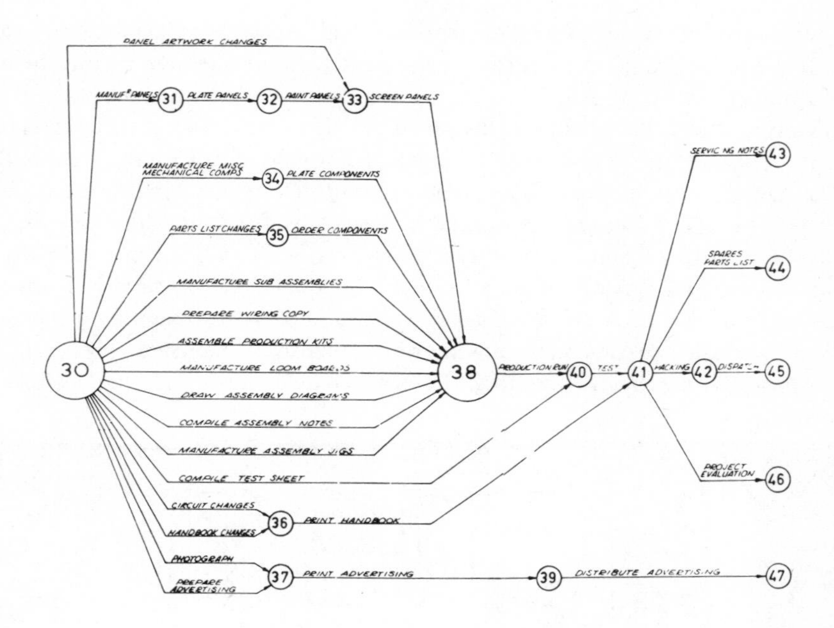

The commercial developer, however, must be able to ascertain that it will be accepted by not a handful of users, but by hundreds. It must also be shown that competition for sales will be acceptable. A satisfactory life of product sales must be assured. Sub-component supply must be acceptable. It must be a reliable design. In all, the maker has to make very significant decisions, decisions that could possibly make or break his enterprise. Fig. 2.17 is a flow diagram of the stages to be covered in the manufacture of a commercial electronic instrument. It takes around two years to complete the path, the prototype maybe only taking two months to build!

Demand for new styles and designs of measuring apparatus is usually in the small production size category so it is rarely possible to make use of the economies of large-scale manufacture. Furthermore, the variations on the central theme that can be required of an instrument principle are great in number and the number of instrument principles vast. This has always been the case. It would seem that it will continue to be the case unless technological forces of economy trend the system designers toward the use of more standard designs. This is happening today in the area of electronics but instrumentation comprises much more than electronics. Transducer elements also need to be restricted in number of available variations if costs are to come down by virtue of dramatic increase in production numbers.

2.9 Instrument designers and builders

The availability of design principles, materials and processes of manufacture cannot produce a new or modified measuring instrument without one vital factor, the minds and manipulative craft of men.

In early times people needing an instrument could usually make it themselves or call upon the artisan skills of craftsmen who would normally create more common things such as jewellery, household objects and military equipment. Galileo, according to Lilley (1951), acknowledged the arsenal at Venice as the source through which he obtained the skills of construction used in his own work. Gilbert in the 1600s used marine instruments making his own magnetic material, 'terrela' globes for his study of the magnet. His book *On the magnet* was the first work written by a scholar for scholars in which experimental methods were extensively reported. Fig. 2.18, from that work, shows a blacksmith forging a magnet held north and south whilst it is hammered.

Fig. 2.18 *Woodcut from Gilbert's treatise showing the simplicity of manufacturing magnets needed for his work*
[Mullineux-Walmsley (1910), Fig.2, p.4]

To begin with, common everyday apparatus was pressed into use for research in the spirit of the enlightened science. The goldsmiths balance, the eyeglass, the ruler, surveyors tools, the quadrant and the clocks featured in early research. These could largely be fashioned by nonspecialist artisans.

The coming of the new experimental attitude to science in the 17th century created a need for more specialised apparatus. Initially, much was developed out of the apparatus of industry. As an example, pumps used for clearing mines of water were gradually adapted to provide greater efficiency at smaller sizes and to work at higher suction pressures.

The simplicity of early scientific apparatus is evident. Eventually such simplicity was seen to be inadequate and that modifications could be made that would enhance the measuring ability of the instrument. One change led to another as further residual errors of operation were established from practice and theory that were then removed by better design. It was in the 17th century that the simple instruments had clearly become complex and sophisticated. A collection offered for sale by Thomas Tuttell, *c*1700 is shown in Fig. 2.19. No longer could any goldsmith make them or any armourer be called upon. Close to the scholars grew the specialist craftsmen who could build instruments to the standards called for.

In 1931 the Physical and Optical Societies held their 21st annual exhibition in England. The opening speaker, Sir Arthur Eddington, had the following to say:

> An exhibition like this brings home to us two things - the debt of the scientific worker to the instrument maker, and the debt of the instrument maker to the scientific worker. I do not know which to place first. I do not think one can be placed before the other; for the instrument maker provides the resources of the scientific worker, and the scientific worker provides the resources of the instrument maker. Whipple (1931).

The craftsmen-technicians joined the designers in the experimental laboratory for co-operative work. Fig. 2.20 shows a contemporary 1884 view of the Experimental Department of the American Bell Telephone Company.

Scholars could detect the likely residual error of the instrument once the craftsman had made it. Craftsmen, on the other hand, knew what was possible from the materials and tools of the day, experiences that the scholar did not always possess. Each complemented the other, as they still do today. The combination of the two would seem the obvious situation to create but that has not always been a simple thing to achieve. All men are not equally talented, the expertise of each needs grouping to achieve balanced co-operation.

Attitudes of the scholar and the craftsman toward each other play a vital part in the development and use of measuring equipment. Because of their intimate knowledge of the operation, craftsmen can often extract more from an instrument than the scholar. Conversely, the scholar can often see how to produce better results because of a better understanding of the fundamental principles involved. It seems too obvious a thing to point out that the two extremes of person do exist and that the team approach is clearly the best to use. But history has shown so often that these two classes of society have not always been able to co-operate to their mutual advantage. Crombie (1969), discussed this relationship from *Antiquity to the Middle Ages*.

Fig. 2.19 *The rise in sophistication of instrument design by 1700 is clear from this portion of the trade brochure of Thomas Tuttell, a London maker*
[Crown Copyright. Science Museum, London]

Fig. 2.20 *Experimental Department of the American Bell Telephone Company, 1884. Technicians and the academics worked together as partners in the development of new ideas*
[Courtesy *Scientific American,* 1884]

The attitude of the instrument user to the instrument maker has waxed and waned throughout time. Today it could be said that in the western world technicians are not given the recognition they deserve. Certainly there is little evidence that instrument technicians have a high place in society.

This is not just a current situation. In one recorded instance of early astronomy the 'observer-in-charge' was not allowed to be physically involved with the instruments (rock structures for star observation).

Lilley (1951) took up the theme of co-operation from the class position held by the humble craftsman as opposed to the snobbish intellectuals (it still exists today). Because of this the scholars often missed opportunities to incorporate the help that craftsmen could have given. He stated that the snobbery began to dissolve in the 17th century. Scholars went out of their way to learn what the craftsmen could contribute. He suggests the barriers fell because of the growing importance in the society of gaining a living from industry as opposed to the previous situation where the wealthy aristocracy did not need to know of worldly things in order to live.

Priestley's view of instrument construction has already been referred to in Section 2.4. It was very clear that he thought that the scholars were giving too little thought to their apparatus in his time. He also criticised the instrument makers for their over emphasis of appearance rather than concentrating on performance of function.

The great bulk of development in science has come from researchers who had an above average understanding of the measuring instruments. That goes almost without saying for it has been shown in Chapter 1 that making a measurement is not merely a case of getting numbers from a system but is a case of obtaining numbers that have true relevance.

The subject of the construction of instruments and the attitudes to their makers throughout time does not seem to have interested many writers. Jones (1967*a*) gives some space to the issue confirming Lilley's above stated views that close and useful co-operation between scholar and craftsman existed in 17th-century Europe. He suggests London became the first world centre for instrument manufacture because of the lack of political stability elsewhere. Certainly London appears to have been such a centre of enterprise and productivity. Whipple (1931) presents a brief history of London makers of scientific instruments. The influence of this activity spread afar. Museums as far as the Eastern block of Europe today have a noticeable proportion of English instruments dating from the 17th to the 19th centuries. London, however, was not the sole and only place where instruments were made. Dutch, German, French, Hungarian, Swedish and, later, American manufacture was also significant. Whipple (1931) mentions how the many Italian companies brought glass thermometer skills to London in the 19th century.

With the introduction of mass production and interchangeable parts made on machines that required little more than the skill of the machine setter–the period around the end of the previous century–the skilled craftsmen of the earlier times began largely to decline in number being exchanged for by the mass produced articles made by less skilled people. Skilled all-rounders who could operate a fine work jobbing shop became harder to find. Pearsall (1974) has published a short account of instrument makers in his work on scientific instruments.

From the period about 1910-1945 T.H. Court and G.H. Gabb engaged themselves in collecting scientific instruments and information about them. Their collections included printed trade cards and advertisements. These card collections are now held in the Science Museum, London, where, along with other similar material, they form a valuable source about instrument makers from the 17th century onward. Those relevant to scientific instruments have been catalogued and published in book form by Calvert (1971).

The cards were profusely illustrated, as Fig. 2.21 shows, with examples of the wares of the instrument firms. Fig. 2.19 is of another card from the collection. Although they often had the patronage of the wealthy and the now famous it must not be thought that instrument makers normally enjoyed any more recognition than other trades people.

Lists of instrument makers are available in Bos (1968), Calvert (1971), Chaldecott (1954, 1976), Pearsall (1974), Pipping (1977) and Ward (1966). Major science museums usually maintain lists of makers for internal research and identification purposes. Holbrook (1972) is a translation from French of an important study of 17th and 18th century instrument manufacture. Instruments of the 13th-19th century are covered in Billmeir (1955). Michel (1966*a,b*) are copiously illustrated works on scientific instruments.

Instrument makers have rarely achieved high status for their instrument work alone.

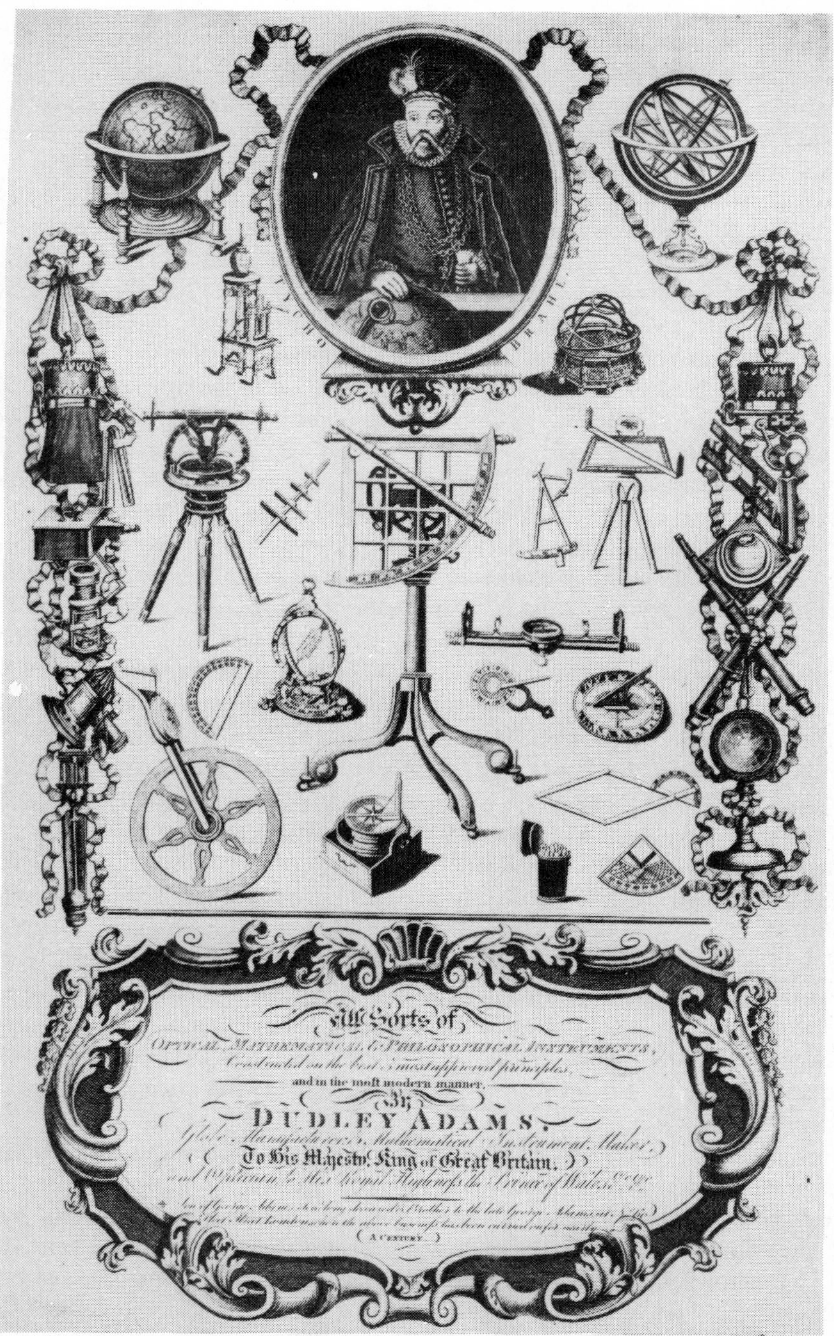

Fig. 2.21 *Trade card from the collection held in the Science Museum, London. These contain a wealth of information about the products of the early small craft firms.*
[Crown Copyright. Science Museum, London]

Some were honoured with Royal patronage, such as the time when George Adams, father of the Dudley Adams of Fig. 2.21, around 1760, was commissioned to make apparatus illustrating the principles of mechanics and hydraulics for George III of England's collection of 'philosophical apparatus'. Such sets of equipment were popular acquisitions of the wealthy of the 18th century. It was quite the done thing to obtain amusement with these on social occasions. The crowning piece was often the electrostatic charge generator for with that all manner of exciting games could be played . . . such as the 'electric kiss'. Fig. 2.22 shows Nollet's 1767 adaption of Hawksbee's design. A large part of the production of the instrument firms in the 17th and 18th centuries was for this highly popularised market. Perhaps that was significant reason for the high degree of concentration on appearance for the apparatus had to grace the drawing room alongside musical instruments and works of art. The task of converting newly discovered physical principles into subjects of amusement led to apparatus known as the Physique Amusante.

As well as wealthy patrons the instrument makers were supported by the Learned Societies who were concerned with the new natural philosophy. Many acquired their own set of philosophical apparatus; many of which have been preserved. Work on these collections has led their modern investigators to gain an appreciation of the lot of the instrument makers of the 17th and 18th centuries. Examples of these studies that have resulted in publication in recent times are catalogues for collections of Billmeir (1955), the American Philosophical Society by Multhauf (1961*a*), the Royal Swedish Academy of Sciences by Pipping (1977), Van Marum's collection - a view is given in Fig. 2.23 - in Haarlem, Holland by Turner (1973), mechanical instruments of Utrecht University by Bos (1968), electrostatic instruments of that University by Lavèn and van Cittert-Eymérs (1967), the George III collection by Chew (1968) and meteorological instruments by Knowles-Middleton (1969*a*).

The University of Utrecht collection is worthy of mention by the way that it became to be preserved. It seems that an enthusiastic laboratory staff member developed a habit of placing the best pieces of equipment used in the teaching programmes away in a private part of his house, which also served as the laboratory. It would appear that no-one knew of this and more than a century passed before the collection was discovered by Dr. van Cittert, the then curator of the museum.

Besides listing the collections, the above-mentioned catalogues include comments. These are valuable for the insight they gave about the use and interest in these collections in their day.

One instrument maker who did achieve a well known place in his own society for his instrument work was John Harrison in the 18th century. A photograph of a portrait is given in Fig. 2.24. Harrison, a carpenter from Yorkshire, spent most of his life developing chronometers. His incentive was a British Government award for £10 000 for a method of determining ships longitude at sea to within 1° for the voyage from England to the West Indies. The award went to £20 000 for a performance figure of half an arc degree. In terms of accuracy of time this meant time-keeping ability to within less than two minutes in six weeks . . . Ward (1970). Harrison achieved far better than called for . . . just 54 seconds error in five months! There was clear evidence

Fig. 2.22 *The statical electricity machine was often the pride piece of 18th century drawing rooms*
[Deschanel (1891), from Nollet J.A. *Leçons de Physique Expérimental*, 1784]

Fig. 2.23 *View looking into the building holding the 18th century van Marum scientific instrument collection, Haarlem, The Netherlands. The central machine is van Marum's large electrical generator.*
[Courtesy Teylers Museum, Haarlem, The Netherlands]

of the award having been won by Harrison and after much delay he was eventually paid in full. Around that time it was not unusual for parliamentary debate to include comment on clocks. It was in 1793 that the English House of Commons awarded Thomas Mudge £3000 for inventing a part of a chronometer, this award being against the expert advice available to the Commons. An account of Harrison's involvement is to be found in Timbs (1860).

Collections of similar technological apparatus were seen, in the 1900s Germany, to be an important part of education. The Charlottenburg Technical High School, a University grade institution, was built over the 1878-1884 period in Berlin. It was a grand establishment in all respects. Twenty museum collections were built into the original plans. These included surveying instruments, kinematic models, mechanical technology, photochemical and physical apparatus. Each collection, it is stated by Dalby (1904) included the relevant instruments of the discipline concerned, examples being building, railways, ships, hydraulics and others. Fig. 2.25 is a part view of the main building.

Although the sophistication of measuring insruments has clearly increased over time it is hard to say whether the cleverness of man has changed since the emergence from the Dark Ages. Bearing in mind that designers always are limited by various factors that restrict their scope it is interesting to reflect on the skills, both mental and physical, that people like de Gennes must have been able to demonstrate. He, in

Fig. 2.24 *Photograph of a portrait of John Harrison who developed accurate chronometers over the period 1728 to 1759*
[Lent to Science Museum, London by W.H.Briton Esq.]

Fig. 2.25 *General view of main building of the Charlottenburg Technical High School of Berlin, opened in 1884. It contained twenty teaching collections*
[Courtesy Newnes-Butterworth, *Technics*, Vol. 1, Fig.1, p.10]

1688, made a preprogrammed peacock that strutted around pecking up corn. Many such creative mechanical works existed.

Cleverness is hard to judge for the term is very subjective and difficult to measure. Perhaps one criterion is to judge the scope and breadth of creative activity demonstrated through a person's endeavour. From this point of view, however, those who have had a greater chance to use resources in abundance could appear to be the most outstanding. Given the state of knowledge existing at the start of the project is an electronic pocket calculator any more outstanding than the mechanical automats of the 18th century? The sobering thought is that looking back we have come a long way through the efforts of many measuring instrument makers. Just how much further is there to go?

The ability of the contemporary instrument maker to produce and improve upon state-of-the-art apparatus depends much upon the degree of knowledge and developed skills that he or she has. Education and training play a crucial role.

Little is known about the education of those persons who produced the instruments in classical antiquity. Presumably it was a case of father handing methods down to son in crafts such as carpentry, metal-working, pottery and the other trade skills. Developments in that period were only minor compared with the changes that arose after 1500 and we do not, therefore, need to give much thought to education of instrument makers in classical antiquity.

As measuring instruments following the time of Francis Bacon were the joint product of the scientist (a term coined in the 1800s) and the craftsman it is necessary to consider the training both kinds of persons would have received.

In the Middle Ages there developed, (see Crombie (1969) for more detail) as a continuation of Ancient world practices or as a spontaneous new arrangement, voluntary associations of people working in similar circumstances. These were called guilds. By the 15th century the craft guilds had become influential in many ways. One factor of their existence, it was hoped, was to maintain the standards of any particular type of guild trade. Possibly the oldest still existing in Britain is the Worshipful Company of Saddlers. It is thought to have gained a Royal Charter from Edward 1 in 1272 but there is no firm evidence to support this claim today. There is also a more modern guild, the Worshipful Company of Scientific Instrument Makers. By the 15th century the Guilds had become narrow oligarchies in which rules of membership and office-bearing were no longer in the best interests of the original purpose of the Guilds.

Guilds lost their legal powers in 1835 but they had ceased to be adequately useful as the sole organisation for maintaining standards well before that date. Today the Guilds of London are still active in the control of courses and examining candidates but there the Guilds do not function as they did in the past.

In Elizabethan times apprenticeships were in force as a method of training craftsmen. The master-apprentice method has stood the test of time but the degree of sacrifice required of the apprentice has reduced markedly to the current time where apprentices learn a trade with little inconvenience compared with the days when they not only worked for the master without earning a wage but had to pay for the privilege.

As time passed it became clear that a more technically trained person was needed who would perform work of a different nature to the tradesman. By 1800 attitudes that would create better technical training were beginning to harden throughout Europe. Indeed some centres of Europe had instituted engineering courses prior to that. Alongside the technician and the tradesman arose the professional man. Engineering and science education gradually crept into the Universities. Britain was one of the last nations of the European area to move in this direction. Burstall (1970) states the first engineering professorship was that of Glasgow University in 1840. Cambridge adopted engineering as a Chair in 1875. A thorough study of science instruction and practice in England has been written by Cardwell (1972). The formation of two American electrical groups has been reviewed by Layton (1976) and McMahon (1976).

Before 1800 instruments could be made by people with a good mechanical trade background. Electricity was not yet compounded with magnetism and statical electricity devices were generally simple in nature.

Those firms that made the mechanical-optical instruments in the 17th and 18th centuries were easily able to adapt to make the electrical pieces. As electricity arose in importance new skills were needed and by the late 19th century the electrician class of tradesman, technician and engineer existed, see Layton (1976), McMahon (1976) and Terman (1976) for some accounts of the developments that took place.

At the early part of the 19th century the term mechanic came very much to the fore. The leaders of the industrial revolution were more usually self-taught engineers of the nonprofessional kind. They were able to practice a purely technological operation but generally lacked the knowledge of science that could provide them with further and deeper understanding. There were, of course, the exceptions.

This began to change around 1810. The Royal Institution of Great Britain was formed in 1799 by a small group of people who saw the need to teach a more scientific approach to the uptake of the new scientific and technological knowledge that was, by then, pouring forth. An 1890s view of the Albermarle Street building is given in Fig. 2.26. An account of how Count Rumford founded the Royal Institution, R.I. as it is affectionately known, is given by Martin (1961). In 1831 the British Association for the Advancement of Science was established to boost a declining emphasis on

Fig. 2.26 *Building occupied by the Royal Institution since foundation* [Martin (1961), p.18. Courtesy British Council]

science, see Howarth (1931). The area of mechanical engineering gained, at first in Birmingham, the professional Institution of Mechanical Engineers in 1847. The Institution of Electrical Engineers followed later in 1871, first as the Society of Telegraph Engineers. Huxley (1882) presented his views of the qualities needed of university and technical people in science occupations in two of his published addresses. In the 1870s electrical training for technicians was formalised, along with many other disciplines, through the formation of the 'City and Guilds of London Institute for the Advancement of Technical Education' in 1879. In 1891 Parliament passed a resolution that enabled customs money to be diverted into technical education. By 1902 'whiskey money' was put to this use throughout Britain.

Interesting is the fact that the Royal Aeronautical Society was founded in 1866, well before the first powered flight was achieved. Institutions in Engineering are discussed in the British Association (1931) study of London and the advancement of science. The development of Engineering Education in Universities was reviewed by Baker (1951). Haslegrave (1951) published an account of the contribution that British Technical Colleges gave to engineering education. Fleming (1951) also contributed a paper on how mechanical and electrical engineers have been trained. Roderick and Stephens (1973) is another study of 19th century technical education in Britain.

It would seem that British technical education was well organised and advanced for the times, by the foregoing reports. This, in the view of several contemporary writers was not the case in practice. The main criticsm was that it had begun too late with too little. Burstall (1970) has discussed the comparative state of affairs in Europe,

United States, Germany and Britain. He points out how engineers were too often taught laborious hand skills rather than expanding their intellectual powers and learning about science. He saw the great engineering enterprises, such as James Watt and Company, closing down because of the lack of follow-through skills that would have kept them in the forefront of technical developments. Many authoritative articles in the first volume of *Technics* of 1904 are concerned with the spirit of inquiry about the state of British technical education.

Scientific instruction had been the subject of several Royal Commissions, Cardwell (1972), in Britain before that inquiry took place in which Strange featured scientific instruments in such a definite way. The Devonshire Commission of Strange, to paraphrase Avebury (1903) were compelled to:

> record our opinion that, though some progress has been made . . . still no adequate effort has been made to supply the deficiency of scientific instruction pointed out by the Commissions of 1861 and 1864.

and that

> . . . the present state of scientific instruction in our schools is extremely unsatisfactory.

An amusing part of Lord Avebury's Address mentioned above relates to a measuring instrument. A young relative of his apparently passed through a great public school with good grades and went to study engineering. He was asked to define a theodolite. He said it was a hater of God.

The Devonshire enquiry was, as Jones (1971*a*) has summarised, certainly a milestone in the history of education for instrument designers and builders but it did not result in much of a change in attitude toward measurement systems design. Some positive good did occur for measurement educators and makers, the Science Museum collection, for instance.

Little of specific relevance to the education of measurement subjects is available prior to quite recent times. It seems safe to assume that instrument designers and makers would have had training in the areas of physics, mechanics, electrics and production methods but not always so. Dew-Smith, Fig. 2.27, the cofounder of the Cambridge Instrument Company in 1880, is described as a 'wealthy experimenter' in a Cambridge Instruments Company (1955) account. His cofounder, Horace Darwin (a son of the famous Charles Darwin), Fig. 2.28, was a mathematician. They produced instruments, through skilled mechanics, that were wide in scope and advanced in thinking. Whitehead (1951) published an extensive review paper on the British contribution to electrical measuring instrument design and development. His account certainly provides evidence that British instrumentation was of high standard and advanced by the 1950s, the end of his review period. Comments predicting the place of semiconductors in his future and a statement on the possibilities in computing and likely uses of thermionic valves make interesting reading from the stand point of the late 1970s.

In the current century we have seen definite interest in measurement education for both measurer and measuring device maker. Specific courses in trades and technical

Fig. 2.27 *A.G.Dew-Smith, cofounder of the Cambridge Scientific Instrument Company in 1880*
[Courtesy The Cambridge Instrument Company Ltd.]

Fig. 2.28 *Sir Horace Darwin K.B.E., F.R.S. cofounder of the Cambridge Scientific Instrument Company*
[Courtesy The Cambridge Instrument Company Ltd.]

grades exist in most countries. Professional courses are, however, still few and far between in the English speaking countries. The European block communities were more advanced in their training for measurement and in their rigorous approach toward instrument construction. This has been the case for a considerable time. They provide far greater emphasis on this area of education. It must not, however, be thought that education can only be provided by obvious structured arrangements. It is a fact that the USA ability at instrument making and application is clearly of very high standard yet these formal courses, at any level, are rare in the instrument technologies. It is this kind of unexpected performance that confounds deliberation of how best to teach instrument technology and science. Established educational systems of the developed countries are harder to modify to suit the changing times than they are in the developing countries. In the developing countries newly created National Measurement Systems are being created in these 1970s. Such schemes generally include training programmes as part of the overall structure of the National Measurement System. It will be interesting to see if this will yield the benefit it appears it should.

In European countries it has long been the practice for teachers to be designers for industry and commerce. The academics of the local University-grade institution will often be Directors, or at the least, co-workers for the local instrument firm. Many developments that come from the European instrument firms were first the subject of research at the University. That has not been so much the case in western tertiary institutions.

At the time of writing, a recent development in the United Kingdom has been the adoption by the British Government of recommendations that measurement science and technology teaching and research should be boosted in the tertiary institutions. This has been given the 'Special programmes' status. A number of Universities have been selected as places that would specialise in this way. It has long been recognised that control cannot be introduced without measurement yet it took many years for western educational authorities to see that the measurement part of measurement and control was lagging far behind the control part. There is now strong evidence to suggest Western countries are gradually accepting this truth and acting upon it.

Many articles have been written in recent years about measurement-education. Some insight can be gained by consulting the Proceedings of an Institution of Electrical Engineers Conference, No. 56, that was held in 1969 on this subject. Since that meeting a number of similar events have been staged by the Higher Education Committee of IMEKO, the International Measurement Confederation. Proceedings of each have been published. Jackson (1975) reviewed technician and craft education of Britain. This, however, underwent further change in 1977.

2.10 Development of the instrument industries

Before the 18th century the quantity and the form of demand for measuring and other scientific apparatus was such that mass production or extreme specialisation did not arise. If anything, the large market for domestic physical apparatus influenced makers in the opposite direction. Patrons and purchasers were more often after an exclusive piece of equipment than one the same as other people might possess. As has already been mentioned earlier, ordinate effort went into appearance. It seems improbable that purchasers who patronised the instrument makers had in mind what Fraunhofer is said to have remarked to a visitor in his time of the late 18th century:

> My instruments, Sir, are made to be looked through, not looked at.

Jones (1967*a*), who reminded modern readers of this quotation, also hastened to say that Fraunhofer's instruments were not so poorly made as would be implied by that event. Fraunhofer was the son of a master glass-worker, and was a working optician for Reichenbach and Utzschneider where he invented many new machines and measuring instruments for improving the manufacture of achromatic lenses. Later he ran his own optical instrument firm.

The specialised needs of the 17th- and 18th-century scientists, as is still the case today, caused instrument firms to work on many different and varied designs to suit many different purposes. The trade cards shown in Figs. 2.19 and 2.21, demonstrate this.

In the 18th century, however, clear evidence exists of some instrument makers making some lines in large numbers of identical forms. This change came about through the rising use of machines of manufacture. Table 2.3 shows the rise of the 'industrial evolution' for in the first half of the 1700s came the improved lathe, indexed gear cutters, vertical drills and boring mills. Later in that century were added the boring machine, lathes close in design to those of today, and a most important development, that of interchangeable parts manufacture.

From the general machine tools developed specialised equipment for making instrument parts. One of these that had considerable impact on precision, was Ramsden's dividing engine. This could produce circular scales with greater precision and at lower cost than the previously used hand-division methods. One of his engines is shown in Fig. 2.29. The original is held by the Museum of History and Technology, Washington DC. With such equipment Ramsden was able to manufacture a geodetic theodolite of unprecedented precision. One of the two 'great theodolites', that he made around 1787, is shown in Fig. 2.30. Pearsall (1974) and Wartnaby (1968) include brief accounts of the dividing engine and the 'great theodolties'. Apparatus, such as the dividing engine, made it possible for people with less skill and patience to manufacture instruments at lower prices whilst at the same time offering better quality of performance. This enabled the market to be widened. Thus can be seen the beginnings of expanding use of measuring instruments in the same manner as better production methods and philosophies enabled technology to be made available to the public at large.

Certain areas of instrumentation, namely clocks, telescopes, surveying and navigational instruments, were the specialisms seen in England and Continental Europe in the 1800 period. In America similar developments were taking place. Hindle (1966) included a guide to instrument makers and manufacture in his general review of early American technology. The many references, cited by Hindle, to published Literature both learned and popular, reveal the extent to which instruments were being made in

Fig. 2.29 *Ramsden's dividing engine, developed to make accurate circular scales in 1775* [Courtesy National Museum of History and Technology, Smithsonian Institution, Washington DC, Photo. 4572-B]

that country. In his opinion instrument-making arose first in conjuction with clock-making, expanding out into surveying and navigational instruments, and then later again into the industrial needs of mechanical gauging and temperature measurement. He saw the telescope as another important development, but that being more for scientific than technological reasons. One famous family of American telescope makers were the Clarks. The history of Alvan Clark and his sons, Alvan Graham and George Bassett, has been traced by Warner (1968). The account includes a list of their instruments made in the period 1844-1897. The 1876 Centennial Exhibition (see Post (1976*a*)), held in Philadelphia would certainly have given some help to spreading the emergent American instrument making firms into other countries. Fig. 2.31 shows the display of apparatus made available to demonstrate how the US Naval Observatory time the transit of Venus. Fig. 2.32 is the 'broken' theodolite used for establishing the time in conjunction with a chronometer.

Pearsall (1974) in his chapter on the materials and manufacture of scientific instruments begins by suggesting that local availability of materials and processes led to countries such as Germany and Holland being centres where instruments using rolled brass were made, ivory designs being the products of Dieppe. Throughout the rise of the instrument industries the material supply factor has been present. In modern times, however, increased global trading and cheaper air-freight has erroded the need

Fig. 2.30 *One of the two 'great theodolites' made by Ramsden c1791*
[Crown copyright. Science Museum, London]

Fig. 2.31 *Timing the transit of Venus. Apparatus shown at the 1876 Bicentennial Exhibition held in Philadelphia, USA*
[Courtesy Free Library of Philadelphia, Post (1976*a*), Fig. 136, p.94]

Fig. 2.32 *Stackpole Brother's, New York, 'broken transit' of 1870s*
[Courtesy National Museum of History and Technology, Smithsonian Institution, Washington DC Photo. 73-7516]

for any particular region to specialise in this way. Instruments made in Britain might use materials from Australia and sell in South America. Conversely, some precision instrument parts made in Australia might use raw materials from Africa and sell to Japan and Europe. Today it would seem that the dominating factors that influence where an instrument product is made are more costs of labour, availability of know-how and the existence of suitable manufacturing plant. Indigenous demand will often entice local manufacture but it is not necessarily a dominant factor deciding where instruments will be made in large numbers.

Fig. 2.33 *Graduation of glass thermometers after the freezing and boiling points were established. Contemporary view in Negretti and Zambra's 19th century manufactury* [Pepper (1874), Fig.148, p.137]

Another example of geographical specialisation prior to the late 19th century is to be found in thermometer manufacture. Whipple (1931) explained how the firms such as Casella, Hicks, Negretti and Zambra (Fig. 2.33) went from Italy to London transferring with them the skills of thermometer making. Traditional London skills, as gleaned from the trade card collection of Calvert (1971) were in optical, mathematical and philosophical instruments. Weighing scales, marine and survey apparatus are also evident. The 61 cards illustrated certainly show few barometers and thermometers.

It is difficult to establish reliably which countries produced the best products from observation of collections now existing. Study of instruments made in the 1700 to 1900 period from many countries shows that each may have had areas of specialisation but the quality varied greatly over and within all countries. A prime difficulty in assessing the performance of past makers from current instrument collections is that collectors tend to preserve the better artefacts in preference to the full range. One gains little confidence in the manufacturing skills of nations from observation of some of the preserved apparatus of past scientists; it often was so simply made that it seems incredible that it was used successfully to make the dramatic discovery, or provide proof, that it was used for. Fig. 2.34 shows the original apparatus used by Otto Hahn to prove nuclear fission in 1938. It is a sobering thought that so many great advances came from such uncomplicated experimental apparatus.

Fig. 2.34 *Original apparatus used by Otto Hahn to demonstrate nuclear fission in 1938* [Courtesy Deutsches Museum, Munich]

Some indication of the capabilities of the various nations involved can be gained by study of the writings of people of the various periods. It seems clear that the Britisher of the 1880s regarded 'Made in Germany' as the general sign of an inferior product and that by 1900 the situation had been reversed by enlightened German policies. The same can be observed about the rise to stature of Japanese instrument products which rose from a post 1945 marketing of easy and quick to produce, inferior articles to a gradually altered industrial programme of production making not only some of the best apparatus available but designs including innovation of value.

Judgement of quality of the various national products is fraught with psychological bias. It would appear that people often express an unfair lack of confidence in the products of their own country even when their products are strongly established in other places. Although the existence of a world leader for a given product often justifies international buying there are many instances of purchases being made on the slimmest of data, being influenced by the quality of the sales literature rather than the product itself. Today so many products are made where the labour costs are low that it is no longer so relevant to talk of made in a certain country. Sewing machines of note are made in India. US instruments are often made in Taiwan and Japan. Australian products could well be made in Singapore.

Instrument marketing is, today, an international trade conducted throughout the world as a global village. Which nation gains the most through instrument design, production and application will, in the future, possibly be more a case of educational policies and government assistance. Left to distinguish one group from the other is what there was in the beginning of instrument design, the individuals and their opportunities to be creative.

In the previous Section 2.8 the process of how an instrument manufacturing industry develops was explained. Variations of the concept stated can be applied to most firms. A few firms will now be considered to demonstrate how instrument-making enterprises have emerged.

Alvan Clark and his sons, referred to above, is the first to be described. Alvan began work in a wagon-maker's shop. He became interested in drawing and engraving and left wagon-making to follow this pursuit. He then acquired an interest in astronomy which he followed up in his spare time by attending lectures given by a member of staff of the firm to which the engraver firm, that he worked for, were sub-contractors. After 10 years he began to emerge as a portrait painter of note. Twenty-three years after he began work he is said to have become involved in telescope-making purely by chance. The occasion brought him into close relationship with his son George and together, in about 1850, they created the company of Alvan Clark and Sons. A view of the factory is shown in Fig. 2.35. George Basset Clark entered the association with a short background in instrument making and repair. At that time Alvan Graham Clark, the other son, was in the middle of a mechanical apprenticeship. Thus soon came together the skills of fine art, glass-working for telescopes, fitting, from the father a knowledge of people of influence and standing and from George Basset appreciation of how to conduct a small business. The Clarks specialised in any form of instrument that was called for, offering a service that continued after that point of sale until the instrument was clearly working correctly. Alvan is said to have had an especially critical eye for detail as might be assumed from his skills at painting. As well as making instruments, the Clarks were often found using them to conduct quality research. Numerous learned honours, including four honorary degrees, were conferred on them. Alvan did not advertise nor seek to promote his firm. Sales were by word of mouth and as the results of their scientific work becoming known.

The business continued to expand and it became necessary to take on other people. They were selected carefully and trained to suit the firm's needs. The partnership was dissolved in 1891 but the name continued for another 60 years. After two takeovers complete dissolution occurred in 1958 thus ending the history of an instrument firm, who like many others, had not been able to adapt to the continually changing demands

Fig. 2.35 *The factory of Alvin Clarke & Sons, Cambridge, Massachusetts, USA [Scientific American,* 1887]

for measuring apparatus. This account is condensed from Warner (1968).

A second account is of a supplier in England. A surgeon, Thomas Ellis of Birmingham, set up a firm in 1817 and nearby, later in 1825, one Philip Harris took-over Evans a chemist and surgeon in the Bull Ring. In 1830 another link produced Harris and Margets, chemists in another shop of the old Bull Ring. Five years later Philip Harris emerged as the firm which progressively expanded first as a Company, then as a Limited Company passing through various structures to become Philip Harris Holdings Limited in 1965. This concern has long dealt with the supply of scientific apparatus for educational purposes. The 1965 regrouping divided activity into electronics, biology, medical and other administrative divisions. This development is typical of many of the older established enterprises that still trade today.

Another, quite different development, is that of Carl Zeiss in Germany. In 1846, a mechanic of the University of Jena formed a firm in his own name, Carl Zeiss. It concentrated on microscope manufacture. This involved, in those times, the considerable difficulty of manufacturing consistent quality lens, combinations which were designed by trial and error to obtain the optical properties needed for the instrument in question. Zeiss realised that a mathematical basis of design would be more efficient.

His theoretical interest attracted Professor Ernst Abbe from the University to work on such problems. Abbe produced formulas for the rigorous design of compounded microscope objectives laying a solid foundation for production realisation and consistency.

Glass was, of course, the key material and this came through combination with Otto Schott, another company, who could produce the kinds of glass having the necessary parameters for realising the lens designs of Abbe. Upon the death of Carl Zeiss there was no interest in the family to continue the firm. Abbe acquired sole ownership. He proved to be as competent a business man as he was a scholar for the enterprise continued to flourish. In 1896 Ernst Abbe passed his considerable assets to a Foundation that he created to continue the firm of Carl Zeiss. He, in doing this, drew up the rules for its future foundation. The Foundation operated as its own owner with no single entity being able to acquire its assets. The Schott Glass Works followed suit and became part of the Foundation later. Abbe was very specific about the degree of research that must be conducted with profits, with the replacement of machines and with social conditions of the employees who formed the Foundation workforce.

The second world war, and the collapse of Germany, seemed to be the end of this Foundation for the workforce and plant were dismantled and separated. All but a small 6% of the workforce were deported from Germany to Russia. The 80 men who went to the Americans, by the terms of the Yalta agreement of 1945, were members of management, scientists and engineers. These men re-established the Zeiss firm at Oberkochen, West Germany. The plant left to the East German authorities was given life again and continued as Carl Zeiss (Jena). The West German development initially lacked just about all crucial machines and information but soon became a strong operation. Today both enterprises serve the world in competition. Staff of the University at Jena again cooperate with the Jena plant. The Oberkochen plant operates in a more remote situation from research institutions but is, none-the-less, well advised by its own research workers who form part of the new Foundation. By being organised as a Foundation the fortunes of the employees are better protected against the possible whims of family ownership or of company take-over difficulties.

Another instrument firm of Europe specialising in optical equipment is Ernst Leitz. In 1849 mathematician Karl Kellner who specialised in optics established the 'Optical Institute' in Wetzlar, Germany (Fig. 2.36). They, too, concentrated on the improved manufacture of microscopes. Close co-operation between maker and user was achieved, considerably aiding improvements to designs. In 1869 Ernst Leitz took over the concern and upon doing so instituted the use of mass-production manufacture as opposed to the former hand-made methods. The range of products was expanded into the many forms of microscope needed for the emerging specialisms of the sciences and engineering professions. In 1935 they introduced the successful 35mm camera known as the 'Leica'. Products were diversified horizontally into the mechanical gauging market known generally, but not correctly, as industrial metrology in the western countries of the world. The coming of the electronic ways of doing things, from the 1930s onward, eventually crept into the design of mechanico-optico instruments. The traditional optical industries of Europe were slow to adopt the electronic instrumentation

Fig. 2.36 *Seventeenth century Wetzlar*

methods. Well established larger firms lacked internal know-how and experience and their product production quantities were too large to change as speedily as could the smaller firms. However, transition was well under way by the 1960s, mostly through co-operation with other electronics groups. Today there is little sign that this hesitancy existed. Surveying instruments including electronic calculators and microprocessors are now marketed and accepted by users in what was once a very traditional field of measurement.

Turning to the industrial use of controlling and monitoring instrumentation the firm of Kent-BBC has been chosen as the example given here. It cannot be said that the development of this enterprise typifies the rise of the instrument industry in general for, as the above examples of the 19th century show, development was possible through many quite different paths.

George Kent, in 1838, set up business in the Strand, London to supply domestic appliances. He not only sold the products of others but introduced lines of his own. One that sold well was a patented knife-cleaning machine. Association with the operation of domestic households and other interests in brass founding led him to enter the market of instruments of measurement in 1883 when he first began to supply domestic water meters. A decade later the firm began to market a newly developed water meter designed for the flow conditions found in water supply authority operations. Fig. 2.37 shows the factory in London c1895. Over time, the product range of the firm expanded in both horizontal and vertical directions through increase in the kinds of sensing equipment supplied and upward into full systems design for controlling plant of all kinds. In 1947 the firm went public. Subsequently management adopted a policy of further consolidation through takeover of many British and overseas concerns that suited the overall profile being developed. In 1974 the George Kent Group of organisations went into association with the massive electrical group of Brown Boveri who took the controlling share percentage. Thus it was that two firms that began in such humble ways have joined forces, via complex paths, to be among the largest instrumentation organisations in existence.

Fig. 2.37 *London factory of George Kent, 1895*
[Kent (1895), p.2]

Fig. 2.38 *Inverse-time, over-current, relay for automatic circuit-breaker release, designed by C.E.L.Brown in 1902*
[Courtesy, BBC Brown, Boveri &Co. Ltd.]

Brown Boveri and Company, on the other hand, began as a partnership between Charles Brown, son of an Englishman, and Walter Boveri, son of a doctor from Bamberg, when they signed their contract in 1890. Both had experience in the design and application of heavy electrical engineering, the then booming new area of industrial development. Their enterprise was different to that of Kent; they set up a factory with a foundry and machine tools to make electrical generators and the like.

As Brown Boveri developed the need to add instruments to the product range became apparent. Brown, in 1902, designed a superior over-current release (Fig. 2.38) that tripped faster for increasing current: the inverse-time over-current relay, Evans (1966). This first design was soon superseded and improved designs were forthcoming as is usually the case in a free-enterprise economic community. Brown Boveri's development into the measuring instrument market, is typical of many companies. A vast number of firms are involved in instrument manufacture but often as only a small part of their enterprise. To gain an idea of the size of BBC try to imagine the value of the expenditure per week on research alone: it is around $4 000 000 U.S. An English language account of the BBC has been published by Evans (1966).

Another company, the General Electric Company (GEC), also became involved in instrument manufacture through needs arising out of the course of its main purpose, the supply and application of the new energy source, electricity. An electrical warehouse in the City of London, named the General Electric Apparatus Company, issued a catalogue in 1887 in which an information source service was offered. This had been offered by the 'H' department; 'H' standing for the name of the man in charge Hugo Hirst, as a way of establishing what products the emerging electrical industry needed in the future. Initially, the company stocked very few instruments; mainly the ammeter and voltmeter.

At that time electricity was supplied on the basis of a fixed charge per lamp installed. A power-used meter was clearly needed and they took up rights to make and market the Aron meter (Fig. 2.39) as the first on a commercial basis in Britain. Other departments were also involved in electromedical instruments. An account of the history of GEC from 1880 to the 1920s is published, Whyte (1930).

Attention is now given to the Cambridge Instrument Company, touched upon in the previous Section. (A number of slight name changes occurred over time but each was substantially the same). This well known English firm was established by Horace Darwin and A.G. Dew-Smith in 1880. Dew-Smith had previously run a small jobbing shop producing instruments for colleagues at the University of Cambridge. It was mainly in the medical field and was almost entirely concerned with novel apparatus that could not be bought or formed from stock items. They started out with three mechanics and two boys. Darwin, influenced by the kinematic design principles enunciated by Clerk-Maxwell, initiated several designs that were able to be made simpler and that performed better by the use of these principles. In 1884 the Company expanded into other quarters in Cambridge, this move enabling the factory layout to be improved in the interests of efficiency. Darwin did most of the design work at home raising a design to the stage of free-hand sketches on squared paper. As he supervised the manufacture very closely these often sufficed for the mechanics to

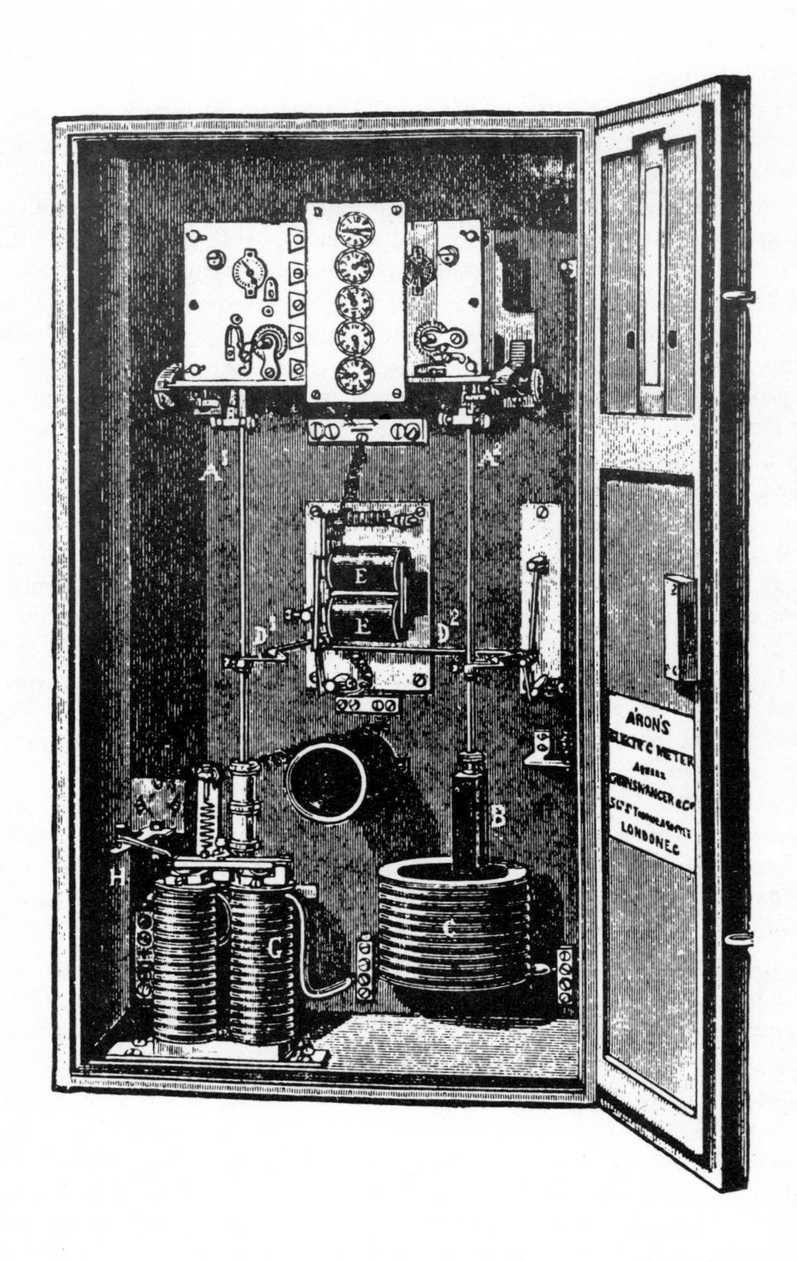

Fig. 2.39 *Aron's electricity meter of 1888, started out as a scientific aid. The GEC firm realised its potential for charging for electricity consumption*
[Whyte (1930), p.159, Courtesy Ernest Benn Ltd.]

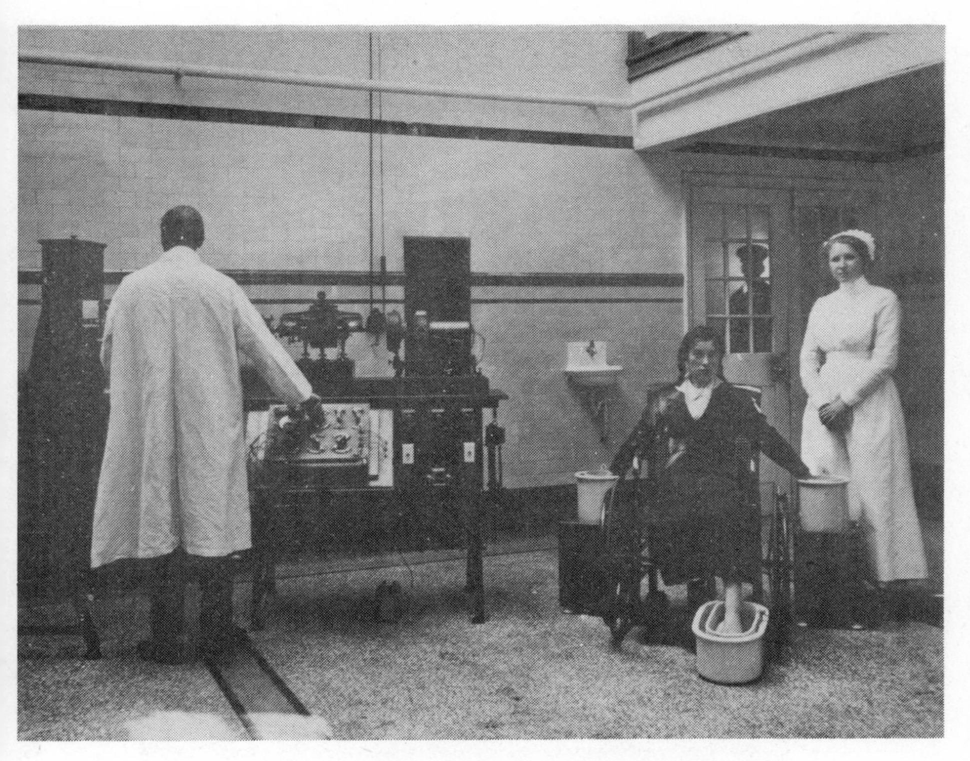

Fig. 2.40 *Electrocardiograph in use in a general hospital, c1920*
[Courtesy The Cambridge Instrument Company Ltd.]

work from. The library of this company contains many volumes of published papers on measuring instruments that Darwin collected together to aid design.

In 1895 the Company moved to a bigger, and again relaid, factory arrangement. The Company became Limited in the same move. At the turn of the century Cambridge Scientific Instrument Company, as it was then called, manufactured a wide range of advanced and unusual scientific apparatus. They marketed the first self-balancing potentiometric recorder using the design of Professor Callendar; their rocking microtome for making thin-sections for biological work is still famous. The designs of many scientists became commercial reality through this firm e.g. Féry's radiation pyrometer, Whipple's indicator (R.S. Whipple joined the firm in 1898), Duddel's oscillograph, Wilson's cloud chamber, Einthoven's string galvanometer and Shakespear's katharometer. Electrocardiograph apparatus, based on Einthoven's galvanometer is shown in Fig. 2.40.

The business of R.W. Paul was amalgamated with Cambridge Scientific Instrument Company in 1919, bringing to the firm Paul's experience and products in the precision electrical instruments area. The name became, until 1924, Cambridge and Paul Instrument Company. In 1922 the company acquired interests in America. Fig. 2.41 shows the interior of one workshop in that year.

Fig. 2.41 *Interior of a workshop of Cambridge and Paul Instrument Company in 1922* [Courtesy The Cambridge Instrument Company Ltd.]

As time passed it became more and more impossible to keep the same number of product lines in the catalogue as were being offered in former times. The 1914 catalogue of R.W. Paul, for instance, offered over a thousand different lines. Many were clearly unique and quite sophisticated products. Some lines were kept in stock but most were manufactured when an order was received. That method of trading could not survive and the Company had to rationalise as time passed. Some designs went to other firms, many were superseded. Today the Company specialises in medical instruments, electron beam X-ray analysers and scanning electron microscopes. This Company has prepared two accounts of its history, the first at its 50th year, Cambridge Instrument Co. (1945) and then again at its 75th year as a trade publication that did not appear in the journal literature, Cambridge Instrument Co. (1955).

The Will of R.W. Paul, who died in 1943, established the R.W. Paul Instrument Fund that is today administered by trustees of the National Westminster Bank, the Royal Society, the Council of Engineering Institutions, and the Institute of Physics. Professor R.V. Jones was elected Chairman in 1960; he has published a paper about this fund, Jones (1973). The fund has terms of reference to provide:

> The design construction and maintenance of novel unusual or much improved types of physical instruments and apparatus needed for an investigation in pure

> or applied physical science and in particular but without predjudice to the generality of the foregoing the advance of knowledge and the arts where a relatively large expenditure may be justifiably risked on experimental apparatus even though it gives no prospect of a financial return Provided always that The R.W. Paul Instrument Fund shall not be used to relieve expenditure in any establishment controlled by the Government nor to relieve any University or other educational establishment of its normal financial obligations.

Some of the former Cambridge Instrument Company products have since become strong lines in the sales of other companies. An example of relevance here is the recorder range that moved into Foster-Cambridge, which is, in turn, also part of the Kent Group mentioned earlier.

Cross-fertilisation of ideas and products has occurred so extensively in the instrument industry that the origins of many designs have become vague or lost completely. The pity is that there does not appear to be much interest in preparing histories of these developments and the designs. Modern sales methods make little use of historical factors and competition, being so great, does not allow the modern instrument firm to put effort into preserving its history and ordering its records. The Cambridge Scientific Instrument Company, for example, does have a collection of its early products and maintains records of early catalogues, documents and texts, including the Darwin archives but little research is done on these. The same is generally the case with most instrument firms that this writer has visited. Constant changes to company structure and ownership over the years, movement of factories from large city to smaller city locations and final closure of firms all add to the confusion about records.

To round out this short set of historical examples, we finish with a short account of how two relatively modern instrument companies developed in the United States, Hewlett-Packard and Leeds & Northrup.

In 1938 two graduates of the Stanford University, USA, Bill R. Hewlett and David Packard, set-up operations to produce a new type of audio oscillator that Hewlett had designed as part of his electrical engineering degree. They began work in a one-car garage at the back of Packard's home in Palo Alto, see Fig. 2.42. Late in the year they promoted this Model 200A unit, winning among their orders a request from Walt Disney Studios to design and build a modification. They received orders for eight units which were subsequently used as part of the sound system in filming *Fantasia*.

The two organised partnership in 1939. By 1940 larger premises were needed and after renting more buildings the firm moved into its own building in 1942. A vacuum tube voltmeter followed. The second world war saw Hewlett in military service with Packard minding the business.

Following the close of hostilities, Hewlett-Packard moved into the microwave measurement field and into generators in 1946. By 1950 Hewlett-Packard had 200 employees, 70 products and $2 million sales. It then followed a growth by diversification, either by purchasing suitable companies or by an internal structure of specialist teams.

This history then goes beyond the scope of this book. Hewlett-Packard (1977) have prepared a short account of the company's history to celebrate its 50th anniversary.

After a brief attempt at school teaching Morris E. Leeds in 1890, entered business

Fig. 2.42 *Hewlett and Packard first set up business in this garage in Palo Alto in 1938* [Courtesy Hewlett-Packard Company]

with James W. Queen & Company, an importer and maker of scientific instruments situated in Philadelphia. Queen's was then the largest instrument firm in America. Most of its imported, and best, equipment came from Germany to be sold to laboratories and colleges. Leeds, then a young man, was largely concerned with its checking and testing. During this time he became aware of his lack of appropriate studies; interest developed within him to make more indigenous designs.

In 1892 and 1893 he attended courses in Berlin, visiting many instrument work shops in Europe and Britain. He was greatly influenced by the quality of Carl Zeiss, then only at Jena, products and its organisation to the advanced ideas of Abbe discussed earlier in this section.

Return to Queen's in 1893 brought dissappointment for hard times had forced a reorganisation. He eventually left Queen's in 1899 to form his own business, the

Fig. 2.43 *Young women producing scientific instruments in Leeds & Northrup production laboratory 1906*
[Courtesy Leeds & Northrup Company]

nucleus being a not-so-flourishing firm owned by a friend, Elmer Willyoung. As well as concentrating on products such as indigenous electrical measuring instruments, he greatly concerned himself with the social aspects of the firm's organisation. One of his aims, stated in an early catalogue, was: 'We have been guided from the first with the definite purpose, to build up an enduring business that shall be founded upon constructions of genuine merit'.

Leeds won a 1931 national competition for the employer-employee relationship of Leeds & Northrup. The firm began life in a floor over a small jewellery shop. It had 25 employees to begin with. Leeds soon moved to a bigger floor over a bakery. He soon began to specialise in certain lines such as bridges, test-sets, resistance boxes and galvanometers.

In 1903 Dr. Edwin F. Northrup, a researcher and theoretical physicist joined as a partner of the firm. A picture of production conditions existing in 1906 is shown in Fig. 2.43.

Certain in-house experience enabled diversification into furnace controls and furnaces. The remainder of the historical development of Leeds & Northrup is to be found in Vogel (1949), the in-depth, well illustrated account concentrating on organisational aspects as well as products.

Space does not permit any more examples to be discussed. Histories of relevance that have been prepared include those of the Marconi Company, Baker (1970); of the General Electric Company of the USA, Hawkins (1950). General Electric (1976) and the Siemens Company Published by Von Weiher and Geotzeler (1977). Development of the analytical laboratories in the USA is the subject of a published dissertation by White (1961). Electrical manufacturers of that same country in the period 1875-1900, have been investigated by Passer (1953). Histories of the Bell Laboratories have been published, Page (1941), Fagen (1975), Mabon (1975). The latter source includes brief biographies of past staff, such as Campbell, Arnold, Black, Nyquist, Shannon and Bode.

2.11 Observatories, laboratories and research institutions

Just as manufacture of an article usually has been carried-out in a special place of industry, so too did the practice of experiment and development of ideas in science and technology gradually be performed in special buildings. Places where observations are made are called observatories or laboratories. The distinction is not clear between the two although the term observatory has been used more for places concerned with study of natural phenomena. A building complex having a structured research programme will often be referred to as a research institution. The content of this Section is about such places, seen in the context of measurement.

Observational occupations began with interest in the movements of the heavenly bodies for religious and other purposes. Instruments of antiquity were extremely basic e.g. sighting rods and marks that helped the observer obtain consistent readings. The Ancient Indian, Sumerian, Babylonian and Egyptian civilisations each built specific astronomical observatory structures. Some were quite huge. Many were little more in design concept than a large sundial, others recorded directions and other provided sighting arrangements. A quadrant in Samarkand had a radius around 600 m.

Observatories continued to be dominantly for astronomical work until the advent of the natural philosophy in the 16th century. The observatory that Tycho Brahe built in 1576 near Copenhagen was one of the most elaborate of its day. It was called 'Uranienburg' – the castle of the heavens. It is shown in Fig. 2.44, as depicted in early woodcut. Its instrument collection was one of the best to be seen in his time. In reality though, all it had was a good range of well divided circles, segments of circles and star position aide-mémoires mounted in various ways to allow the observer to make visual, unaided, sighting of angles between stars and other astronomical parameters. This observatory can be contrasted with the Royal Greenwich Observatory created by Charles II and still operational, Howse (1975). A short history of astronomical observatories is available, Donnelly (1973).

The new method of scientific inquiry that arose following Francis Bacon's time created the need to make observations on more than just the sky. Initially much of the experimental and observational work could be done in the comparative comfort of the drawing room of the stately homes of the wealthy followers of the new philosophy. It could also be done simply by making use of a convenient structure that happened to exist. Galileo dropped various sized balls out of a window of the leaning Tower of Pisa (which was leaning before it was finished). Knowles-Middleston (1969*a*) includes, Fig. 2.45, a print taken from an early manuscript that shows barometric experiments being conducted up the wall of a residential building.

Early meetings of the Royal Society enabled interested people to gather together in a convenient room where they studied a piece of apparatus or specimen on the table before them. This kind of experimental work is still relevant today but greater sophistication of laboratory setup is needed for most scientific pursuits. It was around 1810 that a friend asked Wollaston to show him around his optical laboratory. Wollaston rang the bell and had his butler bring it in. All he needed for his work on optics could be assembled on a small tray.

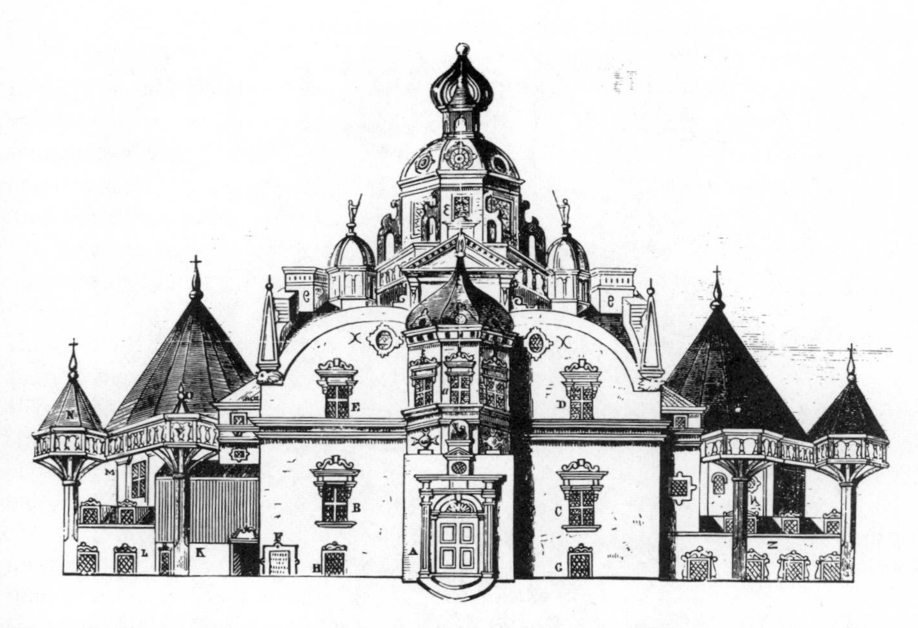

Fig. 2.44 *Early woodcut of Uranienburg—Tycho Brahe's 'castle of the heavens' (1576) the most magnificent astronomical observatory of its time*

The movement toward specialised laboratories and observatories can be traced to be part of the development of specialised institutions of science and, later, engineering. Alchemists had always (see Fig. 2.46) set themselves up in specific laboratories where their equipment was ready to use around them but the modern chemical laboratory as is known today began with the establishment of the workroom (Fig. 2.47) of Liebig in Giessen, Germany. That was in 1842. The term laboratory came down to us from the German term for a chemical laboratory. The Académie des Sciences in Paris, in the 17th century, had rooms where experiments were organised. The famous Greenwich Observatory (which is no longer the place that the famous observations of longitude are conducted now) was established for practical reasons of improving navigational methods in 1675, see Ronan (1975). It, too, had special features included, such as the tall narrow windows shown in Fig. 2.48, to allow certain kinds of observations to be conducted with the instruments of the day.

Laboratories for educational purposes emerged in the 17th century. In such places the collections of philosophical models (Fig. 2.49) would have been put at the disposal of the academics and students. This form of instruction began first in universities, then being adopted by all other levels of education as time passed. Cajori (1962) described the development of laboratories up to the 1920s in more detail than can be given here.

Kelvin, at Glasgow, appears to have been the first to organise a university department in the physical sciences with a teaching laboratory. That was around 1845. Many universities soon followed with the hope that theirs was better than all others.

Fig. 2.45 *Berti's experiment, c1640, on atmospheric pressure was conducted using the wall of a house*
[Knowles-Middleton (1969*a*); from Schott, G. *Technica curiosa, sive mirabilia artis*, 1687, plate 12]

Fig. 2.46 *The first laboratories were for the practice of alchemy.*
[Hart (1924), from painting by Peter Breughel, 1558. Courtesy Oxford University Press]

Some, it seems, however, did not agree that laboratories were a necessity to the practice of physics. It was in 1886 that the newly arrived Professor Threlfall began a campaign to obtain the support of the Prime Minister of New South Wales, Australia to reverse the decision of the University of Sydney not to provide the existing physics laboratory facility to Trelfall. He obtained it eventually.

The science laboratory certainly came into prominence in the middle 19th century in much widened fields. According to Burstall (1970) materials testing laboratories were established in London in 1865, in Munich and Berlin in 1871, in Vienna in 1873 and Zurich and Stuttgart in 1879. Others were established in Sweden and in Russia. It was, in fact, possible in 1884 to call together experts from these to the first international congress on materials testing. The Cavendish Laboratory at Cambridge University was established by Clerk-Maxwell in 1874. Emphasis on physical laboratories was especially strong in Germany. They began to appear in the teaching programmes of American universities from the 1860s. In general, however, Cajori (1962) pointed out the Americans were slow to introduce laboratory work into teaching. Harvard did not have teaching laboratories until well after the 1870s. The Royal Institution in London provided laboratory space for visiting workers in the London building. It was built in 1800 as an assembly of suitable manufacturing machines but it, with the others, soon paled into insignificance by comparison with the laboratories that were to come in the 20th century.

Fig. 2.47 *Organised laboratories for chemistry began with Liebig's at Giessen, Germany in 1842* [Margenau and Bergamini (1966), p.46 from 1842 sketch]

Fig. 2.48 ***The Octagon room of Flamsteed House, Greenwich Royal Observatory had special features built into it to assist observations. From a contemporary engraving by Francis Place.***
[Courtesy The Royal Greenwich Observatory]

Fig. 2.49 *Natural-philosophy lesson taking place around 1769. Note the physical apparatus collections being used*

[University of Utrecht (1977), from Nollet, J.A. *Leçons de Physique Expérimental,* 1769]

The British Association, in 1842, went to the British Government with the proposal that Kew Observatory, which was about to be closed, be used as a place to assemble and test overseas designs of physical instruments. In this the proposers were suggesting, well before time, that a national measurement laboratory was needed. Even today the concept of instrument evaluation by an independent authority is only just beginning to be recognised as worthwhile; instrument evaluation on behalf of clients is conducted by the SIRA Institute in England. An historical account of this development at the Kew Observatory has been published by Howarth (1931).

Probably the first laboratory that could be said to have been established for the common good of science and industry was the Reichsanstalt in Charlottenburg, Germany. This was established under the inaugural Presidency of Helmoltz in 1888. As well as having departments devoted to theoretical research it also had facilities for studying industrial problems.

Of particular importance to measurement was the establishment of the International Committee of Weights and Measures, known as BIPM, in a laboratory, Fig. 2.50, set up in the Pavillon de Breteuil in the Park of St. Cloud, Sevres, near Paris.

Fig. 2.50 *Entrance to the BIPM laboratories where fundamental standards of the international metric system of physical standards are maintained*
[Courtesy Bureau International Des Poids et Mesures, Paris]

This was erected in 1875 to provide the international centre for maintaining the international standards of the metric system that had just then been adapted by many countries under the metric treaty. A history of this laboratory was published on the recent occasion of its centennial, Page and Vigoureux (1975).

It was not long before many of the countries participating in that treaty saw the need to have their own national depository for holding their national standards. The National Physical Laboratory of Britain was established in 1899; R.T. Glazebrook was the first Director. He stated that the object of the NPL was 'to make the forces of science available to the nation'. It began in the Kew Laboratory, mentioned above, moving to the present Bushy House site in 1900. In 1911 a national testing and experimental facility was opened nearby for work on ship models.

In the USA standards of measurement had been introduced during the time prior to 1900 but they had not been given any nation wide co-ordinated control. Cochrane (1966), in his extensive history of the National Bureau of Standards (NBS), called it the era of the laissez-faire standards. Things came to a head when the great fire of Baltimore in 1904 provided strong evidence of the lack of standards. Threads of fire hoses did not interchange as needed at the moment of crisis. Prior to this the NBS had been established but was finding it very difficult to acquire the funds needed. The story of the fire hoses certainly oversimplifies the total political situation, as the official history reveals, but the fire-hose problems did cause a great deal of lobbying to take place. That pressure group, plus others concerned with various consumer trading frauds, helped the Bureau to find its place in the American society of those times.

The role of the NBS then began to widen continuously. During the First World War it carried-out research for the war effort. It performed R & D work on airplanes, radio, rocketry, building, bombs and many other fields, too numerous to mention here. Measurement standards and methods were certainly an important part of NBS activity but by no means the main effort. In 1978 reorganisation took place with the structure becoming a National Measurement Laboratory and a National Engineering Laboratory. This division makes the work somewhat similar to the British system's National Physical Laboratory in England and National Engineering Laboratory in Scotland.

It would appear that part of the rise to prominence of 'Made in Germany' goods during the latter part of the 19th century can be attributed to the formation of the German equivalent of the above research and standards organisations, the Physikalisch Technische Reichsanstalt mentioned above. It also had the mandate to assist industry where possible. In just one decade the effect of work and facilities of this institution changed the state of German manufacturers to a point where Germany had world monopolies in many areas of industrial production.

Today many countries maintain a national laboratory for the work on standards of physical measurement. Developing countries such as Korea, (see Fig. 2.51), Brazil and the United States of Mexico have each recently organised this facility in their national structures. India, additionally, operates a Central Scientific Instrument Organisation that develops common measuring and other scientific apparatus to the stage where the design can be given to industry for commercial development.

In 1946 a new kind of measurement laboratory organisation was formed in Australia.

Fig. 2.51 *Korean Standards Research Institute (K.SRI) established in 1975* [Courtesy Korea Standards Research Institute]

This body, called the National Association of Testing Authorities, or simply NATA, was implemented to give accreditation to the many small to large independent laboratories that were carrying-out testing and measurement for themselves and clients. It was decided that a voluntary organisation that assessed those who wished to be granted registered status would eventually lead to more uniform standards of testing and measurement throughout the country. By 1977 over 1000 laboratories were registered in this way. The concept was adopted by other countries. First the British Calibration Service was formed in 1966; this having far less scope in class of measurement than the NATA scheme for it was restricted to calibration. Later came the New Zealand TELARC scheme. In 1978 the National Bureau of Standards of the USA declared that the NATA concept seemed to be the best model for the USA to follow.

The American Association of Laboratory Accreditation, AALA, is, at the time of writing, to be based upon NATA lines, for it was said that it had stood the test of some thirty years of operation. NATA procedures also had a definite influence upon the operations of the Danish Accreditation scheme.

In the USA, in 1961, standards laboratories formed, with their Secretariat in the NBS, the National Conference of Standards Laboratories, NCSL for short. Of over 500 standards laboratories in the USA, around 300 are members. Membership by other country laboratories is also possible. Membership, however, does not carry with it any accreditation status.

It has become clear from trade and scientific research standpoints that the world is a global village. The importance of measurement is now such that all countries should have similar standards of testing and measurement throughout the whole of daily life and industry, not just at the international physical standard level where first co-operation on standards of measurement developed.

In the same way that professional and sub-professional associations had been created to cater for the special needs of specialist groups such as the physicists, chemistries and engineering divisions there also were formed institutions concerned specifically with measurement. These came into being well into the 20th century. Examples are the Instrument Society of America (ISA), and the Society of Instrument Technology (foundation began in 1943) that later, in 1968, became the Institute of Measurement and Control. Royal charter was granted to the latter in 1975.

By the 1960s over thirty countries had associations that specialised in measurement either as its sole speciality or as a division of it's work. Many were also involved with control as well as measurement.

Out of these, plus other groups interested in the control discipline that emerged after the 1940s, grew the International Federation of Automatic Control (IFAC), the International Federation for Information Processing (IFIP), the International Federation of Operations Research Societies (IFORS), the International Association for Mathematics and Computers in Simulation (IMACS), and the last to appear, the International Measurement Confederation (IMEKO). The latter international organisation is specifically concerned with the various aspects of the theory and application of measurement. In 1977, it had 23 countries as members each being represented by the membership of a nongovernment association of practising measurement workers.

Realising the overlap of the work of these five institutions they formed another grouping called the Five International Associations Coordinating Committee (FIACC). FIACC members do not have assets beyond their membership subscriptions and the generous government support of the Secretariat of each. Action is accomplished through committee work, regional meetings and international congresses or conferences. FIACC has a working link into the UNESCO programmes, thereby relating all five organisations in UNESCO activities.

2.12 Economic considerations: costs and benefits

Measurements are made on physically existing things using physically existing equipment. The instruments must be created from the basic materials using labour of workers. Material and labour resources have never been unlimited or freely available to a would-be user; there has always been the choice of doing something else with them. Priorities must be decided. A measuring instrument does not stand alone; its cost cannot be judged by looking only at the purchase or manufacturing of the apparatus used. In this Section attention is first drawn to the real cost of an individual measurement for too little attention has always been given to this most important aspect of measuring instruments. To provide a balanced economic statement it is thus necessary to consider the worth of a measurement.

It would appear that the cost of the measurements and their worth has not been a subject that many writers have felt they wanted to, or perhaps could, write about. It has, however, recently become the concern of many studies being conducted variously from the makers' viewpoint through to the standpoint of a nations expenditure on measurement. This account begins first by looking at the cost of the apparatus, broadening outward to the cost of measurement to a nation.

The cost of producing an instrument can be broken down into subcomponents prices plus the labour content needed to build the device. To this must be added other factors such as marketing, design development, embellishment, paper work about specifications, use and service, packing, delivery and many more contingencies. Such notions of costing are relatively new in concept being part of the more modern approaches developed as the cost-accounting method. Fig. 2.52 depicts the factors involved in costing an instrument.

In the classical antiquity many of these factors had no relevance. How to use the instrument was usually self-evident. Rarely were more than one of the exact same model made and often the maker was the user so did not need to be concerned with records and sales. As the instruments were made by decorative craft artisans they usually had more embellishment than was needed. The time taken to make a piece of measuring equipment was seldom so important as it is today. Indeed it was not uncommon for instrument-makers from classical antiquity to the 17th century AD to spend, not months on manufacture, but years. Harrison produced his four prize-winning chronometers of the 18th century as the result of devoting most of his life to its task. Originally instrument making was more akin to the fine arts products of today than to the fierce open competition of today's industrial market.

The introduction of production manufacturing methods, such as Ramsden's dividing engine mentioned earlier, started to change this position. Markets widened, production numbers rose and cheaper means of producing the desired performance were realised.

The measuring instrument manufacturing market game has the feature that its players are always doing their best to work themselves out of a job! As soon as a significant new principle or technique is discovered that could cause major sales the manufacturers move to produce it for sale at a relatively high price. Demand grows with reduced prices so means are sought to stimulate demand so that the production runs' sizes are larger. This tactic, if successful, drops the price, floods the market and the manufacturer must again look for another product. The sustained sales of the device that sold so well initially do not usually continue to sell at the volume that first enticed the maker into that market.

Some makers will make adequate profits in this cycle; they will return again to play another round. Many, however, do not get their timing and marketing methods right. They end up with an abundance of superseded product that cannot possibly be sold at even the cost-to-produce price. Companies will go to the wall if this happens too often. Many an instrument firm has poured funds into developing a certain class of instrument only to have others sell better than they. This general picture is, of course, the tale of laissez-faire open marketing. The instrument market has the specific features that the cycle time is only in a matter of months to a year or so, that costs to produce

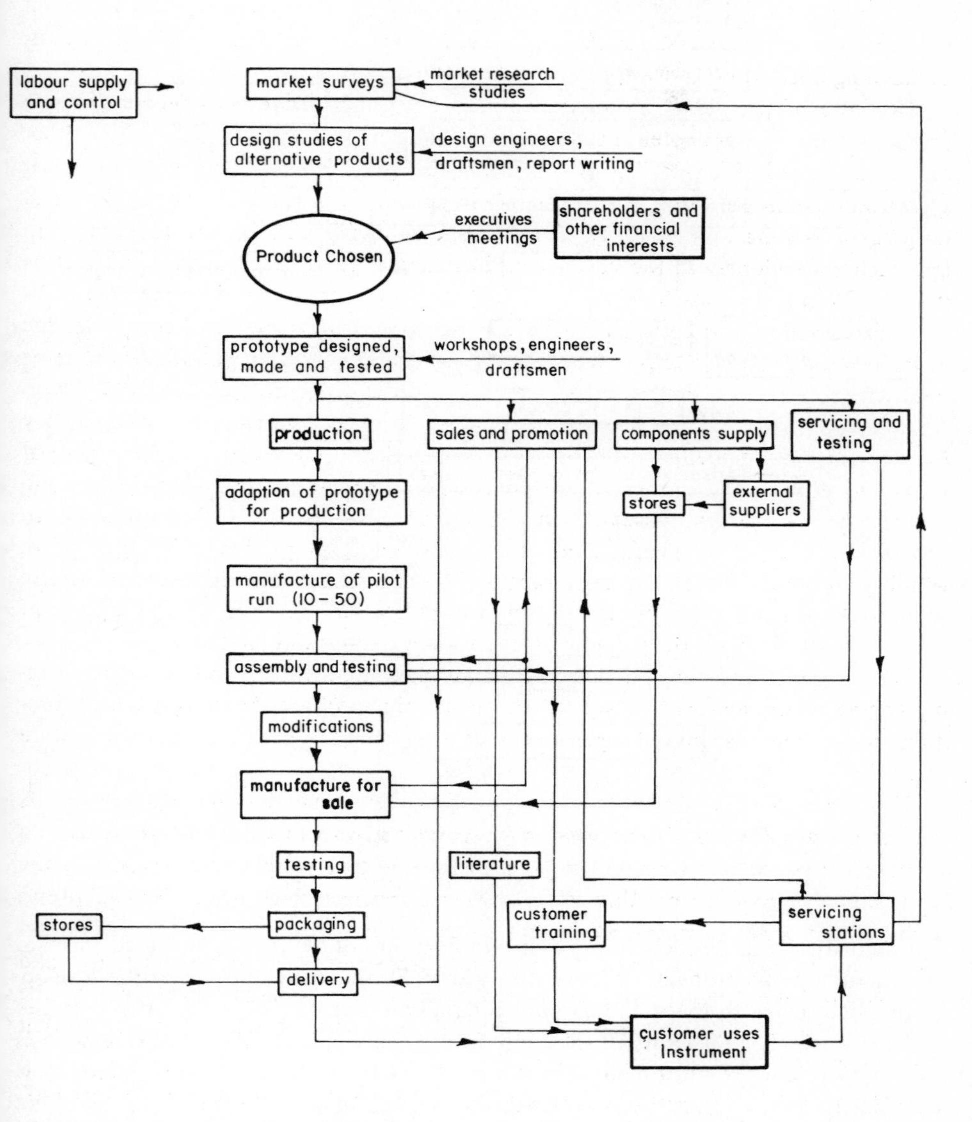

Fig. 2.52 *Factors that must be considered in arriving at the cost of producing an instrument can be seen from this schematic of an imaginary instrument firm*

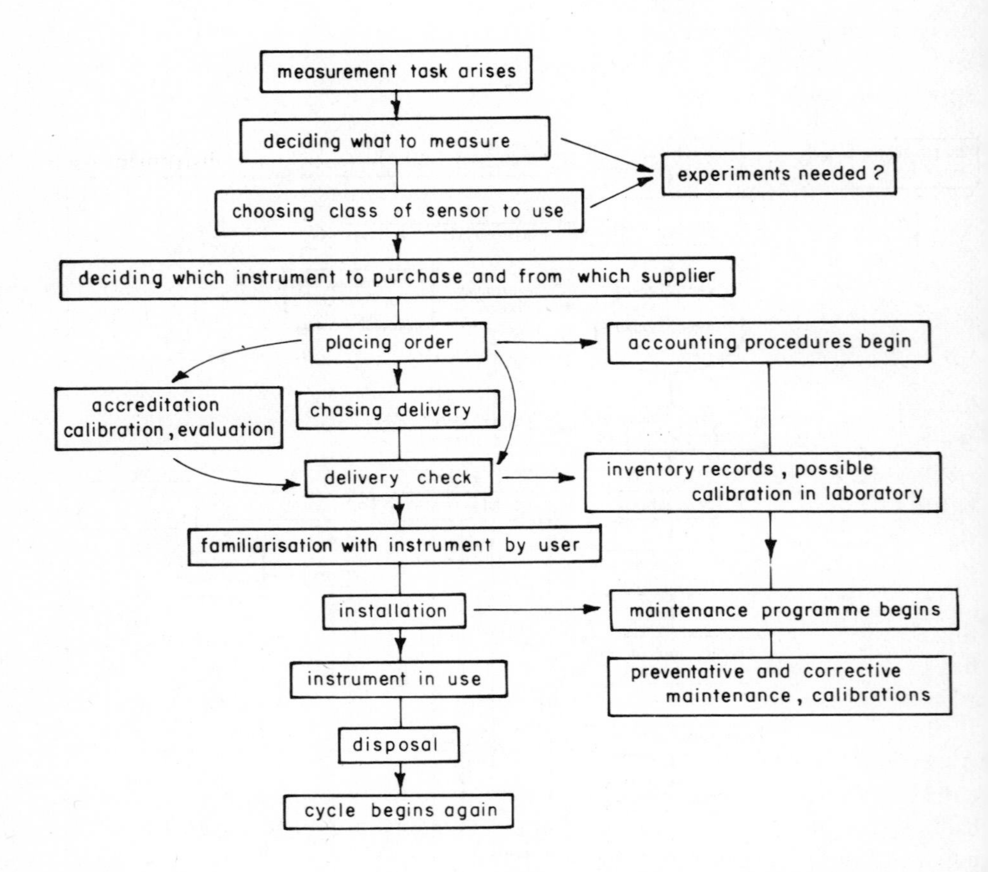

Fig. 2.53 *The cost of a measurement far exceeds the purchase price of the instrument*

fall dramatically as development by the whole group of industry is increased and that the useful life of a product is comparatively long. Furthermore, the greater the development effect made to lower the per-unit production cost the lower the price becomes and the greater the reliability of the component. This would be fine if the volume sales rose appropriately. Unfortunately, sales do not reach optimal levels. High priced, new instruments that exploit the latest methods usually, but not necessarily so, are more liable to breakdown than the mass produced, well developed units. Another feature that the instrument maker has to contend with is the general idea that systems designers consider it is reasonable and normal to call-up specials when specifying the measuring apparatus required. One-offs are the lines that the large firms must produce to maintain customer faith yet they often cost far more than the customer thinks should be paid. People have become used to the idea that their electric toaster works as the designers suggest it should, yet they would not always agree to make the same degree of compromise in using the instrument makers suggestions of what are reasonable basic units to work with. Just as the knowledge about measuring systems has to be restructured for future more efficient use, so must the range and number of marketed instrument variations.

It is generally acknowledged and understood that the cost of the instrument itself can be assessed from a long list of constituent prices as shown in Fig. 2.52. This process is happening everyday in the Board-rooms of instrument-making firms. Management must budget for a manufacturing price to decide sales potential and profit margins. The same, however, cannot be said of users when they assess the cost of a measurement to their organisation or enterprise. The truth is rather startling: the real cost of a measuring instrument is, in reality, many times the purchase price.

Assessment of the real cost of a measuring instrument should be on the basis of considering what costs would have been avoided if it had never been considered. Fig. 2.53 shows the real costs that compose the cost of a measurement. To begin the costing process it is necessary to first include the cost of the time spent deciding what has to be measured. Too often this stage is seen as barely existing for the measurand, the parameter to be measured, appears obvious. In the writer's experience it is clear that many users give too little emphasis to this very basic step.

Having decided the variable concerned it is then necessary to devote time, generating the specifications needed for the method selected. For example, assume the task has been defined as one of measuring temperature of a product because that reflects when it is ready to be transferred. Is temperature really the best parameter to monitor? If so what must be the precision, accuracy, absolute value, shape of sensor and so on? All of these decisions take time, and time is a valuable resource.

The specific instrument principle and the specifications exist; but the task has barely begun for it then is necessary to call for trade information so that a specific purchase can be organised. Tenders are called, or firms contacted, to obtain quotations, data sheets, delivery information, installation details and so forth. At this juncture costing of the real cost of a measurement must not forget to include time spent preparing budget submissions and filing the records.

Eventually, the stage is reached where a purchase decision can be made. That, too, takes up more time of skilled staff. It is unlikely that these above stages can be passed through in less than a minimum of three hours of effort for even the simplest of measurements. Extensive systems take months, even man years, of work to reach this stage.

The instrument is chosen and the order placed; the cost of the ordering procedure should be included. Delivery is awaited. If all proceeds perfectly the instrument arrives without need to make enquiries about late delivery, but more often than desired order advancing adds more to the cost of the instrument.

Upon receipt the wise purchaser will carry out a delivery check to ascertain that it is undamaged, is within specifications, does contain the documentation, which might be yet another cost to consider, and will suit the purpose intended. Such a check, if performed properly can take many hours and will need the use of very expensive test equipment, also placing a cost into the account. If things are found to be wrong more time will have to be devoted getting it corrected through purchase agreements or, which is often quicker and perhaps cheaper, having it repaired in-house.

Overlapping this acceptance stage is the need for the user to become familiar with the instrument. It may need attendance at a training course or be simply a case of reading the operating instructions. Whichever it is, time must go onto the costing sheet. The resources expended on this should not be under-estimated for a complex piece of measuring equipment may require the user to spend much unproductive time learning to use it. Errors in its use in the early stages will also give rise to costs that have to be met somewhere.

Then comes installation. This requires organisation of varied skills, may necessitate inventory records to be created along with an instrument dossier and a spare parts reserve. Plant drawings may need to be updated, instructions issued to operators. Many hours of labour will be absorbed indirectly in getting the instrument into place.

Costs of the instrument continue to mount. To ensure that the instrument is performing properly it will be necessary to have a service programme, a calibration programme and a maintenance programme in operation as well as a possible method of monitoring the instrument if its role is in a key area. To remove an instrument, take it to a calibration laboratory, carry-out a calibration and return it to duty will take a minimum of another two or three hours work. It may take days!

No machine is perfect. Generally, but by no means always, the mean-time-between-failure (MTBF), rate of equipment rises as the cost of the instrument goes up. This means the purchase price can be reduced if the MTBF is shortened. Too often the decision is made to purchase the instrument offered at the lowest figure quoted. This is oversimplifying the purchase decision but the fact remains that most given instruments will need unpredictable service at some time in their useful life. The cost of this kind of service is high for it may occur when labour costs are at their highest penalty rate. Furthermore, the technician may not be familiar with the system and has to learn at the owner's expense.

The instrument has given a useful and satisfactory life but its costs to the user go on. At some stage effort has to be expended to establish a replacement schedule and the choice of its replacement; this can be quite complex. The model used originally will often be superseded by something that may not be quite the same in some respects or it may not be possible to get a replacement at all.

Finally, comes the last of the costs associated with the instrument itself. This is the cost of disposal. It may be possible to recoup some of the costs through sale of the instrument but the opportunity cost of a well used superseded instrument of early vintage is more likely to be a debit than a credit. This cost often does not arise until years after the event for most large complexes tend to store old instruments for many years until they are finally discarded, find their way into a museum collection or are dismantled for use as pieces of some other plant.

Any genuine attempt to estimate the cost of a measurement must include all of these factors. The question posed is one of establishing what resources were tied up by the existence of the instrument that might have been used elsewhere if it had not been considered. A full costing on the above basis gives a satisfactorily quantitative assessment of which instrument to purchase and which way to use to make the measurement.

All of the above costs are concerned only with the choice, purchase, installation and maintenance of the instrument as a working system. The cost of the whole measurement goes on rising for the measuring instrument has to be used to produce meaningful measurements and this imposes a cost. Much of measurement conducted is, unfortunately, not productive. A wrong measurement has often been the cause of disastrous losses to the enterprise. A serious malfunction in the commissioning stage of a new ship can cost the builder thousands of dollars per hour. A malfunction of a weapon of war has a cost that goes without saying.

Thus, this discussion reaches the stage of considering the cost of measurement to a nation.

Studies of the cost of measurement at this level have begun to be given in-depth consideration in the 1970s. Prior to that costing was certainly performed on the annual budgets of such institutions as the national measurement laboratories and similar identifiable groups but little attempt was made to go beyond this small part of the total national cost devoted to measurement.

The term National Measurement System came to the fore when the National Bureau of Standards (NBS), USA, began a serious nation-wide study in 1972. The development of their study can be followed in the NCSL Newsletters from 1972 through to 1978. Minck (1977) defines the NMS, as it has become known, as 'consisting of all national activities and mechanisms that provide physical measurement data to allow the creation of the objective, quantitative knowledge required by our society'. Previously Huntoon (1967) had developed the concept of the NMS in a paper. The NBS study reached the final report stage in 1975 after 151 separate studies had been made. The report contains 27 documents. The two key readings are NBSIR 75-925 'Final summary report study of the National measurement system, 1972-1975' and NBSIR 75-949 'Structure and function of the National measurement system'. The report includes the impact matrix (Fig. 2.54), showing the measurement relationships existing for 25 groups of societal needs. It graphically portrays the dynamics and the degree of such interaction and in doing so, well illustrates the great complexity and widespread use of measurements in modern life-style.

Before leaving the US situation it is appropriate to give an indication of the money value of such a system. One major subsystem of the NMS, that of the Sandia Corporation, spent, in the 1975 period approximately $10 000 000 to maintain ten standardising laboratories. Seed (1973) shows the relationship of the 10 laboratories who are co-ordinated by the Sandia Corporation for the US Atomic Energy Commission. To give an idea of the impact these have on other capital investment he suggests that these ten institutions assist the operations of a subsystem expenditure of around $1 290 000 000.

The ratio of around 1% for impacted funds to measurement funds is typical of estimates made by various bodies. It has been suggested that the economic value of the NMS for a developed technological country will lie in the region of from 0·1% – 5%. Unfortunately such estimates are not based on particularly thorough studies and, at this time, must only be regarded an indicative figures that suggest very strongly

DIRECT MEASUREMENTS TRANSACTIONS MATRIX FOR NATIONAL SYSTEM OF PHYSICAL MEASUREMENTS (March 1976) — USERS / SUPPLIERS	KNOWLEDGE COMMUNITY (Science, Education, Prof. Soc. & Publ.) 1	INTERNATIONAL METROLOGICAL ORGANIZATIONS 2	DOCUMENTARY STANDARDS ORGANIZATIONS 3	INSTRUMENTATION INDUSTRY (SIC Major Gp 38) 4	N B S 5	OTHER U.S. NATIONAL STANDARDS AUTHORITIES 6	STATE & LOCAL OFFICES OF WEIGHTS & MEASURES (OWM's) 7	STANDARDS & TESTING LABORATORIES AND SERVICES 8	REGULATORY AGENCIES (excl. OWM's) 9	DEPARTMENT OF DEFENSE (excl. Stds. Labs) 10	CIVILIAN FEDERAL GOVT AGENCIES (exc. Stds Labs & Reg.Ag.) 11	STATE & LOCAL GOV'T AGENCIES (exc. OWM's & Reg. Ag.) 12	INDUSTRIAL TRADE ASSOCIATIONS 13	AGRICULTURE, FORESTRY FISHING; MINING (SIC Div. A & B) 14	CONSTRUCTION (SIC Div. C) 15	FOOD/TEXTILE/LBR/ PAPER/LEATHER/ETC. (SIC 20-26, 31) 16	CHEM/PETROL/RUBBER/ STONE/CLAY/GLASS... (SIC 28-30, 32) 17	PRIMARY & FAB. METAL PRODUCTS (SIC 33-34, 391) 18	MACHINERY EXCEPT ELECTRICAL (SIC Major Gp 35) 19	ELECTRIC AND ELECTRONIC EQPMT (SIC Major Gp 36) 20	TRANSPORTATION EQUIPMENT (SIC Major Gp 37) 21	TRANSPORTATION & PUBLIC UTILITIES (SIC Div. E) 22	TRADE/INS/FIN/REAL EST/PERS SVCS/PRINT (SIC F-H, bal I, 27) 23	HEALTH SERVICES (SIC Major Gp 80) 24	GENERAL PUBLIC 25
1 KNOWLEDGE COMMUNITY (Science, Education, Prof. Soc. & Publ.)	4 1 / 6 / 2	3 1 / 2 / 1	3 1 / 3 / 1	2 1 / 3 / 1	4 1 / 4 / 2	1	1	2 1 / 2 / 1	3 1 / 2 / 2	3 1 / 3 / 1	3 2 / 3 / 2	2 / 1 / 1	2 2 / 1 / 2	1	1	2 2 / 1 / 1	3 1 / 2 / 1	3 1 / 2 / 1	2 1 / 2 / 1	3 2 / 3 / 1	3 1 / 3 / 1	2 1 / 2 / 2	3 1 / 1 / 1	2 1 / 1 / 1	2 2 / 1 / 1
2 INTERNATIONAL METROLOGICAL ORGANIZATIONS	2 1 / 1 / 1	3 1 / 3 / 1	3 1 / 2 / 2	2 2 / 1 / 1	3 1 / 3 / 2	1			1	1	1			1					1			1			
3 DOCUMENTARY STANDARDIZATION ORGANIZATIONS	3 2 / 1 / 2	3 1 / 2 / 2	3 2 / 3 / 1	3 1 / 3 / 1	3 1 / 2 / 2		3 1 / 1 / 1	3 1 / 2	3 2 / 2 / 2	3 1 / 2 / 1	2 1 / 2 / 1	1	3 1 / 2 / 2	1	2	2	2 1 / 2 / 1	2	2 1 / 2 / 1	2 1 / 2 / 1	2 1 / 2 / 1	3 1 / 2 / 2	1	2	1
4 INSTRUMENTATION INDUSTRY (SIC Major Gp 38)	2 1 / 2 / 1	2 1 / 1 / 1	2 1 / 2 / 2	4 1 / 5 / 2	3 1 / 3 / 2	1	1	3 1 / 3 / 2	3 1 / 2 / 2	3 1 / 5 / 2	3 1 / 4 / 1	3	2 1 / 2 / 1	3 1 / 3 / 2	4	3 1 / 3 / 2	3 1 / 3 / 1	3 1 / 4 / 1	3 1 / 5 / 1	3 1 / 5 / 2	3 1 / 5 / 1	3 1 / 5 / 2	3 1 / 5 / 1	3 2 / 4 / 2	3 1 / 4 / 2
5 NBS	3 1 / 3 / 1	3 1 / 3 / 2	3 1 / 3 / 2	3 1 / 3 / 2	3 1 / 4 / 1	3 1 / 1	4 1 / 2 / 2	3 1 / 3 / 1	3 2 / 2 / 1	3 / 3	3 1 / 3 / 2	2 1 / 1 / 1	2 1 / 2 / 2	4 1 / 2 / 1.	1	2 2 / 1 / 1	2 2 / 2 / 2	4 2 / 2 / 2	2 1 / 2 / 1	3 1 / 2 / 1	4 1 / 2 / 1	4 2 / 2 / 2	4 3 / 1 / 2	4 3 / 2 / 2	2 1 / 1 / 1
6 OTHER U.S. NATIONAL STANDARDS AUTHORITIES	1	1		1	3 1 / 1	3 1 / 1 / 1		1	1	2	1	1	1	1	1							1	1		
7 STATE & LOCAL OFFICES OF WEIGHTS & MEASURES (OWM's)		1	1	3 1 / 1 / 1 R	4 1 / 2 / 1 R		4 / 2	1	1			3 1 / 1 / 1	1	1		3 1 / 2 / 2	1		3 1 / 2 / 1		1	2	4 / 3 / 1		4 1 / 1 / 1
8 STANDARDS & TESTING LABORATORIES AND SERVICES	2 1 / 1 / 1		2 1 / 2 / 1	2 1 / 3 / 1 R	3 1 / 2 / 2 R			3 1 / 3 / 1	2 1 / 2 / 2	3 1 / 3 / 1	3 1 / 2 / 1	1	3 / 2 / 1	1	1	1	2 1 / 2 / 1	3 1 / 2	3 / 1 / 1	3 1 / 2 / 1	3 1 / 3 / 1	3 2 / 2 / 2	3 1 / 1 / 1	3 2 / 2 / 2	
9 REGULATORY AGENCIES (excl. OWM's)	3 1 / 1 / 2		3 1 / 2 / 2	3 1 / 2 / 2 R	3 2 / 2 / 2 R			3 1 / 2 / 2 R	3 2 / 3 / 1	2 1 / 1 / 1 R	3 1 / 2 / 2 R	1 / R	2 1 / 2 / 2 R	4 2 / 2 / 1 R	4 1 / 1 / 2 R	4 2 / 2 / 1 R	3 1 / 2 / 1 R	3 1 / 2 / 1 R	4 1 / 2 / 1 R	3 2 / 2 / 1 R	3 1 / 2 / 2 R	3 1 / 2 / 2 R	3 1 / 2 / 2 R	4 2 / 2 / 3 R	3 1 / 2 / 2 R
10 DEPARTMENT OF DEFENSE (excl. Stds. Labs)	3 2 / 2 / 2		3 1 / 2 / 1	3 1 / 3 / 2 R	2 1 / 3 / 1 R	1		3 1 / 3 / 1 R	1	3 1 / 7 / 1	3 1 / 2 / 1		2 1 / 1 / 1				2 1 / 1 / 1 R	3 1 / 2 / 1 R	3 1 / 2 / R	3 1 / 4 / 2 R	3 1 / 4 / 2 R	2 1 / 2 / 2 R	2 / R	1 / R	
11 CIVILIAN FEDERAL GOV'T AGENCIES (excl. Stds. Labs & Reg. Ag.)	3 1 / 2 / 1		2 1 / 2 / 2	2 1 / 2 / 2	3 1 / 3 / 1	1		2 2 / 2 / 1	2 1 / 2 / 2	2 1 / 1 / 1	3 1 / 7 / 2	2 1 / 2 / 1	2 1 / 1 / 2	2	1	1	2 2 / 2 / 2	3 / 2 / 1	3 1 / 1 / 1	3 1 / 2 / 2	3 1 / 2 / 1	4 1 / 4 / 1	1	2	2 / 3
12 STATE & LOCAL GOVERNMENT AGENCIES (exc. OWM's & Reg. Ag.)	1		1	1 R	2 1 / 1 / 1 R	1	1 / R	1 / R	1	1	1	3 2 / 6 / 2	1	1	2	1	1	1	1		1	1	1	1	1
13 INDUSTRIAL TRADE ASSOCIATIONS	2 / 1 / 1	1	2 1 / 3 / 2	2 1 / 2 / 2	2 1 / 2 / 2		1	3 1 / 1 / 1	2 1 / 2 / 2	2 1 / 1 / 1	2 1 / 1 / 2	2 2 / 1 / 2	2 1 / 2 / 1	1	1	1	2 1 / 1 / 1	1	2	2 1 / 2 / 1	2 1 / 1 / 1	2 1 / 1 / 1	1	1	1
14 AGRICULTURE, FORESTRY FISHING; MINING (SIC Div. A & B)	1		2 1 / 1	1 / R	2 2 / 1 / R		1 / R	1 / R	1	1	2	1	1	2 / 5	1	2	2	2	1	1	1	2	.1		1
15 CONSTRUCTION (SIC Div. C)	1		1	1 / R	1 / R		1 / R	1 / R	1	1	1	1	1	1	6			1 / R	1	1	1	1	1		1
16 FOOD/TOB/TEXTILE/ APPAREL/LBR/FURN/PAPER/ LEATHER (SIC 20-26, 31)	1		3 1 / 1 / 1	3 2 / 1 / 2 R	3 2 / 1 / 2 R		2 / R	2 / 1 / R	3 1 / 1 / 2	1	1	1	3 1 / 2 / 2	1	1	2 1 / 6 / 1		1 / R	1 / R	1 / R		1 / R	1		2
17 CHEM/PETROL/RUBBER/ PLASTICS/STONE/CLAY/ GLASS (SIC 28-30, 32)	2 1 / 2 / 1	1	2 1 / 2 / 2	2 1 / 2 / 1	3 2 / 2 / 2		1	2 / 1 / 1	2 2 / 2 / 2	2 1 / 1 / 1	2 2 / 2 / 2	1	2 1 / 1 / 1	2	1	1	2 1 / 6 / 1	1	1	1	1	1	1	1	1

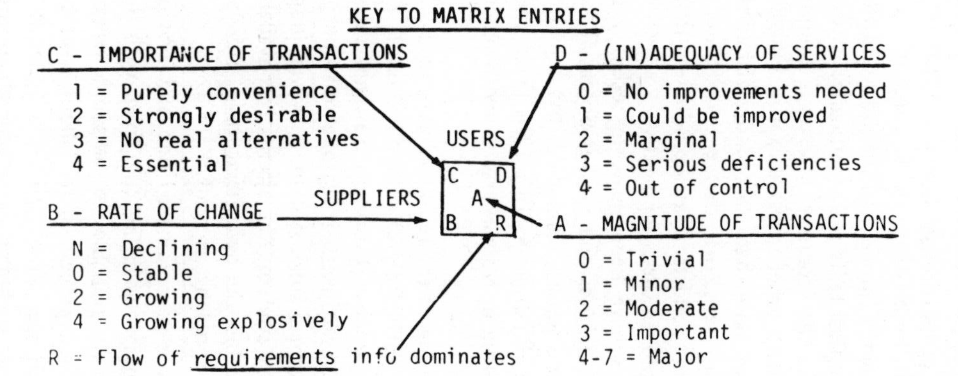

18 PRIMARY & FAB. METAL PRODUCTS (SIC 33-34, 391)	2 1 1		3 2	3 1 2 1 R	3 2 1 2 R		1	3 1 R	2	3 1 2 1	2	1	2	2	1	1	1	3 1 6	2 1 1	1	3 2	2	1	1	1
19 MACHINERY, EXCEPT ELECTRICAL (SIC Major Gp 35)	2 1 1	1	3 1 2	3 1 2 1 R	3 2 R		1	2 1	1	3 1 1 1	3 1 1	1	2	2 2	1	1	1	2	3 1 6	2 1	3 2	2	2	1	2
20 ELECTRIC AND ELECTRONIC EQPMT (SIC Major Gp 36)	3 1 2 2		3 1 2 2	3 1 3 2	3 1 2 2			3 1 2 2	3 1 1 2	3 1 3 1	3 1 2 2	1	3 1 2 1	1	1	1	1	1	2	3 1 6 3	3 2 2 3	2	1	2	2 2 2 2
21 TRANSPORTATION EQUIPMENT (SIC Major Gp 37)	2 1 2 1		2 1 2 1	3 1 2 1 R	2 1 2 2 R			2 1 2 1 R	2 1 2 2	2 1 2 2	2 1 2 1	2 1 1 1	3 1	1	1	1	2 1 1 1	2 1 1 1	2 1 1	3 2 1 2	3 1 6 1	3 2 1	1		1
22 TRANSPORTATION & PUBLIC UTILITIES (SIC Div. E)	2 1 2 1		2 1 2 2	2 1 2 1 R	2 1 2 2 R	3 1 1 2	1	2 1 2 1 R	3 2 2 2	2	2 1 2	1	2 2	2	1	1	1	1	1	1	2 1 2 1	3 1 6 1	2	1	3 1 4 1
23 TRADE/INS/FIN/REAL EST/PERS SVCS/PRINT-PUB (SIC F-H, Bal. I, 27)	1		1	2 R	3 1 1 1 R		2 R	1 R	2	1	1	1	1	1	1	2	2	2	1	1	1	2	7	1	6
24 HEALTH SERVICES (SIC Major Gp 80)	1		1	2 R	4 2 2 3 R			1 R	1	1	1	1	1	1	1	1	1	1	1	1	1	1	1	5	2
25 GENERAL PUBLIC			1 R	2 R	2 1 1 R		1 R		2 1 1 2 R	1 R	2 R	1 R	1 R	1 R	1 R	1 R	1 R	1 R	1 R	1 R	1 R	2 R	2 R	1 R	2 1 7 1

Fig. 2.54 *Direct measurements transactions matrix from study of USA National measurement system*
[Courtesy US Government]

that measurement is vital to a nation and that in-depth studies along the lines of the US investigation should be carried-out for all technological nations. Other countries, such as Britain, have made some investigations of their NMS and as more do so, it will become clear that international agreement on accounting procedures will be needed to introduce standardisation that will enable the figures to be intercompared.

As too few studies have been made by competent observers of the value of measuring apparatus used in a developed country it is difficult to give any more than indications of what proportion of capital is spent on measurement equipment. A recent study in Britain, Finkelstein (1976), came to the preliminary conclusions that in that country instruments form 1% of the output of the manufacturing industry. This alone is not very significant: its real significance is to be found in the impact that this expenditure has on the economy. Measuring equipment has a significant gearing-ratio in its interaction with other invested capital. Finkelstein estimated that measuring equipment constitutes more than 5% of industrial investment. It was also reported, and this will be taken up next, that the general opinion of experts was that this investment brings high returns and facilitates production that was otherwise impossible except by means that employed the equipment.

Many studies have been made in recent years on the economic factors of the measurement industries and market potential for many countries. Reports that are available include series conducted by the Domestic and International Business Administration, Washington DC, USA. This study has researched economic factors of instrumentation in Norway, Germany, Sweden, Belgium, Japan, Brazil, Singapore, France and Australia.

Cost accounting of the resources expended in making instruments, and in applying them within a restricted area of application, is a well developed mechanism for reaching a quantitative cost. Costing a National Measurment System provides more of a challenge but, nevertheless, is reasonable to put into monetary terms for it is a similar exercise to costing a small system. However, judgements about whether to measure or not are made by studying the alternatives available for a given resource useage weighing these up against the benefits accrued. This latter variable, the worth of a measurement, is not so simple to account.

Very little has been done to find and develop means by which management can be guided in budget decisions concerned with the use of new, or existing, measuring equipment. Clearly available are the simple benefits of money saved by implementing a better-controlled process or by labour costs saved by the use of a more advanced measuring technique. More difficult to quantify in an adequate manner, but equally as important, are such factors as general improvement of the standard of a product resulting in increased sales, increased reliability of a product, less wastage in manufacture, potential to produce a product that was otherwise impossible to make, improved performance of the product, improved working relationships between those groups of a nation that impact together in an area of measurement and increased confidence in a nation's products, with subsequently improved exports.

Mention has already been made of the rise to prominence of German products in the late 19th century. This can be linked to improved measurement but not in a quantitative manner. At present the measurers are not yet able to measure the worth of measurement to the same degree as they can assess the cost of measurement. Perhaps this is the reason why many measurement budget decisions generally do not give finance to measuring equipment that is even in the same relation to the proportion of money expended. Very little money goes into calibration, into education and into research of the basic fundamentals of measurement scientific principles and its technology. Fit and forget might be the desire of management but, as with all technology, it is not a reality. Measuring instruments are no exception.

Throughout history there have been occasions when the governing power of nations have considered that certain measurements or measuring instruments would be of great benefit to the nation.

In 1598 the Spanish monarch Philip III, offered a large sum of money to any person who could produce a time keeper that would reduce navigational inaccuracies. Mention has already been made of Harrison who was awarded £20 000 from the British Government who had offered this, in 1714, for a similar reason.

On another occasion James Cook lead the voyage to southern regions to measure the transit of the planet Venus in June 1769. This too was concerned with improving navigational accuracy. Here, it was the Royal Society of London that induced the King and the Admiralty to expend a considerable sum of money conducting this one measurement. There may, however, have been other unrecorded reasons for the voyage.

It is easy to make a case for spending money on measurement for clear-cut scientific, knowledge-seeking reasons. The long-term less quantifiable benefits are, in historical perspective, often seen as more significant than the originally argued case. These are the benefits that mangements would like to be able to cost in money terms.

Chapter 3

Ancient times to Middle Ages: birth of the first instruments

3.1 Introductory remarks

Material provided in the preceding chapters sets the scene for investigation of the historical development of the science and technology of measuring instruments. The following chapters trace this development in a chronological order passing through four defined eras that can be identified as such for the purposes, more of exposition rather than for the fundamental epochs that can be associated with the same dates.

These eras begin with a time span that is much longer than those that follow: that of classical antiquity. This covers the beginnings of traceable civilised man on earth (around 3000 BC) through to emergence from the Dark Ages where the birth of the experimental sciences occurred around 1500 AD.

The second era defined here covers the 300 years following 1500 AD. These years can be thought of conveniently as the age in which experimental science, with its necessary instruments of growing complexity, became firmly established.

In 1800 the low-impedance source of electricity - the Volta primary cell battery - was invented beginning a new era in which electromagnetic method was discovered and applied to produce much more complex and sophisticated sensing systems than had been seen previously. This era is seen here to extend until around 1900 after which the knowledge of electronics found its first applications to communication and other kinds of instrument system. The 20th century was, indeed, an exciting one for, in its early decades, instruments became so firmly established that there arose the birth of the modern instrument manufacturies.

Electronic technique developed slowly at first but the arrival of the transistor in 1948 enabled many long-established concepts to be realised. Since the 1940's, the acceleration and sophistication of measuring systems has grown faster and wider in scope than at any other time in the history of measurement. Because it is difficult to trace historical developments as trends and proven usefulness until some time after the event this account ends around 1950.

To cover such a large period the account must, by necessity, be restricted by many constraints on the limit of detail that can be given. It is the intention that the reader gains an impression of the full account over time, an account that does not seem to have been attempted before.

The aim throughout has been to provide more than a catalogue of the instruments used in the various times by creating a linguistic and pictorial image of the underlying principles and the depth of understanding that might have existed at a given time. It is, of course, just one view tempered by the limited experience at the disposal of that one person.

Although a number of works have accounts of historical measurement concepts and practices they invariably tail out in the early 19th century where the knowledge needed to appreciate many of the then introduced methods lies in the domain of electrical engineering. Those works that do carry through into modern times tend to concentrate on the more obvious kinds of measuring apparatus, such as theodolites and other easy-to-appreciate instruments. The bulk of general works on the historical aspects of measurement concentrate on what has become known as the 'weights and measures' class. These are discussed later in this Chapter. It has been the intention here to put this class, and the many others, into a total perspective, the rationale being that all instrumentation has common basics such as were explained in Chapter 1.

A certain amount about the historical development and application of measurement can be gleaned from study of the works on the general aspects of the history of science and technology and from contemporary explanations of science. These also provide information about the background factors that must be appreciated when considering any particular measuring method. Example works are Lodge (1893), Marmery (1895), Royal Society (1924), Neuburger (1930), Farrington (1936), Sherwood Taylor (1936), Hogben (1938), Butler (1947), Needham (1954), Singer *et al.* (1954), BBC (1958), Klemm (1959), Cajori (1962), Pledge (1966), Africa (1968) and Crombie (1969). Works that are specific to any particular discipline are also relevant. Reference to these will be found in appropriate later sections.

The above remarks apply to the development of measuring apparatus over all times. Attention now narrows down to that period chosen as the first period of development i.e. classical antiquity to the Middle Ages.

It is generally agreed that it was around 3000 BC that civilised, and advancing, forms of man's existence first emerged. We must not completely ignore the older peoples that are known to have had organised lifestyle as early as 30 000 BC but these peoples, to the best of current knowledge, had no obvious forms of measuring instruments. They made do with their natural senses using little technological backup to assist everyday routines. Fig. 3.1 illustrates, in a simplistic manner, the approximate periods during which the ancient civilisations of interest here were in a state of dominance.

The introduction of instruments of measurement went hand-in-hand with the adoption of technological aids for, as was discussed previously, the tools of work are often easily adapted to be tools of information. It has been said that the story of civilisation is largely the story of weights and measures: the statement should be broadened to allow for all measures made, not just those of daily trade.

The period of interest in this Chapter spans approximately 4500 years. The periods following it span just 450 years. Yet in the latter period the advances made in measuring instruments make those of the former period seem insignificant, perhaps almost

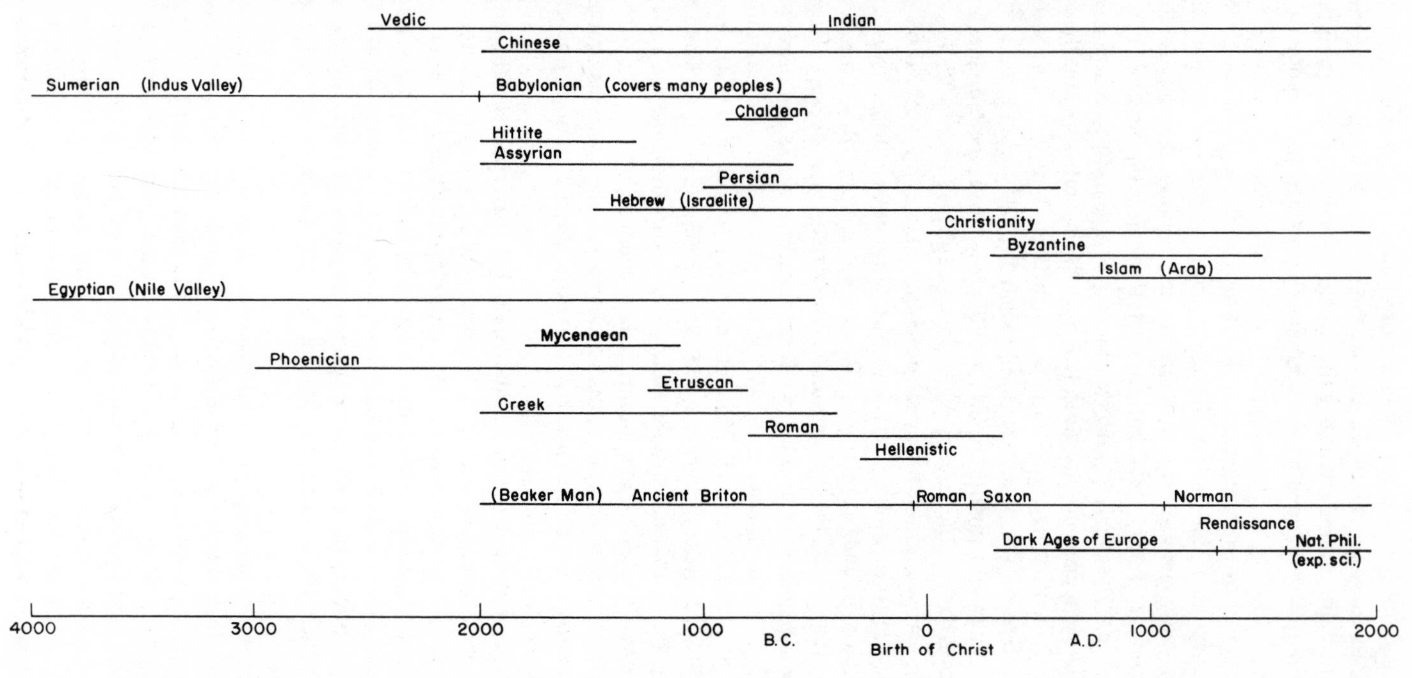

Fig. 3.1 *Occurrence of the ancient Nations*

trivial. Nevertheless, to ignore this birth period would be wrong for many of those early primitive instruments still suffice today. Many are still used with little alteration. Man has the ability to select what is worthy of retention by virtue of proven value.

In classical antiquity measuring instruments developed predominantly for purposes of controlling the daily routines of trade, work, government and religion. There is little evidence to show that the Ancients used measuring instruments to seek knowledge. Evidence that a new attitude eventually emerged, arose from such work as the measurement of the circumference of the earth by Eratosthenes around 250 BC. This important event is discussed later. Measuring devices apparently developed from intuitative foundations, with practice improving the design as trial and error; and chance, showed that it was possible to do better.

Throughout this long period, instruments gradually became more sophisticated, but even at their most complex they were not (at least, to the standards of our time) difficult in concept or in execution of manufacture. Perhaps the main reason for the slow development can be attributed to the kind of science that was practised in those times.

It is clear that the Ancient Egyptians, Assyrians, Chaldeans and other civilisations on the Middle-East areas were beginning to build up knowledge by some deliberate programme of work. Their work, however, was often far from correct (as must be some of today's knowledge!) and it did not seem to be commonly available to the common peoples. Each of these peoples appear to have taken up the knowledge and technologies of earlier civilisations by trade or by occupation. Their knowledge, especially that to do with trade and commerce, spread to the Mediterranean civilisations who, in turn, adopted many of the methods and apparatus of measurement. It passed from the Greeks to the Romans and then to the new Middle-Eastern peoples who carried it, with more modification, back to the European peoples. Thus it was that ancient knowledge came through the Dark Ages to the European nations where modern science flourished.

Of this knowledge it was that of the Greeks that dominated. They were a well organised society having time to concentrate and order knowledge. Earlier peoples had probably only a casual interest except when it was reinforced by a specific need, such as the need to know of seasonal times, how to cure ills and when astronomical events would occur for religious purposes.

Unfortunately, the Greeks were dominantly interested in speculative ways of science. They generally thought – there were the few radicals who saw the shortcomings, – that all of the knowledge needed was already provided in the mind. All that was required was to order and reveal it. Thus philosophical questions were generally more interesting to them than observational ones. The quotation given from Aristotle in Section 1.1 shows the practical outcome of such biased views. Many of the so-called 'truths' handed down to the peoples of the 1500s were not truths. Many could have been verified as false if simple observations had been conducted to test them.

Developments of those times, in all probability, arose via the technological path of emergence in which scientific method played no part. Academicians were not in close contact with the practitioners. The division of classes and attitudes created the situation where instrument developments must have come predominantly from inventive processes, rather than from deductive paths in which gathered knowledge of principles was applied to create technological apparatus.

Although this period can be generally seen as lacking a scientific component the issue must not be closed with the satisfaction that this simplistic assessment is entirely correct. It must be realised that there is much that is yet to be rediscovered about these times. A century ago the oldest civilised man was said to have existed about 3000 BC. Now it has been shown that the date must be pushed back to 30 000 BC times. Many of the instruments of the past were not made of materials that would endure the passage of time. Consequently, those that have been found in recent times are those that have survived. Weights and measures are well represented for they were manufactured from durable materials such as stones and metals. Very few wooden measuring instruments have survived. Many of those displayed in today's museums are reconstructions. Much of the evidence is only available from verbal descriptions, pictures found in tombs, on statues and as a small part of other memorials. Authors of those times as is still the case today, were seldom interested in the niceties of measuring instruments.

It is clear, however, that we cannot expect to find startlingly new discoveries for there is no evidence that any particular group was advanced greatly over the whole of peoples of those times.

This period has interested many modern day scholars as will become evident in the later discussions of the ancient weights and measures.

Instruments of the times preceding the 2nd to 3rd century BC were clearly for use as part of the routine of living; they provided the means to regulate life. It is seen from the translated works of the writers of Greek and Roman times that although measuring instruments and methods were used they played little more than a minor role and would have been regarded as common place to their makers and users, as would have been such ordinary items as, say, a spoon. No ancient author appears to have given space to the philosophy of measurement although several references to weights and measures, to counting and to surveying and astronomy exist. Specific instances such as these will be covered in the respective sections following. It is also interesting to note that much of what we would measure today as absolute quantities referred to a standard of the unit concerned was achieved in those times without need for standards. Taxation, the quantity of goods that could be sold to foreigners, the division of estates and such like were settled by the simple expediency of dividing the whole into fractions. Thus it was that a harvest was taxed by giving, perhaps, one-twelfth to the state. Money exchange was well developed and in such cases the need for weights and measures arose as the mechanism by which the necessary conversions were made.

Greek mythology possesses an important character named Prometheus. Legend has it that he invented weights and measures.

In carpentry a key skill was the use 'of rule and measure' see Bury (1952) for translations of the original Greek from Plato's *Laws* Book 1. In Book 5 of the *Laws* Plato (*c*427-347 BC) wrote about the universal nature of numbers and their divisions, and that the 'numerations of the coinage system of dry and liquid measures, and of weights should all be of the right size and consistent with another'. Such writings as these provide evidence that the need for standards was apparent to all.

As well as appreciating the physical meaning of the term 'measure' early peoples also saw that there was something symbolic in the concept of the balance. There exists numerous passages where one factor is weighed against another. The King was weighed against the silphium herb as Fig. 1.10 shows. Justice was seen to be associated with the weighing scales. A famous part of Aristophanes' (*c*448-338 BC) '*The frogs*' is about a debate between two poets attacking each other's work. The solution is provided by weighing their words on symbolic scales to establish which is the heavier.

The Greeks had a number of words that had measurement meanings. A 'canon' was a rule, or a standard for judgement. From this we have today the 'Canon' of the Church and the 'canonical expression' of mathemetical terminology. Metrology also was derived from works of that time. As today, they too used the terms measure and weight to describe both physical and symbolic measurements.

One word that they also possessed in common usage, and one that has not been pressed into service to describe some aspect of measurement today, is the letter group, stathmon, stathmos, stathmo, stathmere and stathmeon. These variously described the carpenters rule, the plummet or plumb-line, the weight of wool, the standard weight that the people voted should be set up, estimates of distances, certification of full measure, weighing as used in the metaphorical sense, a balance for weighing and weights and measures. A word 'stathmology' was proposed by the writer as the alternative name for what has become known today as measurement science, see Sydenham (1976). A company in the USA that specialises in measuring-instrument design operates under the name of Statham, apparently for the same reason.

Another word that the Greeks used was 'organ'. Aristotle published his ideas on logic in a set of treatises called the *Organon*. They used this word in connection with the organs of sight and hearing. The Latin word mensio, dimensio from which 'measurement' has come, meant 'a measuring'.

It was also very apparent that standards of land survey were important. Plato, in his *Laws*, Book 8, stated his idea of this importance.

> No man shall move boundary-marks of land, whether they be those of a neighbour who is a native citizen or those of a foreigner . . .
> realising that to do this is truly to be guilty . . .

3.2 Science at Alexandria

In the last centuries of classical antiquity, around 200 BC, the immense city of Alexandria was created at the mouth of the Nile river. As well as its famous Pharos lighthouse there were built a Library and Museum. Bonnard (1961) described this great institution. It had, around 50 BC, no less than 700 000 volumes (each a roll of papyrus). He sees the establishment of the Library as the time that scientific research began to acquire the factors that we now recognise as forming sound and scientific method.

Aristotle's idea was that science is the fruit of collaboration, advancing through the collective endeavour of generations. Scientists each provided links of genius that carried on end to end, this being unlike the creations of the fine arts where past experience counts for much less. Aristotle had embarked upon a 'patient enquiry, carried out in the spirit of respect for facts' Bonnard (1961). The Library and Museum was akin to the University that came later, for there the scholars were not all of one 'school' but developed ideas in their own way. The aim was to collect facts, classify them and then explain them.

Out of this great institution came such people as Euclid (geometry), Hipparchus (trigonometry and astronomy) and Eratosthenes (mathematics and geography). To Eratosthenes is given the place in history of being the first (but was he?) person to use measurements and scientific deduction to measure the circumference of the earth. He created what so many great intellectuals had found difficult to do before him, the ordered scientific experiment in which an hypothesis is tested and applied.

This great place was, however, destined to be destroyed; around 150 years after its establishment it was sacked and fired. Other similar events followed over time. It was refurbished, razed, rebuilt and was burned again in 273 AD. By 640 AD it may still have existed but we have no positive evidence that it did.

From the vast collection of works in this place we have inherited but a few. Certainly the existence of these institutions allowed men the time and gave them the facilities to consider deeper aspects of the potential of science. From it we can certainly trace the first use of measuring instruments as efficient tools of knowledge.

3.3 The Dark Ages (*c*300 AD - *c*1300 AD)

The break-up of the Roman Empire began a period in history in the European and Middle-Eastern areas where the knowledge that had been amassed lost much of its value through abuse by those that inherited it. The bulk of it was destroyed but even so a considerable quantity passed into the civilisation of the Middle-Eastern peoples. There it was, at first, ignored or taken up into the life style of those people where it was obviously applicable. These new custodians practised a quite different form of science; they did not, at first, have the stability of settled life form and did not have the interest in preserving the facts that had been made available to them. The result was a period in history where little knowledge of their existence became available to modern historians through the medias of the written word or pictures. Thus it became known as the Dark Ages. These peoples, at least in the European areas, did little to advance the body of knowledge. Many of those involved had no concept of what they had inherited in their invasions so did not take measures to preserve what they found unless it suited their day-to-day needs. The main invaders were the barbarians of Germanic origin who constantly fought for control of each other's property. They had little time to devote to learned activities.

It was in the Eastern regions that knowledge survived best, for there the remnants of the Roman Empire continued somewhat intact under new rulers. It is now well recognised that the Islamic peoples of the East acted as a bridge across which the

Graeco-Roman learning was passed to the Western world, as Graeco-Arabic, in the Medieval and Renaissance times. This link came about through the Arabs becoming interested in translating the Ancient's literature into their own language, Arabic, and subsequently this, in turn, was translated by Christian Church scholars into European tongues.

After the scattering of knowledge and downfall of the Roman Empire by the barbarians in the 5th - 7th centuries AD a surprising event occurred. Religious fervour in the East provided a catalyst that fused many tribes together to become a powerful nation. By the 8th century AD the Mohammedans had become the new intellectual leaders of the known world of the day. A thirst for knowledge arose which, they discovered, could be largely satisfied by gleaning it from the ancient sources. From the Greeks and Hindus they acquired literature that they translated into their own Arabic language. Favourite topics were chemistry, astronomy and mathematics, each of which already had a well developed attitude toward instruments of measurement and observation.

This period of history has not yet had enough attention by modern Western historians but where it has been studied it has been shown that Islamic science did, to a small degree, contribute to the knowledge taken up.

To understand their contribution it is necessary to appreciate that Islamic science was not practised in the same way as was Western science. Islamic science is intricately bound up with the religion of the Islam. The writings in the Holy Book of Islam, the *Koran*, are seen as a guide to the way to seek information, where to look for it and how to apply it. The written word was, therefore, important and the Arabic language was consequently well known and widely spread. Book-making was a central form of activity in this period.

A brief introduction to the ways of Islamic science and technology has been provided by an exhibition held at the Science Museum, London in 1976. The guide book, by Harrow and Wilson (1976) is a useful source of fact about the way that science then was practised and about the scientific instrumentation that these peoples built and improved upon. According to those authors Islamic science anticipated what modern-day scholars are just beginning to realise; that Western science is not entirely adequate and that the Islamic way may offer a better alternative. Islamic science offers the theme that harmony with the world is also needed. Such does not readily occur within the Western form. An illustrated account of Islamic science is available, Nasr (1976).

Islamic science inherited knowledge from the Romans, the Greeks, the Persians and the Indians. Add to that its own contribution and the result was a major chapter in the interpretation available. It gave the body of knowledge the arabic numerals, improved astronomical beliefs and instruments; it integrated different points of view of the various national schools of thought and also added the dimension of mysticism. Out of this grew alchemy that would lead, later, to the more correct chemistries. This period brought about a new interest in instruments for they now assumed in the eye of the user a magical component which made them more important and more worthy of attention to performance and, especially, decoration.

Of this period one cannot help but wonder just what the true significance of it will be for as more studies are made available in the English language, the more the surprises that result. There may have been a Dark Age period in Europe but it was not so dark in the East.

Simultaneously with the rise of the Islamic movement there occurred the growth of the Christian Church in Europe, its foundations being in Rome. It was characteristic of this religion to send missionaries out into the unenlightened areas to spread the gospel. It was also characteristic that these missionaries in their monasteries spent much of their time on scholarly matters. They translated the Ancient books, mainly into Latin for that was their official language. Through these endeavours the Greek knowledge came to Europe via the Roman interpretations.

The Romans did little to increase or improve the works of learning but they at least provided a valuable link. Contact between the Christians and the Mohammedans led to the monks realising that other sources of knowledge existed. More translations from more languages were carried-out in monasteries. Despite the instability of the Dark Ages in Europe the monks' work remained relatively safe, for monasteries were regarded as sacred places. The work of the monks gradually mounted into a sizeable body of knowledge which, at first, no one made use of. The monks did little to apply this information so it was passed on again without being tested against reality. Indeed, it was regarded as so perfect that it was worth more than people's lives to disagree with its facts. The Church, as well as being custodians of knowledge, assumed the role of protector.

The resultant state of stagnation of progress in science gradually became alive as men of the 11th to 15th centuries gradually introduced changes of attitudes into the Churches' understanding of the knowledge it has acquired. As this enlightenment developed, so did the realisation that much of the so-called truth expressed by the Ancients was, in fact, wrong. The knowledge was slowly put to the practical test and it was then that the measuring instrument acquired its first real substantial importance. The new natural philosophy that was to break out around the 15th century onward could not be practised without constantly improved instrumentation. Measuring instruments for daily use became overshadowed by the much greater growth of those for scientific purposes. A new sequence of instrument development emerged by which new forms of measuring instrument were just as likely to emerge from a prototype, constructed first for use in an obscure scientific pursuit, than it was from daily routine requirements.

Allied to the Church Institution was the establishment and growth of another important and less restricted kind of institution of learning, the University. Intellectual revival showed signs of awakening in Europe when a long-standing centre of learning in Paris lead to the foundation of the first University in 1101 AD. This began a wave of interest in learning as a pursuit for all those who could acquire it. By the end of the 13th century over 20 Universities (Fig. 3.2) had been established in England, France, Spain, and Italy (as we know these regions today). The main thrust of these was toward understanding of the Aristotelian works. In general, no effort was put into verifying them or even into questioning them with the exception of a very few people who

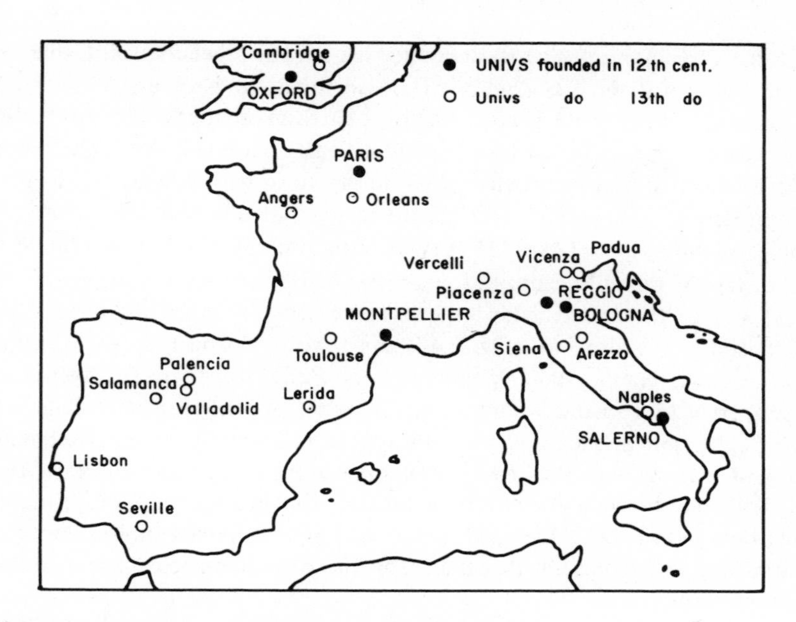

Fig. 3.2 *University towns at the end of the 13th century AD* [Hart (1924), redrawn version]

were prepared to fight the static unchanging attitudes that prevailed. Although situated in many different countries, these Universities were able to act as a concerted body for they each made extensive use of the Latin language.

It is difficult to provide much information about the instruments of the Dark Ages for little was recorded. If it was recorded it has not survived and, as with the ages before, the actual instruments used were mainly made of perishable materials. Today our legacy includes only those that were made of durable brass, iron and stone plus the sources of pictorial and written description such as stained-glass windows, illuminated manuscripts, features in monuments and verbal descriptions in the Arabic literature. A selection of these examples is introduced in the later sections of this Chapter where specific classes are discussed. Reviews of the science of this time are available in Marmery (1895), Hart (1924), Crombie (1969) and Cajori (1962).

3.4 Renaissance: rebirth of learning and the arts (14th and 15th Centuries)

Toward the end of the Dark Ages the 'radicals' had begun to erode the stoic attitudes and had started to show that the 'truth' inherited from the Ancients was not always to be relied upon. Indeed, some of it was so wrong that man's concept of his existence and his world had lead him in quite the wrong directions. Men such as Roger Bacon (*c*1214-1294), St. Thomas Aquinas (1225-1274), Robert Grosseteste (d 1253) and later Copernicus (1473-1543), Tycho Brahe (1546-1601), Bruno (1550-1600) who was burnt at the stake for refusing to denounce his scientific beliefs, and Galileo (1564-1642) who also suffered the rigidity of closed minds to his ideas of science, each helped bring about the much needed changes of attitude that gave birth to the Renaissance. It was a period in which learning and art again became fashionable, proper and in which intellectual human activity could be practiced with adequate patronage and freedom of action. Sherwood-Taylor (1950) relates instrument development with the happenings of those times. Grant (1978) is about physical science in the Middle Ages.

By the start of the 16th century, the number of Universities had swelled to over 50 (Fig. 3.3) and to these were added specialised places for practising science, such as Brahe's 'Castle of the Heavens' (Fig. 2.44) and the Royal Courts where both thinkers and doers were given the stability and support needed for them to advance science once again. At least 15 Courts were recognised as major centres of renaissance activity by 1600. Scholarship and learning were but part of this rebirth but the combination of this form of interest with the arts created institutions and centres towards which all men of skill and learning desired to gravitate. The introduction of printing with moveable type began about 1440. This, too, had a great impact on learning, enabling communication to become much easier. The rise of modern science and technology had begun in earnest.

The major part of the history of instruments of measurement lies in the 350 years that started in the Renaissance. What preceded this period was minor in concept, in manufacture and in application when compared against the advances that were to follow.

One prime difficulty arising with any treatment of a complex multivariable subject such as the history of measuring instruments is that an ordered scheme of explanation is required to suit the serial input characteristics of the human being, yet the information is not generated in this way. Measuring instruments can be discussed on a basis of use of a principle, such as the thermoelectric effect; on a basis of the application to which a varied group of devices are applied, such as the measurement of temperature; or it could be treated as the class of societal use, such as industrial, scientific or medical instruments and so on. Add to these possibilities the fact that there exists hundreds of physical principles, that can be used in combination in thousands of different ways, for millions of varied applications and a feeling of total disorder can arise. To make matters more disordered, historical study also introduces the time factor.

This treatment is arranged to cover chosen periods in which each is further divided to suit the characteristics of measurement in the period. In this first period it is quite straightforward to first give the general history of the period and then deal with several salient classes of device on a mixed variable and principle basis. This is so because the measuring instruments for this 5000 year period changed very little and were made in very few forms. They conveniently divide into the groups – time measurement, weights and measures, astronomy and surveying with two other divisions being needed for the miscellany of the rather more unusual physical variables, such as temperature and electrical parameters and for mechanical devices including calculators and feedback mechanisms.

Fig. 3.3 *Courts and universities provided the work place for scientists and artisans of the Renaissance*
[New Caxton (1969), Courtesy Istituto Geografico de Agostini, Milan]

3.5 Time interval and time of day

Probably of all of the measured variables that man quantifies that of time is the most used yet possibly the least understood. We perceive that the entity involved has changed value with the passing of events but there is no concept of absolute time. All that can be measured is the notion of time interval, the duration of sequences and events.

Time interval is measured by either comparing the duration between two events against either a natural cyclic motion or a man-made timing device. The most stable natural phenomenon, and the one most obvious to men since the earliest time, is the cyclic behaviour of the heavenly bodies. Thus the rising of the Sun or the zenith of stars provided early man with the means to pace life. From this source developed the hours of the day, the calendars of the year and greater periods. It is also possible to construct equipment that passes through changes of state which can be used to define an interval of time. These are man's clocks.

In early time, clocks were inferior time-keepers to natural means and it was not uncommon practice, even until quite recent times, to use the sun to set the man-made clocks. Today, however, clocks have been devised that make use of the atomic resonance phenomenon (which is, in reality, a natural standard of a less obvious kind) that are better time-keepers than the stars. Indeed they form the basis of the unit and they are used to detect discrepencies in astronomical movements from our imperfect models of the Universe. Although the variable time is a philosophical enigma, it does have the distinction of being the best defined measured variable at present. Current physical standards can maintain the world's clocks within fractions of millionths of a second per day.

Clocks began in classical antiquity either as devices for observing the Sun and stars, – in themselves this kind had no inherent time-rating characteristic, and as those mechanisms that provided some means of pacing an interval by employing a suitable physical principle.

Natural time-keeping

Natural time systems are based on astronomical principles. They are tied in to the rotation of the earth, as that decides when the sun presents each day, and to the motion of the earth around the sun, for that decides the seasons. Time recorded on a clock that uses internal time generating principles follow the motions of the stars. This is known as sidereal or mean time. The time scale indicated by the sun is not exactly the same as mean time; this is called apparent time. Universal time is a relatively modern device (origins in around 1880) that keeps the whole world linked together to a common datum time system that is paced to the mean time as established at Greenwich in England. This is called the Greenwich Mean Time. It is actually elsewhere in England but is referred to the zero longitude, time-line defined at Greenwich. From 1971 the prime standard of time was decided by 'the Caesium clock', not by astronomical means.

Ancient peoples did not have to contend with global time systems; it was sufficient for them to establish time for their local area and to maintain the seasonal calendar to assist annual control of life. The measurement of long periods is covered in the subsection on astronomy given below.

The need therefore for natural time-keeping was for devices that would provide means of observing the transit of interest, such as sunrise, maximum sun height and sunset and some method of dividing the passage of the sun into a scale that divided the day into suitable time divisions. Early astronomical time-measuring instruments, therefore, consisted of sights or markers that either enabled the observer to decide when the Sun passed a given datum point or, alternatively, cast a shadow onto a divided scale. The stars were also used to establish time and in that case only the sighting method was practical as no shadow can be cast.

Sighting mechanisms have been found from the later periods of Ancient Egypt. A simple end-split palm leaf stalk has been found that was used for sighting, Singer *et al* (1954). It was called a 'merkhet'. That, shown in Fig. 3.4 dates from around 600 BC; it was the property of the astronomer-priest of the god Horus in Upper Egypt. The same instrument was adopted by the Greeks who called it a 'horologos'. It served a dual purpose for it was also used in laying out buildings. The merkhet shown is made of bronze inlaid with electrum, a gold-silver alloy. This example is one of the few original time-keeping instruments that have survived the times. It is in the Science Museum, London. It is reasonable to assume that similar, less well made, instruments were in use before the time of this one.

In the Greek language, 'hora' means time and 'logos' means telling. Hence the name they used for the merkhet. Today the name 'horology' is used to describe the art of making clocks and measuring time. Passage of time has given it the specific meaning for mechanical forms of clock which do not use electrical techniques – see De Carle (1965). Numerous works have been written on horology and electrical horology and, as is to be expected for they did not advance the art very much, little space is given to the time-keeping instruments of the Ancients. The review by Ward (1970) and descriptive catalogue by the same author, Ward (1966) provide a good introduction to what is known. The catalogue provides useful insight about extant apparatus. Leach has reviewed primitive time reckoning in Singer *et al.* (1954).

Shadow devices cast a shadow of the sun's rays onto a scale formed into the instrument. This group includes the well known flat plate sundial, the shadow clocks of Egyptian origin and the spherical bowl sundials as were used by Eratosthenes to compute the circumference of the earth.

According to De Carle (1965) the word dial is derived from the word for day, which is 'dies'. He suggests, reasonably enough, that man soon realised that the shadow of the sun, formed by a stick or a tree, would have been found empirically to have a systematic path and that man would have subdivided the shadow path into divisions. A refined instrument developed from this principle was certainly in existence by 1000 BC. A shadow clock of T form shown earlier in Fig. 1.7, survived virtually intact from that period. It has been restored according to modern interpretation of the evidence; the original did not have the crosspiece. The T-shaped shadow clock had to be turned around at the high noon so that the shadow then fell again on the scale inscribed on the longer bar.

The sundial also developed out of the shadow stick principle but they were more sophisticated than the T-shaped shadow clock for the gnomon (the shadow stick) and the graduations were set to suit the latitude that they were used at. It also allowed all-day measurements without the need to move the instrument at noon. All manner of

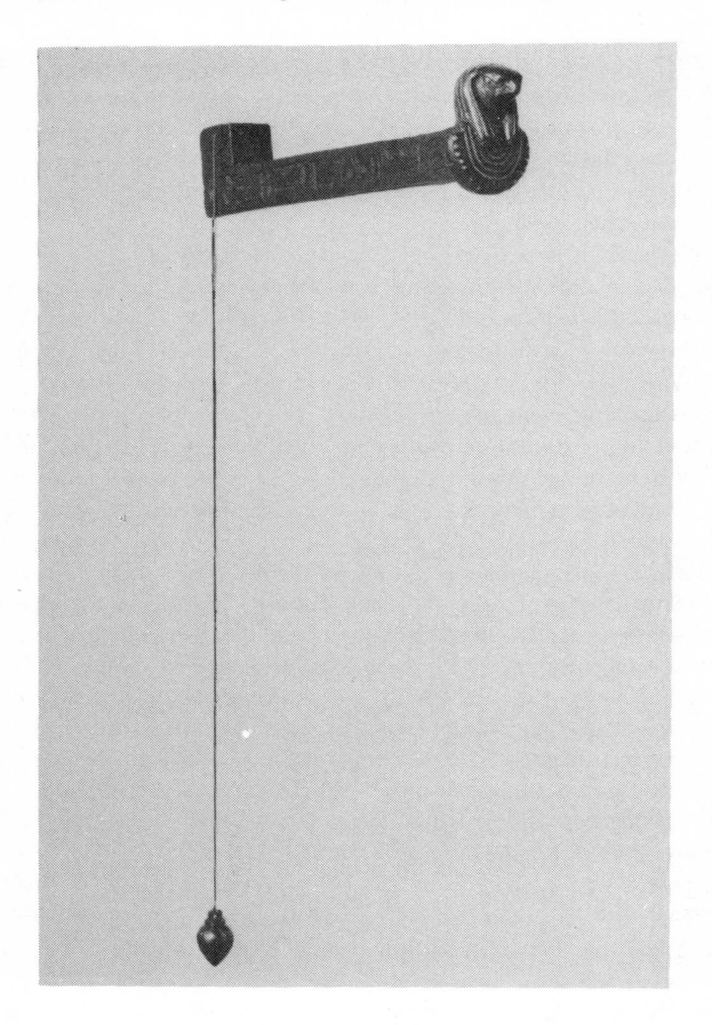

Fig. 3.4 *Merkhet from Egypt, c600 BC*
[Crown copyright. Science Museum, London]

gnomons were used e.g. sticks, strings in tension, blades. Marmery (1895) states that Anixmander (610–547 BC) introduced the gnomon into Greece and invented the sundial! The dial was used in forms of a flat plate, and later spherical curves. The hemispherical bowl with a vertical gnomon is called a 'polos', Farrington (1936). With the addition of an armillary sphere it could be used as a star clock at night (see Section 3.7). A Roman sundial made of carved limestone is held in the British Museum. It has a curved scale of spherical shape and is graduated to allow for the changing solstice over the year. This design is said to have originated with Berosus, a Chaldean astronomer of *c*300 BC. The sundial is mentioned in Isaiah 38, 8, suggesting it was in use as early as 2000 BC:

> Behold I will bring again the shadow of the lines, by which it is now gone down in the sun dial of Achaz with the sun, ten lines backward.

Practice of those times required the light of day to be divided into twelve hours. Subsequently as the season changed the duration of an hour varied. For this reason, many early clocks had a series of scales divided into twelve to suit the various times of the year.

Another early Egyptian shadow clock used two pegs inserted vertically in holes along a horizontal stick. Supported between the two pegs at their top was a fine member that cast a shadow on the stick, Singer *et al.* (1954).

Sundials are still in common use in modern times and they are capable of indicating time to within minutes if properly set up and if the sun shines. Gunther's (1935) handbook of the Museum of the History of Science of Oxford, gives considerable detail about sundials (and other old instruments) for that museum has a renowned collection. Gibbs (1976) is a study of extant masonry sundials of Greek and Roman origin.

At night the stars were used to define the time. Sighting devices, such as the merkhet, served the dual role as time measuring instruments and as tools of astronomical observation.

As the astronomical methods were not always usable, owing to weather, other techniques were adopted that made use of the motion within a piece of physical apparatus.

Man-made mechanisms

The earliest man-made clock to provide time by implementing a physical principle appears to be the water clock of Ancient Egypt. In their more advanced forms they became known in Greece as 'clepsydrae', a term now used to describe all water-clocks.

The oldest extant water-clocks date back to about 1400 BC. A cast of one is shown in Fig. 3.5. The bowl was filled with water which escaped through a small hole formed in the side at the bottom. The level in the container – graduations were marked – indicated the time that had elapsed. Designers of the time appear to have appreciated that a cylinder of uniform diameter did not provide a falling level that was constant with time (pressure head reduces as the level falls resulting in the water escaping more slowly with time). They apparently realised that the conical shape gave improved linearity. The true shape would be a curved cross-section but the other errors of this simple system would probably have made this more complicated shape an unnecessary refinement to strive for.

If the water flows into the container at a steady rate then the height of water in the cylinder is a linear indication of time that has passed. A constant input rate was easy to realise if the source of flow was allowed to overflow a small reservoir containing the discharge orifice, thereby maintaining a constant head on the orifice. Fig. 3.6 shows this design arrangement.

Fig. 3.5 *Cast of the water-clock found in the Karnak Temple, Upper Egypt. The original was made of alabaster c1400 BC*
[Crown copyright. Science Museum, London]

Yet another method is to float a small empty container, having a hole in its bottom, in a container filled with water. The floating container will gradually fill and sink in a systematic manner. Such timers were certainly used in classical antiquity, probably to regulate the time allowed for water to be taken from an irrigation supply channel.

It has been suggested, but apparently without much supportive evidence, that the sand-glass may have evolved from the water-clock, the link being that sand could be used where water was not available. The hour glass is, however, more likely to be an instrument of later times, times when clear glass could be worked to the degree of precision needed. Sand-glasses are certainly known to have been in use by the 3rd century AD.

Water-clocks have fascinated many historians and although few clepsydrae have survived, we know quite a lot about them from descriptions in writings. They became more like the clock, as we know it, in the time of the Greeks. It was then the custom to time the speeches of orators in courts and senate by the water-clock. It is said that a clepsydra was, therefore, a central item of the public buildings of the Greek and later, Roman times. For all the worth of this assumption no actual clock or illustration seems to have survived.

The best known water-clock to have attracted the attention of historians, is that of Ktesibios (*c*260 BC), pictured in Fig. 1.26. This was described verbally in Vitruvius' *De architectura*. Mayr (1970) has researched the material on the development of one class of water-clock at depth; the Ktesibios clock is of importance for another reason. It appears to be the first man-made device to make use of automatic feedback control. Instead of using the wasteful overflow method of maintaining a constant flow into a cylinder, like that shown in Fig. 3.6, it used a float valve to regulate the flow rate. Mayr's interest was more for this feature than for the value of the clock as a time-keeper. The original written descriptions, however, leave much in doubt as to the exact nature of the Ktesibios and other water-clocks – many interpretations of this clock's description exist. Fig. 3.7 is a clock built early in this century to one, very imaginative,

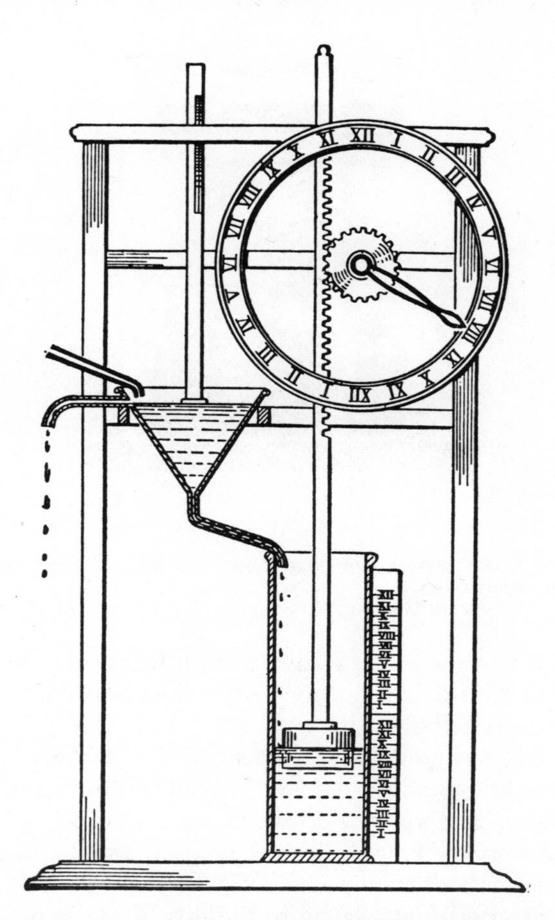

Fig. 3.6 *One form of clepsydra, according to De Carle (1965); Constant flow is provided by the overflowing feed reservoir. All existing pictures and constructed water-clocks of this type are reconstructions; none has been found*
[De Carle (1955), courtesy Teach yourself books, Hodder & Stoughton Ltd.]

Fig. 3.7 *Water-clock of Ktesibios; over-imaginative reconstruction once made for the Deutsches Museum, Munich*
[Courtesy Deutsches Museum, Munich]

interpretation of the contemporary accounts. When Julius Caesar was invading Britain, he recorded that he wished he had a good water-clock available to him to enable him to use the tides more effectively.

Farrington (1936) has provided an account of how Empedocles (*c*490-*c*430 BC) used a clepsydra to investigate the nature of atmospheric pressure. The device that he described (from the ancient works) was little more than a metal cylinder having a small hole in a conical bottom. Empedocles did not, however, appear to have used the clepsydra as a timing device but only as an airtight container.

A translation of manuscript relating to the role of the clepsydra in the Hellenic courtroom is to be found in Botsford and Sihler (1915).

Hogben (1938) suggests (p.230) that the Ktesibios clock also had a geared function, driven by a water wheel, that caused the length of light-day scale to be rotated with the seasons. This feature was incorporated into the clock made by Speckhart for the

Deutsches Museum, shown in Fig. 3.7 – refer to Neuburger (1930). Other reconstructions were made in 1567 and in 1587 by Säulen – both were illustrated in contemporary reports, pictures of which are held in the Bildstelle of the Deutsche Museum.

Water-clocks continued to be improved by the Mohammedans who inherited the Graeco-Roman knowledge. The Chinese also built magnificent clock machines on this principle. One needed seven men to keep it working, it was so great in size. Mayr stated that few authors have studied the original manuscripts from the technical point of view and, therefore, historians have, as yet, made little contribution to our modern understanding of the full degree of sophistication built into early clepsydrae. His chapter on the Islamic work is definitive and most enlightening. A reconstruction, that by Wiedemann and Hauser c1918, of the clock of Pseudo-Archimedes (the name given to its 10th century or earlier maker by Mayr) is shown in Fig. 3.8. Many versions of clepsydrae were described in the times of Islam that coincided with the European Dark Ages. Cajori (1962) also refers to Islamic work.

Clepsydrae of later form used a mechanism to convert the float level in the cylinder into indicated time appearing on a clock face. It has been thought for some time that these may be generically linked to the mechanical clocks of the 13th century but, until recently, such a link had not been established. Work by Needham and colleagues, Needham (1954) and Needham (1965), has revealed that the Chinese had described, in 1172 AD, a very sophisticated astronomical water-clock of 1088 AD. It used water to drive it yet did not rely on the rate of water flow to provide a time-generating mechanism. A model of this clock has been built (Fig. 3.9), it is described briefly in Ward (1966). This clock is an important link, Needham (1970), for it is the first known use of the escapement principle to control the rate of a clock. Needham (1965), Section 27, devotes considerable space to this clock. Early development of gears and clockwork is reviewed in De Solla Price (1974).

It is conjectural as to whether the knowledge of such Chinese clocks was available to Europeans, but certainly the use of escapements in Europe came into use at the end of the Dark Ages. The first all-mechanical clocks – see Ward (1973) – in which some form of cyclic time-keeping principle was used appeared about 1280 AD. The oldest extant clock of this type dates from around 1400 AD but reliable records show that many existed before that. Early mechanical clocks struck the hours only – the latin word 'clocca' meant bell – and were usually made for use in public buildings such as churches. They derived energy to overcome the frictional losses from a falling weight, used the verge and foliot escapement and were very large. Many exist today in cathedrals and churches throughout Europe. Fig. 3.10 is a modern picture of the Dover Castle clock that was manufactured around 1550. These designs were used with little change from around 1300 to 1600. Fig. 3.11 shows manufacture of smaller weight-driven clocks around 1570.

The foliot method of regulation was not capable of precise rate control and for 300 years was not improved upon. In the early 1600s the pendulum was added to the methods and timekeeping moved into a new era.

It has been suggested by Marmery (1895) that the Arabs measured time by pendulum oscillations and invented a balance clepsydra operating on a new principle for

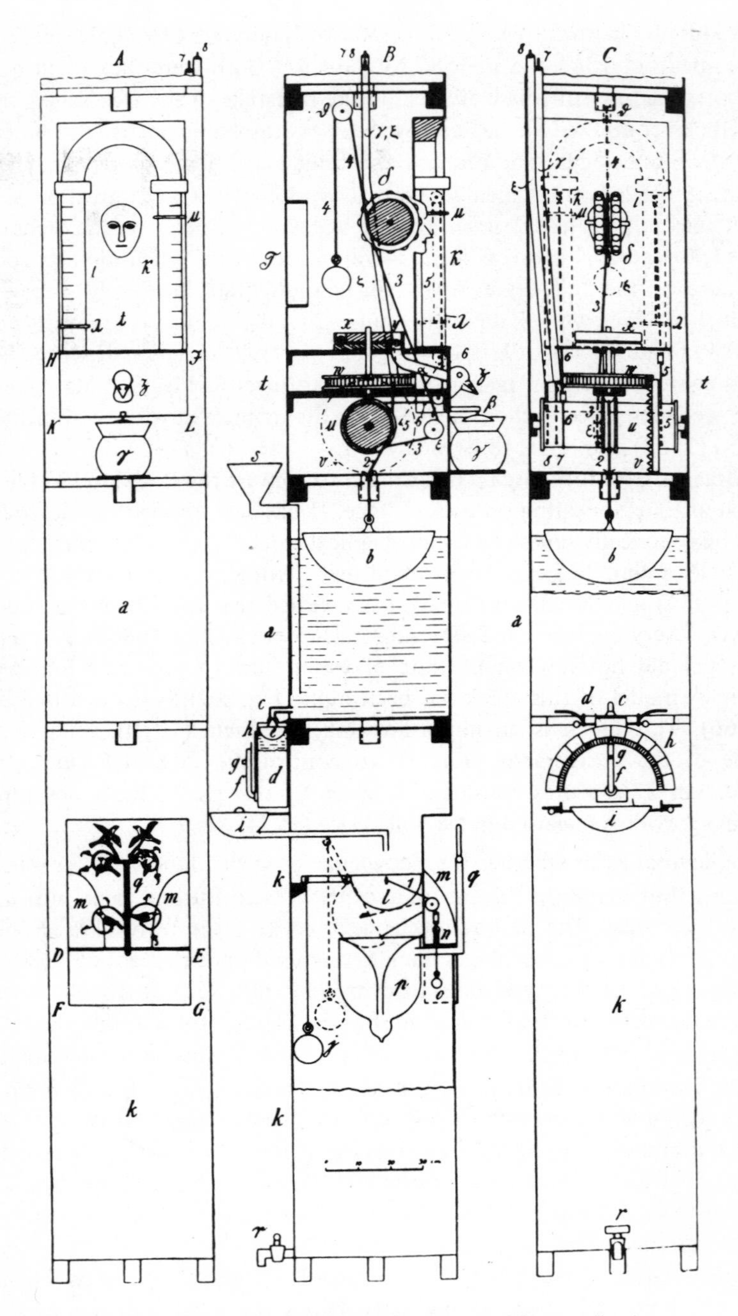

Fig. 3.8 *Water-clock of Pseudo-Archimedes, 10th century AD or earlier*

[Mayr (1970), Fig. 18, p.30 from Wiedemann, E., and Hauser, F. 'Uhr des Archimedes und zwei andere Vorricht ungen', *Nova Acta. Abh. d. Kaiserl. Leop-Carol. Dtsch. Akad. d. Naturf,* Vol. 103/2 (Halle a/Soale 1918)]

Fig. 3.9 *Water-clock escapement reconstructed from 1172 AD. Chinese sources. It was the first known use of an escapement to control the energy release of a clock*
[Crown copyright. Science Museum, London]

Fig. 3.10 *Dover Castle striking clock using a falling weight for power with the verge and foliot escapement. These were first built around 1300.*
[Crown copyright. Science Museum, London]

its time. He did not, however, provide evidence to support these statements but as the Chinese of the same period are known to have had balance clepsydra, Needham (1965) Section 27, it would appear to be a reasonable statement.

Before leaving the subject of time-measuring instruments of classical antiquity, mention is made of another interesting and ingenious method that was not destined to be adopted but, nevertheless, formed part of the history of the development of the clock. Fig. 3.12 shows a replica of a Chinese fire-clock of around 1000 BC. Its operation is self-evident. As the flame burned inside the dragon's body, it moved along horizontally. At equal time intervals it reached the next overstrung cord burning it through. That released balls onto the dish making a pleasant, time-indicating sound. The Chinese also built clocks driven by sand running onto a 'water' wheel.

Needham (1965) Section 27, includes an extremely useful chart (Table 59) showing the relationships formed between numerous named workers of Europe, Babylonia,

Ancient Egypt, China, Islam and India on the subject of mechanical clocks. The chart covers 250 BC to 1872 AD. It reveals how the Chinese had the verge and foliot escapement reintroduced back to their peoples in the 16th century AD from Europe, without the realisation that they may have been the source of the idea arising in Europe 200 years previously.

Fig. 3.11 *Manufacture of early mechanical clocks. c1570*
[Klemm (1959) p.32 from Stradano *Nova Reperta,* 1570]

History of horology has always had a considerable following; numerous books and papers have been published. Introductory bibliographies appear in De Carle (1965) and Ward (1970). Kurz (1975) has traced the master-to-pupil links from early times for the Near-Eastern developments. The exhibition guide by Harrow and Wilson (1976) lists Islamic clocks. Needham (1965) Section 27, covers Chinese clocks. National, and to a lesser extent local, museums that deal with science subjects invariably have sections devoted to horology, such is the general public interest in this kind of ingenious mechanism that the lay person can appreciate with little formal training.

3.6 Weights and measures

As was the case for the variable time, other measurements of classical antiquity were largely concerned with ordering the daily routine of a stable life style. As trade

Fig. 3.12 *Replica of c1000 BC Chinese fire clock*
[Courtesy Science Museum of Victoria]

developed to a point where straight barter was unsatisfactory, the need for means of exchange arose that would enable assets of a more universal kind to be transferred. Settled peoples invariably had taxation systems by which the rulers carried-out their public works. Currency was introduced generating a need for measurements as the weight of precious metal was often involved. It, therefore, became necessary to determine weights and volumes.

Construction of the buildings and other man-made structures generated the need for measures of length, angle, area, flatness and the like. Allocation of water from the main irrigation channel into the land-users ditches required more measurements. This collective group of measurements has become generally known as 'weights and measures'.

The origin of weights and measures lies in obscurity but it is certain that they developed gradually through intuitive trial-and-error processes. These were brought into use as primitive man realised what he could do within the limits of resources available to him.

In a tomb at Naqada, Egypt, that dates from around 4500 BC, archaeologists found a red limestone beam-balance of about 90 mm length complete with a set of stone weights. This appears to be the oldest instrument associated with weights and measures that has been located. Bailey (1977), however, mentions sets of weights from 7000 BC but does not provide details of his source of information. The actual age of the first manufactured weights and measures apparatus is here of little importance. It is sufficient to realise that most early civilisations independently saw the need for similar devices. In all probability, many balances were made of wood for that material was more easily worked. They, however, would not have survived save in exceptional conditions of preservation. The first known picture of a balance is in a stone relief of a Vth Dynasty tomb at Gizeh, *c*2500 BC. It is of jewellers at work.

The first measures made use of individually chosen standards that the user adopted from articles existing in the immediate natural world. The need for units and standards must have been very obvious for so many different peoples adopted similar means to control their measurements. Trade would have forced those of a group or community to choose a convenient article that they each agreed to use as the standard for the unit in question. Units of length were based on the lengths of parts of the body i.e. the width of the thumb, the width of the palm of the hand, the forearm length, the outstretched arms and paces, length of a foot and so on. The Egyptian cubit, a length unit, used the palm as a modular element. It appears to have its origins around 3000 BC, first as marks scribed on wood and stone slabs and then, later, as more formally declared stone standards. A similar unit has also been found, dating from around the same period, east from Egypt. Perhaps the close relationships of internation trade speeded up the transfer of techniques: it seems reasonable to presume so.

It is also clear that sharp dealers realised that they could manipulate their local standards to suit their own gain. It was common practice for a dealer to have two sets of standards, one for buying and one for selling. The Bible contains many passages about this practice i.e. the use of diverse weights. The same unfair practice was often used by tax collectors and tax payers. Rulers, too, were often paid dues and tribute according to biased units, Ellis (1973).

Thus it became necessary to introduce what is now called legal metrology. Standards of mass (weight), length and volume were set up by many early governments. No longer was it legal to use the nearest grain of wheat as the 'weight' to put in the balance scales when measuring the weight of gold. The Babylonians had a system of weights and measures that was traceable to central standards. A statue of Gudea, ruler of Lagash in the 3rd millenium BC has been found that has a graduated length scale carved into a tablet sitting on the knees.

The Greeks also had declared standard weights. The word 'stathmeon' had a meaning as the 'weight that the people had set up'. Stathmo carried a meaning 'to certify as containing full measure'. 'Stathmeon' was sometimes used as 'the public weight under authority'.

Internal and external trade in Roman times necessitated the establishment of a state-guaranteed system of weights and measures as well as for the coinage, see Durant (1944). Barter gave way to exchange of goods for precious metals, this in turn led to official currency.

The existence of such legal weights and measures did not, however, mean that all nations of those times had basic trading standards by law; and having the law did not mean that trade was always in accordance with them. It must be remembered that the standards of those times were indeed perfect as far as their absolute value went for that aspect was defined by the choice of artefact. The practical difficulty lay in transferring them into local standards throughout the length and breadth of the countries concerned. Even comparing the legal standard against local standards would not have been possible to any real measure of accuracy for the only apparatus in use was crude. Weighing, at least, had the advantage of being able to provide quite high precision comparison from quite simple balances.

Considerable evidence exists to show that early Chinese culture saw the importance of having legal standards, and of approving them at a certain time of year to allow for temperature effects. Chapter 14 of the 'Book of the Lord Shant' *c*340 BC condemns 'reliance on private appraisal and speaks of the folly of trying to weigh things without standard scales, or forming an opinion about lengths in the absence of accepted units . . . '. Needham's works contain this and many other references to the Chinese attitude toward the importance of measurement in practice and as a philosophy.

Lessons learned by the early civilisations had to be learned over again by the barbarians who inherited the European areas. The story of British and Continental weights and measures is full of oddities and absurdities. The text of the 1225 AD Magna Carta contained this legal statement about trading standards:

> (35) There shall be standard measures of wine, ale, and corn (the London quarter), throughout the kingdom. There shall also be a standard width of dyed cloth, russett, and haberject, namely two ells within the selvedges. Weights are to be standardised similarly.

It hardly constituted a tight specification for the creation of a national traceable legal measurement system. It was the Worshipful Company of Grocers (incorporated in 1345 AD) that had 'custody of the Great Beam' in medieval England.

Bailey (1977) quotes an early description of a 1554 unofficial international intercomparison, by Hasse, of Russian measures against English ones. Edward 1, in 1303, ordained 'that throughout all our kingdom there be one weight and one measure, and that they be marked with the mark of our standard'.

The main trade variables that needed measurement were, therefore, weight and distance. Coupled with time measurement the Ancients had realised the basic requirements of what later was introduced as the SI unit system. From these three base units can be derived most others that were to be needed for many centuries. They could be used to realise area, volume, flow (but even the Romans did not realise that the velocity of water from a pipe was a parameter of the discharge rate), density and possibly force. There is, however, no evidence that supports that this process of derivation from base units was appreciated until quite recent times.

Each of the many nations adopted an incredible variety of standards, defining countless units for the three variables, time, length and mass. They have been well researched and reported by writers already mentioned, Ellis (1973), Klein (1975) and Skinner in Singer *et al.* (1954). Other works on weights and measures of classical antiquity are Woolhouse (1856), Harkness (1890), Ridgeway (1897), Hallock and Wade (1906), Petrie (1934), Bendick (1947), Moody and Clagett (1952), Berriman (1953), Rush and O'Keefe (1964), Kisch (1965), Skinner (1967) and Zupko (1968).

Interests in this area are catered for by the International Committee for Historical Metrology. This Committee, with its origins in 1955, officially became part of the International Union for the History and Philosophy of Science in 1974 having its' own section granted in the International Congress held at Tokyo in that year. Tasks set in 1975 included listing institutions interested in historical metrology, listing people interested in this area of history, editing a bibliography of historical metrology and organising the First International Congress of Historical Metrology for the autumn of 1975. In 1976 the Committee staged a conference, on the theme 'The role of measurement standards in human civilisation', in Budapest.

The 1975 conference concentrated on old measures, the metric system and problems of studies in this area.

Determination of mass

Weighing techniques were first used for transactions involving precious materials, mainly gold. To begin with, there was little need for weighing of bulk commodities for they could be apportioned by simple division of a number of items or by the use of quite crude, indirect, forms of measures. Precious metals and stones, however, needed a more precise method owing to their higher intrinsic value per unit mass.

The oldest weights yet found are those from Naqada, Egypt that were mentioned earlier – see Skinner in Singer *et al.* (1954). They were made from limestone and were formed as small cylinders or as cones having convex bases. The smallest represented 0·5 beqa, which is equal to around 6·5 gm. Weights have been found by the thousands. Until *c*1450 BC they were made of highly polished stone. After that time the use of metals became common but such weights were not as durable as stone and fewer have been found. The shape of weights ranged from simple cubic forms to very complex shapes, usually of birds and animals. Fig. 3.13 shows Naqada weights, those from El-Armana *c*1350 BC, from Babylonian sources *c*700 BC, and other metal weights from early Assyria and China. It was common practice to mark the weights with the sign of the authority issuing them. In some cases, for reasons of intrinsic value, weights made from metals, such as lead, became used as coins. The denominational size of weights increased over time to extend the range of capability to cover heavier objects of the later to be entered, non-barter trade. One 'sleeping-duck' weight, from 2350 BC, weighs 60·55 kg.

Considerable detail is available in the published literature. Skinner, in Singer *et al.* (1954), provides a succinct introduction to ancient measures. That by Ellis (1973) is written for the popular readership. Works such as Berriman (1953), Petrie (1934) and Ridgeway (1897) give greater detail. Needham (1962) Section 26, provides insight into weights in use in early China. Sanders (1947) provides an illustrated history of weigh-

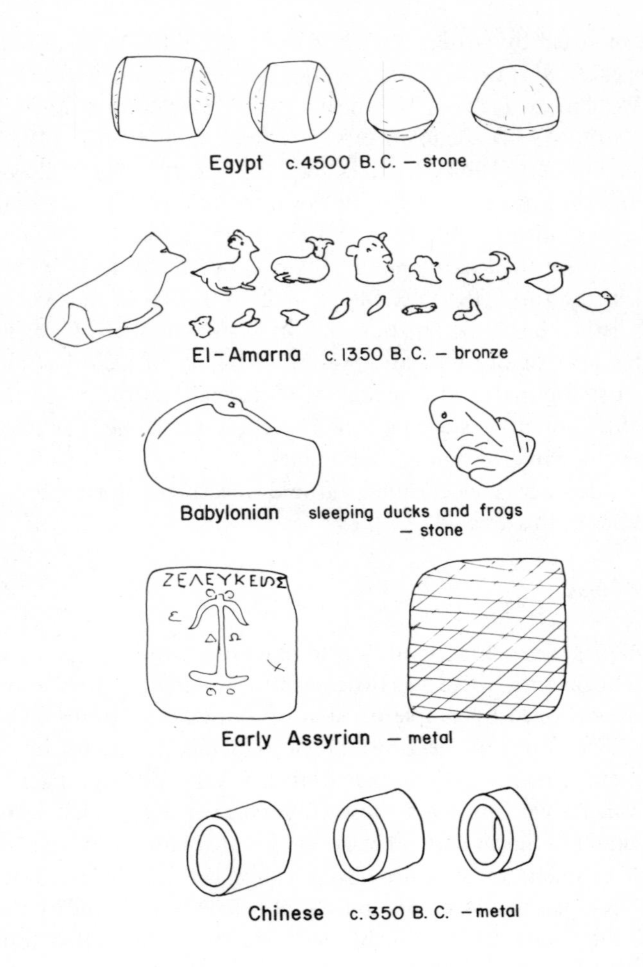

Fig. 3.13 *Shape and form of early standard weights used with beam balances*

ing using information from the extensive collection of W & T Avery Ltd.

Attention is now given to the balances used to make the comparison between the standard weights and goods.

Balances began as a beam having equal lengths each side of the central pivot. Pans were hung on each to hold the masses being compared. The whole unit was held up by hand. The Romans later introduced the so-called steel-yard method in which the arms each side of the pivot are of unequal length, the longer one being able to balance the shorter by using a lesser size balance weight. Spring balances were first used by al-Khazini *c*1140 AD.

The sensitivity of an equal-arm balance can easily be calculated from basic principles of mechanics – refer to Walker (1887), Brauer (1909), Dittmar in Thorpe (1937), or

Felgentraeger (1932). Sensitivity, the amount of deflection that results for a given degree of weight difference between the two pans, is decided by the length of the arms and the mass and arrangement of any existing pointer arm that rotates at the centre of the beam to indicate differences from exact balance. Study of available illustrations of balances of the ancient world shows that they did not normally incorporate the pointer feature, but it seems reasonable to assume that they did use some kind of judgement method. Perhaps they simply sighted the beam to a convenient member behind it, or looked for movement of the beam in a particular direction.

Error of comparison is influenced by the length of the arms on each side of the pivot and the effective mass of the arms. These factors can introduce greater turning moment into one side of the balance arm than the other. This, however, was not a serious error for simple hanging of the balance without weights in the pans would show up any difference. It could then be adjusted out by addition of any convenient weight. Furthermore, the masses could be reversed to establish the difference. The mean value was the correct weight.

Additional sources of error arose from the stiffness, to rotation, of the pivots. Early balances invariably used what we now call flexure pivots. They were made of leather, cords or fibre strings. Some later metal designs incorporated wide knife edges. The pivots could well have limited the sensitivity of early balances for a string in considerable tension can be quite stiff to bending. Sanders (1947) illustrates the development of beam-end attachment methods.

Fig. 3.14 shows designs taken from photographs of reconstructed original equal arm beam-balances and from early contemporary pictures and carvings. Note the many different forms of end attachment of the pan strings, of the central pivot arrangement and of the shapes of beams. Today we would criticise many of these designs as being insensitive and prone to error. It must be remembered, however, that they were used to compare weight as a balanced comparison of like masses – now called a null balance – a measurement operation that did not require errors to be as tightly controlled as they must be for balances in which the pointer arm is used to decide differences.

Skinner in Singer *et al.* (1954) estimated that the balances from 5000 BC to 1500 BC had resolutions ranging from 0•13 gm with 6•48 gm being in each pan (1-in-42 parts), to 1•94 gm when weighing 130 gm masses (1-in-67 parts). The Ebers papyrus tells of medical recipes needing a weight of 0•71 gm for an ingredient. Sensitivity was improved so that by around the 4th century AD the first example was improved by a factor of two and the later case by a factor of six. Measurement of weight to 1-part-in-400 for a 130 gm mass was not adequate for the use of the druggists and gold merchants of the following periods and the balance underwent considerable improvement. Balances of today can easily discrininate to 1-in-100 000 parts. By the 1500s the sensitive balance began to change considerably in design arrangement. Pledge (1966) includes a table of the discrimination of weighing instruments over history; it was compiled using actual balances. The older, simple devices have continued in use until current times for they are as adequate now for some sectors of routine trade as they were then. The history of the sensitive chemical balance has been studied by Stock

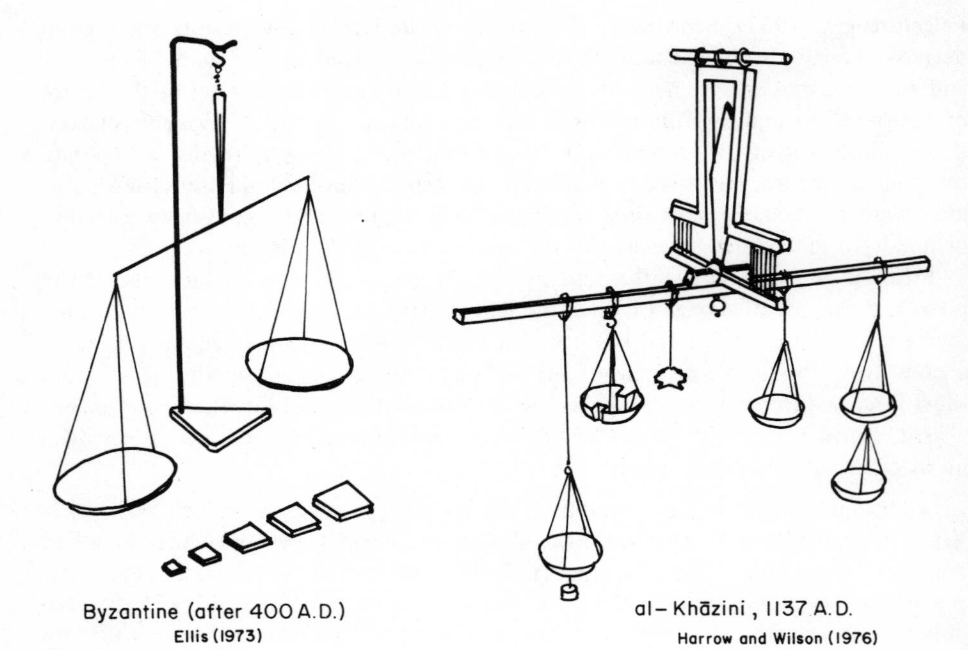

Byzantine (after 400 A.D.)
Ellis (1973)

al–Khāzini , 1137 A.D.
Harrow and Wilson (1976)

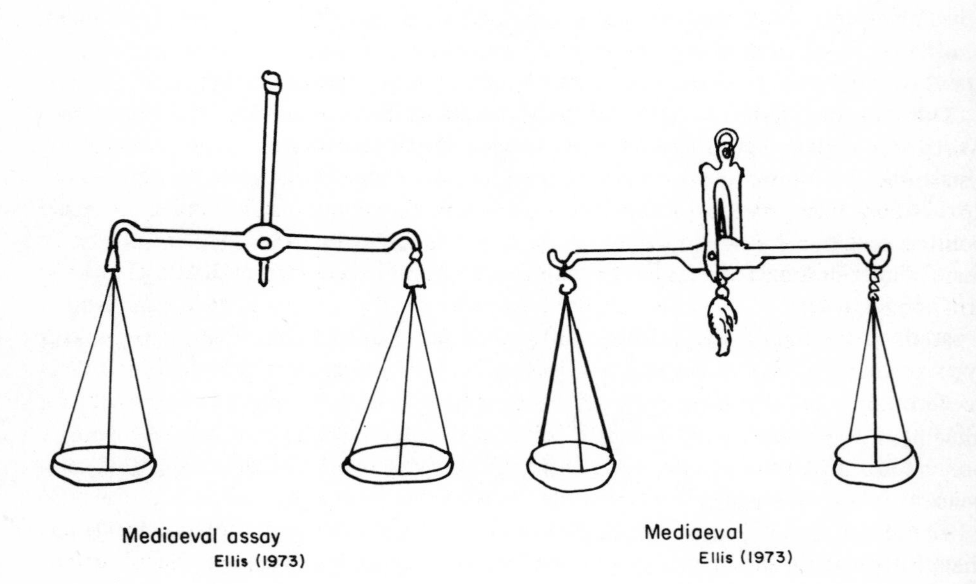

Mediaeval assay
Ellis (1973)

Mediaeval
Ellis (1973)

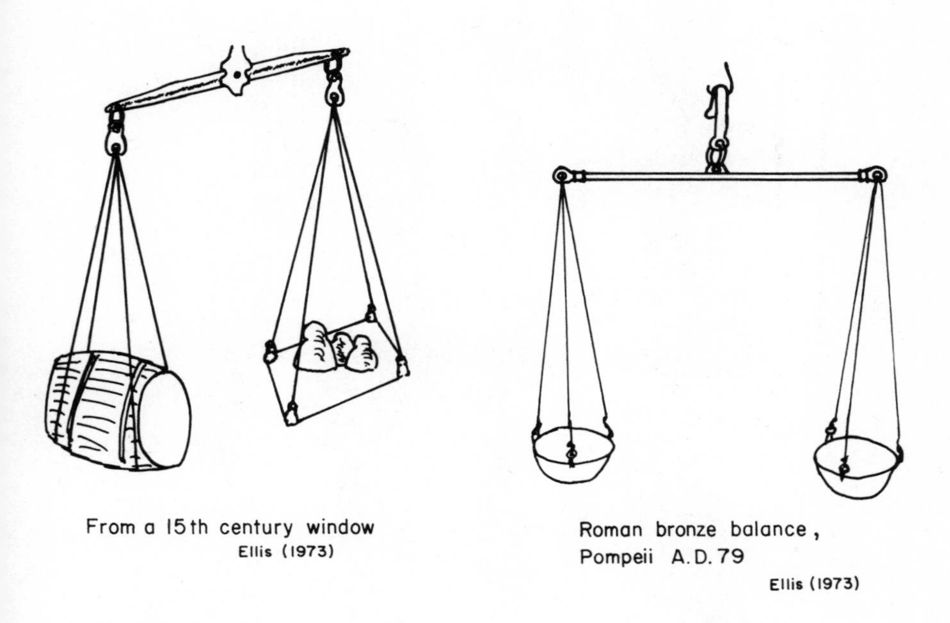

Fig. 3.14 *Designs of equal-arm balances to 1500 AD*

(1969) but his account begins in earnest at the time where this Chapter ends.

The Romans, perhaps influenced by the laws of levers that were enunciated by Aristotle in the 4th century BC, introduced the unequal arm balance – the 'steel-yard', also called the 'bismar' in Northern Countries. The Chinese, Needham (1962) Section 26 has explained, also used the steel-yard design in wood or ivory forms. He also pointed out that it was surprising to see that this form appears to be the first used there. The oldest extant Chinese balance dates to the 4th century BC and is of equal-arm design.

Fig. 3.15 is a fine specimen found at Pompeii and therefore predates 79 AD. In this type of balance the runner weight is slid along until balance is achieved. Another version, Fig. 3.16, implements the lever principle in another way; it is also known as a 'desemer'. The pivot is slid along between the mass and unknown weight suspension points until a position of balance is found. Weight is then read off at the point of suspension.

The advantage of this design is that no additional loose weights need be carried and that the overall weight of the apparatus is greatly reduced because the standard set of weights for the equal arm balance reduces to one weight that is at least a fifth the weight of a full set. It is a very convenient form, easy to carry, very robust and all parts remain together for safe-keeping. They are, of course, still in use today. The well developed degree of understanding that the Romans had of the principle involved can

Fig. 3.15 *All metal steel-yard balance found at the Pompeii site*
[Crown copyright. Science Museum, London]

be gained from reading the one-page debate on the steel-yard that has been translated by Hett (1955). This is part of works regarded as being written by Aristotle or, as is the case for this particular passage, by someone of the same school: see *Mechanical problems – 20*. Needham (1962) Section 26, refers to the Chinese having a similar knowledge at a similar time.

Characteristic of their approach to life the Roman's advanced the balance in respect to its practicability. They do not appear to have improved its potential for fine work

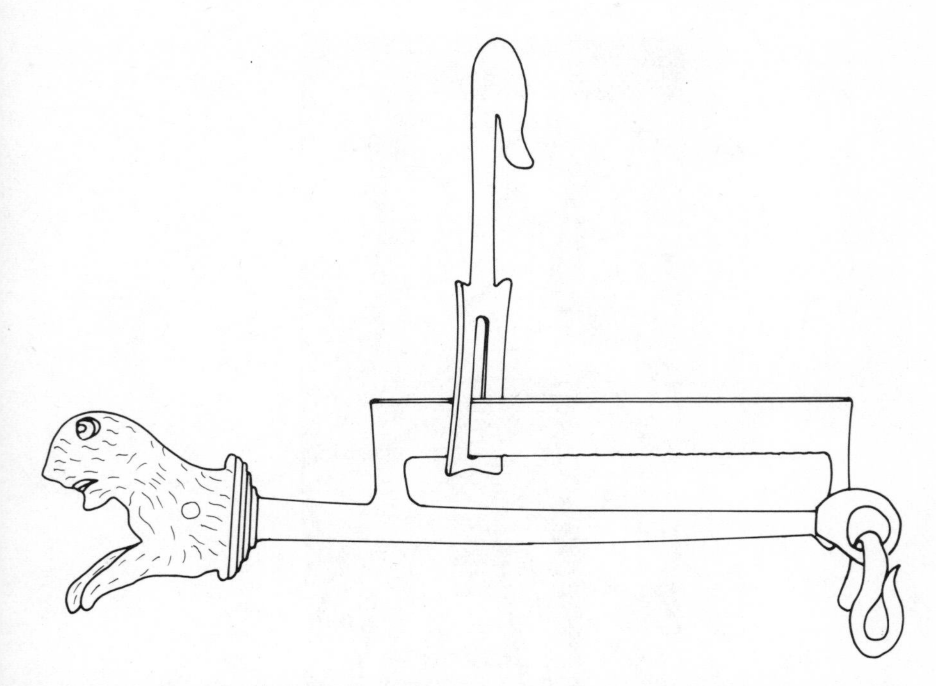

Fig. 3.16 *In this form of the Roman steelyard the balancing mass and the load points are fixed. The pivot is moved to find balance*
[Neuburger (1930), Fig. 261]

at high precision. They did show that all-metal designs were practicable. To do so would have been more possible for them than it was for those that preceded their time, because skills and availability of iron and bronze supply and working had improved in their time.

Then began the Dark Ages. Least affected were the Eastern regions of the former Roman Empire which became known as the Byzantine Empire. There, balance construction continued to be improved. Reports about all-metal designs for weighing to high resolution have been discovered in the literature of the Islam. Ellis (1973) provides a picture of a Byzantine balance that exhibits a long, light beam and which has an error indicator pointing upward inside a datum-marker frame. This kind of balance became dominant from this time and remained in use until the 16th to 17th century AD where often the pointer was more generally placed to hang below the beam. Fig. 3.17 shows the manufacture of balances in medieval times.

The Islamic world invented some advanced balances. There has already been mention of the spring balance of al-Khazini. He was a physicist who taught that mechanical problems could be handled by using the concepts of centre of gravity and equilibrium conditions. A book entitled *The book of balance of wisdom* was completed by him around 1121 AD. His realisation that air also exerted a bouyant force on the pans and components of the beam balance was a significant advance. From this,

Fig. 3.17 *Medieaval manufacture of scales*
[Ellis (1973). Courtesy Wayland Publishers Ltd.]

he designed a five-pan balance that is illustrated in Fig. 3.18, see Harrow and Wilson (1976) and Cajori (1962). The five-pan balance shown is a model built to a modern interpretation of records. Assuming that it has been modelled correctly we can draw several possible conclusions about the designers understanding of some important points of design, they may, of course, have instead arisen from empirical experience. The balance hangs from a set of rings that allows the whole unit to hang vertically. This would help the automatic establishment of the balance zero position but, as shown, the rings could have been improved upon for they would have exhibited considerable friction. Use of a multiple flexure suspension gives the required degree of rigidity in the transverse direction whilst allowing rotation in the axial direction. For null-balance working the flexure system would have worked well but it probably was not realised that the use of this simple suspension arrangement provided a centre of rotation of the beam assembly that was not well defined, nor lying near the axis of the beam. The balance would have given a non-linear indication for deviations from zero but that mode of use seems unlikely as no scale is provided. The balance shows that al-Khazini had a well developed approach to the scientific design for its various features indicate that it was realised from knowledge of basic principles of mechanics and fluids.

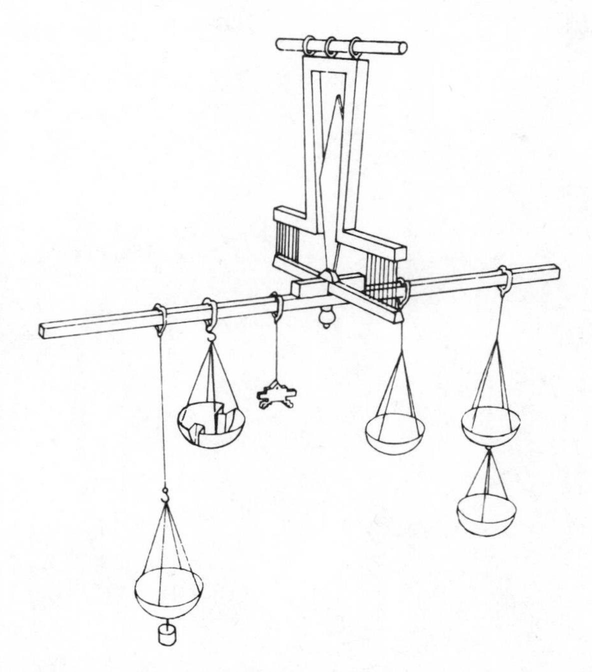

Fig. 3.18 *Al-Khazini's, c1121 AD five-pan balance; allowance is made for the buoyance of air* [Harrow and Wilson (1976), p.28]

The most prolific use of balance during this time period must have been for the studies in alchemy. The Arabs gave us the word alchemy from 'al-kīmia', but the practice of seeking ways to convert base metals into gold had been in existence well before this time. Alchemy, however, apart from sowing the seeds for more scientifically-based chemistries that were to follow later, did not help advance scale technology. The practice of alchemy did not make much use of analysis of substances. It was more a case of weighing out portions of various substances to make up a mix that was then processed. This did not need good balances. Sensitive balances developed primarily for reasons of trade in precious materials and in medicinal uses. The 'carat' unit of weight derives from ancient standards based upon the weight of seeds. The Arabs gave us the carat from their word 'qirat', the seed of the coral tree. Today it is now 0·2 gm, Klein (1975).

The use of such a poor standard of weight did not restrict the sensitivities of Islam balances being raised above the older world performances quoted earlier here. Holmyard in Singer *et al*. (1956) tells of coiners at a Muslim mint weighing, in 780 AD, to 0·003 gm. His quotation from early accounts reveals the early use of enclosed scales and sophisticated procedures of use:

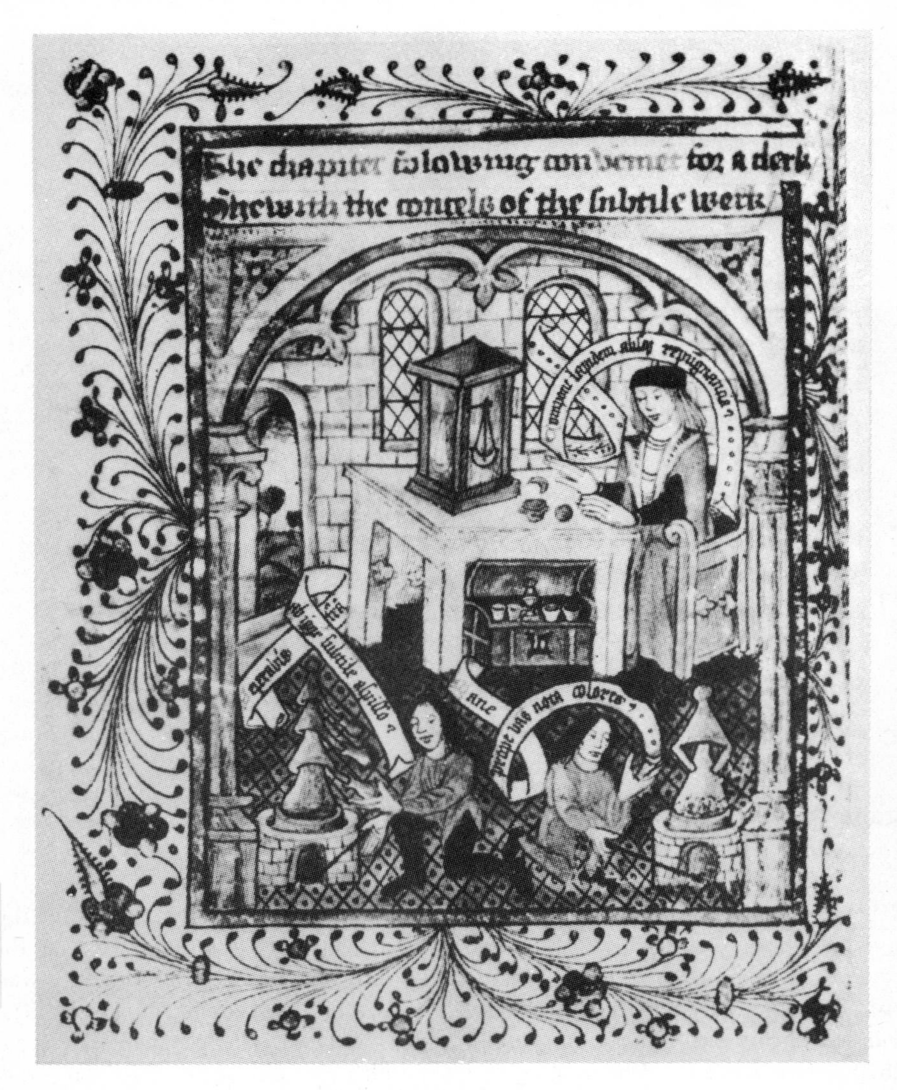

Fig. 3.19 *Earliest known illustration of an enclosed balance*
[Norton, T. *Ordinall of Alchimy* c1477. Add. 10302, f.37, Courtesy British Library]

> To reach such accuracy it was needful to use the finest chemical balance, with closed case, double weigh the glass weights against each other, to read a long series of swings of the balance.

The earliest actual illustration of a balance that is enclosed appears in Thomas Norton's work of *c*1477 AD, entitled *Ordinall of alchimy*. A facsimile print of this was published by Arnold, London in 1928. Fig. 3.19 shows the illustration of part of Norton's alchemy laboratory that features the balance.

There is clear evidence of balances being used for scientific studies throughout the Middle Ages of Europe as well as in the Islam world. Klemm (1959) has given one instance in his study of the rise of western technology. It relates to the works of Leonardo da Vinci as compiled by Richter (1939). After writing about how to construct a sail-like wing of a stated size da Vinci wrote:

> And if you wish to ascertain what weight will support this wing, place yourself on one side of a pair of balances and on the other place a corresponding weight, so that the two scales are level in the air; then if you fasten yourself to the lever where the wing is and cut the rope which keeps it up, you will see it suddenly fall; and if it required two units of time to fall of itself, you will cause it to fall in one by taking hold of the lever with your hands; and you lend so much weight to the opposite arm of the balance that the two become equal in respect of that force; and whatever is the weight of the other balance, so much the wing will support as it flies; and so much the more as it presses the air more vigorously.

The relevant sketch made by da Vinci is given in Fig. 3.20.

The design of balances had been improved greatly by 1500. There was, however, still much that could be done, and had to be done, to create measuring instruments that would give the new generation of rising chemistry researchers the detection capability that they needed. Out of what appears to be a period where this degree of improvement did not evolve, it is surprising to see that the first reported balance that has such capability occurs around 1547 with the report of Ercker's scales in a contemporary book. This instrument has been reconstructed at the Science Museum, London. The logical question that this raises is where are the other balances that must have been made? Perhaps further research of medieval sources will, one day, unearth more information about this significant change that came about.

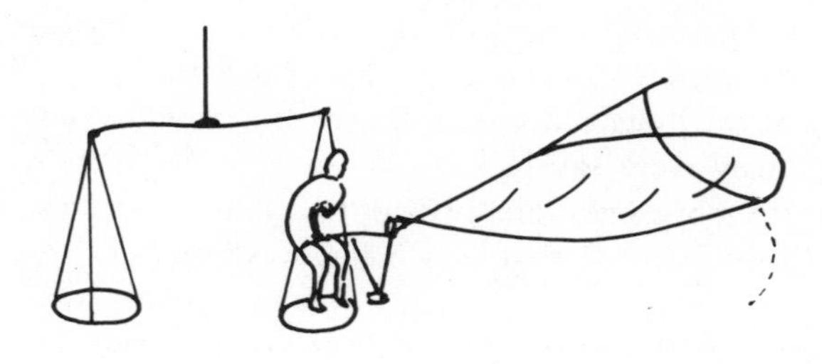

Fig. 3.20 *Leonardo da Vinci's sketch on how to use a balance to measure the lift forces of a wing*
[Richter (1939) from Leonardo da Vinci's sketches]

Lengths and distances

Daily routine of the earliest peoples would have developed their concept of distance. It seems they must have comprehended two kinds of distances. First were those to do with their personal needs, such as lengths that arose as they made a building or piece of furniture. Second were those distances that they were not generally called upon to realise as a quantity, but would have appreciated existed, the distances to the horizon, the extent of land, the height of mountains and the size of the heavens being examples.

Standards were needed for the units of the first kind because quantification was essential. Generally, the citizen of the ancient peoples would be concerned with lengths ranging from the smallest he could see, the width of a fine line, to the largest that he might have to experience, that of a journey. It would have been quite rare for him to have needed to measure long distances much beyond a few hundred kilometres and then never to great precision and accuracy.

Short distances were given units chosen from immediately availability objects that could serve as convenient standards. Parts of the body were an obvious choice; everyone had a set of standards with them when needed. Standards were often taken from reasonably invariable natural products, such as the length of seeds.

In all cases of length standardisation the standards chosen were arbitrary, as is the case for all physical standards of man. The variability of such standards was great. Attempts were made many times during the histories of the ancient peoples to set up the legally authorised length standard but the reality was that difficulties of transport, of standards comparison and of getting people to cooperate usually resulted in the situation where although the central standard did exist, a local one was that used. With no traceable link from the national authority down through local substandards (in the standards sense the word substandard does not mean that it is inferior to a degree of less usefulness but only that it is the standard of a lower grade than the ultimate one) it is easy to comprehend why so many different lengths were in force across a country. Trade was complicated by this but the practical difficulties of providing a fully traceable system were insurmountable in those times. Indeed man has only had such a concept in almost complete global force for less than a century.

Today many of the units and standards used appear ridiculous to us. What must be realised is that they were satisfactory, at least, to begin with. Once established, however, it became more and more difficult to overhaul them into a coherent system. It was not until very recent times that a common system was possible over the whole world.

Today the standard of length is defined in terms of the wavelength of light emitted by a well defined radiation source. It is based upon a natural phenomenon that is known to be adequately stable and reliable for the most precise needs that arise. So today we, too, still use a convenient natural principle, just as did people of earlier times.

Klein (1975) has shown that numerous units for each of the commonly needed measurement parameters existed and he has discussed the great confusion that this brought. Here space permits only a few to be examined. The cubit, foot and some others are now discussed to give an impression of the problems and the state of understanding that early peoples had of their length standards.

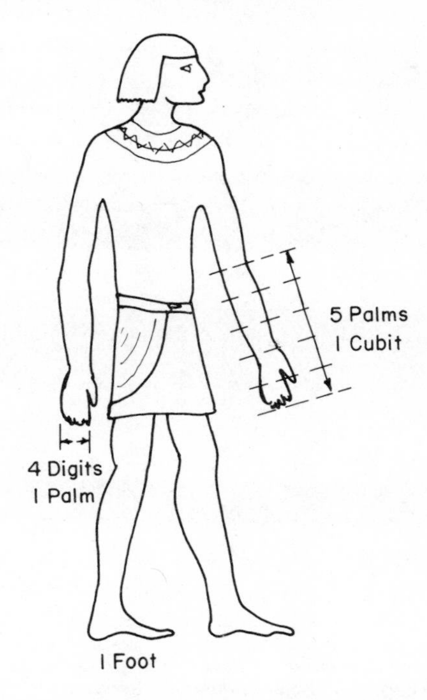

Fig. 3.21 *Relationship of the cubit length unit to parts of the body*

It is pure conjecture that the first length units were based upon the length of the arm, the foot, the fingers, the palm width and so on, but there is evidence that the cubit arose this way for the Egyptian symbol for the cubit is a picture of a forearm. Furthermore, subdivisions and multiples fit a sequence that suits typical body part sizes. Fig. 3.21 shows the relationship between the cubit, and the palm and the digit. (Some confusion might arise when reading about the cubit for the Royal cubit was equal to seven palms, not five.)

Many cubits were used by many nations in the area. The following, taken from Berriman (1953), shows these and their relative standard sizes:

Cubit	*Metre*
Roman	0·444
Egyptian 'short'	0·450
Greek	0·463
Assyrian	0·494
Sumerian	0·502
Egyptian 'royal'	0·524
Talmudist	0·555
Palestinian	0·641

Fig. 3.22 *Stone and wood cubits of Egyptian origin, 2400 BC to 1st century AD* [Crown copyright. Science Museum, London]

It is an interesting fact that the size of the Egyptian cubit has been shown to have remained within a 5% variation over the period from 3000 BC to 1850 AD. This is quite unexpected for the length of a human forearm varies well beyond that figure. Considerable study has been made of the cubit. Skinner in Singer *et al.* (1954) has discussed cubits in particular.

Many cubit standards have been found. They are made from wood or stone. Fig.3.22 shows several held at the Science Museum, London. The cubit was subdivided into equal intervals and then again each minor division was again divided into pure fractions ranging from a half downward through the progression 2, 3, 4, 5 . . . . to 16. This may seem curious but their number system expressed fractional quantities as the successive sum of unit fractions that all added together to sum to the value needed, for example, 2/29 was expressed as 1/24 + 1/58 + 1/174 + 1/232.

Being made of stone or wood they would have been subject to temperature expansion error; this would not have shown up in their methods of intercomparison which did not have optical aids to help transfer the interval from one cubit to another.

The cubit was taken up in Europe and other areas. It was still in use in some countries in 1966. A world study of weights and measures made by United Nations specialists – see Klein (1975) – revealed this.

Distances greater than one cubit were expressed as multiples of the cubit with some decimal values being given other names. The 'atour' was equal to 20 000 cubits. Dimensions of the land of Egypt were 20 atours in the Lower area and 86 atours was the length of the Upper Region.

The inch-foot system can be traced to likely origins from the width of the thumb for the inch and from the obvious length of a human foot for the foot unit. Klein (1975) tells us that the word 'inch' comes from the latin word 'uncia' meaning one. This passed into Old English as 'ynche'. The Romans used the basic unit of a 'digitus' which was also based on the width of a finger.

In the Roman system 16 'digit' equalled one 'pes' which was the latin for foot. From the foot came the next largest unit the 'pace'. In Latin this was 'passus' and it was equal to 5 pes. Their system went on through higher units to the 'milliare' which was equal to 1·476 km.

The foot unit appears in many other civilisations, sometimes even alongside the cubit. Klein (1975) has discussed the differences that existed which were great. He shows that there existed numerous units of around the 300 mm length.

Three feet equal a yard in the old Imperial English system. The word 'yard' meant a measuring stick in the Anglo-Saxon tongue. Also tied into this tale Klein (1975) suggests that the word 'yerde', a spar of a ship, was how a person appeared when measuring out a rough yard of material by holding it from the fingertips to the nose with the arm outstretched. From armtip to armtip with both arms oustretched, is derived the fathom that, therefore, equals two yards.

Possibly the most quoted standardisation procedure for arriving at the legal German foot comes from the 16th century method of defining the 'right and lawful' rood (rute) of the Germans. In 1575 Jakob Kobel suggested that the variability of one person's foot could be averaged out to arrive at a more uniform standard by taking the average of the length of the left foot of the first sixteen persons to emerge from church at a given date. A cartoonist of the same period drew a parody of this (Fig.1.11) suggesting what could happen to the range and kind of people that might make up the group. It includes a person with a wooden left leg, a knight with long curly armour on his foot, people with deformed feet, and so on. Even so this odd collection would still have averaged better than taking any single one as the official measure.

The 'ell', referred to in the clause of the Magna Carta quoted earlier, was to do with cloth measurement. The range of ells was great. The shortest appears to be the Dutch one at around 300 mm and the longest, that of the French, was almost double that value.

Such was the state of length units by the Renaissance period of Europe. Standards of all measurements were little better than they had been 2000 years earlier!

Attempts were made to improve things in medieval Europe. Several suggestions tried to remove the use of units based on parts of the human body and replace them with standards made from physical objects, such as a bar of iron. Edgar (959-975 AD), King of England, established 'the measure of Winchester' which, according to Ellis (1973), implied the existence of a physical length bar. A brass-rod length standard exists from the 12th century.

Toward the end of the period of interest of this chapter we find a 14th-century treatise suggesting how the confusion might be cleared up and order once again created across the whole of the standards of weights and measures:

> 'By consent of the whole realm of the King's measure was made so that an English penny, which is called Sterling, round without clipping, shall weigh thirty-two grains of dry wheat from the middle of the ear; twenty pence make an ounce and twelve ounces make a pound and eight pounds make a gallon of wine and eight gallons of wine make a bushel of London.'

Even today we still purchase many goods on the basis of such mixed units.
So much for the problems of standards and units of length. Attention is now given to apparatus used to measure length and distance.

Basically it is reasonable to assume that the tradesmen of the day would have made measuring sticks by scribing scales on them using what they regarded to be a more superior scale or by dividing their own using geometrical principles. Such instruments have been found. For longer distances they would have counted paces for crude needs, such as the distance to a land feature. They would have tried to do better by making sticks which were multiples of their basic foot, yard or cubit, setting these end to end. Evidence exists of the use of knotted cords: the priests of the Vedic age of Ancient India excelled at laying out altars to given specifications using them. These were the ancient forerunners of the highly successful metal survey chains introduced in England after the Renaissance period.

On the Nile river were mounted 'nilometers' that recorded the height of the all-important flood as it occurred each year. Most instruments were simple; noteably sophisticated measuring devices began to appear around the time of the birth of Christ.

A most advanced device for measuring long road distances was the hodometer. It is important, not just because it was able to indicate distance travelled and perhaps record it, but because it was among the first of the known uses of geared mechanisms in the Ancient world.

Two, apparently independent, developments of the hodometer have been recognised, in China around 260 BC and in Alexandria around 60 AD. It does seem that the former was the original use and that the work in Alexandria may not have been an independent original invention. Furthermore the Chinese also produced the famous south-seeking chariot that used differential gearing to keep its figure pointing in the same direction irrespective of the motion of the chariot on which it was mounted, more of that later in this Chapter. Such a construction leads one to think that the Chinese certainly had the earliest understanding of geared mechanism.

The account of the hodometer by Needham is most authoritative and detailed covering, in the main, Chinese sources. One reconstruction, built as a model, is featured on a 1953 postage stamp of China. It is shown in Fig. 3.23. The two puppets strike the drum at regular intervals of distance. One can imagine the Imperial processing with this instrument as a feature of the colourful parade.

The other hodometer development is that of Heron (*c*60 AD). A later version of the time, described by Vitruvius (*c*192 AD), had the additional feature that it dropped balls into a collector to indicate the coarse values of the distance run. This is pictured in Fig. 3.24 as a reconstruction given by Bailey (1923). In modern measurement

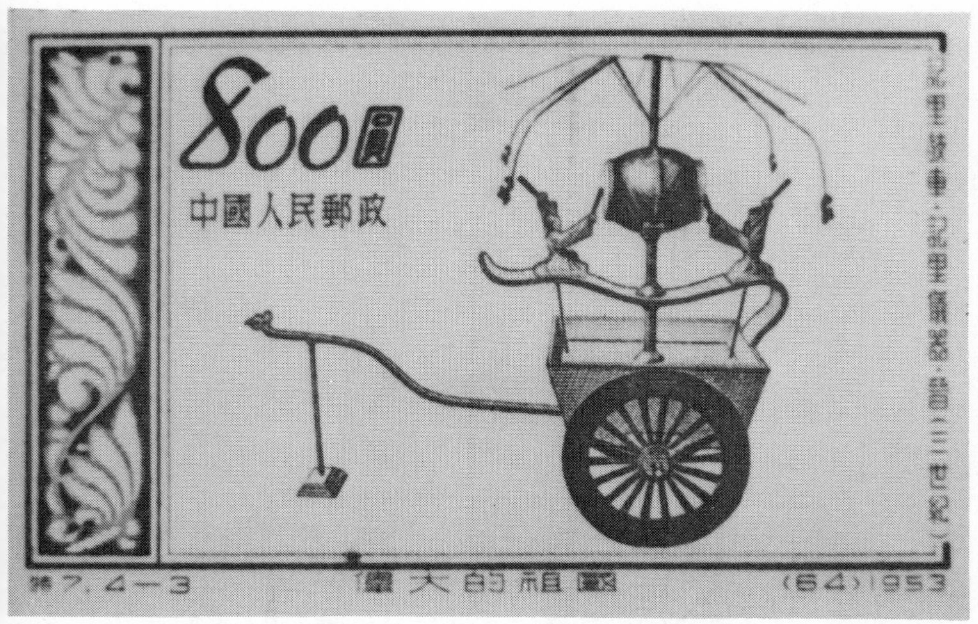

Fig. 3.23 *Copy of Chinese hodometer of Tsin dynasty 300 AD: as reconstructed by Wang Chen-To, and featured here on 1953 postage stamp*
[Photo. Science Museum, London]

parlance, this hodometer recorded distance run in analogue-digital format. The early form is shown illustrated in Neuburger (1930).

Knowing that the Chinese built the hodometer it is a little less surprising to learn of another metrological tool that they invented for measuring small distances. It is a sliding caliper used for gauging size up to around 150mm. Fig. 3.25, drawn from Needham (1962) Section 26, shows a copy of this device made for the late Professor W.P.Yetts. The L-shaped bar with the finger hole and the Chinese letters upon it slides relative to the other part secured to the lower bar by the slot and the right-hand end, rectangular, ferrule. It is clearly the forerunner of the vernier-caliper that made its entry into the western world many centuries later. A scale, not clearly shown in the picture but drawn in Needham's other illustration in his work, is formed on the frame. This measuring instrument was made, from the inscription upon it, in 9 AD. According to Needham, Leonardo da Vinci sketched an instrument like this in his time which was at a much later date.

Distances to the stars, of the circumference of the Earth and others lay in the domain of the astronomers and 'scientists'. It is more appropriate to discuss their measurement in the next Section on astronomical instrumentation.

Measures derived from length, mass and time

A modern measurement system having standards for the three units, mass, length and

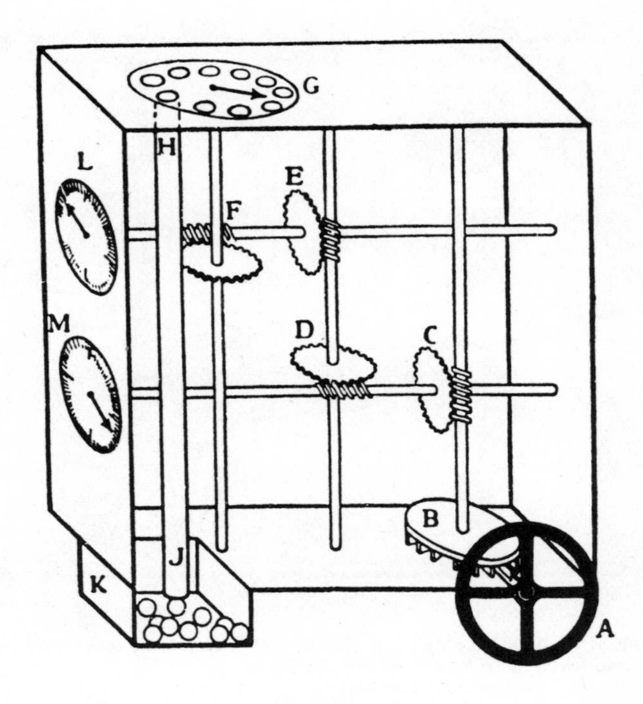

Fig. 3.24 *Hodometer described by Vitruvius, c192 AD. This featured analogue and digital recording of distance travelled*

[Bailey (1923), Fig.20. Courtesy Oxford University Press]

Fig. 3.25 *Sliding caliper made in China in 9 AD*

[Needham (1962), Section 26, Fig. 287, p.45]

time allows derived variables to be generated from them by suitable combination. Thus it is clear to us that area needs measurement of two lengths, volume of three lengths, density needs mass and three lengths to be measured and so forth. In the ancient times, however, such methods were not always possible in practice or even realised as existing. The concepts of area and volume were certainly understood by the Egyptians and those after them, but appreciation of density was much slower to form in the minds of the ancients. Measuring apparatus usually develops from an understanding of what the variable is that is involved so there was little need in early times for instruments to measure density, flow, acceleration, force and other variables.

Area was calculated by the use of length measurements coupled to a basic understanding of elementary geometrical relationships. The Babylonians, for example, see Farrington (1936), divided the total shape into right-angled triangles, rectangles and trapeziums of appropriate sizes, adding up the total after calculation of each.

Volume of large structures such as ramps were amenable to the same method and little other possibility existed because of their rigid immovable nature. Measurement of the capacity of smaller volumes could, however, be achieved without recourse to length measurements at all. Capacity measures were needed for wine, corn, medicines and more. A mosaic has been found showing corn being measured in Roman times, Fig. 3.26.

Fig. 3.26 *Measuring corn in a standard capacity measure. Fom a mosaic of Ostia.*
[Courtesy Place des Corporations, Ostie. Photo. P.Ronald, Editions Arthaud, Paris]

Standards for liquids and other flowing substances, such as seeds, were measured using convenient receptacles elected as the capacity measure. It would not have been easy for early peoples to form such measures as length determinable containers. No attempt was made to make them deriveable from the length standards. The exception that is established was that of Ancient China where red millet seed was used to link volume to the weight standards.

Early capacity measures found include a copper cup from around 1400 BC of Egyptian origin, pottery jars from most civilisations and horns. The Egyptians had the unit of capacity called the 'apet'. It was based on wine jars and was used as early as 2000 BC. The early Greek peoples used a small capacity measure to arrive at the amount of wine to be added to a given quantity of water. These were not legal measures but were similar to what was used as a standard.

Weight, or more formally mass per unit volume, is called density. The ratio of the density of a substance to that of water is the specific gravity. These, therefore, can both be derived from weighing and from volume measurements. Without a volume derived from the basic length standard it is impossible to relate the densities of various substances unless the capacity relationships are known. The relative value, however, specific gravity, does not get caught up in that tangle and can be measured reasonably easily with simple apparatus.

The ancients had a good appreciation of the variation in specific weight of different substances for that is evident from direct experience. The first person to give this concept a numerical basis appears to be Archimedes. The account of how he tested the quality of gold in a crown varies. Two versions are given here as stated by Cajori (1962).

In the first, according to Vitruvius, he took a piece of gold and silver that each weighed the same as the crown he wished to test for its gold content. For these and the crown he determined the volume of water displaced by each. From these three measurements of volume he calculated the proportion of gold in the crown. According to another author he did it another way, using the method of apparent loss of weight of each when hung in water. This gives the same result.

From this kind of research Archimedes compiled his work 'Floating bodies'. According to Needham (1962) Section 26, the Chinese did not appear to have knowledge to match that of Archimedes until after his time.

Today the simplest instrument for testing the specific gravity of a fluid is the familiar hydrometer. According to Cajori (1962) this device originated in the 4th century AD being described in full by a letter of Bishop Synesius. It was a 'hollow, graduated, tin cylinder, weighted below.' Its original use was to test the quality of water to ensure that it was not hard; such was considered unwholesome for use in medicines. Cajori stated that there seemed to be no good evidence for attributing this idea to Archimedes.

The knowledge about fluids of Archimedes and his followers did not, however, go much beyond that mentioned above. It was to be some 1500 years later that such fundamental issues as the pressure exerted by a fluid would be resolved. The Greek scientists ran out of time as a nation and it was then up to the Arabs to further such work.

Marmery (1895) gives credit to Alhazen, *c*1100 AD, for improving the hydrometer and for preparing accurate tables of specific gravities after the previous work of Abur Raihan. Already mentioned is the work of al-Khazini on the five-pan beam balance that was used for this area of study.

3.7 Astronomy and surveying

In this Section are grouped the measurement methods and instruments used in connection with position finding and control of shape and size of large structures. The fields that fall in this class are surveying of sites and of land of large extent, astronomy and navigation. The techniques that each use overlap to a great extent.

Position of an object having extended shape in the three-dimensional space can be defined by quantifying up to, but no more than, six positional measurements. There are, for an extended object, six degrees of freedom in movement that could occur. To tie position down we can specify three lengths and three rotations with respect to a chosen positional datum framework. Those degrees of freedom that are restrained as constant values need not be measured. Thus for a point object lying on a line there are no degrees of rotation and two of the three translations are restrained. This leaves only its position along the line to be determined. This is the case of a one-dimensional measurement, a single length measurement. If the object in question is a point lying in a plane then two-dimensional variables must be measured to determine the position of the point in the plane. It is also possible to locate position of a point in a plane by the use of triangular relationships in which case an angle and a length can be used instead of two lengths. The need nevertheless is still for measurement of two positional variables. The same applies to position in space.

Thus position measurement can be involved with the need to measure angles. Indeed for large-scale position determinations it is often more practical to measure angles instead of lengths using just one length to establish the actual size of the triangle involved. In astronomy it is not possible to directly measure many of the distances involved and angular methods must be employed with a baseline established, if possible, to tie down the absolute magnitude of the triangular relationships.

The Ancients measured long lengths with hodometers, (later called waywisers), pacing, knotted cords, and by baseline plus angle methods. Another main metrological parameter of interest then is angular measurement.

Before dealing with those devices that were used to determine the magnitude of angles it is first necessary to dispense with instruments that were concerned with the generically more basic parameters – level, vertical, flatness and straightness. Each of these concerns the detection of deviations from a stated position definition for a given set of constrained dimensional variables.

A level line is a line that has all points in it lying on the same line and additionally that is coincident with the geoid of the earth at the place of interest. That there exists variation in the smooth curvature of this geoid and that the geoid is curved, not flat, does not concern us here for these understandings would scarcely have been observed in the works of the Ancients. A level plane is the two-dimensional form of a level line.

Fig. 3.27 *Scene from a tomb at Thebes, c1450 BC, showing the use of strings to establish straightness*
[Singer *et. al.* (1954), Fig. 313, p.481. Courtesy Oxford University Press]

It has been established that the Egyptians used string lines, Fig. 3.27, to test flatness and straightness and to establish a straight line. They also used communicating water trough levels to establish a level plane, such as was needed for, say the base of a pyramid. Such practices have been known since the earliest times.

The vertical is defined in a similar way with the exception that this time the vertical is perpendicular to the horizontal level line or plane. Again deviations of departure from this line (or plane) were detected with strings using the well known plumb line or a square to define the vertical.

In both cases above the measurement problem is to define the orientation of the line or plane and to detect departures from the defined line or plane. This was done in early time using the eye to see distance variations from the ideal, or to detect angles that indicated such displacement departures. With no optical devices available to assist the eye – lenses can be first traced to no earlier than around the end of the Islam period preceding the Renaissance – the instruments used were designed to provide definition of the datums making up an angular excursion.

Angular resolution of the unaided eye is approximately 1′ of arc. This means, in radian measure, that the eye can resolve an interval at a distance that is in the region of 1/7000 of that distance. It is unlikely that lengths could be measured to this degree of accuracy in those times implying that there was considerable scope for measurement of such objects as stars for the eye, at such an embryonic state of astronomy, would not have been the limiting factor on the precision of star movement measurements. Considerable scope would also have been possible for mapping the star fields; the resolution of the eye made a vast number of heavenly bodies observable. Another factor in the favour of the early civilisations was their global position in clear-sky climates.

The situation was not so favourable on land for surveying techniques would have been limited by the severe turbulance of the lower atmosphere that were brought about by the wide temperature variations that occur during a tropical day. The eye again, however, was able to resolve to approximately the same limit as these effects allowed on well chosen days. Today surveying and astronomical measurements of position can be carried-out to around a part in one million, but the improvement is the result of the application of optical aids to sighting devices and to the use of non-human sensing methods such as electromagnetic radiation devices that have no natural counterpart in man's collection of senses.

The eye can resolve an interval of distance of approximately 50-100 μm when it is placed at the optimal distance of about 300 mm from the marks defining the interval. This capability can be used to sight error of straightness between a taut line and a test surface, or levels of water in a communicating trough. Through such methods the angular detection limit of the eye can effectively be enhanced. For example, detecting water levels to 1 mm over a 100 m trench distance implies an angle of 1-in-100 000 parts is observed.

Instruments of the period from genesis to 1500 AD were characterised by their total lack of optical magnifying elements. The development of surveying methods and astronomical instrumentation was one of the struggle to make better sights and records of position with only the unaided eye available as the primary sensor.

Angle is theoretically defined in terms of two lengths expressed as ratios. From the theoretical point of view angle has no need for units and, hence, no need for standards. In practice, however, it is convenient to work with units of angle such as degrees, minutes, seconds, grades and radians and to declare certain pieces of physical apparatus as having angular magnitudes somehow expressed in their construction. A divided circle provides a working standard for degrees and anyone caring to form their own circle will arrive at the same magnitudes within the error of division. The fact that angle can be expressed this way for convenience does not prove that angle has dimensions: it has no physical dimensions. In radian measure of angle the size of an angle is expressed as the ratio of two lengths. The circular degree system derives from a circle that is divided accordingly.

A circle is easily created to quite high precision using only a piece of string and a fixed central pivot place. It is easily divided into four quadrants by using comparison methods involving only a fixed length stick. From there we could look for a simple explanation of why the quadrant is divided into 90 degrees or 360 degrees in the full circle. Many theories exist but each seems to have reasons for suspecting it. We are sure that it was the Babylonians who divided the circle, this being done to give them a quantitative way of recording where stars and other movements were seen in the heavens. It is also certain that they used a sexagesimal number system based on 60 states in each digit position (compared with ten states for each position of the decimal system). Perhaps 6 times 60 was a good reason for 360 degrees in a circle? As they could resolve finer angular intervals than a degree, it was divided further into 60

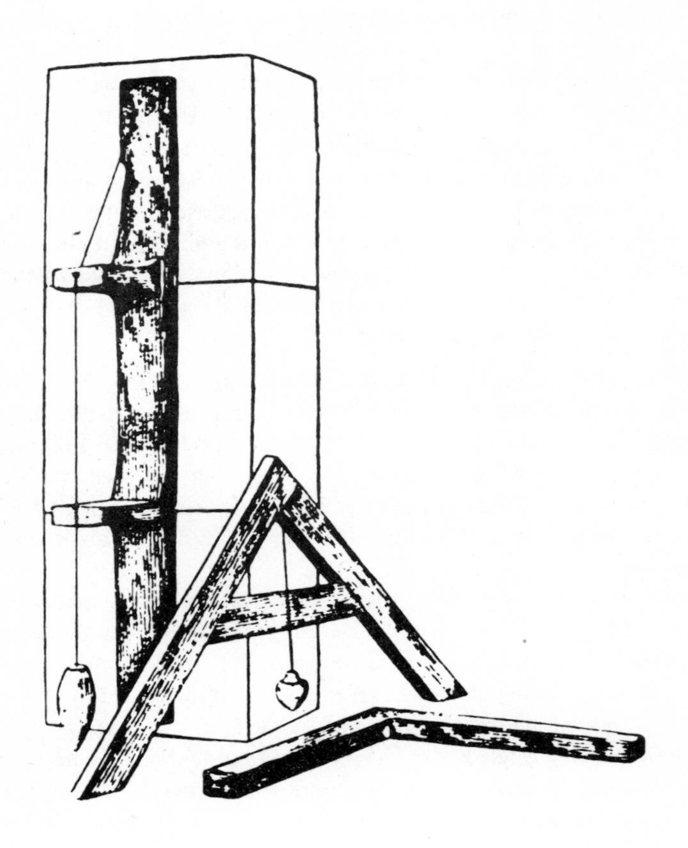

Fig. 3.28 *Mason's measuring instruments from Thebes, Egypt c1100 BC*
[Singer *et. al.* (1954), Fig. 314, p.481. Courtesy, Oxford University Press]

minutes (60′). Averaging of values coupled with the fact that some observers could resolve to better than a minute of arc and other reasons meant that the minute needed further subdivision, hence the 60 seconds (60″) as the next divisor from the sexagesimal system. Klein (1975) discussed reasons for the 360° system.

Survey work and setting up for astronomical observations often needed the quadrant to be made as a self-contained measuring tool. In its simplest form it was the square made of wood tied, or doweled, together. Fig. 3.28 shows the range of measuring instruments that were at the disposal of the Egyptian mason. Instruments like these were handed on from civilisation to civilisation right down to today. It is clear that the Romans and the Greeks did little to improve on these basic instruments, there was no need for change, they served their purposes well. Inscriptions and written descriptions from Greek and Roman times show that they, too, used the square with its plumb-line (called a plummet), the pair of compasses, the caliper and small hand squares.

Fig. 3.29 *Egyptian groma from Fayum, c150 BC, as restored for the Science Museum, London* [Crown copyright. Science Museum, London]

Another variation of the square combined with plumb lines was the 'groma'. A restored Egyptian groma is shown in Fig. 3.29: it dates from around the 2nd century BC. It is made from palm leaf centre stems lashed together at right angles. From the end of each arm hang limestone weights on fibres that define verticals. Thus, with this instrument, the user was able to set up two planes that were vertical and square to each other. A more elaborate design has been reconstructed from contemporary literature; that by Bailey (1923) is shown in Fig. 3.30. Neuburger (1930) gave an account of a development of the groma, called the 'chorobates'. His reconstruction, shown as Fig. 3.31, was made from a description by Vitruvius. Neuburger explained that there exists evidence that shows that Vitruvius also knew of the glass tube, air-bubble, kind of water-level.

With such basic measuring instruments early civilisations were able to build magnificent structures such as the Pyramids, the Acropolis and the city of Persepolis. Today the same devices are still used in much the same way although in quite recent

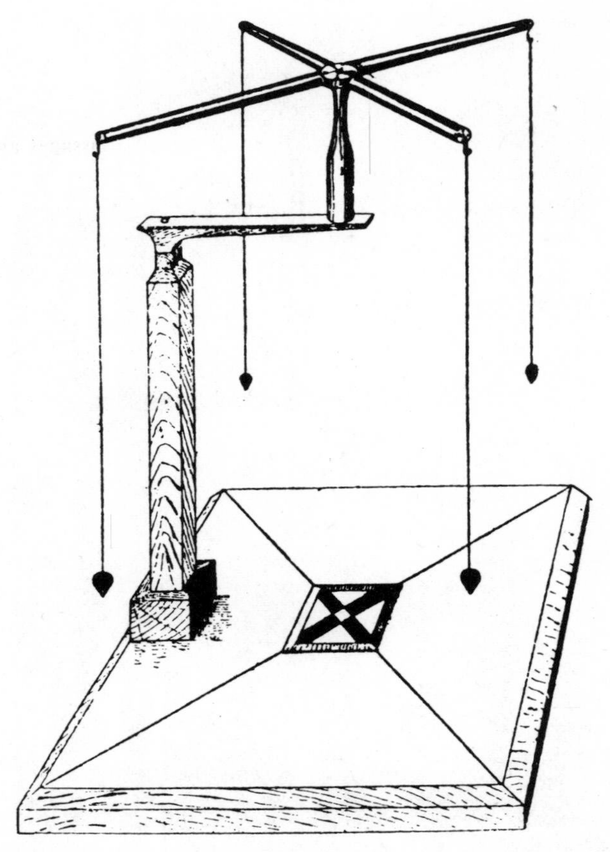

Fig. 3.30 *Roman form of the groma; reconstructed from original sources*
[Bailey (1923), Fig. 18, p.298. Courtesy Oxford University Press]

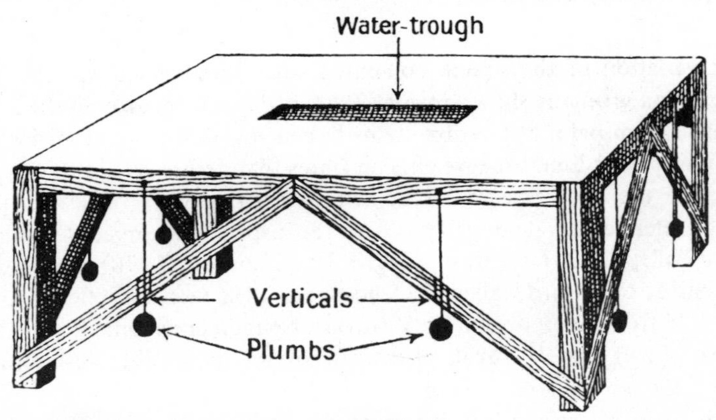

Fig. 3.31 *Greek chorobates; reconstruction by Neuburger*
[Neuburger (1930), Fig. 535, p.394. Copyright R. Voigtlander's Verlag]

times electronic forms of surveying instruments are making a positive impact into this area.

Probably the most advanced surveying instrument of the ancient times was the 'dioptra'. A detailed description of this instrument was written by Hero of Alexandria. Cohen and Drabkin (1948) provided translations of various passages and explanation of this device, the forerunner of today's theodolites. Neuburger (1930) states that H.Schoene had 'recently constructed one' according to the early accounts. Reference to Fig. 3.32 well shows the sophisticated use of instrument principles to provide constrained motions and rotations in a manner that was clearly well in advance of other instruments of its time.

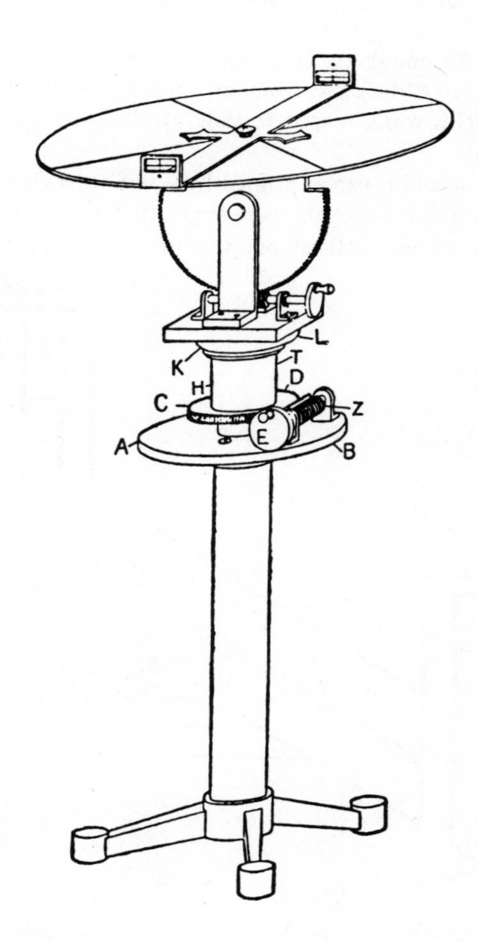

Fig. 3.32 *The dioptra of Alexandrian times; as reconstructed from Hero's descriptions* [Cohen and Drabkin (1948),p.336. Courtesy Harvard University Press. Copyright 1976 L.C.Rosenfield]

The Cohen and Drabkin account describes accessories that were available for use with the basic dioptra (sometimes spelled diopter) to give it water-tube level capability, see Fig. 3.33. The attachment, as reconstructed by these authors, allowed the levels of the water at each end of the brass communicating tube to be directly compared with the sighting datum lines by arranging them to be in simultaneous vision by the observer. It is, however, doubtful if the dioptra was made to the precision shown in Fig. 3.32 (and elsewhere in the literature) but the principles involved are extremely ingenious for the time.

Contemporary accounts said that the dioptra solved the problem of driving tunnels from both ends simultaneously but there appears to be no evidence to say they were actually used for that purpose. They certainly were used for astronomical observations, Cohen and Drabkin gave space to that application. They also commented on the tunnelling use.

The ingenuity does not end there. To go with the system was a target on a graduated staff. Again, according to Cohen and Drabkin, these had a disc target formed with a black and white half. This was arranged to slide up the staff moving a pointer with it that lay alongside a scale.

The contemporary account also explained how this surveying system was used to conduct several classes of surveying measurements, such as how to measure horizontal distance without going to the farthest point, how to measure the depth of a ditch

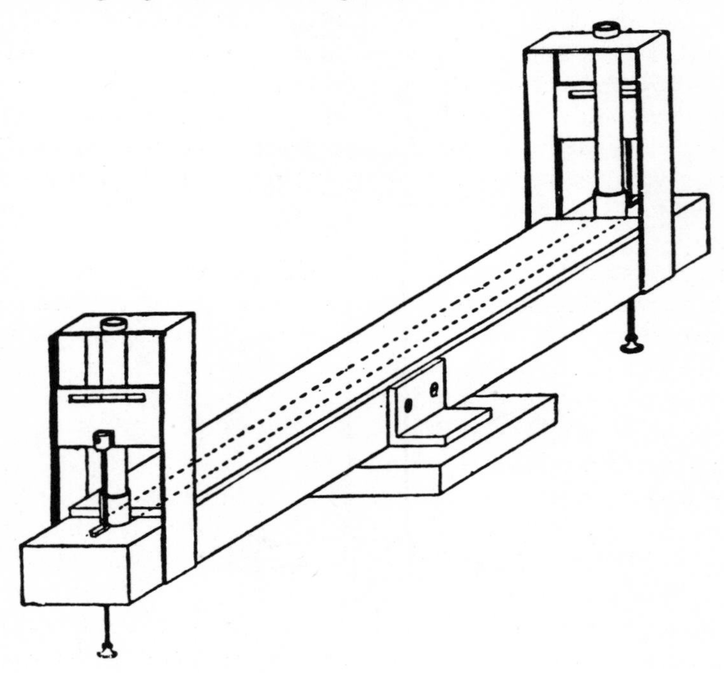

Fig. 3.33 *Levelling attachment for use with the dioptra*
[Cohen and Drabkin (1948), p.337. Courtesy Harvard University Press. Copyright L.C.Rosenfield]

without entering it and, as already mentioned, how to drive a tunnel from both ends.

Principles and their instruments discussed here in survey work were also directly usable for astronomical use.

Astronomical instruments and apparatus

Since earliest times man has been an avid observer of the heavens. This need has arisen for reasons of religion, for navigation, for exploration and conquest, for agricultural seasons and eclipse prediction, to tell tides at sea, and for scientific needs as part of man's then increasing awareness that basic knowledge was worth seeking. Numerous works have been compiled on the history of astronomy, the main theme generally being on the ideas that were in vogue and how they were gradually improved upon. Here we draw upon such sources to extract information about the instruments and other apparatus that were used up until the 15th century AD. Table 37 of Needham (1959) Section 20, shows the comparative development of astronomy in Europe, Babylonia, China, Arabia and India.

Despite the lack of optical magnification in those times, this period produced a vast array of useful devices. Very few new principles have been added since then for this range included most of the possibilities. Today's instruments are still based upon principles of instrument construction and implementation that came from the period. What modern man has added is a greater and wider range of sensitivity, and some additional methods such as radioastronomy, that were not possible in those times owing to lack of knowledge and technological capability.

Astronomical apparatus of this period divides into those instruments used to make observations and those used to assist setting up observations and for recording findings. The two groups were used hand in hand. One kind was used to find new information, the other to record it in a manner that would give appropriate retrieval for subsequent observations.

All of the instruments of observation were primarily for observing the shadow cast by the sun or for watching the transit of stars, and the like, at night. (Some were later adopted to measure angles in field survey). These began as the simple gnomon placed vertically in the ground, reading the shadow cast on a horizontal plane or stick. It is clear that this procedure required careful control of the length of the gnomon, its geographical position and its orientation. To preserve precision in such measurements the Chinese used a 'gnomon shadow template' – see Needham (1959) Section 20, – to improve upon ad-hoc measuring sticks and unreliable length measures. Templates were made of a durable material such as jade or terra-cotta. Extant templates date from 164 AD. By 500 AD there were, in China, bronze instruments in which the gnomon was combined with the plate.

Sundials, already partially covered in the section on time measurements, developed from the crude shadow clock of early Egypt (refer to Fig. 1.7) through the flat-plate kind, the upturned hemispherical 'scaphe' form used by Eratosthenes that, in turn, developed from the earlier Mesopotamian 'polos', to the armillary spheres that arose out of the Greek armilla.

The polos form was found only in Mesopotamia. It comprised a hemispherical bowl placed to catch the suns rays. Hanging above it was a small ball that cast a shadow into the concave surface. The path of the shadow was used to establish time of day and the time of year.

To this the Greeks added a skeleton sphere to replace the hemisphere. This formed the armillary sphere. It was made up of rings added to suit the need. These rings, arranged as concentric spheres, provided the positional information about places of interest in the universe. Needham (1959) Section 20, provides an account of Western and Chinese armillary spheres. Their design, although basically the same at the fundamental level, varies widely; some were most intricate and very complex. Fig. 3.34 shows one of the earliest design arrangements, that used by Ptolemy. The size of armillary spheres ranges from desk-top devices to those having circles of more than a metre in diameter. Many were so well built that they survived hundreds of years in the open.

Needham (1959) Section 20, has traced how the armillary sphere instrument went through a significant transition. He suggested that a 'torquetum' (an astronomical instrument, once described aptly as 'a collection of astrolabes set at different axes') designed by Kuo Shou-Ching, Fig. 3.35, constructed around 1270 AD, was one of the earliest uses of the equatorial mount later adopted for the mounting of telescopes.

Portable sundials and night dials were also developed. One design, known as the nocturnal in the 16th century AD, is shown in Fig. 3.36. These, like the armillary spheres can be used to plot positions or alternatively to locate them. Early forms were made with a template that fitted concentrically with the sighting hole. Needham (1959) Section 20, has explained how these were used. He also traced their development from as early as possible (1000 BC). Full appreciation of the operation of these and many other astronomical instruments requires a good understanding of the star fields and of astronomical theory. A definitive work on the early development of astronomical theories is that compiled by Neugebauer (1975).

Probably the largest instruments in the 'sundial' class were the giant observatory structures erected to cast shadows from gnomons as high as 20m onto 50m long horizontal scales levelled by water troughs. These structures were built large to increase resolution. They also had the advantage that they were a very permanent record of shadows; transits marked on their scales were reasonably certain of lasting over centuries. Fig. 3.37 shows the first of several complexes that were erected throughout India in the early 18th century, Bholanath (1970). Such structures were common in the far-eastern countries. With such large scales, the resolution of an observation was raised to a part in 100 000 or more, but this did not mean that the accuracy was increased to the same degree.

We have seen how simple instruments gradually were given scales so that quantification was available to an improved degree with time. Interwoven into this development were devices that combined a sighting tube with a divided circle – those for direct observation of a chosen area. In the entrance to the Hall of Physical Sciences of the Museum of History and Technology at the Smithsonian Institution in Washington DC USA is a reconstructed 2nd century AD Greek courtyard as it would have been set up

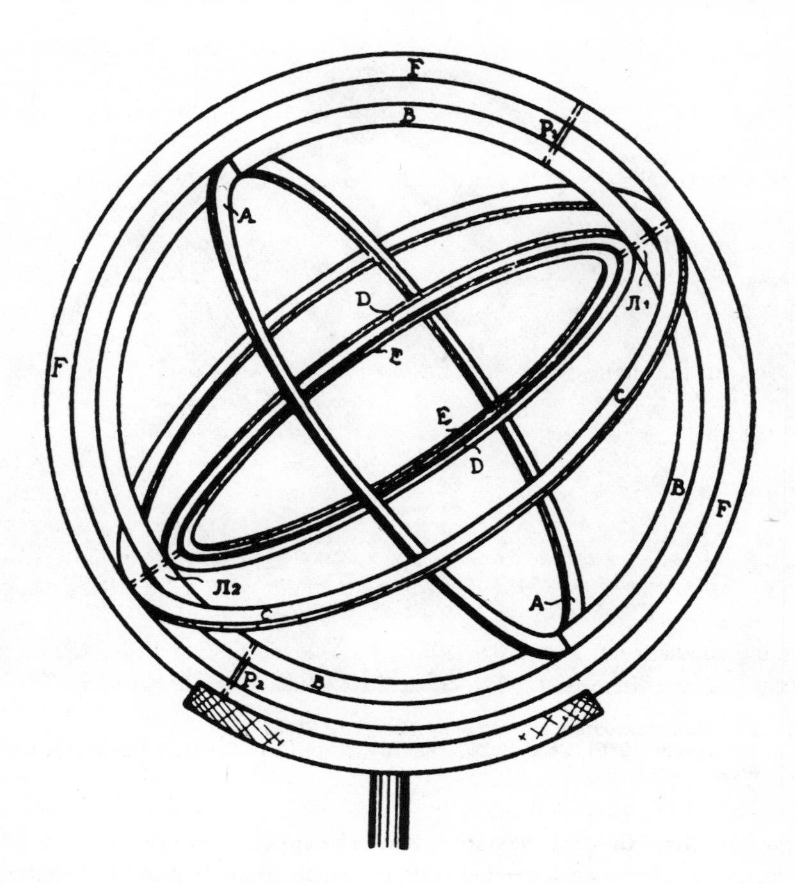

Fig. 3.34 *Reconstruction of Ptolemy's armillary sphere*
[Needham (1959), Section 20, Fig. 154, p.341. Courtesy Cambridge University Press]

for astronomical observations. From Fig. 3.38, a view into this courtyard, the observer can be seen to be sighting the heavens with a sighting stick using a block at each end to define the sight line. Having established a certain line in space he then would have recorded its direction, presumably as the lengths of the two moveable timbers. Simple apparatus like this was widely used in the earliest times of astronomy. The sighting 'tube' would have been in many forms, like that shown above or as a tube proper. The virtue of a tube was that it excluded extraneous background radiation enabling the eye to resolve fainter objects.

The obvious improvement was to pivot the sighting tube, turning it about a graduated scale. Many such instruments have been shown to have existed. Thus was begun the line of development that would eventually lead to many direct observation instruments for example, the telescope with its scales, the sextant, the theodolite, the plane table apparatus and the armillaries and torquetums already discussed above.

Where instruments used a quarter of a graduated circle they were called quadrants. Octants and sextants are strictly those instruments having the appropriate sector size

Fig. 3.35 *Equatorial torquetum by Kuo Shoy-Ching c1270 AD*
[Needham (1959), Section 20, Fig. 164, plate L III. Courtesy Cambridge University Press]

but in the later times the term 'sextant' came to be applied to all types.

The quadrant comprised a divided scale on which rotated, about the centre of the radial scale, a sighting bar called the 'alidade'. They were either portable types, usually made of brass, or they took the form of permanent structures built into the fabric of an observatory. Each served the purpose of enabling angles between two sight lines to be measured. Portable kinds were used for astronomical, surveying and, later, artillery aiming uses. Some were used with the scale plane held horizontally, perhaps attached to some form of levelling arrangement; others had a plumb line attached to allow them to be set up with the scale orientated in the vertical in the manner of a plummet. Such instruments were used by Guillaume de St. Cloud to determine solstitial altitudes of the sun and the obliquity of the eliptic in 1290 AD, Crombie (1969). Well made, precision, portable quadrants became generally available by the 2nd millenium AD.

Add to the vertical form of quadrant a method to establish its scale relative to the vertical (or horizontal) and a 'rete' or 'spider' that lay over a changeable plate containing positional information about the positions of the stars and the result is the astrolabe. The name derives from the Greek for 'to take a star'. Early forms were also of spherical form.

Astrolabes were extensively used throughout the Western and the Eastern world. They originated, as such, in the Greek era and continued in common use until around

Fig. 3.36 *Nocturnal of about 1700 AD*
[Crown copyright. Science Museum, London]

1670 AD in Europe and later still, in Arab countries, into the 19th century. The Oxford University, Museum of History of Science, contains close to 100 astrolabes from many countries; Persian, Egyptian, Moorish, Spanish, Italian, Flemish, French, German and English examples are held, ranging in age from around 1000 AD to the last made in England, a paper instrument, refer to Gunther (1935). A fine specimen from the Science Museum, London is shown in Fig. 3.39.

The first description of a scientific instrument, to be written in the English language, was that by Chaucer, *Treatise on the astrolabe*, written in the 14th century, Crombie (1969). Astrolabe design represents the epitome of sophisticated instrument design and execution for the period up to 1500 AD. Their intricacy of scales and their exquisite workmanship represent the height of instrument ability that man could achieve in those times. The astrolabe was called the 'mathematical jewel'. In 1512 AD

Fig. 3.37 *Astronomical observatory, New Delhi, 18th century. It was built in the tradition of giant observatories whereby resolution was increased without using optical magnification*

Fig. 3.38 *Reconstruction of third century AD Greek courtyard set up for taking astronomical observations*
[Courtesy National Museum of History and Technology, Smithsonian Institution, Washington DC Photo 74-2075]

Johannes Stoffler published a 'Treatise on the astrolabe and on surveying'. He was so highly regarded that his prognostications, about a meeting of three planets close in space that would bring a great wet, caused one man of note to build a Noah's Ark (In the year concerned, 1524 AD, it was drought that came, not rain!).

Although the astrolabe was a most ingenious device and could be made to quite tight tolerances by the 16th century it was, nevertheless, a very crude instrument in terms of performance. That the early explorers had only these and crude clocks to guide them on their epic journeys must remain a source of astonishment to us today. The mariner's form of astrolabe is more correctly called the astrolage. It was simpler than the astronomical model and was the forerunner of the sextant of today. Pearsall (1974) although concentrating on instruments that can be collected today and, therefore, are not as old as the period of this chapter, provides an introduction to the various forms of astronomical and surveying instruments. Wynter and Turner (1975) provide a more pictorial insight showing the beauty of many of these pieces of apparatus.

One other important observing instrument that made a considerable impact on navigation was, of course, the magnetic compass. The magnetic properties of the iron-ore, lodestone, were known to the Greeks. Needham (1962) Section 26, traces similar knowledge back into Chinese history. The first use of lodestone or a magnetised needle to sense direction for purposes of navigation is not easy to establish for although there exists a considerable amount of literature on the subject, it cannot all be interpreted as showing that the magnet was used in this way. The difficulty is that practical use and the experimental scientific interest progressed closely together. Suffice to say that the earliest written accounts in the medieval literature of Europe appeared around 1200 AD, refer Crombie (1969), in such works as *De naturis rerum* by Alexander Neckam. In the Chinese literature works of 1000 AD, and before, there is considerable evidence to suggest that floating needles may have served as compasses as much as a thousand years before that. Fig. 3.40 shows the floating and the dry suspension forms of *c*1050 AD Chinese origin magnetic compasses. De Solla Price (1959b) includes discussion of the origin of the magnetic compass.

Widespread opinion suggests that the compass was in general use to assist navigation much later than this, dates given varying from the 13th century AD onwards. The *Epistola de magnete* by Petrus Peregrinus, 1269 AD, described improved forms of wet and dry magnetic compasses. Gunther (1935) stated that a magnetised needle supported by a piece of wood was used by seamen in the eastern Mediterranean in 1242 AD. The significance of the early name for the magnetic compass the 'calamita', which may have come from the Greek word for frog, has been discussed by Needham (1962) Section 26. As pointed out by Pearsall (1974) the compass did not change much during the many centuries following its general adoption in the European areas of the world.

To complete, what can only be a brief visit to the observing instruments of early astronomy and surveying, we now consider the cross-staff. This instrument was introduced in 1342 by Levi ben Gerson, possibly from earlier work of Jacob ben Mahir from a century before. In its early form it was a central bar having only one sliding

Fig. 3.39 *Obverse of a French astrolabe, late 15th century, that has geared cog-wheels*
[Crown copyright. Science Museum, London]

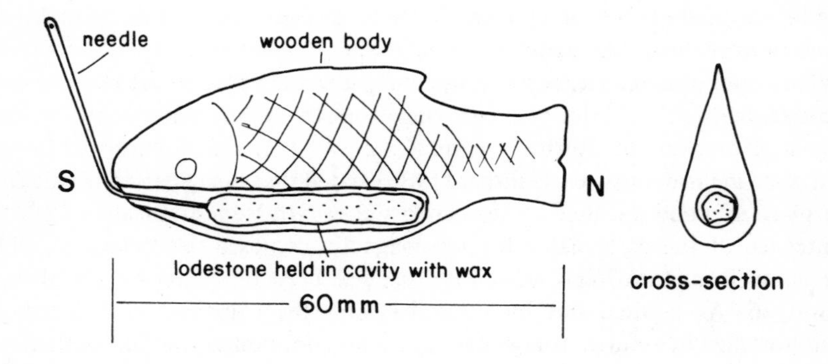

Fig. 3.40 *Cross-sections of floating, wooden fish and dry pivoted turtle shape magnetic compasses of Chinese origin c1150 AD*
[Needham (1962), Section 26, Fig. 324, p.256 and Fig. 325, p.257. Courtesy Cambridge University Press]

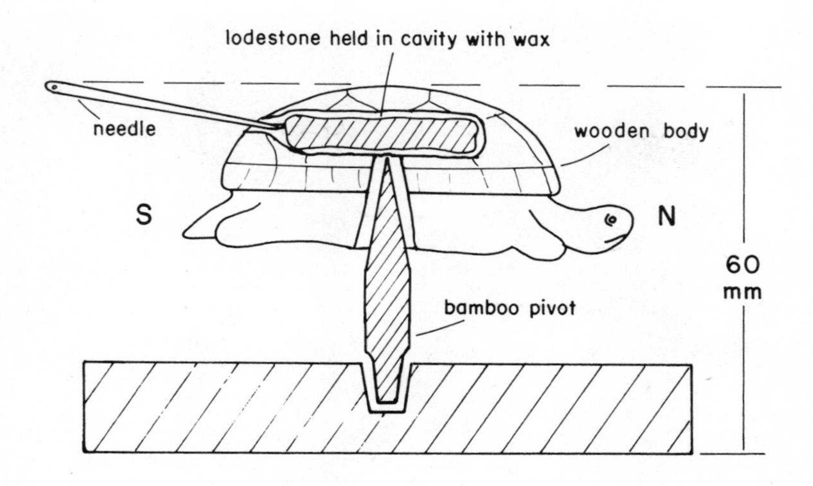

Fig. 3.40 *(continued)*

crosspiece. The central bar was engraved with a scale that enabled altitude to be read off after the end of the stick and the tip of the crosspiece were adjusted until the sun was seen coincident with these two while the cross-staff main-shaft was held horizontal. It was used at noon to 'shoot the sun' and was also used to observe the stars. Fig. 3.41 shows a modern replica based on an instrument made originally by Thomas Tuttell. A picture of this appears in Tuttel's trade card, shown earlier as Fig. 2.19. That illustration clearly shows that the instruments of the period of this chapter were still of significant importance many centuries later. In it can be seen the cross-staff, quadrant, plummet, astrolabe, armillary, square, magnetic compass and instruments that are to be considered next, those for storing information about the heavens and for modelling the celestial movements, i.e. globes, orreries and planetaria.

It was apparent to the early astronomers and surveyors that they could derive benefit from physical models representing what they observed in the sky and on land. The first of these were the star and land maps that began as crude sketches showing the salient positions on plane and spherical formats.

Maps have been found from Babylonian times that show that their concept of the Universe was that the Earth was a disk floating in an ocean with Babylon in the centre. A map formed on a clay tablet has been found of the Sumerian city of Nippur. It was accepted that a spherical space contained the sky, but the celestial maps were sometimes produced as spheres flattened out, the planisphere. A Babylonian one exists, in part, from *c*1200 BC. Later forms were produced in stone, formed in bronze for use by navigators or printed on silk. Needham (1959) Section 20, contains illustrations of different types. Astronomical records have been found on clay tablets and papyrus . . . Singer *et al.* (1954).

Another development, that may have its origins again in Babylonian times but certainly existed in the time of the Greeks, was to put the information on a solid

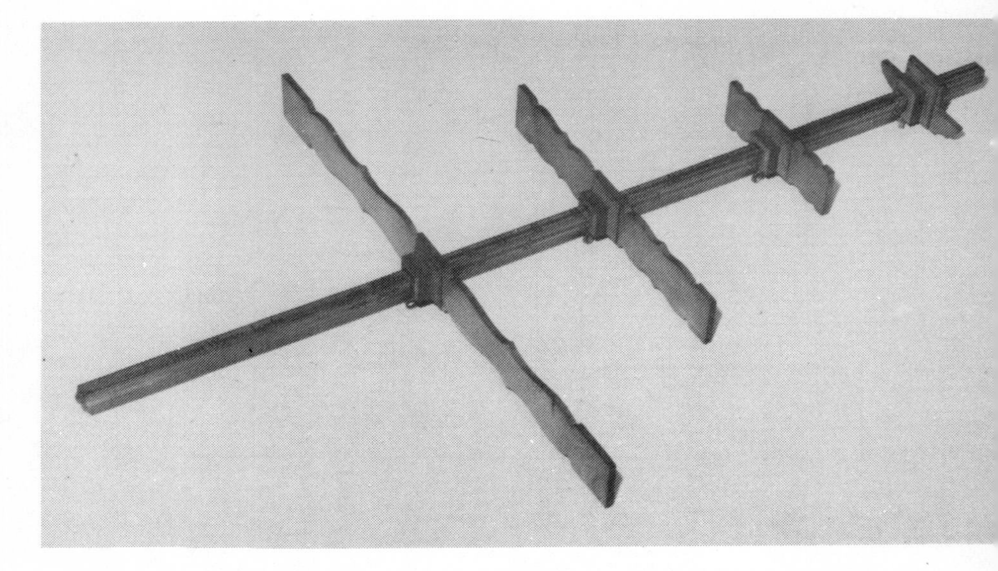

Fig. 3.41 *Replica of cross-staff originally manufactured by Thomas Tuttell* [Courtesy Harriet Wynter Ltd.]

sphere i.e. the celestial globe. Needham gives information of sources about globes that date them back to a possible 350 BC and certainly to 270 BC. These were often pivoted for rotation and were made of stone or copper. The terrestial globe came much later for it first had to be resolved if the earth was flat or spherical.

A logical development from the simple celestial globe was the addition of a clock-work, water or weight-driven drive. Needham mentions many such drives of Chinese origin around 1000 AD. One step further was to form a multilayer model in which the planets and the stars of interest rotated as they appear to in the heavens. Thus developed the planetarium in its mechanical form (optical projection, of course, came only in recent times). These could be used to demonstrate the motions of the heavens for purposes of demonstration and for 'research'. They also were able to provide calendrical information. The model would be set to the current known state of the universe and then run on at scaled up speed to decide just where, say the moon, would be in so many days of years.

Evidence exists that suggests that Archimedes devised a planetarial device. It showed the motions of the sun, moon and five stars. Cohen and Drabkin (1948) contains translated passages about this contrivance.

In 1900 AD sponge divers – see De Solla Price (1959*a*, 1974) – found a corroded metal device that subsequently, by the use of modern X-ray and gamma-radiographic methods coupled with considerable scholarship, has been shown 'to be a calendrical computing device in the tradition of planetarium construction founded by Archimedes', Fig. 3.42. It was named the Antikythera mechanism after the place where it was found. It is thought to have been built in Rhodes around 80 BC. The mechanism uses epicyclic, differential gearing. It does not appear to be the same unit as that of Archimedes, Cohen and Drabkin (1948), but was certainly a kind of astronomical computing mechanism. It demonstrates the use of very advanced mechanical principles for its time, so advanced that the Antikythera mechanism stands out as an incredible and almost unbelievable step in progress.

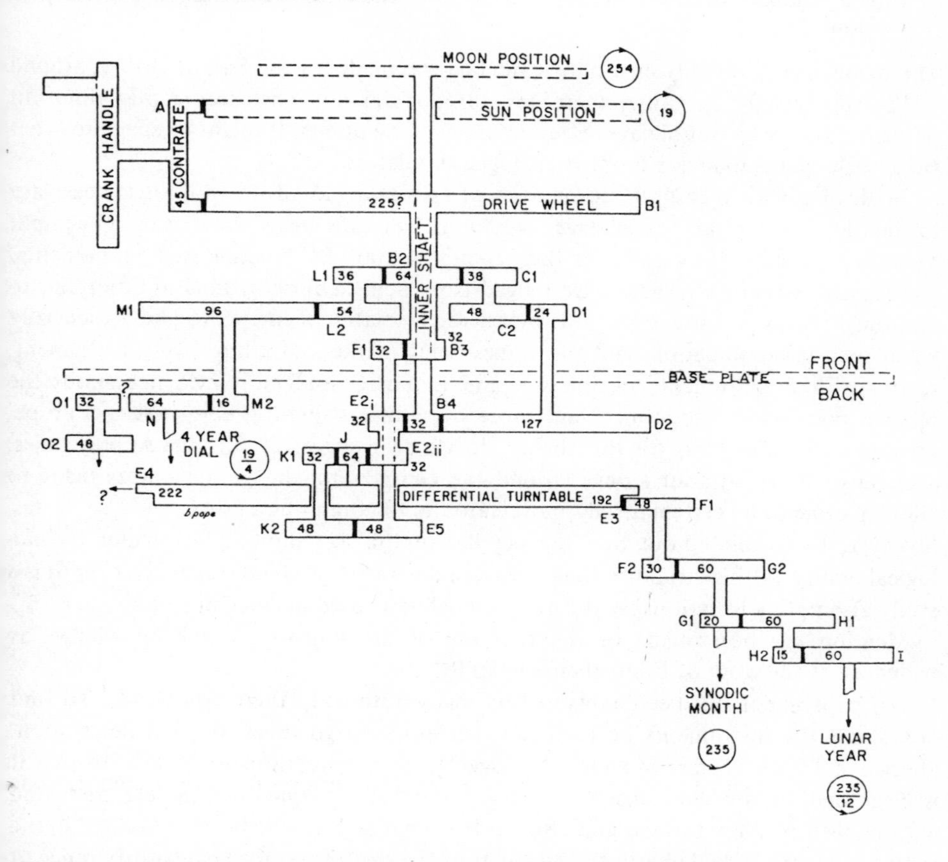

Fig. 3.42 *Gear train of the Antikythera mechanism – first analogue simulator?*
[De Solla Price (1974), Fig.33 p.43. Courtesy American Philosophical Society]

This basic model was followed over time by dynamic armillary devices and then by what became known as the orrery in the 18th century AD. Gunther's (1935) description went as follows:

> 'An orrery is a machine for representing by wheel-work the various motions of the heavenly bodies. This machine differs from a Planetarium in that it exhibits diurnal as well as the annual motions of the earth, the revolution of our moon, and sometimes the rotation of the sun and of certain planets on their axes. The larger instruments, called 'Grand Orreries', exhibit, moreover, the motions of the secondary plants, but more particularly those of Jupiter.

> By the orrery it is possible to represent all the vicissitudes of summer and winter, of spring and autumn, of day and night, the risings, settings, and culminating of the heavenly bodies, together with their constantly varying altitudes and azimuths, their right ascensions, declinations, and amplitudes, their conjunctions and oppositions, their transits and occultations or eclipses. The orrery is therefore a machine of the first importance as a means of education in planetary motions.'

The name was adopted from a time when one was made for the Earl of Orrery, around 1712 AD. Strictly speaking, therefore, orreries did not exist before that time but popular usage does not always differentiate the fine points. It might be suggested that these early planetaria were the first analogue simulators.

In dealing with recording mechanisms of astronomical information, it is necessary to mention the famous Stonehenge structure. Over time many theories have been put forward to explain the reason for the peoples of 2000 BC erecting such monumental and massive structures. These have ranged from being a burial ground of Chieftains to sacrificial places and other religious practices. The latter seems more likely especially when seen in combination with the stones being markers of a huge, very permanent record of the movements of the sun and moon. Considerable evidence about the relative position of the stones, and those of similar structures elsewhere in Europe provide a plausible basis for this theory. It has been suggested that the Aubrey holes, a circle of holes without stones around the central Stonehenge area, were made to allow a stone to be systematically moved around to count out a 56-year eclipse. It has, however, been pointed-out that the peoples of this age showed far greater technological ability in this structure than they demonstrated elswhere. Lorenzen (1966) is a study attempting to rationalise the dimensional units used in Ancient times.

Mention has been made of the first use of instruments to seek knowledge, as evidenced in the work of Eratosthenes *c*250 BC.

His experiment has been explained by many authors in their own words. To find out about the instruments he used it is necessary to go to an original description. Sherwood-Taylor (1945) provides a passage from a translation by Heath. In that it is clear that Eratosthenes made use of a scaphe, an upturned hemisphere graduated with concentric rings around and also up the bowl and in which was mounted, in the centre, a gnomon. He deduced the diameter of the earth from the angular difference of the sun's shadows between two places some 1000km apart – see Fig. 3.43. It is left to the reader to obtain detail from Sherwood-Taylor (1945) or from a modern interpretation of the event by Hogben (1938).

It would be incorrect to regard Eratosthenes as the only person of his time to use instruments in scientific pursuits. Hogben (1938) explains many other measurements deduced from astronomical observations.

Aristarchus, for instance, used the 'parallax method' to realise the ratio of the diameter of the sun to that of the moon. Hipparchus also conducted studies aimed at similar findings. The precision of the parallax method, however, rests on the ability to observe the half size of the moon's face. Without optical magnification the method is not capable of a reasonable answer. Many mistakes were made by many observers. Somehow, however, they managed to obtain values that were similar, thereby reinforc-

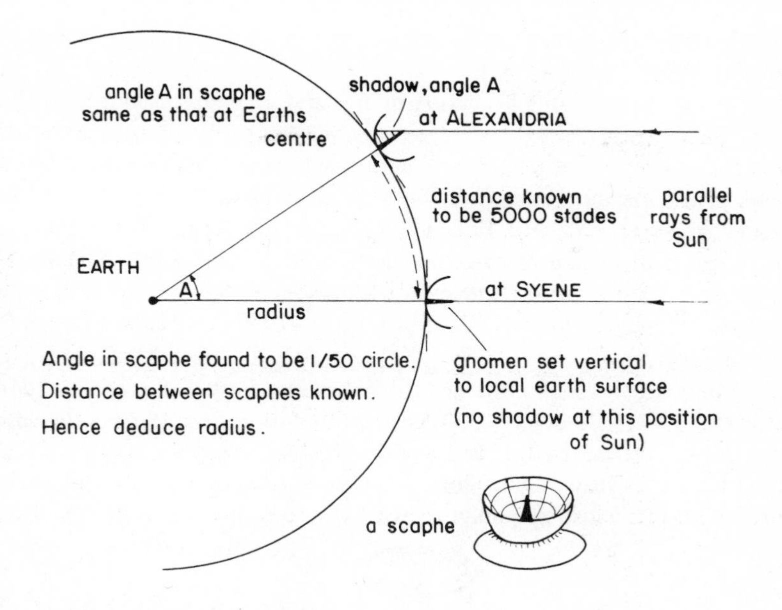

Fig. 3.43 *How Eratosthenes measured the diameter of the Earth in c250 BC* [Sherwood-Taylor (1945), Fig.15, p.35]

ing the validity of some of the wildly incorrect values.

It is always necessary to remember that scientific advance usually hinges on often seemingly unimportant work of others. Such was the case for the Greek astronomers. Eratosthenes was not alone in realising that instruments could be used to seek knowledge. Draper and Lockwood (1939) provides background to the development of astronomy as it progressed from simple to complex phases of instrumentation.

3.8 The beginnings of other instrumentation

So far we have only covered the dominant uses of instruments of early times, those that were involved with the main streams of need. When the many pieces of apparatus, even up to the Alexandrian times, are reviewed collectively, as has been done here it becomes clear that there existed a greater capability in inventiveness and craftmanship than many writers of the past have given the ancient peoples credit for. Furthermore, it has not all been described, by far. To finish this review of early times attention is now devoted to other areas of interest which produced requirements for measuring instruments.

Mechanics

In the field of mechanics and hydraulics wheels, gears (spur, worm, epicyclic), levers, wedges, flexible hinges, engraved scales, inclined planes, pulleys, capstan wheels, intermittent motion devices, elastic elements, siphons, hydraulic presses, pneumatic cylinders, steam turbines, double-acting pumps and pressure cooking are all evident in antiquity. Such principles were used to assemble many inventions, some have been mentioned already. Authors who have assembled specific statements about the technical capabilities of this period include Burstall (1970), Klemm (1959), Neuburger (1930),and Singer *et al*. (1956). Cohen and Drabkin (1948) and Hett (1955) are useful sources of relevant original writings that have been translated into English. Needham (1965) Section 27, in covering Chinese developments, provides much about others in the course of his explanations. De Solla Price (1959*b*) discusses fine mechanical devices in ancient times.

Many works cite the coin-in-the-slot the holy water dispenser of the 1st century AD, shown in Fig. 3.44. Neuburger (1930) gives a translation from Heron's *Pneumatica* but no illustrations have survived of this device. This was, however, a fairly simple device compared to the feedback controlling mechanisms used to regulate the level of oil in oil-lamps, in an inexhaustible goblet of wine or water, in other wine dispensers and in

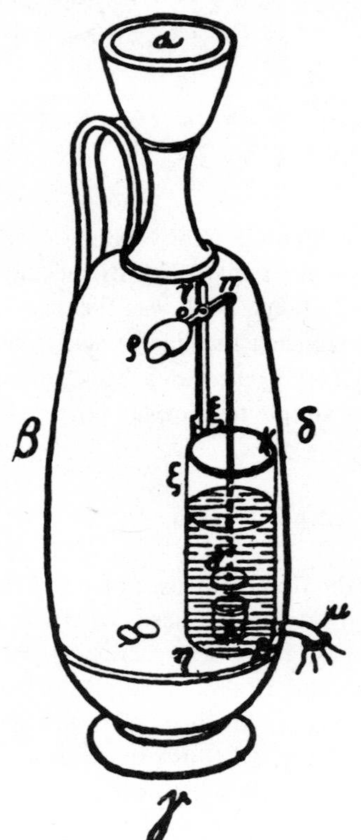

Fig. 3.44 *Automatic dispenser of holy water by Heron of Alexandria. This version from* [Neuburger (1930), Fig. 264, p.206. Copyright R. Voigtlander's Verlag]

clepsydrae. An intensive study of the original literature, such as it exists for much was already secondary by the Middle Ages, on the origins of feedback control has been carried-out and published by Mayr (1970). His work contains many illustrations of such devices made in early Greek and Arab times.

A very spectacular mechanical device, often erroneously confused with the hodometer already mentioned and with magnetic direction sensing, is the south-pointing carriage of Chinese origin. Needham (1965) Section 27, gives considerable detail about these mechanisms. He ascribes their date of origin as around 200 AD. Within the strict context of this book these are not measuring instruments but they do, as have other nonmeasurement devices discussed here, contain principles that are directly relevant to instruments of later times. These carriages appear to have contained differential gearing (the exact nature of which is uncertain) arranged so that, regardless of the direction of movement of the carriage, the figure maintained its arm pointing south or in whichever direction it was first set. Practical limitations would have given rise to error with continued movement but tests conducted on a reconstruction, Lanchester (1947) (Fig. 3.45) shows his arrangement, proved that the device could have been very

Fig. 3.45 *Second century AD, Chinese, south-pointing carriage. Model as reconstructed by Lanchester*
[Crown copyright. Science Museum, London]

Fig. 3.46 *Cardan suspension used to stabilise the sights of a navigator. A proposal by Besson, 1567 AD*
[Needham (1965), Section 27, Fig. 474, p.230, from 1567 AD etching. Courtesy Cambridge University Press]

effective on reasonably smooth surfaces. An account of 500 AD specifically indicated that these were not toys for amusement but practical devices to assist envoys keep track of direction when journeying across great open plains. It pointed to the South because the Chinese in those times used that direction in preference to our North convention. Many were made over the period 200-1100 AD, later versions being of bronze and often combining the hodometer principle to measure distance travelled. The south-pointing carriage was certainly the forerunner of automatic direction guidance controllers.

In 1550 AD Jerome Cardan described a method of suspending an object in which gravity forces are used to keep the object hanging in the same vertical direction. Cardan did not claim this as his own device (but it took his name) and it certainly had been in use at least as early as 180 AD. (see Needham (1965) Section 27) in the form of the suspension ring system of Ting Huan. Today these are known as gymbals and are used, for example, to support ship's compasses and chronometers and to transport sensitive photodetectors that must be kept from inverting. Fig. 3.46 shows an interesting suggested use put forward just later than the period of this chapter.

The origins of mechanical calculators can also be traced to these times. The word calculate stems from the Latin for 'pebble'. This connotation was drawn from the early use of the sand table in which troughs drawn in sand were used to hold pebbles as digits in each significant position of the number system. The abacus or counting-frame came next; They were known to the early Egyptians, Romans and Chinese. In some parts of the current world the abacus is still used as a rapid and effective means to perform calculations at unexpectedly high speed. Mention has already been made of the several calendrical computing devices.

Seismology

The ancient world had a vivid and direct experience of earthquakes and volcanoes and many theories were put forward for their origin. A brief history of seismology has been published, Wartnaby (1957). Earthquakes have always been a source of consternation to the Chinese and with the skills of invention shown in early times it is not surprising to learn that they invented the first seismic activity indicator.

Late in his lifetime (78-139 AD) Chang Heng constructed a seismoscope which he called the 'earthquake weathercock'. A picture of the reconstruction according to Li Shan-Pang, is given as Fig. 3.47. The device would drop a ball into the appropriate toad's mouth according to the direction of the seismic wave source. Exactly what was

Fig. 3.47 *Chang-Hêng seismoscope from the second century AD – cut away to show interior* [Crown copyright. Science Museum, London]

inside the instrument is unknown. Li Shan-Pang reconstructed this instrument with a hanging pendulum that released a ball according to which way it was displaced but Imamura opted for an inverted pendulum mechanism in his reconstruction. As with much of the instrumentation of antiquity we may never know the exact details. Needham (1959) Section 24, presents the arguments as a comparative study of depth. He reaches the conclusion that similar instruments were still being made up to the end of the 1st millenium AD, at which time they appear to have probably been lost to man.

Meteorology

In Greek times meteorology contained the study of celestial matters, meteors for instance, as well as weather conditions. In Athens a wind direction indicator was constructed around 100 BC to become the first weather-vane. However, according to Cajori (1962), it was unlikely to have been a common instrument for neither the Greeks nor the Romans had words for it. Rain gauges were introduced much later. A Korean one dates from the 15th century AD; Chinese devices were used in 1442 AD and evidence suggests well developed knowledge existed earlier in 1247 AD. They also had snow gauges in that period.

Considerable use was made of the hygroscopic nature of natural materials such as feathers, charcoal, wool, and cereal crop seeds to give an indication of what is now called the relative humidity of the air. Details of a specific instrument that could be called a hygrometer, however, first date to later in the 15th century. Knowles-Middleton (1969*a*) describes the working principle of models by Leon Battista Alberti and by Nicolaus of Cues from that period. Both used a water-absorbing substance (wool or sponge) placed in one side of a beambalance. Cajori (1962) gives a quotation about the device by Nicolaus of Cues. This principle was still in use in the late 18th century . . . see Fig. 4.45.

The earliest device that might have served as a thermometer of ambient temperature would seem to be the air thermoscopes of Hero of Alexandria and Philo of a similar time. Philo's device, reconstructed in the Science Museum, London, was made of a leaden globe that contained water. This is shown in Fig. 3.48. As the water temperature changed it caused an observable flow of water. A description is given in Chaldecott (1976) along with that of the also reconstructed, more complex, but similar, device by Hero. The important feature about these early devices was that they were also disturbed by the ambient air pressure. Little more is available about thermometers until they were re-established in the same form by Galileo 1592 AD. Knowles-Middleton (1969*a,b*), Biswas (1970) and Frisinger (1977) each give accounts of the historial development of meterological instruments.

Optics

Astronomy certainly had some influence in the scientific study of optical laws but did not, in early times, result in much knowledge becoming available about fundamentals.

Fig. 3.48 *Reconstruction of the thermoscope of Philo*
[Crown copyright, Science Museum, London]

It is more likely that casual curiosity about simple elements, such as mirrors and rock-crystal lenses, sparked interest in such matters as the laws of reflection and refraction.

Looking-glass mirrors have been found from the Egyptian times and on. Burning glasses were appreciated as such, but apparently not as magnifying elements. It is said that a lens was found in the ruins of Ninevah. Aristophanes, in a play performed in 424 BC, included a conversation about a 'fine transparent stone with which fires are kindled' and which 'though at a distance, melt all the writing (in wax)'. Chinese literature contains mention of bronze mirrors, the earliest being Needham (1962) Section 26, in 672 BC. The oldest extant mirror from China is dated at 6 AD.

Hero's *De speculis* and *Optics* by Ptolemy gradually expanded the laws of reflection and refraction but rather imperfectly. They held the view that light was a substance that went from the eye of the observer to the subject. Their work was extended notably by Alhazen, a millenium later in the Arab world. He established that the rays of incidence and reflection lie in the same plane and that rays bunched together concentrate the heat from the sun. The first study of the eye which resulted in a drawing was by him. Some of his terminology is in use today e.g. the cornea, retina, vitreous humor.

Spectacles were first thought to have been used around 1300 AD in Italy. Magnifying glasses of rock were likely to have been used in 1100 AD China. Alhazen is said, Marmery (1895), to have pointed out the magnifying property of lenses.

By 1500 AD enough of the basic concepts, of what are now called 'Gaussian optics', were understood allowing the lens to emerge as a magnifying element (by Lippershey in 1608 AD) from which would be learned the laws of simple ray-tracing procedures. It is thought, by some, that Roger Bacon (from a passage dated to 1269 AD) experimented with a crude telescope. Following is the passage, plus Sherwood Taylor's (1945) footnote about it.

> And when we wish, things far off can be seen as near, and *vice versa*, so that at an incredible distance we might see grains of sand and small letters and the lowest things may appear high and *vice versa*,* and hidden things be seen openly and open things be hidden, and one thing may be seen as many and *vice versa*, so that many suns and moons may be seen by the artifice of this kind of geometry . . .
>
> Bacon 1269 AD

* It does not seem improbable that Bacon in experimenting with lenses made a crude Keplerian telescope of two convex lenses: this would magnify and invert as he asserts. He probably had polyhedral lenses which give many images, just as he describes.

Sherwood Taylor (1945)

Whereas it may, one day, be proven that the optical magnifier was used prior to the time of Lippershey it is most unlikely that it will be shown that such instruments were commonly known and freely available. As well as existing as a practical device they had to be promoted to a point of acceptance. Such a process is most unlikely to have gone unchronicled in some way.

Electricity

Instrumentation of the 20th century AD relies mainly on mechanical techniques to provide structures and mechanisms plus electrical techniques to convert the energy forms into a universal regime, that of electricity (including electronics which is becoming increasingly difficult to distinguish from electricity), for transmission, conversion, processing and computing purposes. Other physical principles are, of course, used but they are less dominant.

It is, therefore, a little surprising to learn that electrical fundamentals form one of the newest bodies of knowledge, coming into real prominence only in the late 18th century AD and after. It is indicative that prime sources about physics and engineering in antique times, such as the works of Needham and Singer, give almost no detail of ancient understandings of electrical phenomena. The simple fact is that little was known to those people.

It is possible to trace knowledge of statical electric charge back to Thales of Miletus (640-546 BC). He discovered that amber, when rubbed with a suitable material, could attract small things. From that event the Greek word for amber, electron, came into use to describe the effect. That scant finding, however, was not expanded upon until the 18th century AD, when the two types of charge (now called positive and negative) were discovered by Dufay, a garden director of Louis XV. Morgan (1953) suggested that electricity was not learned about because it did not readily appear to the early people, but this is not true. They did observe lightning, they did know how to use the torpedo fish to administer electric shocks for therapeutic uses and they did perform surgical work on tissue with metals. They, however, did lack means to observe its properties and none were forthcoming for several centuries. What might, one day, be resolved, is why, with the same means as were available to 18th-century scientists, electricity was not revealed. It would seem that no artefact exists today that was concerned with the measurement of experimentation of electricity before the early 18th century AD. Collections of electrostatic instruments have their earliest artefacts dating around 1730 AD. Chipman (1964) summed up the situation in the opening paragraph of his paper on the earliest electrical instruments. Anything before 1800 AD would have to have been electrostatic in nature. As there were no enunciated principles in either a theoretical or a practical form there could be no instruments. He was apparently satisfied to begin at 1800 AD. Knowing that the ancients had knowledge of attraction and repulsion by charged bodies it is surprising that a simple device, such as pith-balls on strings, was not used even if only as a toy. Perhaps it was? It is not possible to provide an illustration of an electrical apparatus of the period in question for none appears to exist.

The 20th-century entry into electronics, however, owes its origins to more than just charged amber in antiquity. Notions about the composition of matter began then. Atomic theory had advocates in Greece. It is thought that Leucippus first originated the theory. His work is, however, indistinguishable from that of Democritus of Abdera (*c*460-370 BC) who 'taught that the world consists of empty space and an infinite number of indivisible, invisibly small atoms'. The theory, however, had to be rediscovered by Dalton centuries later and lay dormant until then. Still (1944) provides historical background to the early times of electrical-science.

Thus we come to the end of the account of measuring instruments and similar apparatus that were devised and created by men of times up to 1500 AD. It has been shown that the range is far greater than might have been expected and that by the Alexandrian times man was almost at a point of breakthrough into developments that were then delayed by 1500 years. It has been said by many historians that the ancients were not good scientists for they either, (if Greek) placed too little attention on observation or, alternatively (if Roman) placed too much emphasis on just using known knowledge. Perhaps it has been shown here that these ideas need revision in the light of those advances that we now know about. There is, however, need to be cautious about the interpretation of ancient accounts. Many of the designs that are stated in modern popular expositions give the impression that the reconstructions, and most of the devices of this period are those, are absolutely correct. There is room for doubt in many instances: this must be recognised.

It is clear that much of today's instrumentation had its origins in this period. The notable exceptions are instruments or their features, that are based upon electrical principles. Needham (1962) Section 26 has the following passage on the fly sheet to his work:

The balance and the steelyard, the square and the compasses, are fixed in a uniform and unvarying manner. Neither the people of Chhin nor Chhu can change their properties, neither the northern Hu nor the southern Yueh can modify their form. These things are for ever the same . . . . A single day formed them; ten thousand generations propagate them.

Huai Nan Tzu
2nd century BC

Chapter 4

Experimental science becomes established: Middle Ages–1800 AD

4.1 Emphasis on observation

The period preceeding that now considered can be regarded as one where sensing was performed using only the natural senses given to man. With a few exceptions, the seismoscope of Chang-Heng is one, no use was made of aids to enhance or modify man's natural senses in the observation process. Early man discovered those effects that did not require additional apparatus to interface him with the phenomena of interest.

A second epoch emerged when apparatus began to be devised to enable man to observe the physical world at levels of resolutions that cannot normally be seen without devices to alter the scale of the parameter and in energy regimes differing from those of sight, hearing and touch.

Thus the arrival of the telescope, and soon after the microscope, in the early 1600s AD (hereafter 'AD' will be omitted; there will be little need to refer to BC dates) heralded the start of a new era in observation, one in which man was given greater potential to witness the physical world. Through optical magnification he was able to observe smaller-magnitude effects, meaning he could observe more distant heavenly bodies, smaller stars and look into the finer structure of matter. Providing the eye with a magnifier, Fig. 4.1, having a gain of ten can increase, subject to adequate quality of image, the capture of objects in the field of vision of three-dimensional space by a nominal 1000 times. A reasonably simple and unsophisticated instrument provided the key for man to enter previously unknown domains of knowledge.

A similar situation was to arise around 1800 when electrical sensing gained a firm footing through the development of the indicating instrument.

Over the three centuries of interest here the new method of scientific enquiry became established. Natural philosophy became the accepted way to practise science, one that is still current today. We have previously seen how observation plays an important part in scientific and technological endeavours. Many writers have suggested that observation was not given due emphasis in the ancient approach to science but it has been shown in the previous Chapter that man had been developing instruments over the period at a gradually increasing rate and that it would seem that this continual progress was halted at a time (Alexandrian decline) when even greater expansion

JAMES AYSCOUGH,

OPTICIAN,

At the Great GOLDEN SPECTACLES, in *Ludgate-Street,* near St. PAUL'S, *LONDON,*

(*Removed from Sir* ISAAC NEWTON'S HEAD *in the ſame Street*)

AKES and SELLS, (Wholeſale and Retail) SPECTACLES and READING-GLASSES, either of *Brazil*-Pebbles, White, Green, or Blue Glaſs, ground after the trueſt Method, ſet in neat and commodious Frames.

CONCAVES for SHORT-SIGHTED PERSONS.

REFLECTING and REFRACTING TELESCOPES of various Lengths, (ſome of which are peculiarly adapted to uſe at Sea;) Double and Single MICROSCOPES, with the lateſt Improvements; PRISMS; CAMERA OBSCURA'S; Concave and Convex SPECULUMS; MAGICK LANTHORNS; OPERA GLASSES; BAROMETERS and THERMOMETERS; SPEAKING and HEARING-TRUMPETS; with all other Sorts of Optical, as well as Mathematical and Philoſophical Inſtruments.

Together with Variety of MAPS, and GLOBES of all Sizes.

Fig. 4.1 *Optical aids provided gain for the eye, extending man's visual acuity. Trade card 1750 from the Science Museum, London collection*
[Photo. Science Museum, London]

had just become possible.

Most advances in instrumentation of the 16th to 18th centuries had their genesis in those early times. The technological breakthroughs that realised the telescope had been prepared by work on optics by many investigators. It appears that the development of the measuring instruments themselves was a steadily widening process having an occasional quickening when a new and powerful principle was discovered and implemented. The dramatic effects in science and technology were more to do with the application of the instruments to problems that had very significant outcomes than to the design features of the instruments.

This period is characterised by widespread adoption of scientific apparatus to a widening range of uses. The basic needs of living, such as weights and measures and time-keeping continued to be important and to be improved. In addition, there now emerged an interest in seeking and applying new principles about phenomena that had no obvious pragmatic use and that would not normally be experienced in the normal course of life.

As it was, in antiquity old ideas were improved upon where needed, new ideas were discovered and applied and the scene set for yet more discoveries and implementations. Knowledge became more easily recorded and transmitted to a wider range of the populace owing to the invention of the moveable-type printing press and to social changes. More leisure time (for some, at least) enabled knowledge to be won at a faster pace. The rise of patrons provided the means for more people to practice the new science and to participate in the arising interest in technological matters.

The result was a broadening of the need for instruments as well as a widening of the types. Special purpose apparatus, built initially in small numbers for a specific test or experiment, became in demand by others. An instrument-manufacturing trade developed and flourished with the help of a not unsubstantial popular market component. The seeds of mass production and machine manufacture of assemblies developed in this period.

A patent system emerged to protect inventors and manufacturers. The earliest was in Venice where the letters patent was established in 1474. An English patent system began in 1561.

Materials needed to build better instruments gradually became more prolific and of better quality – refer to Table 2.2 in Chapter 2, but the demands of instrument-makers did little to influence their supply sources. More significant was the development of the machine tools – refer to Table 2.3, that enabled precision parts to be made in the durable metals. Mechanical ingenuity continued to flower in this period.

Improved machining, coupled with the discovery of many more laws of nature, enabled the range of scientific instruments to be expanded. Optical aids were coupled with mechanical devices to produce previously unknown designs. The dynamic pendulum was married to the mechanical clock to enhance its performance. Rust, a London instrument-maker, coupled a static pendulum to the sextant to provide an artificial horizon, Calvert (1971). Disciplines began to cross-fertilise resulting in a wide range of sensing and measuring hybrids.

It was also realised in this period, sometimes from practice but more often from

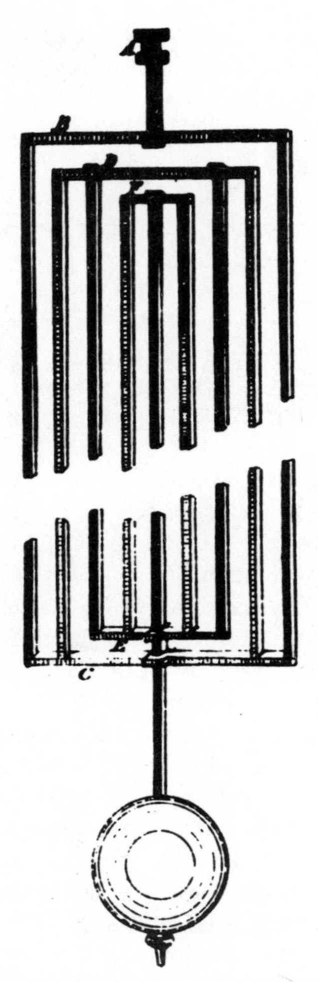

Fig. 4.2 *Suitable combination of brass and iron member provided a pendulum that was less affected in length by ambient temperature changes. 'Second-order' instrument errors were reduced*
[Godfray (1880), Fig.7, p.40]

theoretical understanding, that influencing effects often commonly referred to as 'second-order effects', limited instrument performance. These caused a measuring instrument to be inaccurate and imprecise. It was realised that second-order effects could be reduced by incorporating into the basic instrument design appropriate corrective design features. An example of this is the grid-iron clock pendulum invented by Harrison in 1726 and pictured in Fig. 4.2. Made of brass and steel elements, suitably arranged, the length of the pendulum varied far less with temperature than either a steel or a brass pendulum. As the rate of a pendulum clock is influenced by pendulum length this measure improved time-keeping.

Measuring instruments for testing, as opposed to knowledge-seeking in the purely scientific search, appeared in this period developing rapidly from crude to workable devices. Da Vinci's idea for testing the tensile strength of a wire was rudimentary compared with the testing machine introduced two centuries later by Musschenbroek around 1720. The two are contrasted in Fig. 4.3. Timoshenko (1953) is concerned with the development of strength aspects of materials.

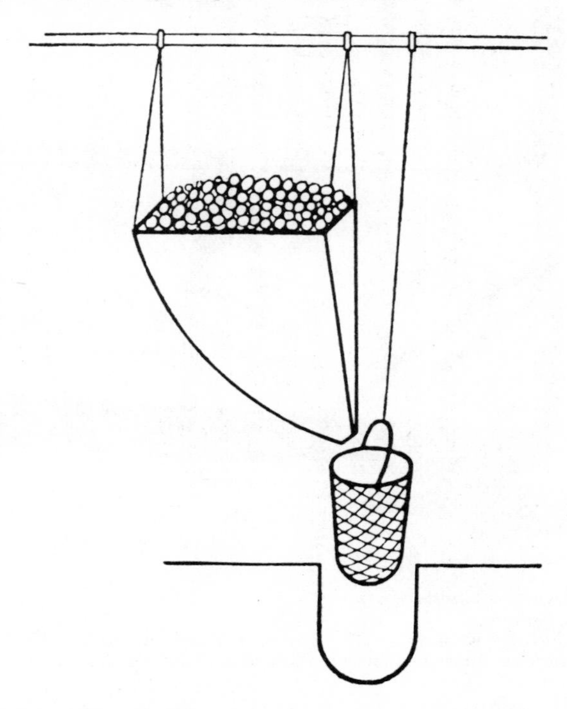

Fig. 4.3 *Tensile testing apparatus*
a da Vinci
[Timoshenko (1953), Fig. 10, p.6. Courtesy McGraw-Hill Book Co.]

Commercial needs were an influencing factor in the development of some instrument lines. The clock and surveying instruments were in high demand. The expansionist exploration activity of the European countries created demands for improved clocks and navigational devices. Later in the period the same need still remained for although the instruments had been significantly improved by then and exploration into the unknown was of less significance, demand continued for reasons of international trade and sea travel. Communicational needs arose for instruments replacing the earlier incentives. Horses were no longer fast enough for message carrying; inventors sought means of providing faster communication. There developed the concept of information signal transfer over an energy carrier medium, that had its beginnings in antiquity as chains of fires and then as semaphore stations (first established by the Arabs). Almost commercially workable electric systems were in existence by

Fig. 4.3 *Tensile testing apparatus*
(*b*) Musschenbroek
['Leids Fysisch Kabinet – III', Boerhave Museum, Leyden, Netherlands. From van Musschenbroek 'Physical experimentales et geometricae dissertationes' *c*1740]

the end of the 18th century.

With the need for information transfer was imperceptably linked the development of the understanding of magnetics and electrics. By 1800 these two were ready to merge from their original standpoints as distinctly separate studies to provide the principles and hardware for the creation of efficient and highly commercial telegraph systems. The electromechanical telegraph did not just happen, it evolved gradually. Its effect was revolutionary but its design evolutionary.

Feedback can be recognised in hydraulic devices described in Ancient Greece. Systems, in which the sensing and the actuation function can be clearly discerned were developed for other regimes in this period. Drebbel's incubator, shown earlier in Fig. 1.27, shows the use of feedback to maintain constancy of temperature. Other regulators – see Mayr (1970) – of this period were for water level, pressure in a pressure-cooker, velocity of windmill operations, direction of windmills, sail regulation and steam engine control. There is, however, little evidence of wide-spread adoption of the concept of feedback until the later part of the period in question here when steam engines were fitted with speed governors just prior to 1800. Punched card control, as opposed to data storage only, has its origins in the loom, Fig. 4.4, built by the French mechanic Falcon in 1728.

Instruments for surgery and other medical operations, although not the subject of this study, are linked because they enabled mechanical ingenuity to be developed that paralleled that of measuring and musical instruments. It was from physiological experiments that the electromagnetic era, that was to follow, arose. Work on animal physiology of the late 1700s apparently in another unrelated path of learning, linked into electricity around 1800.

This period, then, was one in which electricity and magnetism found their way into scientific apparatus and into measuring instruments. Conditions were being created whereby powerful and widely ranging measurement devices would be able to be built in the following century.

Fig. 4.4 *Falcon's loom of c1728 was the first to use punched cards for aiding control*
[Crown copyright. Science Museum, London]

This period saw the refinement of mechanical manufacture from hand-made objects that were relatively crude in precision (but not in style and aesthetic beauty) to machine produced components that were considerably closer to a nominal tolerance because of the principles of manufacture adopted. Design based on understanding of the physical principles involved in a mechanical system was introduced as scientists and engineers gradually discovered more laws of relevance.

Optical instruments were invented and constantly improved in this period. The telescope, originating for use in astronomy, found application in other uses where the eye needed gain. Optical methods also enabled man to learn more about the general laws of radiation.

The ever-widening availability and the new style of scientific enquiry provided expanding chances for new principles to be discovered that would enable more measuring instruments to be developed. Many new transducer principles were realised and their path to practical application begun in these times.

This was also an era when the need to record measurements as an automatic process had its genesis. Although remote electrical registration was barely possible by 1800 it was practicable to implement mechanical methods of producing a trace or marks to record the magnitude of a measured variable with time, position or other variables. Work on mechanical recording methods undoubtedly assisted the electrical methods of later times.

These five areas of development are the subject of this Chapter. They are selected as appropriate divisions for they provide suitable threads of explanation for the major developments of the period 1500-1800. In a general history such as this work is, it is not possible to trace many individual developments and it is the intention that the reader would follow up any particular interest by reference to the citation given.

Literature of and about this period is prolific, especially in the primary journals. In general, however, historians and other researchers have not concentrated on measuring instruments. Wolf (1935) and Sherwood-Taylor (1950) are sources specific to the 16th and 17th centuries that introduce instrumentation. As has already been said information about instruments had to be gleaned from works on the general history of science and engineering, from catalogues of instrument collections and from works on specific subjects. A good introduction to the available literature is that by Knight (1975). Specific works on instruments of the later part of the period of this chapter have been listed in Section 2.9, where instrument-makers were examined. Wolf (1938) covers science and technology in the 18th century.

4.2 First foundations of electricity

It was Gilbert, in his 1600 work *De Magnete*, who coined the terms ‘electric force’, ‘electric attraction’ and ‘magnetic pole’. He called materials that could be made to attract small objects ‘electrics’ and those that could not be made to do this ‘non-electrics’. Gilbert carried the study of magnetism and statical electricity ‘out of the realm of wild superstition placing the two subjects into a framework of scientific understanding'

At first thought, magnetism and electricity appear to be unrelated physical effects but even in these early times the two are linked because of their power to attract small objects. Gilbert investigated the earth's magnetic field, which he was able to study by way of connections with Drake that enabled measurements to be conducted across the known world. Gilbert has appropriately been called the father of electricity and magnetism, for he studied these in a systematic manner laying down the foundations for others to build upon.

From Gilbert's times onward there developed several instruments for observing components of the earth's field and directly, from his work, developed the electrical indicating instrument.

Lodestones were commonly known in Gilbert's times and were the source of means to magnetise iron to form needles of magnetic compasses. These objects were of great value usually being mounted in highly decorated, precious-metal mountings.

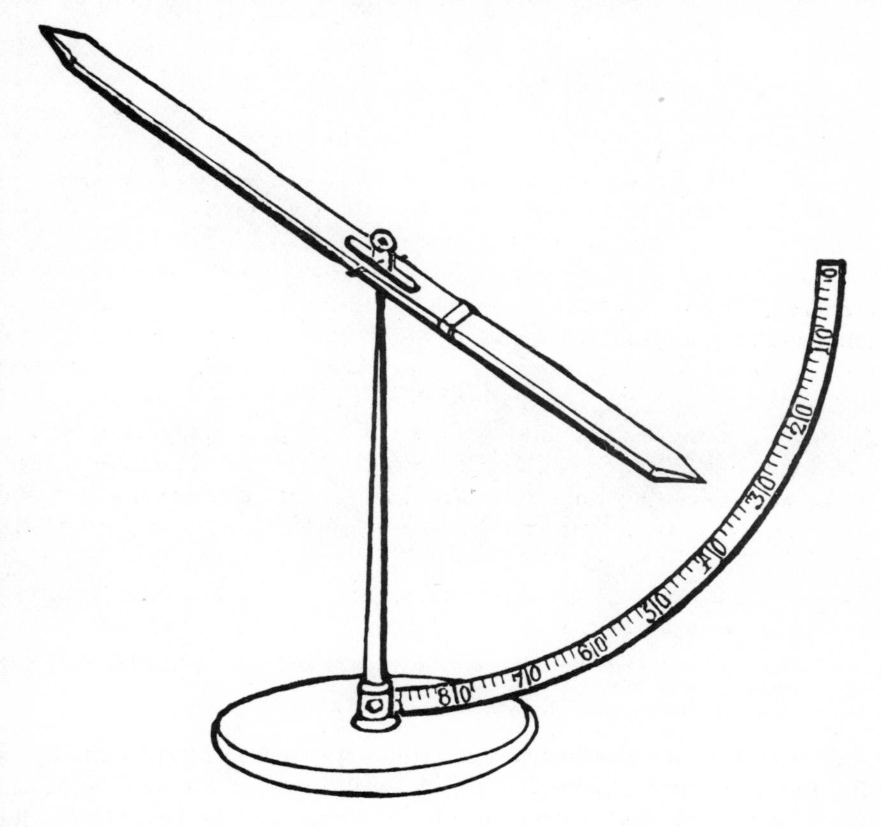

Fig. 4.5 *Simple dip-needle capable of quantitative measurement*
[Pepper (1874), Fig.322, p.351]

Fig. 4.6 *Dip circle of the late eighteenth century made by Nairne to the design using frictionless bearings by Michell*
[Courtesy National Maritime Museum, London]

A 1750 terrella (a spherically-shaped lodestone) was able to lift 17 times its own weight 160 years later, Gunther (1935). A small lodestone exists in the National Maritime Museum, Greenwich which is said to be that used by Drake to touch up his compass needles on voyages around the world.

As the magnetic compass played an important part in navigation it was to be expected that its users would soon discover that the earth's magnetic field strength varied from place to place and that this variation varies with time. They also discovered that the direction of maximum force dipped into the ground by varying amounts. As time passed, these effects were studied using special purpose instruments but initially people in Gilbert's time used instruments that were available for general sale, such as a mariner's compass.

The dip needle began as a simple needle, Fig. 4.5, pivoted on a horizontal axis. It soon acquired a vertical circular scale divided in angular degrees. Dip was first described independently by Robert Newman and Gerhard Mercator in 1576. In the early part of the 17th century magnetic surveys were made for iron-ore deposits in Sweden; Tasman made declination measurements in the Australasian region. Dip varies at the most by 90° from the equator to the poles so dip-needle instruments had to be carefully made to eliminate constructional and scale errors. By the late 1700s they often used agate knife-edge bearings to reduce the rotational frictional torque. Another intriguing design rolled the needle-spindle on the intersection of two much larger freely rotating wheels thereby multiplying the rotary drive effect of the needle so that the actual spindle torque appears much reduced to the needle shaft -- called a frictionless bearing. A unit thought to have been the one that Cook was instructed to use each day of his third voyage of 1776-1779 is shown in Fig. 4.6. This represents the well-developed mechanical design capability existing by 1800. It is around 250mm diameter and is on exhibition at the National Maritime Museum, Greenwich. This bearing arrangement is attributed by Turner (1973), to Sully (1680-1728). Turner (1973) contains several illustrations of demonstration frictionless bearing assemblies.

The azimuth compass was gradually improved. By 1800 it had been given a horizontal scale, reduced friction pivot, optical magnifying glass, gymballed suspension and other devices for use at sea, a vernier reading method (after Pierre Vernier 1580-1673, who invented the method), reduced rotational-moment cards and needles that enabled the card to respond more rapidly than the slower motions perturbing it during time at sea and in improved means for sighting the direction of the object of interest. Fig. 4.7 is and 1801 etching of compass constructions.

The azimuth compass was also married to an octant having an optical telescope sight-tube in one instrument variation. An example exists in the Van Marum collection at Teyler's Museum. Turner (1973) includes illustrations of magnetic devices of the collection.

Declinometers were azimuth compasses in which the range was restricted to a small rotation on the vertical pivot. They were used to measure the variation of the magnetic field direction with respect to the geographical North. A design pictured in Pipping (1977) has a simple optical system that enabled the declinometer to be aligned with respect to the meridian by observing the sun. The combination of optical techniques with previously mechanical devices became common practice in the period 1500-1800.

In 1784 Coulomb (1736-1806), a French military engineer, announced the law of force of attraction between two magnetic poles. It now bears his name. He also showed that the same law applied for electric charges. The style of apparatus that he used,

Fig. 4.7 *Etching of compass construction released by J.Wilkes, 1801*

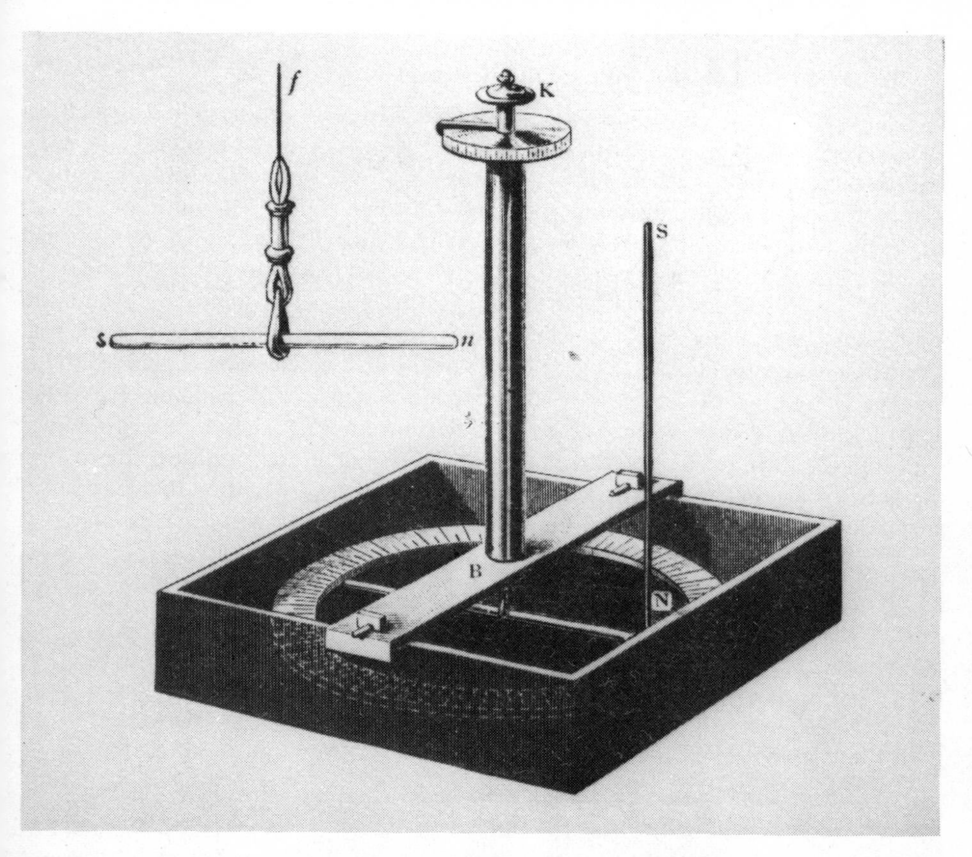

Fig. 4.8 *Form of torsion balance used by Coulomb in his investigation of the forces of attraction* [Mullineux-Walmsley (1910), Fig.18, p.25. Courtesy Cassell Ltd]

the torsion balance, is shown in Fig. 4.8. This design of balance is credited to Rev. John Michel in Glazebrook (1922), Vol. 3. The degree of twist that has to be applied to the torsion fibre to bring the magnet (or charged beam in the electric force use) back to a datum position is a measure of the force of attraction. The apparatus, therefore, enabled relative forces of attraction or repulsion to be compared in a quantitative manner. The torsion fibre was made of silver or platinum. In the electric attraction version the beam had a small disk of gold leaf, or a gilt pith-ball, attached to the end of an insulating material beam made of gum lac or glass.

Although known as Coulomb's Law, it was discovered a little earlier by Cavendish but his work, unfortunately, was not published until well after that of Coulomb. Cavendish, as well as establishing his place in the development of chemistry, also worked on electrical subjects. It is clear that he anticipated later work of Faraday and Ohm. Commenting on Cavendish's work, Maxwell said in 1879:

> .... we have the sole record of one of those secret and solitary researchers, the

> value of which to other men of science Cavendish does not seem to have taken into account, after he had satisfied his own mind as to the facts.

This failure to communicate his work of the late 18th century, meant that electrical science had to wait more than 50 years before the work that Cavendish had completed was repeated by others and disseminated for general use. Quantification of the laws of attraction enabled instruments to be designed on a more rigorous basis.

Cavendish used the law of attraction between two bodies, which Newton had established and which has the same form as the laws of magnetic and electrical attraction, to 'weigh' the earth in 1797-8. His apparatus, Fig. 4.9, comprised a deal-wood lever, some two metres long, on which were attached small lead balls at each end. The whole was suspended by a metre-long, fine, silvered copper wire connected to the middle. The period of oscillation of this torsion system was 7 minutes. Two other larger lead balls were positioned to cause deflection of the beam which was observed with well mounted telescopes. Observation of the period of oscillation, combined with the geometrical and mass constants, enabled Cavendish to quantify the constant of gravitation G. This result enabled the average density of the earth to be assessed. His

Fig. 4.9 *Model of apparatus used by Cavendish to 'weigh' the earth*
[Crown copyright. Science Museum, London]

experiment has been repeated many times by other workers over time, the latest method being to make use of laser interferometry of torsional systems having a nominal 100 m long suspension fibre. More details of early work on *G* are available in Glazebrook (1922), Vol.3.

As the laws of attraction between masses, charged bodies and magnetic poles are of similar form, there gradually developed the concept of potential fields to describe them in a general theory. To this day we are still, however, not much the wiser about the fundamental nature of these fields. We understand their laws and observable effects yet do not know what they are as a 'substance'.

Before moving to the electrical aspect of instruments of the period to 1800 it is now appropriate to make brief mention of the genesis of what is generally known in European and Scandinavian countries aptly as 'teletechnique'. . . telegraphy, telephony, television and other forms of information transmission systems. It was in this period that evidence exists that proved that people were thinking of how they might use 'action-at-distance' effects to transmit information between two points. The earliest ideas are found in magnetic form but they came to workable fruition, as we shall see later in this Section, in the electrical form.

The earliest account seems to be that of Baptista Porta, a Neapolitan, who published a book in 1558 in which he wrote:

> owing to the convenience afforded by the magnet, persons can converse together through long distances.

his ideas matured; later he wrote:

> I do not fear but with a long absent friend, even though he may be confined by prison walls, we can communicate that we wish by means of two compass needles circumscribed with an alphabet.

His concept of telemetry is clear but the practice somewhat fanciful. It is doubtful if experiments were made to attempt to utilise this principle. Strada, in 1617, expressed his ideas:

> And thus by several motions of the needle to the letters (inscribed on the compass dial) they easily make up words and sense which they have a mind to express.

More is available about Strada's ideas in Timbs (1860). Even if the ideas were not workable Strada's concept of sympathetic movement between two mechanical members placed at a distance from each other was in the right direction – the first workable telegraphs of the early 19th century did just that. An account of the development of teletechnique has been published by Sydenham (1975). It is but one of numerous works on this area of communication. The subject is taken up in more detail in the following two chapters on later times for the introduction of the methods found acceptance because of the availability of more appropriate technology and principles than were in existence up to the end of the 18th century.

It is often the case that measurement devices develop out of need to quantify the magnitude of an available energy source. Such was the case in the development of electrical indicating devices. The simple generation of statical electricity by rubbing amber developed into the 'electrical machine' for generating large quantities of charge for purposes of serious study or as the source of amusement. Otto van Guericke probably built the first electrostatic generator in 1663. These evolved as rotating glass or sulphur, cylinders of spheres holding the hand or a suitable pad of material against the rotating member so as to charge it up. Fig. 4.10 shows Watson's design being used for games of amusement. Later, around 1760, glass disks were commonly used. The efficiency of these purely frictional machines was low and they were improved upon by the 'influence' design that appeared around 1865. In this design the charge is effectively amplified and swept away for storage in a Leyden jar, capacitor bank.

Fig. 4.10 *Dr. Watson's frictional electrical machine, being used here for amusement. The 'electric kiss' was a favourite game*
[Pepper (1874), Fig. 198, p.221]

Many of these machines still exist in a workable state. Details and illustrations are to be found in Chew (1968), Gunther (1935), Multhauf (1961*a*), Pipping (1977) and Turner (1973). Dibner (1957) and Lavèn and Cittert-Eymers (1967) give the historical development of the various types. Educational accounts are to be found in Pepper (1874) and Mullineux-Walmsley (1910).

These machines could supply very large electrical power levels. Finn (1971) published his findings of the electrical outputs of several 18th century electrostatic machines. The Plate machine of the Van Marum collection, shown earlier in Fig. 2.23, when used with 100 Leyden jars to store charge, could generate 30 000 joules at a potential of 330 000 volts (1 watt is an energy rate of 1 joule per second). It was the biggest of its time. Another design is shown in Fig. 2.22.

Coulomb's method of measuring the force of charges for investigative reasons was usable as a means to measure the voltage (also called tension, potential, electromotive force) difference existing between two parts that are insulated from each other. The Coulomb apparatus became known, when used for electrical measurements, as the electrometer or electroscope (the former seems the preferable term for those instruments that provided a quantitative measure). Coulomb's apparatus, however, was not the first electric charge-indicating device for Gilbert had made use of the magnetic-restoring moment of magnetised needle instead of the torsion fibre in what he called a 'versorium' form of electroscope.

Electrometers of many forms were devised. Qualitative, electroscope, forms are shown in Fig. 4.11. They began as pith balls on strings (Canton in 1735) that deflected apart as they charged, and as thin gold leaves that operated on a similar principle having its origin in the 1786 work of F.R.S. Bennet, 1750–1799 – see Lavèn and Cittert-Eymers (1967).

For rigorous study some form of scale was added to enable the deflection to be given an identifiable number, being a measurement of tension.

Quantification in electrometer design can be traced back to around 1785. The Van Marum collection contains an advanced version, after the style by Lane of 1766, in which a micrometer screw enables the length of the spark gap to be measured, thereby yielding a measure of the voltage available at the onset of spark-over. These were used, for example, to give rough quantification to shock levels in early electromedical practice. This collection also contains a unit in which weights are used to decide the force exerted between the two charged balls. Dial outputs were often preferred by investigators for they could observe the trend of the signal.

Precision mechanical components are a feature of the Maréchaux micro-electrometer made just before 1805. According to Turner's description this was an attempt to make a precision measuring device out of the gold-leaf electroscope. One of the leaves was replaced by a screw-displaced sphere that could be moved toward the hanging leaf by known distances. The screw used has 50 turns to the Rhineland inch and was graduated into angular degrees of a 101mm diameter disk rotating with the screw. It was used to observe the diurnal variation of atmospheric electricity. This design, Pepper (1874), was originated by Hare, of Pennsylvannia. Another form in the Van Marum collection is that of Kinnersley made *c*1785. It made use of the energy of discharge to heat air trapped under a column of indicating liquid causing a level change. It is confusingly called the electrical thermometer. The design originated in America around 1761.

Many experiments relied on deciding the onset of the flash-over of sparks. To assist observation of this event optical devices were used to magnify the effect. Unfortunately, the harmful effect of massive charge such as flows from a storm cloud to ground, was not understood. Rickman was killed repeating an experiment of Franklin's in which a wet line was sent up by a kite into the sky. It was common practice then to use the human senses to detect the presence of charge.

Electrometers have continued to be of use in high-voltage measurement to this day, as will be seen in later chapters. Indicating instruments that used electromagnetic, as opposed to electrostatic, principles are generally known now as galvanometers but that name was first used by Bischof for an electroscope of similar design to, or possibly

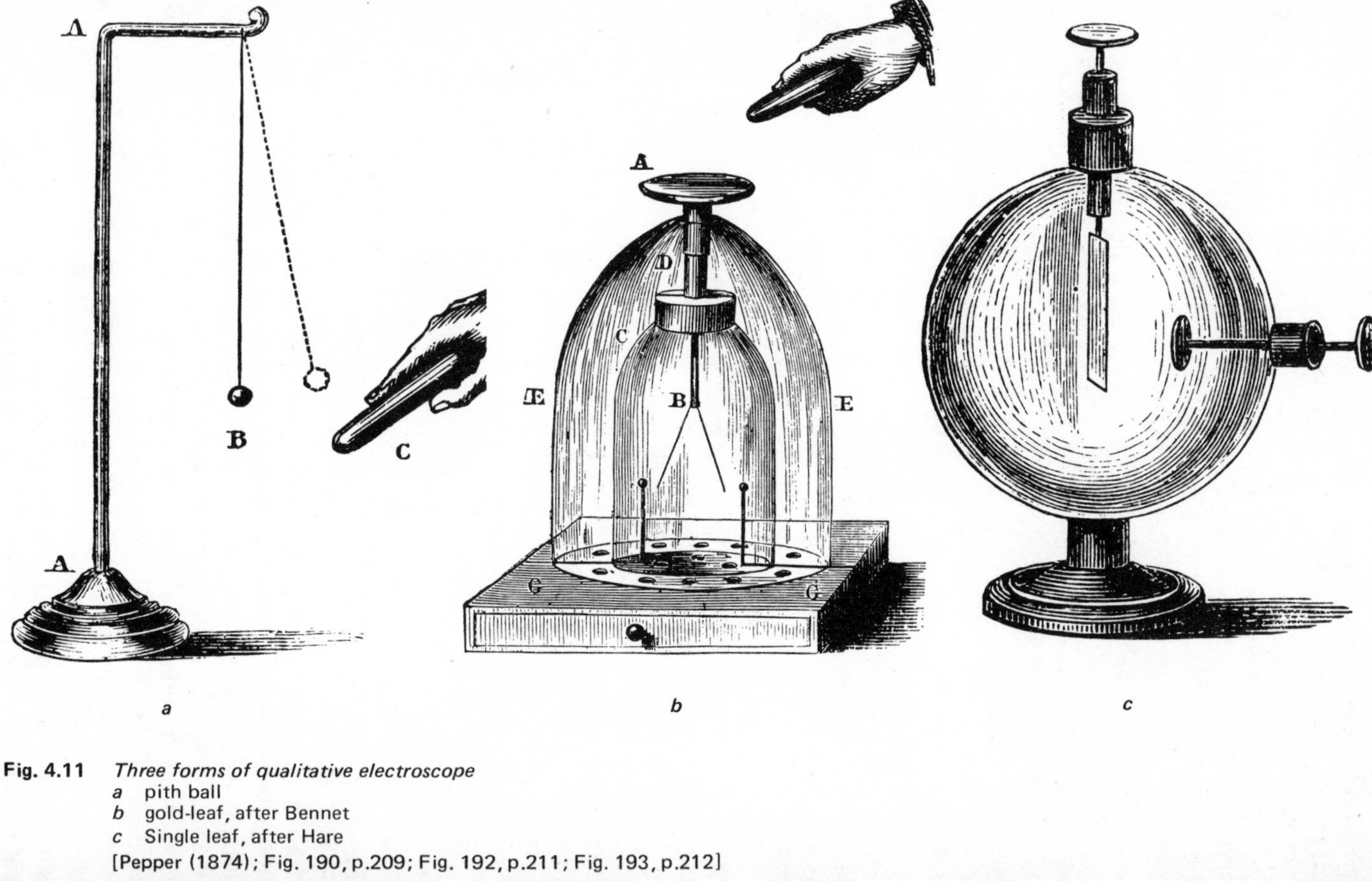

Fig. 4.11 *Three forms of qualitative electroscope*
a pith ball
b gold-leaf, after Bennet
c Single leaf, after Hare
[Pepper (1874); Fig. 190, p.209; Fig. 192, p.211; Fig. 193, p.212]

predating, that of Maréchaux, mentioned above, see Chipman (1964). Bischof's was reported in 1802. His instrument is illustrated in Fig. 4.12. Note the use of a vernier scale instead of a micrometer screw.

Out of the general study of statical electricity were born many of the basic electrical components now used in instrumentation. The Leyden jar has already been mentioned. It was invented independently by von Kleist in 1745 and by Musschenbroek in 1746. They tried to fill a water-filled bottle with electricity but they unfortunately held the bottle and the filling conductor, thereby receiving a substantial electric shock. The name Leyden jar came from Nollet's use of this storage device. Initially they were filled with water; later copper filings were tried. It eventually became obvious that the requirement was for two separated conducting sheets; foil was placed on the inside and outside of a glass bottle. Fig. 4.13 shows a battery of Leyden jars arranged with a Lane unit and an electrometer on the top. Lavèn and Cittert-Eymers (1967) estimated that the capacity of a similar bank using 15 jars is 0·02 microfarad and that it could withstand 100 000 volts, delivering 100 joules (capacity of course, depends on foil area and glass thickness). The early electrical systems were well able to supply large energies but only by virtue of the very large voltages that characterise static electricity systems. This was a severe drawback that held up the application of electricity to practical tasks until a more suitable form, the chemical cell was developed in 1800. More of that later.

Another, rather humble, component of electrical science is the insulated wire. That too, took some time to emerge. Initially the reasons why electricity went where it does were not understood. In 1730 Steven Gray (also Grey), a member of the Royal Society from 1732 and a pensioner of Charterhouse, began to be interested in electricity. To him we owe the discovery, Fig. 4.14, of what materials to use for conducting and for insulating electrical charge. He used silk threads to support a dampened packthread line, thereby conveying charge over a distance. Thus was invented the concept of insulated wires and busbars. He proved conclusively that the colour of objects, and other such parameters, were not the basis of charge flow control.

Gray's work triggered others to improve the knowledge and it was not long before reliable experiments could be carried-out to seek the speed of transmission of electric charge and the limits of the distance over which it might be conveyed. Timbs (1860) stated that Gray sent it over 250 metres. He described the experiment by Watson that transmitted charge through a large body of water to the other side. This was also reported in Priestley's *History of Electricity*. In 1747 the distance was up to 8 kilometres as the result of a Royal Society sponsored experiment (the cost, ten pounds five shillings and sixpence). The paper reporting this event was also significant for it contained the first statement, by Franklin, that there existed two forms of charge, negative and positive. Finn (1976*a*) looks into Franklin's electrical work.

The speed of transmission appeared to be instantaneous in experiments; we now know it travels at the speed of light. The history of the measurement of the speed of light is discussed by Bray (1930). As lines were lengthened no difference could be detected but having no theoretical understanding of this, investigators continued to be amazed and seek a value for its velocity for many years after this time. Lardner (1862)

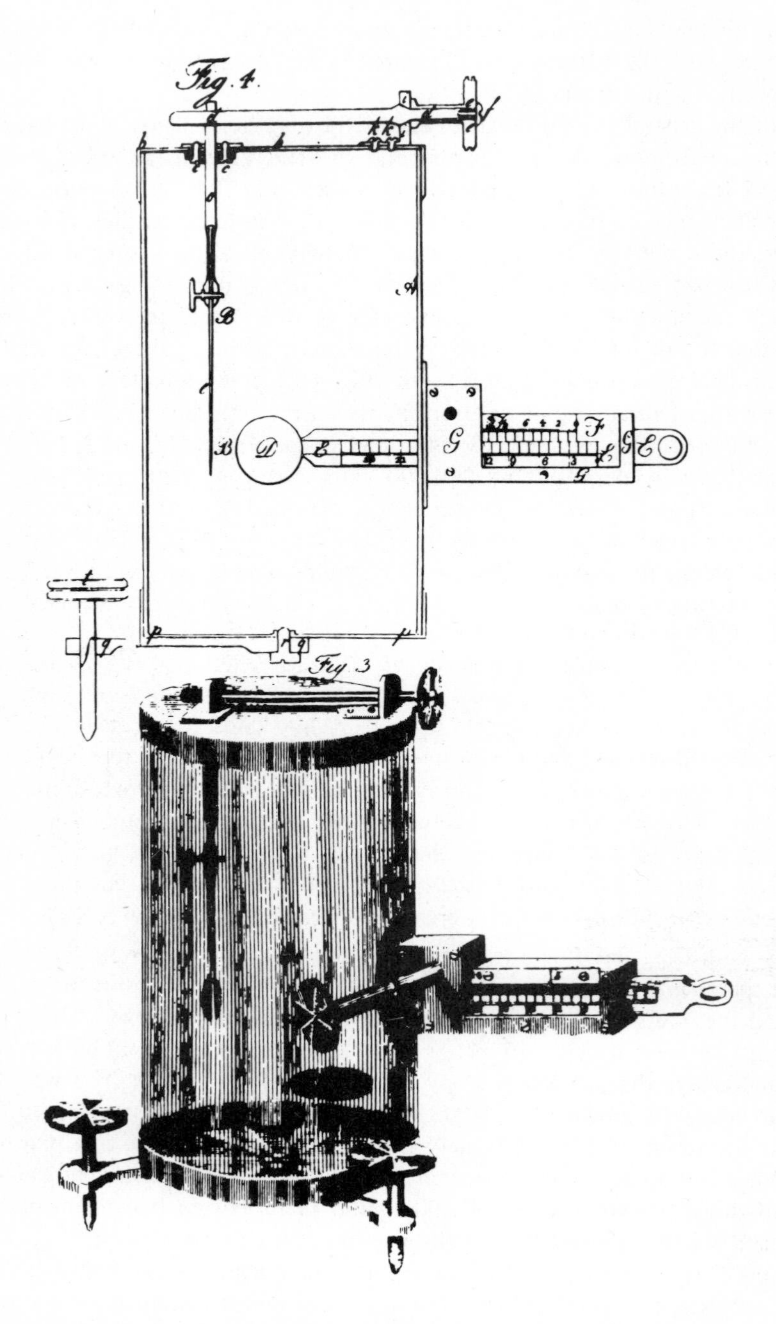

Fig. 4.12 *Bischof's electrometer of 1802, that he called a 'galvanometer'. The name was adopted for electromagnetic indicators instead*

[Chipman (1964) Fig.2, p.126. From *Medical and Physical Journal*, 7, p.529, 1802]

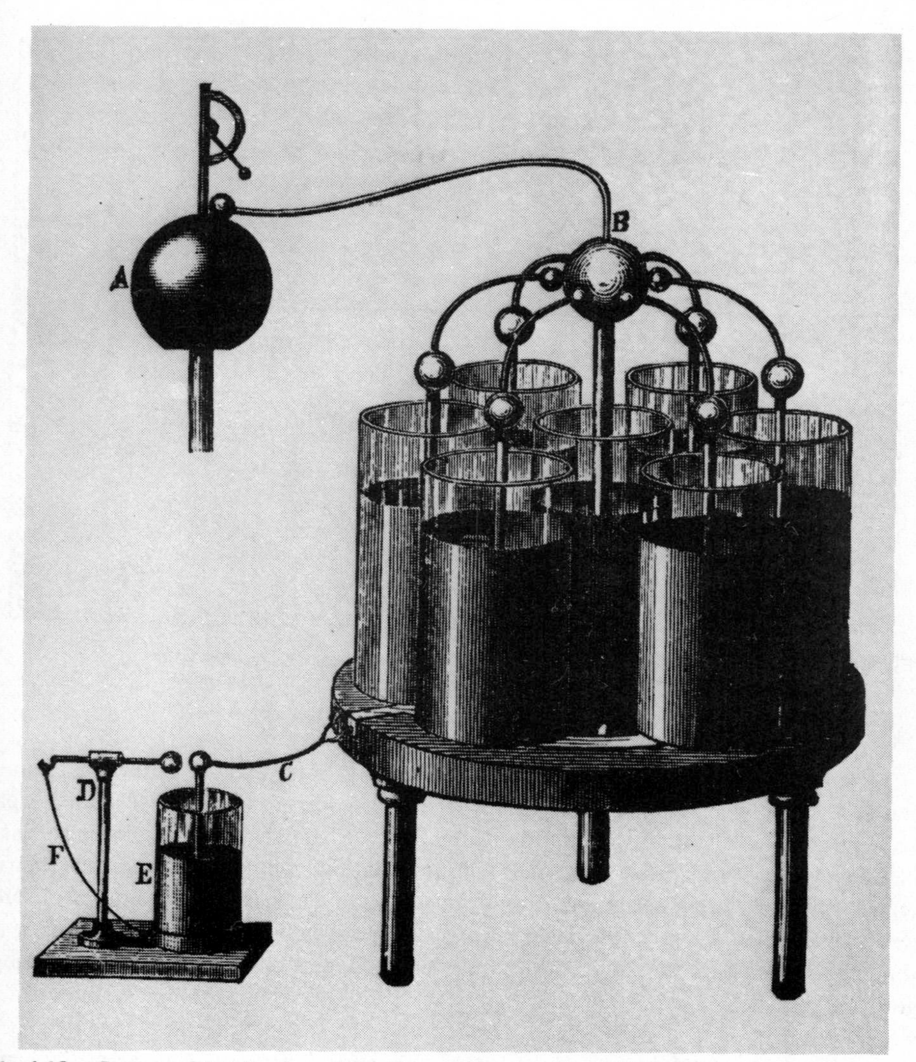

Fig. 4.13 *Battery of Leyden jars with a Lane unit-measuring jar (lower left) and half-circle scale electrometer (upper left)*
[Mullineux-Walmsley (1910), Fig.104, p.118. Courtesy Cassell Ltd.]

Vol. III, in his verbose description of the telegraph, marvelled at the speed at which messages could be sent. As the result of a controlled experiment over a wire link of 700 km round-loop he concluded that it was possible to transmit 19 500 words per hour over a distance upwards of 200 kilometres. In his time the velocity of electricity was still not known. Wells (1856) quoted the speed estimates in his time; it was often greater than the velocity of light. Deschanel (1891) asserted, even then, that it has no definite velocity, apparently confusing long-line integration effects with the speed of electricity.

Fig. 4.14 *Cruikshank's study of a lecture to the general public by Stephen Gray*
[Morgan (1953), p.25. From 17th century etching by Cruickshank]

Means to convey messages more rapidly than by horse existed in the semaphore stations. These were supplanted by the electric telegraph around 1850. The most well known semaphore system was that implemented by Chappe, in France, during the last decade of the 18th century. Fig. 4.15 shows a station. A typical message was transferred through twenty-seven stations, from Calais to Paris, in three minutes. There were several people, however, even before Chappes' system was implemented, that thought that electric communication systems could be built. Wireless systems using ground conduction had been investigated in the 1750s. Transmission line systems using static electricity source signals can be traced to an article in Scots Magazine of 1753, signed by C.M. Many writers suggest that this person has not been identified but Frith and Rawson (1896) states it was Charles Marshall. The article was entitled '*An expeditious mode of conveying intelligence*'. It described a multiwire system, one for each letter of the alphabet terminated in a small ball having under each a piece of paper. As a particular wire was electrified at the sending end the paper would be attracted at the receiving end indicating the letter sent. This system was never built as far as is known but other people, after a pause of interest, implemented the concept in several forms. One method, by Lomond in 1787, used only one wire and coded the information to be read out from the respective movements of pith-ball indicators. Timbs (1860) states that a working model of this telegraph was witnessed in France in 1787 with which Lomond sent two or three words to his wife in another room.

According to Voigt's Magazine (stated by Timbs) an electric spark telegraph was

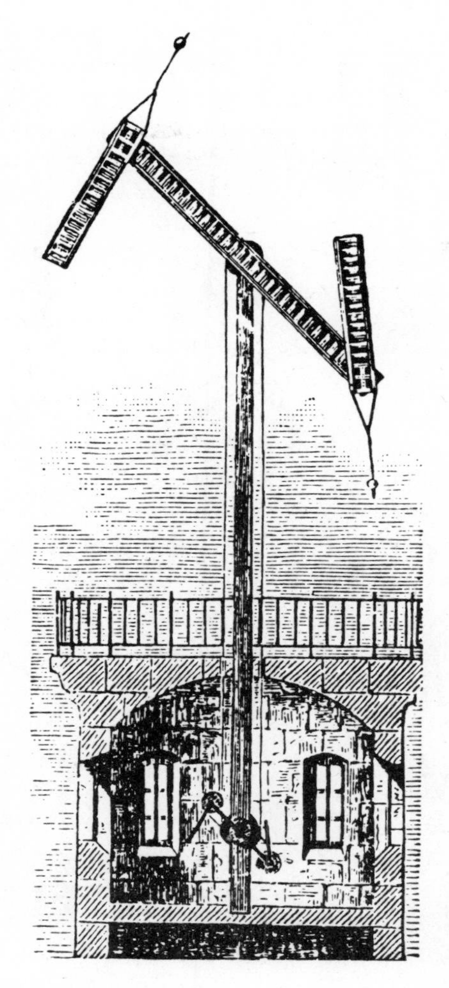

Fig. 4.15 *Chappes' arm semaphore c1790*
[Frith and Rawson (1896), p.29]

built by Dr. Salva of Madrid 1798 that worked over a several kilometre distance. A telegraphy system that is still extant from this period is that of Sömmering, a Professor from Kassel, Germany. Success of a Napoleonic takeover of Munich in 1809, because of the Chappe system giving information about the movements of the Austrian troops led to Sömmering being commissioned to construct a 'system of telegraphy'. He built a prototype, but it was never used for public utility. It remains complete, today, in the Deutsche Museum, Munich. Fig. 4.16 is an engraving of this apparatus from Frith and Rawson (1896). His method used the decomposition of water by electricity to generate gas at the position of the letter sent. The gas rose to be collected under the balanced inverted spoon causing the spoon to tip up, dropping a ball into the gong, warning

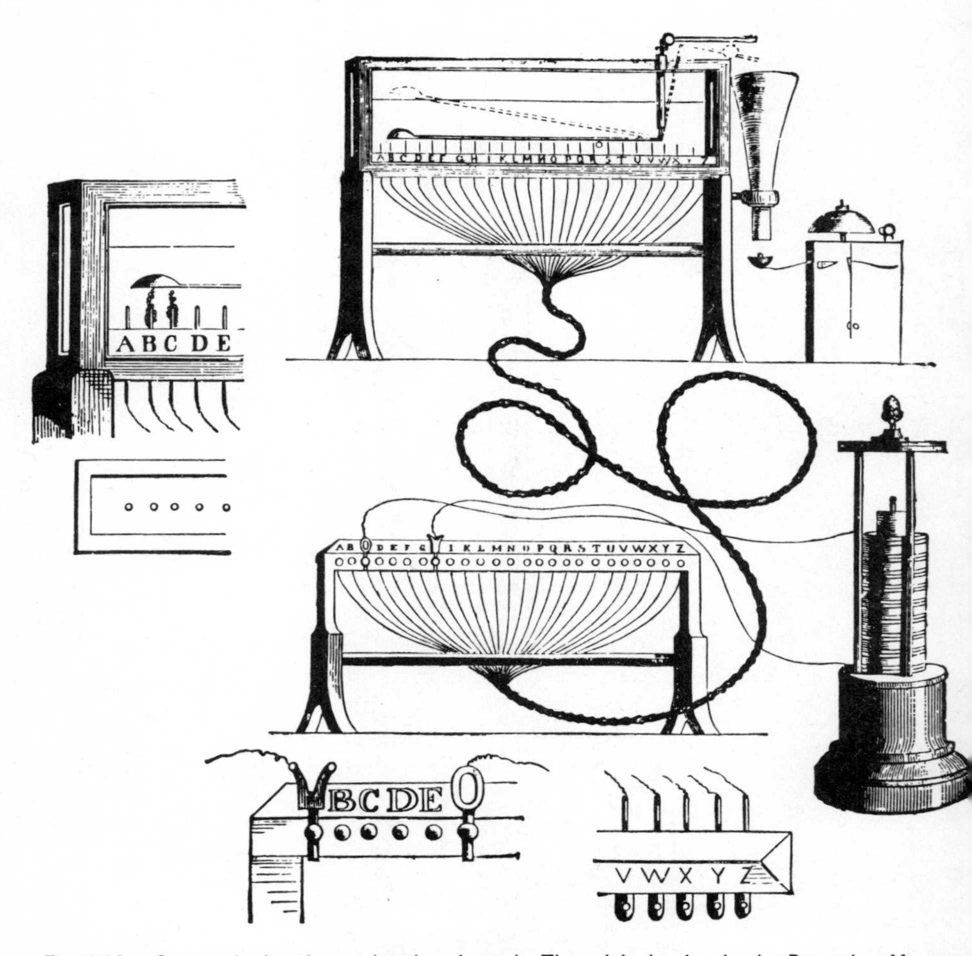

Fig. 4.16 *Sommering's galvano-electric telegraph. The original exists in the Deutsches Museum, Munich*
[Frith and Rawson (1896), p.31]

arrangement. The operator would then observe each letter sent by the appearance of bubbles on appropriate electrodes.

It is appropriate to include this telegraphy system in this chapter even though it was designed after 1800 for it represents one of the last attempts to construct a telegraphy system without the availability of electromagnetic concepts. Techniques originally exploited for communication have found widespread application in measurement systems, for they, too, require information transfer.

The above descriptions show that although electricity was slow to be investigated by the Ancients it was soon at a level of understanding equal to other disciplines by the end of the 18th century. There is more of relevance yet to report.

Several important electrical output transducer effects became known during this

Fig. 4.17 *Experimental apparatus used by Bergman c1766 to investigate pyroelectricity*
[Pipping (1977) dustcover. From Bergman, T., *Trans Royal Swedish Acad. Sci.*, 1977. Courtesy Royal Swedish Academy of Sciences]

time. In 1703 the Dutch brought tourmaline from Ceylon and it was noticed that when heated over a fire it attracted ash. In 1756 this was shown to be a pyroelectric effect by Wilke and Aěpinus. Lavèn and Cittert-Eymers (1967) describes several artefacts of this kind and age. It was also known from this time that Iceland spar becomes electrified if pressed with the hand retaining the charge generated for some time after. This is the piezoelectric effect. The transactions of the Royal Swedish Academy of Sciences contains a paper by Torben Bergman on pyroelectricity. An illustration from it was used for the dust-cover of Pipping (1977); it is given here as Fig. 4.17.

We have already touched upon the electrolysis of water. This was first accomplished, according to Chipman (1964), by Nicholson and Carlisle in 1800 but Mullineux-

Walmsley (1910) credits it to Trostwyk in 1789. The water had to be made a little acidic to produce the effect for, in modern terms, an electrolyte is needed. They noticed it by accident when a connection was wetted to improve induction. Knowledge of this effect was extended much further by Faraday at a later date, when he took the state of knowledge from the qualitative stage to the quantitative state. Conversion of electrical energy to thermal was also well known by 1800 for many an experimenter had fused small gauge wires.

From the instrumentation viewpoint, probably the most significant discovery of a physical effect in the whole history of measuring instruments was the invention of the voltaic, also called galvanic, source of electricity. The invention of the primary electrochemical voltaic cell by Volta in 1800 provided a much needed source (although those of the times were unable to realise the long-term significance) of cheap and simple low-voltage source. It enabled electricity to be researched and studied in a previously unknown manner.

This development began when Luigi Galvani, Professor of Anatomy at Bologna, was working on nerves of frogs legs in 1786 – Fig. 2.12. Owing to the prior work of John Walsh, in 1773, who decided that the electric eel contained in it 'animal electricity' Galvani drew the conclusion that the reason that a leg twitched when he was operating, or when hung, was because of animal electricity in it. History suggests that his wife made the observation and that, anyway, it had already been seen by Swammerdam several years before, Pepper (1874). Just who discovered the effect matters little to this account but it is true that Galvani's work was the event needed to bring this effect to the attention of others.

The actual moment, as reported by Galvani in his 1791 paper, which translates as 'Commentary on the effect of electricity on muscular motion', Foley (1953) and also in Batten (1968), went as follows:

> I dissected and prepared a frog . . . . and placed it on a table, on which was an electrical machine . . . . When, by chance, one of those who were assisting me gently touched the point of a scalpel to the medial crural nerves of this frog, immediately all the muscles of the limbs seemed to be so contracted that they appeared to have fallen into violent tonic convulsions. But another of the assistants, who was on hand when I did electrical experiments, seemed to observe that the same thing occurred whenever a spark was discharged from the conductor of the machine. He, wondering at the novelty of the phenomenon, immediately appraised me of the same, wrapped in thought though I was and pondering something entirely different. Hereupon I was fired with incredible zeal and desire of having the same experience, and of bringing to light whatever might be concealed in the phenomenon.

Galvani was fortunate in some ways for a frogs-leg was probably the most sensitive detector of electricity in his day. What Galvani did not appreciate fully was that the convulsion was due to the scalpel and another metal object forming an electrochemical cell with the frog body fluids. Galvani's original apparatus of 1784 is held in the Deutsche Museum. It was Volta who reasoned this as the cause and from this deduction he built chemical cells, stacking them on top of each other, or side by side, Fig. 4.18, to form a battery of cells. His battery used silver-zinc disks separated by

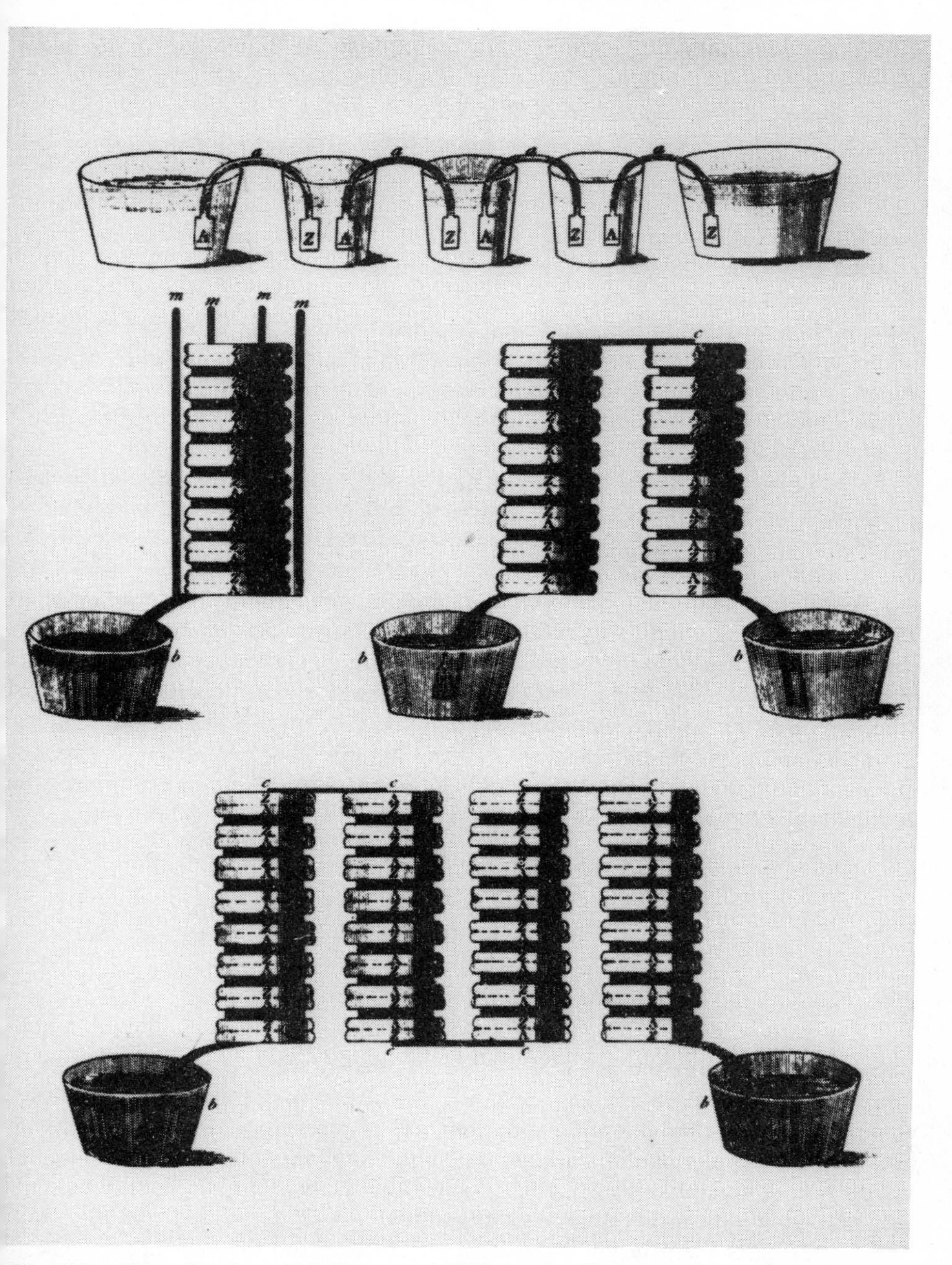

Fig. 4.18 *Illustration from Volta's paper of 1800 showing his 'crown of cups' and 'pile' forms of voltaic cell*

[James-King (1962*a*), Fig.3, p.236. From *Phil Trans. Roy. Soc., London*] salt-water moistened paper. Volta's paper on this battery, Volta (1800), was given good circulation and it produced rapid influence on the work of others.

To appreciate the importance of this simple invention we need to compare the statical and galvanic sources available around 1800. The source impedance of the statical source was very high, it was generally unable to supply a steady current (the term, current, was not used until 1820), it was expensive to produce by comparison and had to be 'topped-up' periodically or operated by an assistant. The statical source supplied only high, kilovolt, tension that dropped away as the experiment proceeded. The high voltage required extremely good insulation of the conductors and components. In a nutshell, to use the statical electricity method was akin to investigating laws of electricity and physical effects with a supply of 30 000 volts or higher. Imagine trying to construct a simple amplifier or sensor with these levels with the technology of the time! The importance of Volta's work was recognised by naming the unit of tension after him . . . the volt.

By 1800 then, there was thought to be two kinds of electricity and that electricity and magnetism were in some way related, but just how remained unsolved for another two decades. Mathematical formulation of the capacitor was understood, the concept of voltage existed, ideas of current were forming and electricity was seen to be related to other phenomena. The voltaic cell enabled these relationships and quantities to be related bringing about the electromagnetic era of the 19th century. The clues for this had already been noticed. Sparks from statical generators had been observed to magnetise compass needles doing this with either polarity of magnetic pole. Iron held in a water tank in which was placed an electric eel could be felt to quiver and compass needles deflected when the fish was stationary, Chipman (1964). Many instances in which electric sparks caused magnetisation were recorded but two decades were to pass before 'the penny dropped' revealing the link between electricity and magnetism.

4.3 Further development of mechanics

Burstall (1970) groups the period 1500-1750 as that 'towards the industrial revolution'. His account explained how scientific foundation was given to a previously empirical way of designing and making objects of technology. From his study of both power and information machines we extract the latter information amplifying it with further examples. Stanley (1901), in an introductory chapter, reviewed mechanical design of instruments up to the 19th century.

Table 2.4 gives the dates of formulation of principles and theories for mechanical subjects, including hydraulics and pneumatics. It shows the great expansion of the theoretical understanding of the design of static and certain classes of dynamic machines. Laws of elasticity enabled spring balances, dynanometers and electrometers to be built that provided quantification of forces. Theory of beams provided rigour to the design of the highly stressed member of an instrument. Laws of fluid flow provided the basis for transducing flow into displacement. Laws of pendulums gave designers a new principle for time-keeping.

As is always the case in the development of science and technology the new laws were discovered with the instruments built from the knowledge available at the time. Guericke, in 1661, measured the pressure of the atmosphere using simple mechanics to create a test situation shown in Fig. 4.19. The illustration suggests that the force of air-pressure was thought to be bearing downward. If it had been realised that it operates from any direction he need only to have hung the weights from under the upturned cylinder, thus simplifying the experimental apparatus. A replica of this apparatus has been made at the Deutches Museum. Another use of the balance was by Sanctorius who, around 1610, proved the body lost weight through perspiration by weighing himself.

With emergence of laws of applied mechanics also arose, through the spread of

Fig. 4.19 *Guericke's apparatus for measuring the force of atmospheric pressure*
[Klemm (1959), Fig. 13, p.160. From Von Geuricke, O., 'Experimenta Nova', 1672]

Fig. 4.20 *Polhem's models, 1729*
[Courtesy National Museum of Science and Technology, Stockholm]

printing, the use of drawings and models to depict 'ingenious design principles'. Ferguson (1977) has expanded the value of visual images as was mentioned in Chapter 1. One example of systematic model making in the period of this chapter was the 1729 'mechanical alaphabet' of Christopher Polhem (1661-1751), a Swedish engineer. He grouped mechanical principles with vowels as the 'five powers' of the lever, wedge, screw, pulley, and winch having some seventy-five other models as the consonants of his alphabet. Fig. 4.20 shows some of the remaining models in the National Museum of Science and Technology, Stockholm. They are made of wood and embody principles that were implemented at the time in a range of materials.

In the 18th century 'mechanica' was a field of knowledge about the various forms of machine. Many teaching collections contained large numbers of models of principles, many of which are clearly related to instrument use. The mechanical instruments of the University of Utrecht Museum have been catalogued, Bos (1968). The frictionless bearing referred to above appears in this collection along with dynanometers, the 'static enigma' balance to be discussed below, instruments to show the equilibrium of forces acting in different directions and intriguing mechanical relationships in which cones roll up hill and one that proves that the fastest descent down a track is not a straight path but a cycloidal route. Other collections having similar apparatus are the George III collection of the Science Museum, London, Chew (1968), and the van Marum collection, Turner (1973). The value of these educational aids lost much of their relevance in the early twentieth century: visual aids are only just beginning to become recognised again. These models from the 17th and 18th century embody principles that are still as relevant to modern mechanical design as they were then. Publication of encyclopaedic collections of mechanisms continued into the next century, Robison (1822), Barlow and Herschel (1848).

Around 1769 Greek engineer, Count Carburi designed a metal track, having vee-grooves in which were placed metal balls, to move a million and a half kilogram block. The van Marum collection, Turner (1973), contains models of this apparatus which may be the first use of ball bearing 'slides'.

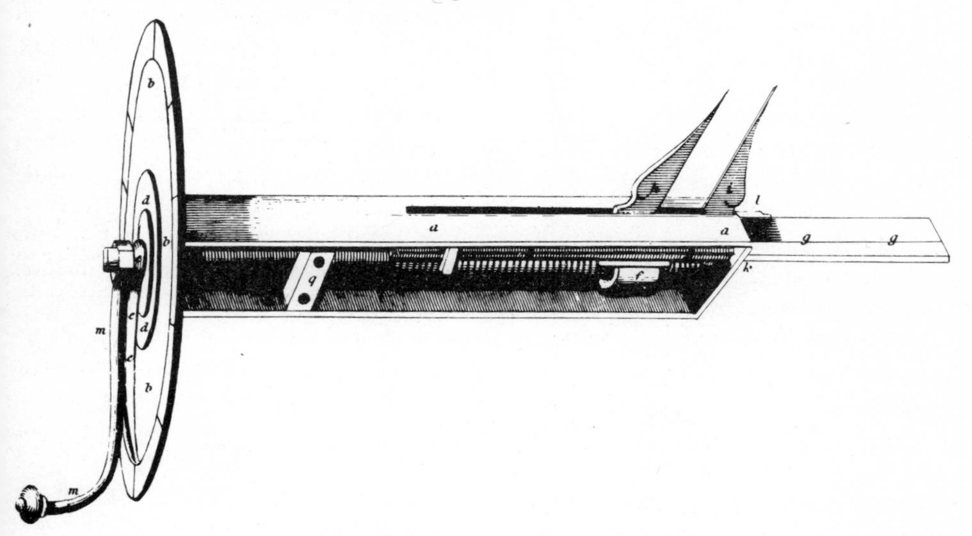

Fig. 4.21 *1639 etching of screw micrometer by Gascoigne*
[Photo. Science Museum, London]

In this time mechanical devices of greater complexity and refinement evolved. Pearsall (1974) suggests that angles could be read to an arc-second by the introduction of Ramsden's, dividing-engine, produced circles. That was around 1787. As well as radial scales there became available linear scales with improved means for reading them. The screw micrometer was originated by Gascoigne, Fig. 4.21, around 1640 being reported by Hooke in the Transactions of the Royal Society in 1667. The vernier scale *c*1640 was applied for both linear and angular use in dividing a scale into finer divisions than could be conveniently ruled. The need for scales arose in many ways . . . for verniers, for the disks of a micrometer barrel, for spring balances, in theodolites, quadrants and in mathematical instruments.

Clock design called for better pivots, more complex arrangements of components and greater reliability. Experience learned in this and other fine mechanics used assisted the manufacture of extremely complex automata and calculating devices.

The use of elastic elements as design components was evidenced from early times as the power unit of cross-bows and other war machines. Perhaps it was Hookes' work on elasticity of *c*1665, see Keynes (1960) for illustrations, that prompted the appearance of the elastic member as the basis of force measuring instruments. According to Burstall (1970) the two forms shown in Fig. 4.22 appeared in a published account of 1694. Their exact origins seem to be unknown, but they show that considerable development occurred. Details of these forms, including the enigma about the actual operation of the left hand unit, are given in Benton (1941).

The 'proving ring' used today as the force-to-displacement transducer of force transducers can be traced to this time. Many accounts describe the dynanometer of M. Regnier. Fig. 4.23 is a contemporary etching released in 1803. The University of Utrecht Museum collection, Bos (1968), contains two early 19th century units modelled on Regnier's design. There is also a unit, containing an internal spiral spring, for measuring the force of a fist blow. As it was difficult in those times to calibrate this dynamic instrument it was calibrated with a qualitative scale . . . fort-ordinaire-FORT-très fort-extraordinaire. The W. & T. Avery Museum collection contains a sector spring balance, Fig. 4.24, that uses a similar principle. It dates from the middle 18th century, Sanders (1947).

The use of the torsion elasticity has been discussed above in the work of Coulomb and Cavendish. Despite the general knowledge of spring elements for force production and measurement there is little evidence in these times of the use of elastic elements finding their way into instruments as suspensions and pivots, the so-called flexure pivots. Precision weighing balances had evolved close toward using agate-bearing beam ends by the early 19th century, the early cord pivots having disappeared altogether.

According to Sanders (1947) it was the Turnpike Act of 1741 that brought about the design and development of the platform weighing scale. The Act required toll-gate locations to erect 'any crane, machine or engine, which they shall judge proper for the weighing of carts, waggons or other carriages'. This was for toll charging purposes. From this came the scale of John Wyatt, illustrations of which appear in Ellis (1973) and Sanders (1947), and many more designs. He used a compound lever system in which the position of the load on the scale platform did not alter the reading. It is still used widely today. This design was slow to take on at first and was, in fact, first patented in America in 1831 whereafter a patent was applied for in England only to be rejected as already being in common use.

In 1669 de Roberval devised the so-called 'static enigma' mechanism, Fig. 4.25,

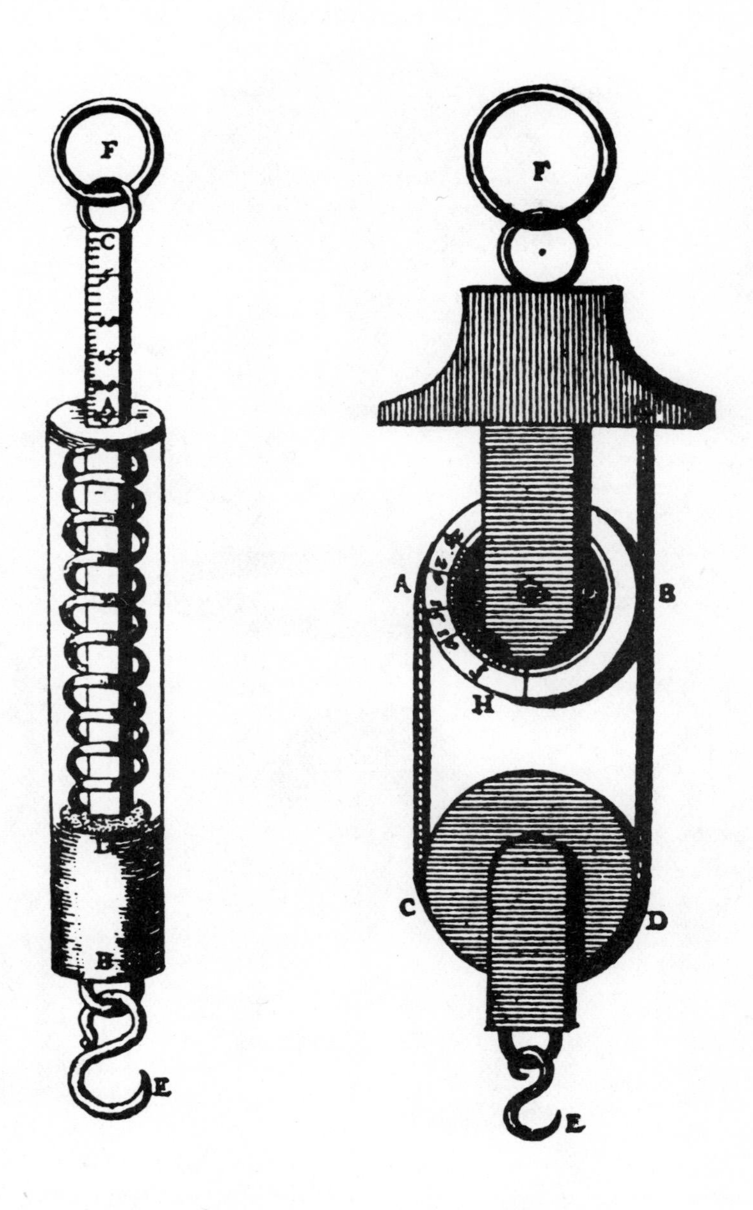

Fig. 4.22 *Spring balances, 1694. Left: coil spring. Right: rotary clock spring in upper pulley*
[Benton (1941) plate VII. From 16th century etching. Courtesy Newcomen Society]

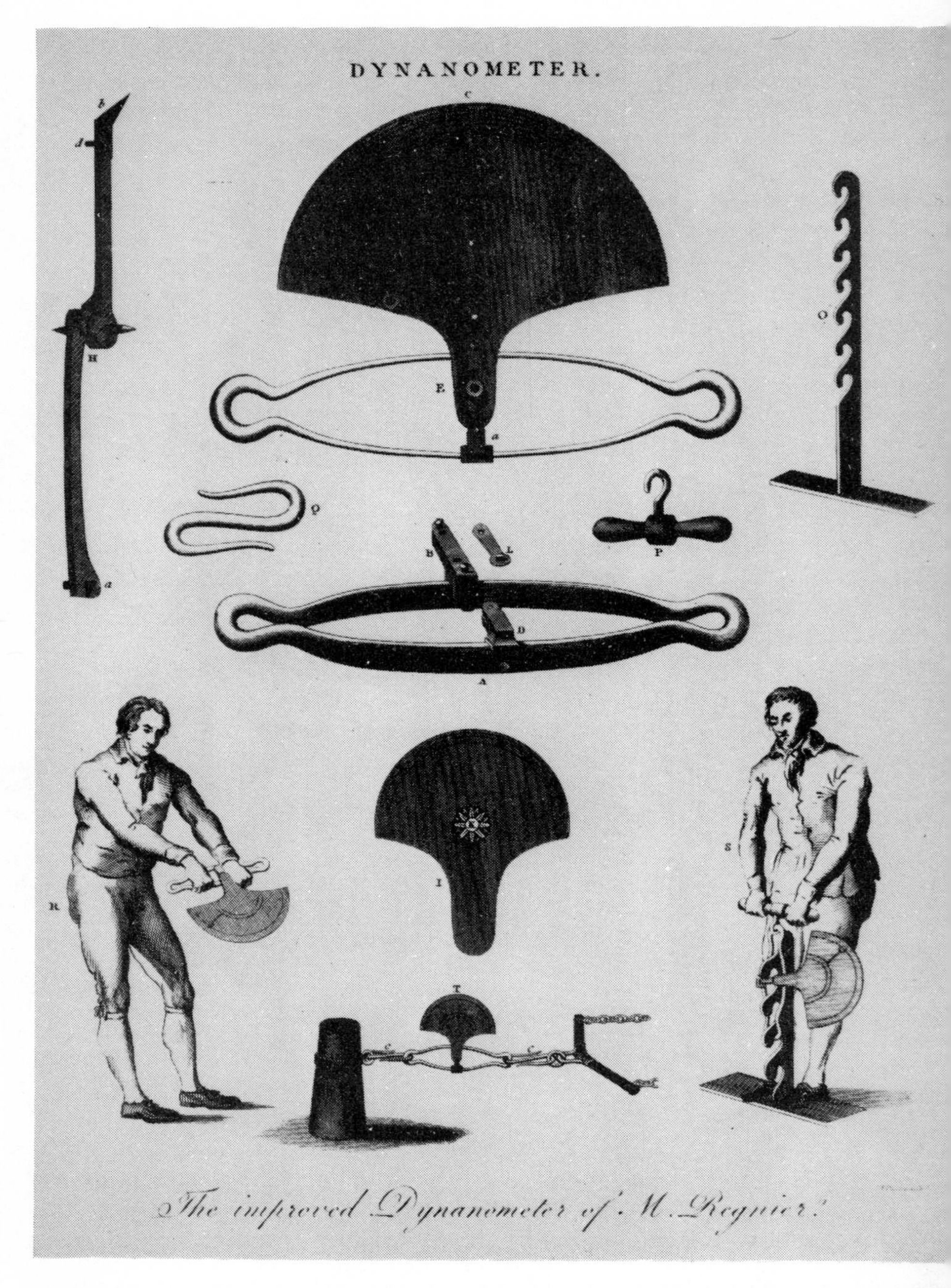

Fig. 4.23 *Print of Regnier's dynanometer; published in 1803 by J.Wilkes.*

Fig. 4.24 *Sector spring balance, middle 18th century*
[Courtesy W & T Avery Ltd.[

reporting it in 1670 in Journal des Scavans. Regardless of the position of the loads on each side they will remain in balance if they are of the same mass. This was very puzzling to the mathematicians of his time, hence the name. This principle was eventually put into use in weighing around 1800, enabling scales to have secured pans that were easier to use than hanging pans. A demonstration unit is illustrated in Bos (1968) and Turner (1973).

Clocks were a main stimulus of mechanical design activity; along with falling-weight drives were added, from *c*1511, the spring element power source. This required means to adjust the drive torque as the spring rand down. for this the fusée, Fig. 4.26, was invented by Jacob Zech around 1525, De Carle (1965), Ward (1970). This device derived uniform torque output by gearing the spring drive drum to the output drive shaft via a gradually reducing radius, cord-connected, spiral. Another method was the stackfreed mechanism in which the friction used to brake the spring was reduced as the spring ran down. Both devices called for considerable precision of manufacture. The oldest surviving spring-driven clocks date back to 1540. Ward (1970) states, as a foot note, that existing evidence suggests a spring-driven clock, with fusee control, was made in 1430.

Fig. 4.25 *Roberval's 'static enigma' balance, signed by J.V. Wijk.* [*Utrecht University Museum. Photograph Jac. P. Stolp.*]

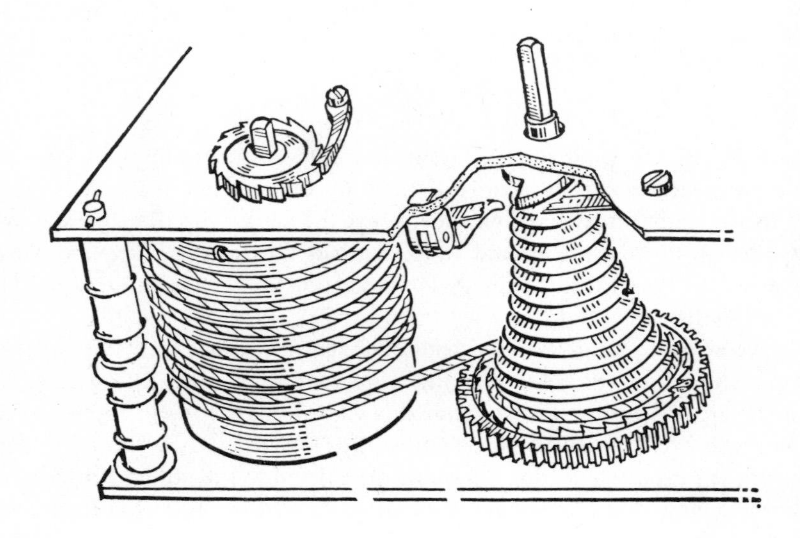

Fig. 4.26 *Fusee mechanism to maintain constant torque as spring runs down; invented 1525* [De Carle, D. (1965), Fig. 5, p.24. Courtesy Teach Yourself books, Hodder & Stoughton Ltd.]

Another aspect needing improvement in clock design was control of rate. The foliot mechanism was superseded when the discovery was made that the swinging pendulum had a period that was not, to the first order at least, dependent on the mass of the pendulum, but on its length. Galileo is said to have realised this from observations of the chandeliers at the Cathedral chapel at Pisa. He did, for certain, perform experiments that led him to state that the period of vibration of the 'simple' pendulum varies as the square root of its length. He also used a pendulum to measure elapsed time. Lenzen and Multhauf (1965) have described the early work with pendulums in another field of use, for pendulums were of interest then for more than timekeeping. Until quite recent times (1960s) the pendulum was the prime method of determining the absolute force of gravity and for a considerable period the consistency of the period suggested, Fig. 4.27, that a pendulum of a certain length might be used to

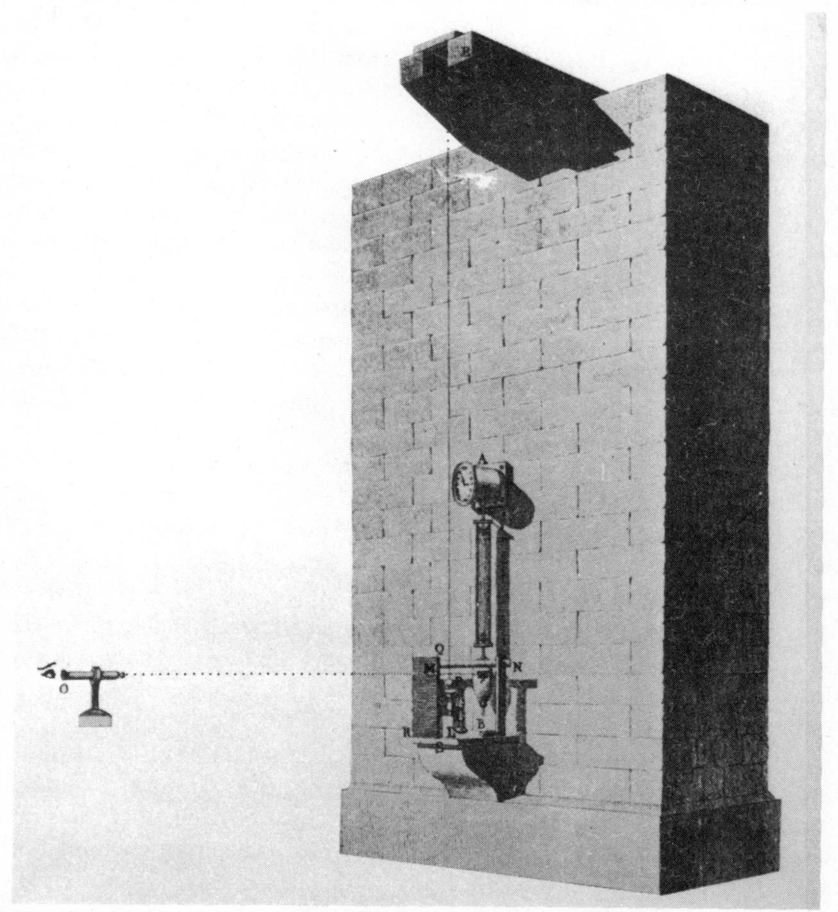

Fig. 4.27 *Pendulum apparatus used by Borda and Cassini in 1792 to determine length of the seconds pendulum at Paris. The method of coincidences was used to time the period of the longer pendulum according to the period of the clock pendulum*
[Lenzen and Multhauf (1965) Fig.6, p.311. From Mémoires publiés par la Société Francaise de Physique, 4, plate 2]

define the length unit by setting it's length so as to swing at a stated period, the so-called 'length of seconds' pendulum. It was from these less known applications that a considerable body of knowledge arose about the relationship for swinging pendulums.

Huygens, in 1657, invented and patented the pendulum clock later setting down his theory (in *Horologium oscillatorium* 1673) of the compound pendulum. This was three decades before Newton published his laws of motion and gravitation. The use of the pendulum for gravity determinations, and as a possible length standard, went hand-in-hand with pendulum clocks for the best method of timing the gravity pendulum swings was to use the 'method of coincidences' in which the two pendulums were observed together, counting their swings until they came into visual coincidence. The verge and pallet escapement was modified into the anchor design by Clement *c*1680. but Hooke claimed the invention in 1666.

The design of clocks reached a pinnacle in this period with the Harrison's chronometers, Ward (1966, 1970). His four designs, from 1737-1759, are on working display in the National Martime Museum, Greenwich. Made large, by modern size standards, they incorporate incredible care of execution coupled with design ingenuity to compensate for second-order effects. They can only give rise to wonderment to the observer. Fig. 4.28, of his No. 1 and No. 4 versions, shows the gradual reduction of size. The watch shaped units, however, are much larger than are used today. Two other models after Harrison's design are also shown, Kendall's copy K.1 as used by Cook, see also Howse and Huchinson (1969), and K.2 used by Bligh on the Bounty.

It seems reasonable to assume that out of clock-making skills and methods flowed the necessary ingredients for other complex mechanical systems. Frankel (1977) is a recent overview of time-keeping technology. Study of extant mechanical automata from this period proves conclusively that many other people had the skills of innovative design and execution that were embodied in Harrison's clocks.

Timbs (1860) states that the name 'automaton' derives from the Greek meaning 'self-moved'. He described several well known devices from earliest times through to his day. Time has mixed myth with the truth and many of the accounts must be considered with caution. They include an artificial eagle of 1470 and an iron fly. The Emperor Charles V had many automata made; armed men on horses and wooden sparrows that reputedly flew. Louis XIV had a small coach and horses made that moved around the table with its coachman whipping the horses. The Memoirs of the Academy of Sciences of 1729 include a report of a set of automaton actors presenting a pantomime.

The most famous must surely be the duck of Vaucanson, (1709-1782). It resembled a living duck that ate corn, exhibited anatomically correct movements, drank water and passed excrement (but not as the result of a digestion process). Others imitated this device. Timb's list of automata built as humanoids and animal representations was extensive. It seems difficult to believe many of the claims made in reports. However, several of these machines are still extant proving that they were, indeed, made and did function as described.

An outstanding example is the writing machine made by Friedrich V. Knaus in 1760 (in the Technological Museum of Industry, Crafts and Trades, Vienna). This

Fig. 4.28 *Chronometers by, and after Harrison. Rear: Harrison's No. 1, 1737, Centre front: Harrison's No. 4, 1759, Right front: Kendall's copy, K.1 and left K.2*
[Courtesy National Maritime Museum, London: on loan from Ministry of Defence — Navy]

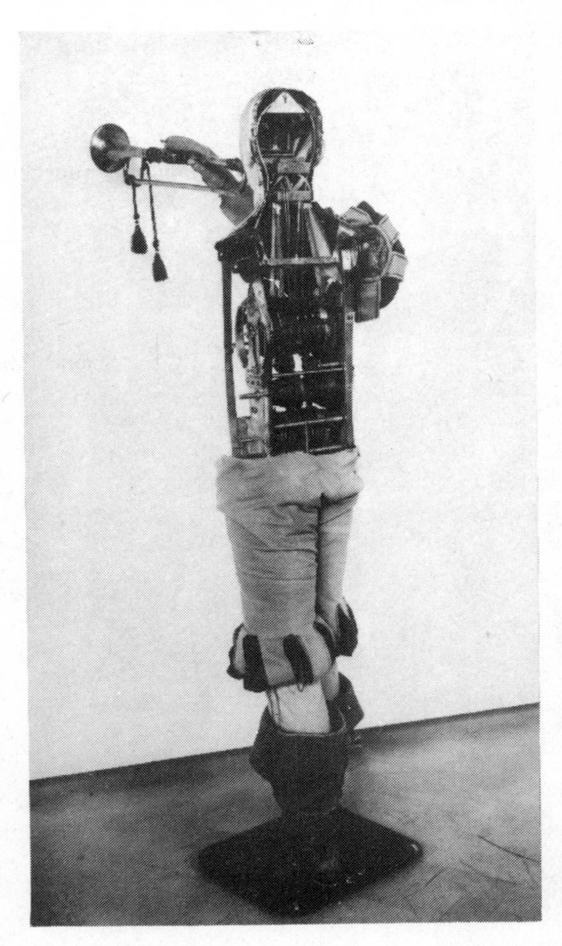

Fig. 4.29 *Rear view, showing internal construction, of Kaufmann's automaton trumpeter, 1810* [Courtesy Deutsches Museum]

comprises a lady figure holding a quill ready to write on a sheet of paper. On command it writes flowing script. The required movements can be pre-programmed into its internal workings via a rotating cylinder having pins projecting from it where needed. Another fine example is the trumpeter of Friedrich Kaufmann *c*1810, Matschoss (1925) that is in the Deutsches Museum, Munich, Fig. 4.29. Musical instrument players were popular automata subjects . . . piano-forte players, other trumpeters and violinists. Maelzel, early in the 19th century, built a whole band of 42 wind instrument players called the Panharmonica. The pursuit for human replication also encouraged research into aural sound production in this period. Many inventors investigated how to make speaking automata but with limited success. Hillier (1976) provides some information about automata and toys.

Automation also began to be applied to manufacturing machinery. Falcon's loom,

Fig. 4.4, used a belt of cards having holes punched where needed to enable the boy operator to rapidly select the required weaving combination for each row. In 1804 J.M. Jacquard developed the idea further eliminating the need for a drawboy to set the card. His machine is usually that referred to as the forerunner of punched card control, an incorrect statement in the light of Falcon's Loom and punched paper-roll controlled musical machines of the 18th century.

Calculation by machine also became reality in this period. Mechanical aids for numerical calculation date to the earliest times in simple forms such as the abacus. In the early part of the 17th century more complex instruments began to appear. The early development is described in Pugh (1975) in the updated catalogue of the collection of calculating machines and instruments held by the Science Museum, London. Trask (1971) provides a brief pictorial overview of the development of calculators. Murray (1948) is a study that explains, in depth, how mechanical devices can be constructed to carry out both analogue and digital computations. Eames and Eames (1973) takes up the historical development of computing equipment in considerable detail from the period *c*1830 onward.

The first calculating devices of the European tradition were the rods, or 'bones', of Napier from 1617; they were made as individual rods and as cylinders set up in a box. They operated as the forerunner of logarithm scales on the now, fast-fading, slide-rule. Their manufacture required a reasonable grade of precision manufacture and scale engraving. According to Pugh (1975) a description of a calculating machine is known to have existed from 1623 but the oldest extant apparatus dates to 1642 when Pascal constructed an adding device. Fig. 4.30 shows details of the manufacture involved. Pascal made some fifty of these machines holding a patent that prevented other calculating machines of any form from being constructed for a considerable period. An original Pascal calculator is held by the Conservatoire des Arts et Métiérs, Paris. Some other museums have replicas.

Later, Leibniz, using a different technique of carry, Fig. 4.31, made a significant contribution by inventing the idea of multiplying by consecutive addition and constructing a model that did this in 1694. Moreland, 1666, also invented a calculator described as,

> new and most useful instrument for addition and subtraction of pounds shillings, pence and farthings; without charging the memory, disturbing the mind, or exposing the operator to any uncertainty; which no method hitherto published can justly pretend to.

It was very small, measuring only 75 by 100 by 6mm. Also important was that his account of this calculator described a more advanced one that could also multiply.

As can be readily realised the main problem facing would-be makers, even until the end of the 19th century, was not a lack of design principles but the need to be able to make precision gears and other small mechanisms. In Moreland's time, each gear would have been made by hand, cutting it from a blank of metal. The first commercially viable mechanical calculator, stated Pugh (1975), was that of Thomas de Colmar, called the Arithmometer. Principles embodied into these early calculators form the basis of current instrumentation although the current trend of this 1970 decade is to replace

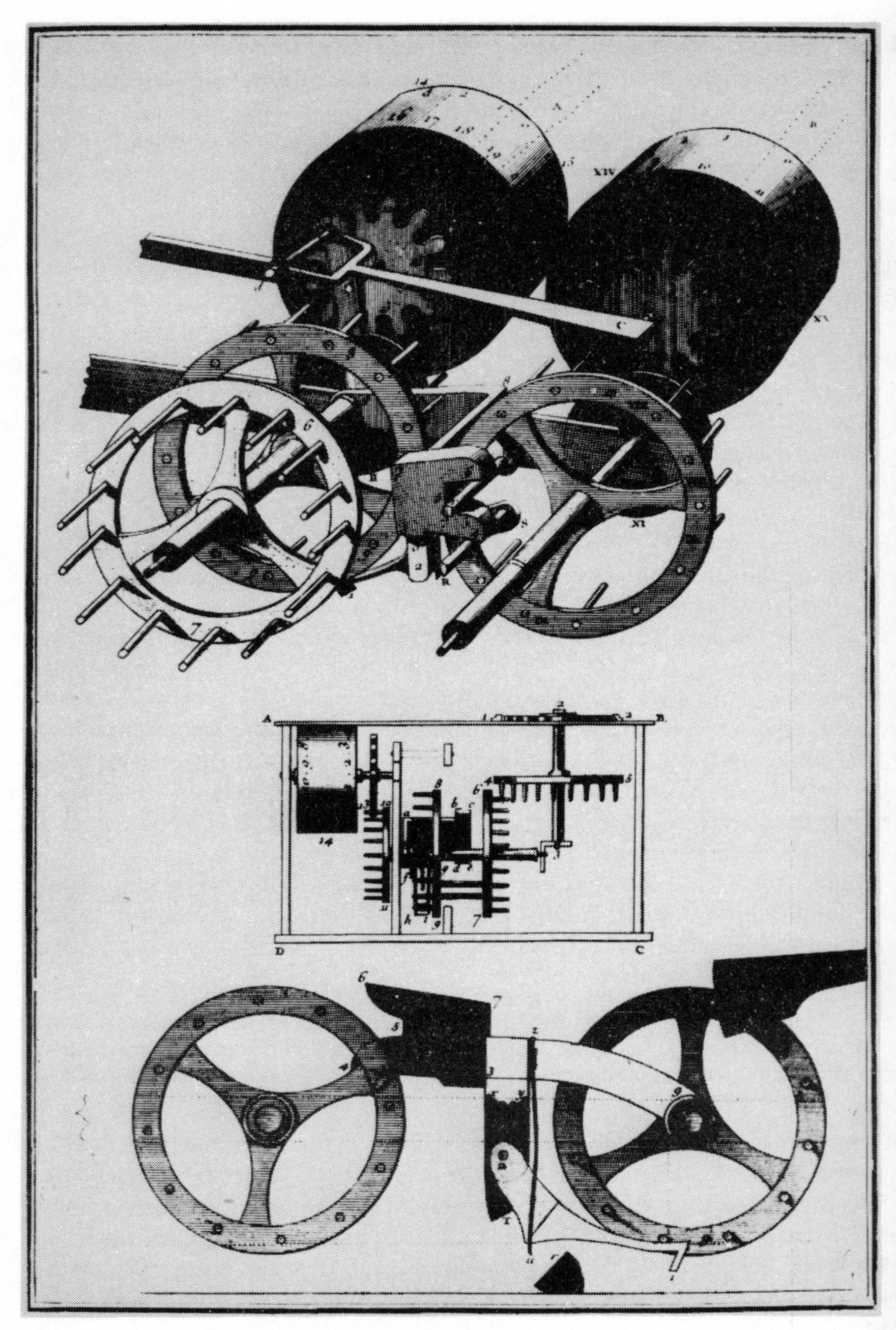

Fig. 4.30 *Internal construction of Pascal's calculating machine*
[Trask (1971) p.65. From 17th century engraving]

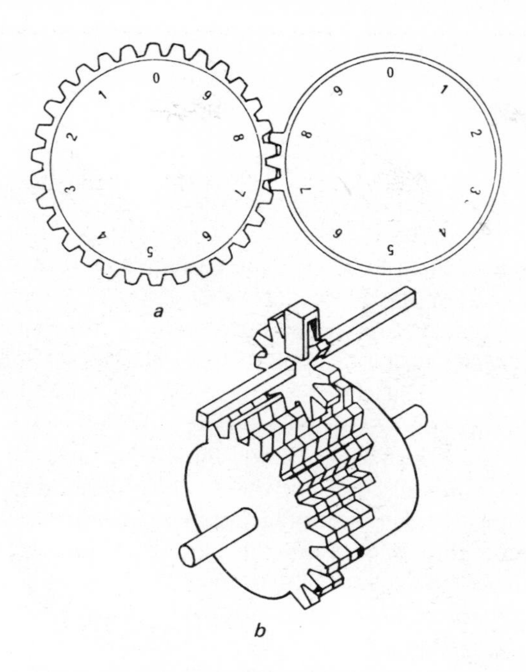

Fig. 4.31 *Carry mechanisms of 17th-century calculators*
a Pascal's stripped-gear carry
b Leibniz's stepped-wheel carry

mechanical calculating mechanism with electronic alternatives. The familiar trip and distance-run odometers of the car dashboard, that use Pascal's carry mechanisms, will soon be gone ousted by solid-state electronic counters.

Summation of revolutions – the Pascal and the Leibniz carry principles – found their way into devices such as the waysider. The hodometer, Fig. 3.24, of Roman times was replaced by more elegant and commonly used distance measuring devices. These usually had a mahogany wheel of diameter (around 800mm was common) to suit the gearing and the unit of distance for which they were constructed to measure. The wheel drove a small clock-face indicator through worm and bevel gearing. Examples are described in Bos (1968) and Pearsall (1974) and one appears in the middle left of Fig. 2.21. Waywisers of this general design are still used today to transduce distance, and hence velocity and acceleration, for the testing of motor vehicle performance.

Closely related is the pedometer, Wartnaby (1968). These were invented in the mid-sixteenth century. A cord, attached to the leg, was used to count paces by a ratchet action on the rotating counting shaft that displayed the total as dial position. One shown in Wartnaby (1968) has four decades giving a 1 to 10 000 paces range. This design was eventually replaced by a pendulum-driven arrangement in which the need for an attachment to the leg was not needed. Mechanical skills, married to optical aids enhanced land survey giving it a new level of precision, Richeson (1966).

It would seem that by 1800 there was little room to improve mechanical design.

But, as will be seen in the next chapter, the introduction of the mass-produced, interchangeable component-part philosophy yielded a major change to the mechanical aspect of instrument design.

4.4 Blossoming of optical instruments

The foundations of optical theory can certainly be traced to before 1500 but it was from around that date onward that worthwhile principles were revealed and quantified such that designers could build observing and measuring instruments that had impact. The 'new philosophy' called for observation and any device that could allow the investigator to see where others had not previously been able to see found application. Improved mechanical skills and designs enabled eye piece micrometers to be fitted giving a measuring scale and unit to the qualitative observing instrument.

Table 2.4 gives dates of the discovery of the dominant laws of optics. Reflection, refraction, colour composition, lens theory, polarisation, wave and corpuscular theory were each worked upon in this period. From these laws came numerous optical devices based, at first, on simple principles and construction. They ranged from the very familiar telescope and microscope to projection systems, spectroscopes, heliostats, eyepieces, spectacles and 'camera' apparatus.

Fig. 4.32 *Descartes diagram showing his ideas of visual feedback control.* [Sherwood-Taylor (1945), Fig. 35, p.112. From Descartes *De Homine,* 1677]

The first available optical sensor was, of course, the eye and this was the subject of study by many investigators – now termed opthalmology. By the 16th century the internal structure of the eye had been studied by dissection methods and the various parts named. The print given as Fig. 4.32 gives an appreciation of Descartes' concept of the eye and how it is used to control position of the hand by feedback. Descartes was the first to state that the human body is a deterministic machine mechanism. The debate still continues today as to whether there is more to the body than can be observed to function in a deterministic way. In his time considerably less functions were thought to be determinable than we understand are today. Descartes subscribed to the then current 'soul' theory. In his control system, a small 'soul' (the pear shaped object) was retained as the centre of 'intelligence', possibly to appease followers of the theory. From the instrumentation point of view his diagram shows a developed understanding of optical ray theory and feedback control principles.

There exist many contemporary prints about man-made optical devices, often with diagrams of their relevant optical theory. Schneiner's apparatus for projecting sunspots, Fig. 4.33, is a fine example from 1630.

The lens has its origin extending well before the invention of the telescope and several documents have been found containing passages that can be interpreted to be about lens shapes and their action on rays of light. The introduction of the optical system for magnification and other similar geometrical gain transformations, had considerable impact on the design and utility of instruments then in existence.

Just who was the designer of the first telescope is still in doubt but it is generally attributed to be a Dutchman, Hans Lippershey. Cajori (1962) provides a short account of the relevant events explaining that the situation was quite complex. Gunther (1935) states his belief in a section on telescopes. Many people, already then making spectacle lenses came close, if not actually to, the invention of the method for magnifying distant objects. The first maker of spectacles is also contested – see Needham (1962) Section 26 – but it does seem certain that they were used to correct long-sightedness in the 13th century. Pepper (1874) states that they were supposed to have been unknown by the Ancients and that it was either Alexander da Spina, a monk at Pisa in *c*1300, or Salvinus Armatus, a nobleman from Florence who died in 1317 and who has the claim inscribed on his tombstone, but Needham dates invention to before 1286.

It is not particularly important, from the long-term point of view, who the inventor was and in all probability there were many who had the kind of knowledge and practical skills needed to make such instruments. By the 1500s the manufacture of spectacles was commonplace, Fig. 4.34, but that did not mean that they were as widely used as they are today. Short-sighted correction came three centuries later. Lippershey was a spectacle maker. It is said he stumbled upon the use of two lenses to obtain telescope magnification when testing spectacle lenses.

Many accounts exist on the invention of the telescope and its early development. A definitive history is that by King (1955). Timbs (1860), Sherwood-Taylor (1945), Cajori (1962) and Thoday (1971) each provide short accounts. Howse (1975) describes them in relation to the Greenwich Observatory. It seems Lippershey, upon discovering in 1608 that he could see distant objects as though they were closer, immediately

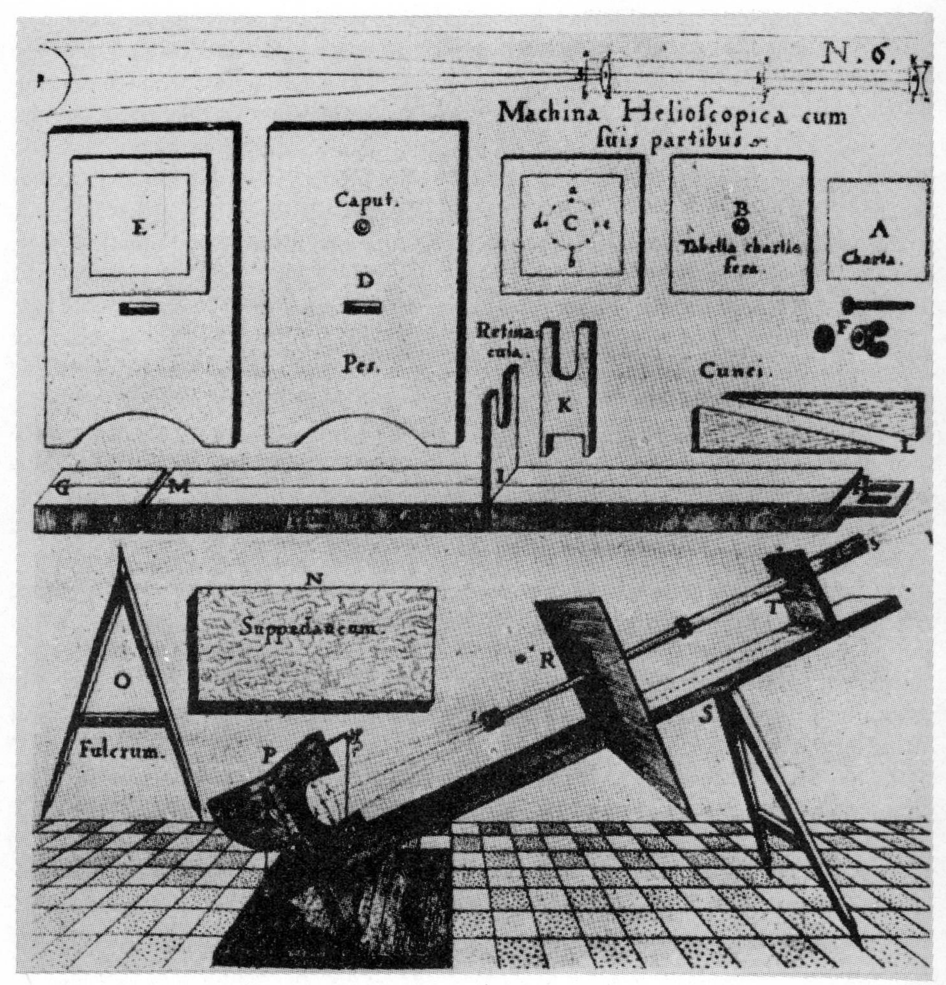

Fig. 4.33 *Scheiner's apparatus, 1630*
[Sherwood-Taylor (1945), plate IX. From 1630 publication by C. Scheiner]

constructed a telescope using two convex lenses mounted in a tube. He presented this to Prince Maurice of Nassau who helped spread the news of this invention. Galileo heard of it in May, 1609. He studied it for a day and then constructed an improved version using a plano-convex object glass and a plano-concave glass as the eyepiece. Both were mounted in a lead tube. He published details of his constructions and procedures of design less than a year later in *Siderius Nuncius*. The oldest existing telescopes are two of Galileo's. They are in the History of Science Museum, Florence. Lippershey also constructed a binocular telescope in 1609.

The first telescopes provided very poor quality images but continual development gradually brought about improvement as each error source was understood better. At first the gain was only three diameters, but by 1610 Harriot, an Englishman, had one

Fig. 4.34 *16th-century woodcut of spectacle making*
[New Caxton (1969) p.4491. From 16th century woodcut. Courtesy Istituto Geografico de Agnosti, Milan, Italy]

working with fifty times magnification. Galileo worked on the optical polishing and design to raise his magnification to similarly better figures.

The long focal length lenses used for spectacles were a fortunate choice (perhaps that is why they were stumbled upon for telescope use) for they exhibit less of the classical defects (chromatic and spherical abberation) than do more powerful lenses having greater curvature. Early telescopes always had a confusing coloured image due to chromatic abberation of single element lenses when transmitting wide-band visible radiations. This defect was overcome by Chester Hall *c*1740. The first refractor that was free from chromatic abberation was made by Dolland in 1758.

Timbs (1860) includes a short article on Guinard's glass for achromatic telescopes. He traced the developments of the 17th century revealing how the excise difficulties imposed on manufacture of flint-glass by the British Government caused that country to lose it's earlier technological lead in both supply and telescope design. Enterprise then went to Guinard of Neufchatel who operated free of such controls. By 1790

Guinard had succeeded in making disks up to 0·3m in diameter. Guinard was persuaded to move to Munich to work with Fraunhofer. He there obtained the patronage of Maximillian Joseph, King of Bavaria. French artisans obtained knowledge of his techniques and added high-quality telescopes to the then available German designs. England was thus left behind because of shortsighted policies. Such was the impact of the better production of mere glass.

The Galilean telescope provided an erect image, but could not yield a wide-angle of view with high magnification. Kepler, in 1611, used the convex eyeglass instead of a concave one. This inverted the image, which was of no real consequence for astronomical work, and increased the usefulness of two lenses. The Ramsden eyepiece, a development from other work, was added to the optical stock of basic components in *c*1770. This improvement made use of an additional lens in the eyepiece construction.

Refracting telescopes of those early times must have been very tedious to use and study soon resulted in the reflecting alternative. In 1663 James Gregory proposed his 'gregorian' telescope that used two concave mirrors. It was not built at the time. A

Fig. 4.35 *Newton's second reflecting telescope. Copy of the original in care of the Royal Society, London*
[Crown copyright. Science Museum, London]

little later Newton, 1668, built a working reflector using a slightly different arrangement in which the image is brought out of the side instead of viewing it through a hole cut in the centre of the primary mirror. Fig. 4.35 shows a copy of the second telescope that he made in 1671. A little later, in 1672, a Frenchman named Cassegrain, proposed a gregorian style in which the secondary mirror is convex. Gregorian and cassegrain designs used parabolic primaries and the newtonian a spherical mirror. Thus were founded the basic arrangements of many reflecting telescopes in use today. Over time methods to make and 'silver' better mirrors were improved but this aspect of development slowed in the 19th century whereupon photographic methods and, today, photoelectric methods of image observation have become the areas of improvement to telescope technique.

Refractors were initially inferior to reflecting units but design gradually improved and by the 18th century it was again possible to build telescopes that were as useful as reflecting types. The triple-lens object glass came into being in 1763. The size of a telescope is usually judged by the diameter of the primary mirror, or the object glass in a refractor. Length is not necessarily an important factor in deciding the resolving power, magnification, angle of field and brightness of image. Telescopes in the 17th century were often built with focal lengths of over a hundred metres. The so-called aerial telescopes comprised an object glass unit – for an illustration see Thoday (1971) – that was held on the top of a high pole using a tight silk line to provide alignment with the eyepiece that was mounted on the ground. Huygens had one having a nominal 70m focal length. With it he discovered the first satellite of Saturn in 1655. This system was presented to the Royal Society, London.

By the end of the 18th century the largest telescope then in existence was that of Herschel. It, built in 1789, had an aperture of around 1·23m and focal length of 12·3m. Fig. 4.36 shows the structural detail of this great instrument that took £4000 from King George III's coffers. Herschel made his own mirrors, including 430 parabolas alone. He made many for sale to supplement his income. To give an idea of what was possible with the knowledge of his time and within the life span of one man, Herschel also catalogued many thousands of stars, discovered 806 double stars, counted the stars of 3400 gauge fields and estimated the brightness of hundreds of stars. This was made possible with the telescope. He also developed new theories of stars and investigated the nature of infra-red radiation. Telescopes by Galileo, Newton, Herschel and the Earl of Ross are each discussed in Timbs (1860).

Just as the telescope made a great impact on observation, so did the microscope. This invention is ascribed to either Zacharias Joannides or, perhaps, to Cornelius Drebbel by Cajori (1962) but again, as with the telescope, it is doubtful if one man could be accorded the claim for the microscope developed out of the simple, single-lens, magnifiers that existed in Ancient times. It is clear, however, that the compound design having two lenses, came into existence around 1590. Timbs (1860) and Pearsall (1974) both provided brief accounts of the historical development. Timbs discussed a unit, made by the Jansen's – Dutch spectacle makers of Middleburg – that they made in 1590 and presented to the Archduke of Austria. It was said to be two metres long. Pearsall stated that the first illustration of a magnifying glass in use appeared in 1592.

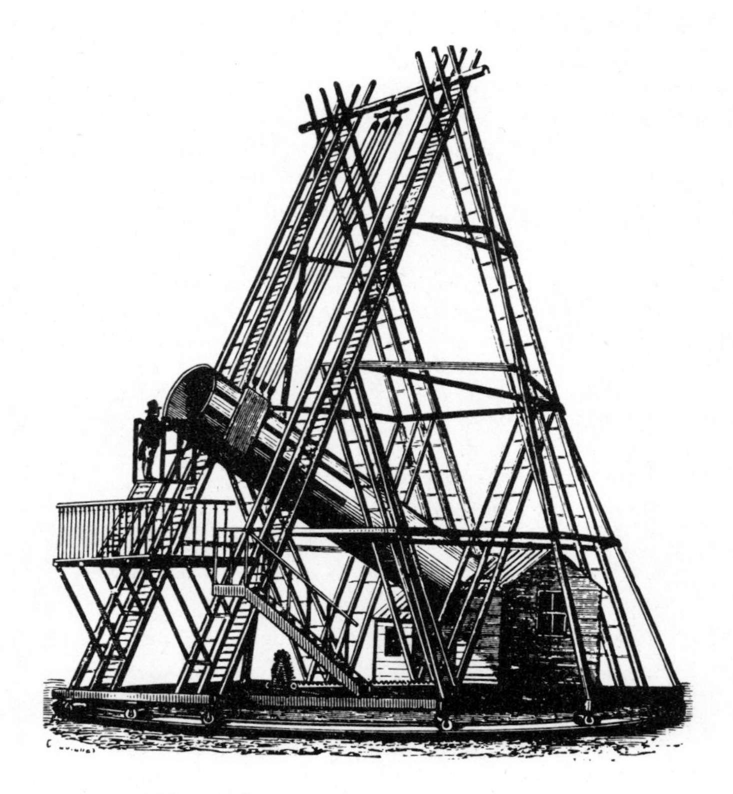

Fig. 4.36 *Herschel's '40-foot' telescope*
[Lodge (1893), Fig.83, p.283]

Fig. 4.37 *Original Leeuwenhoek microscope c1650.*
[Utrecht University Museum. Photographer Jac. P. Stolp]

They certainly existed before that time. Galileo had one that he claimed could magnify 50 000 times; he was quoting areal, not lineal, magnification. It was Hooke, *c*1660, who added the third lens forming the design that lasted to this day.

Microscopes existed at the start of the 17th century but it was the work of van Leeuwenhoek that brought the relevance of this instrument to the attention of the many. Throughout history there are always those who, due to circumstance or method of presentation, cause the work of others to be taken up, the so-called 'gate-keeper'.

Leeuwenhoek's microscope is simple indeed, using only a single lens. It is 7cm long Fig. 4.37 shows the original unit that is commonly known because of so many museums around the world having replicas of the original existing in the University of Utrecht Museum. His design made optimal use of available knowledge and skills. By using a surprisingly small spherical lens (one writer describes it as about ball-point pen, ball size) he was able to reduce abberations to a workable value. He left 26 of these to the Royal Society, London, but all have vanished. Their magnifying powers, according to Timbs (1860), varied from 40 to 160 and they performed so well that the compound method was 'laid aside' for several years. These tiny pieces of glass were each hand-ground as a double-convex lens. One existing today in the University of Utrecht Museum has recently been tested, Philips (1975). It can resolve to 690 lines per millimetre, which is roughly equivalent to the resolving power of a good quality modern 4mm objective. Its' magnifying power 275 times.

In 1665 Robert Hooke published 'Micrographia' in which he described his personal microscopy system. In doing so, he showed how the basic compound microscope was combined with several additional aids to make it a workable system. The prime requirement of microscope work was, and still is, that the object has to be held in space relative to the optics with a high degree of positional stability. Leeuwenhoek's design, see Fig. 4.37, provided movement of the specimen across the field of view, and also into it, to allow focussing of the image. Hooke, as can be seen from Fig. 4.38, went further by adding a source of light to illuminate the subject material.

Hooke's unit must have been tedious to set up for all positional adjustments were devoid of fine screw drive controls that were capable of keeping the image in the same place. The screw barrel, according to Pearsall (1974) was introduced as a more effective means of focussing by Marshall late in the 17th century. Gunther (1935) states how, in 1770, Benjamin Martin did this, work that eventually led to the use of the standard Royal Society thread in 1859. Pearsall (1974) provides a short introduction to the various improvements to the microscope. In 1758 Dolland began to market his achromatic lenses for telescopes but it was not until 1824 that these became available for microscopes. The inversion objective did not appear until the 19th century.

Telescopes and microscopes, being functional over a long enough life to become working collectors pieces, and because of their high popularity as objects of antique value are well documented. Their 'popular' exposure is such that it might be thought that they are all there was to the application of optical methods; many other applications existed. Collections of microscopes are held in numerous museums, from the small to the large. Published catalogues of relevance include Gunther (1935),

Fig. 4.38 *Hooke's diagram of his microscope system*
[Philips (1975), from Hooke, R., *Micrographia,* 1665]

Rooseboom (1956), Multhauf (1961*a*), Frison (1965), Chew (1968), US Army (1967), Warner (1968), Thoday (1971), Turner (1973), Ronan (1975), Wynter and Turner (1975) and Pipping (1977). Pearsall (1974) is a useful introduction to the varieties of telescopes and microscopes. Historical aspects can be studied by reference to Hogg (1867), Frey (1872), Van Heurck (1893), Disney (1928), Clay and Court (1932) and Bradbury (1967).

The availability of the telescope, Fig. 4.39, provided the means of improving the definition of the sight tube used in earlier surveying and astronomical measuring devices (Compare Fig. 4.39 with Fig. 3.38). The sextant, evolving out of the astrolabe, steadily acquired optical aids that improved sighting accuracy and allowed a more functional arrangement by using mirrors, intensity filters and the ability to sight the horizon and the point of interest simultaneously. The magnetic compass, the theodolite, and all manner of laboratory apparatus were instruments that benefited by optical magnification. Basically, these were all improved by the addition of geometric gain. Pledge (1966) well shows how important the microscope was to scientific advancement. Spencer-Jones (1972) is an account of 18th century astronomy that reveals the value of the telescope to science. Another example is its use in geodetic measurements, Butterfield (1906).

From the study of telescopes and microscopes image defects and from astronomical experiments there was gradually gained a knowledge of other optical phenomena that would eventually provide other less obvious applications of optical technique. Bray (1930) provides brief histories of these discoveries when explaining the general knowledge of optics.

Galileo was among the first to attempt to measure the speed of light. A little later, Römer, in 1676, explained his theory about the infinite speed of light to an, at first not convinced French Academy of Sciences. Following on, in a similar study of the apparently unexpected appearance times of astronomical bodies, Bradley, at Oxford, by chance verified Römer's work. Huygens, in 1678, presented his theory of light. Thus began a long debate about wave or corpuscular theory that is still largely un-

Fig. 4.39 *The '8-foot' mural quadrant of middle 18th century preserved in the Quadrant room, Old Royal Observatory, Greenwich*
[Courtesy National Maritime Museum, London]

resolved today. What is important from the point of interest here is that knowledge of the fundamental nature and speed of light was gradually building up to the point where interference would be explained ready to become part of optical instrumentation of the 19th and 20th centuries.

It was in this period that the spectral nature of light began to emerge from such works as Newton's on the dispersive power of prisms. Discovery of diffraction effects, by Grimaldi (reported in *Physico-mathesis de lumine*, 1666) laid the way to realising that optical instruments had fundamental physical limitations imposed on a design by the parameters of geometry, design layout and materials used. It became clear that any principle of measurement can only provide a given level of resolution and accuracy, natural limits existing to prevent improvement beyond certain achievements.

According to Cajori (1962), Newton had himself set up experiments that must have exhibited the different refractive index of various materials, later to be the basis for achromatic lens design and that he also, in experiments with the solar spectrum, must have seen the 'Fraunhofer' lines. In both cases he apparently failed to either observe the effects or to realise their significance. The first observation of spectral lines (from luminous gases) was made by Melvill in 1752 but his work was little known until the next century. It was then that spectroscopy began.

The majority of optical work of these three centuries was concerned with visible light and its spectral wavelengths. Here though, were the beginnings of the realisation that other 'radiations' existed. According to Jones (1972) the first clues to the existence of the infra-red portion of the then unknown electromagnetic spectrum range (only visible light had been observed) can be found in Newton's *Opticks*. There Newton reported an experiment that showed how heat could be transferred across vacuum, whereas sound could not. This required the use of reasonably sensitive thermometers and good vacuum technique, which by then were reasonably commonplace.

Little other interest in this apparently obscure observation arose until Herschel, in 1800, in the course of observing the sun, began to realise that darker filters (darker to the eye) did not necessarily filter the heat in proportion to their apparent visual darkness. He devised an experiment, Fig. 4.40, in which a prism (which unfortunately itself acted as a filter) was used to disperse the wavelengths into a spatial spread along which he set up sensitive thermometers that had been blackened. Jones gave the characteristics of these thermometers as having a response time of around five minutes and a sensitivity of, perhaps (in the units then used by Herschel) $0{\cdot}25^{\circ}$F. The need to allow for ambient temperature changes led Herschel to use two thermometers at each position, one in the invisible I-R radiation, the other being shielded. Jones suggested that this was probably the first use of the differential measurement technique. Today the use of a standard, blackbody (or near-black) source in the field of view of long wavelength sensors is standard practice as a means to control error. Herschel's experiment unfortunately possessed several factors that prevented an accurate study to be made. Allen and Maxwell (1944) describes these and how other workers gradually improved measuring apparatus in the field of I-R detection.

Thus it was that, although the first great impact of optical technique was simply to add gain and allow images and rays to be guided where needed the methods used, in turn when coupled with other sensing variables, led the way to new and more profound principles that allowed yet more varieties of instruments to be built . . . and so on. By 1800 optics had already begun to become recognised as being part of heat and radiation theory. Two, apparently distinctly different classes of phenomenon had begun to be proven to have fundamental laws of the same origin. In a half-century it would be seen that magnetics, electrics, optics, heat and mechanics were all inter-related. Optical technique would expand into an ever-widening range of knowledge seeking and controlling devices. Still more instruments would be constructed encapsulating, as a permanent record, man's knowledge of physical laws in hardware packages.

Fig. 4.40 *Artist's reconstruction of Herschel investigating the thermal effect across the radiation spectrum*
[Photo. Science Museum, London]

4.5 Growth of the instrument range

Optical devices, such as the telescope, and mechanical devices, such as the clock, might appear to be the major advances in the instrumentation of this period. They were not, however, by any means, the only developments for as new physical laws were being discovered they, too, were gradually finding their way into the implementation of previously unknown sensing and measuring equipments. Space does not permit more than a fleeting explanation of these additional fields. This section concentrates, therefore, on several groups that are today commonplace, relegating the remainder (those that have come to the writer's knowledge) to a table containing information that enables further detail to be obtained elsewhere.

As soon as some measure of degree is added to a qualitative measuring set-up it becomes necessary to generate a unit for the variable in question. With the expected, haphazard, development of understanding of an emerging new variable goes a parallel degree of confusion over the units to use. Early problems with weights and measures units have already been discussed in Chapter 3. Rapid expansion of transduction devices in this period brought with it much difficulty about units.

Although attempts were made to bring about widespread use of common units, at least for the traditional measurement variables, little progress was made in this period. To the confusion was added the need for more units for what became common measurements, temperature, atmospheric pressure, density and velocity, being some of the newly arising variables calling for units. Klein (1975) provides considerable information about the problems that arose as these areas gained maturity.

Throughout the foregoing part of this Chapter we have already touched on several embryonic transduction processes e.g. the pendulum to measure the force of gravity and to define the length standard, electrochemical reactions converting electric current into gas generation (which became, for a short period around 1900 the way to measure current used in electricity useage), pyroelectricity and the dynamometer. At least forty energy conversion effects were implemented as measuring and controlling transducers in this period, some taking many different physical forms. For example a thermometer of this period is used to measure temperature by converting some effect of temperature into an observable displacement. This particular kind of conversion stage was implemented in dozens of different ways. Hence from the earliest times in transducer development the problem of classifying the various branch developments has been extremely complex.

Temperature and heat

Measurement of temperature began for needs of the range that normally was experienced in everyday circumstances. The first devices were the thermoscopes of the Ancients mentioned in Chapter 3. Development began again when Galileo made a thermoscope around 1592. Chaldecott (1976) provides detailed information about this and other instruments in the course of describing the collection of the Science Museum, London. Galileo's thermoscope was made from a long thin glass tube having a bulb at one end. This was heated and placed upside down into water or wine held in another vessel. Due to the reduced internal pressure produced, the liquid would rise up the tube a small way. Air remaining in the, now upper bulb, changes pressures with temperature thereby indicating temperature variations as changes in the level of liquid in the tube. It was, therefore, also effected by barometric pressure and did not respond singularly to temperature effects alone.

The earliest known thermoscope having a scale of degrees and, therefore, giving a quantitative measure, is credited by Chaldecott to Telioux who wrote about his work in 1611. Von Geuricke also constructed a thermoscope of unusually long length in about 1660. The term thermoscope is now used more for devices measuring changes in temperature than for those giving absolute value to the degree of heat present.

The term 'thermometer' was first used by Leuréchon in 1624 and whereas any temperature measuring device would generally fit its definition it has become to be used more for instruments having the range of the familiar mercury-in-glass thermometer.

Atmospheric pressure dependancy was effectively eliminated in the 'Florentine' design first made *c*1641. It used similar principles to modern units but was made, see Fig. 4.41, from a much longer length of tube. Indication marks were fused into the tube wall. They also took a straight form which was more prone to breakage, Knowles-Middleton (1969*a*). The 'Florentine' design was much superior to anything before it but, even so, lacked consistent and repeatable basis of construction and calibration. Each thermometer made was an isolated unit having its own unique scale. Alcohol was used as it could be obtained with greater purity and because it possesses a higher coefficient of thermal expansion. Mercury was in use by the early 1700s. This was shown by the extensive work of de Luc (1727–1817) who came to this conclusion whilst researching barometers, Whipple (1972).

The liquid-in-glass thermometer underwent gradual improvement to become the thermometer that is so familiar today (but may now soon be displaced by a more robust and less expensive solid-state equivalent). Improvements eventually led to the manufacture of thermometer units that had traceable calibration, a common unit and an agreed scale. Knowles-Middleton (1969*a*) states that two 18th century thermometers had 18 different scales provided!

Table 4.1, adapted from Chaldecott (1976) with additions, gives some idea of the confusion that arose with scales that used different fixed points. Note the use of a reverse scale. The basic problem was that the nature of heat and, therefore, the notion of temperature was slow to be understood at a fundamental level. Each investigator created his own scale and fixed points using what, to him, seemed to be a reasonable 'normal' range and fixed points that were convenient. The boiling and freezing points of water were obvious choices. To these were added such apparently stable points as body temperature. According to Pearsall, thermometers were not made on a commercial basis until around 1750. This would have assisted standardisation.

It must be left to the reader to follow the development of the humble sealed thermometer, with such works as the Knowles-Middleton (1966, 1969*a*, 1969*b*), Pearsall (1974) and Chaldecott (1976). Numerous examples are extant in the scientific museums. Catalogues by Gunther (1935), Knowles-Middleton (1969*a*) and Turner

Table 4.1 *Comparison of some scales used for thermometers: from 1600 to present (As many as eighteen scales were in simultaneous use in the mid-eighteenth century)*

As marked on a George Adams mid-eighteenth century thermometer frame (Chaldecott, 1976)	Newton	De Lisle	Fahrenheit	Reaumur	Centigrade (SI Celsius)	Original Celsius	Kelvin Absolute
Just freezing	0	160	32	0	0	100	273·16 (IPTS-68)
Winter	1	154·5	37	2·3			
Spring and Autumn	3	146	48	7·5			
Air at Midsummer	6	132	63·5	14·5	No fixed points defined by IPTS-68 in this interval		
Hottest external parts of human bodies	12	104	95	29·5			
Water just tolerable to the hand at rest	14·5	92·5	108	35·5			
Water hardly tolerable to the hand in motion	17	81	121	41·5			
Melted wax grows stiff and 'opake'	20	67·5	136	48·5			
Spirit of wine just 'boyles'	25	43·5	163	61			
Water begins to boyle	33	7	204	80·5			
Water boyles vehemently	34·5	0	212	84	100	0	373·16 (IPTS-68)

Fig. 4.41 *Reconstruction of Florentine thermometer constructed by Gonfia c1657*
[Crown copyright. Science Museum, London]

(1973) provide illustrations and useful statements giving depth to the knowledge about each artefact listed.

Sealed, filled thermometers were not the only kind invented in these times. Bimetallic devices using the differential expansion of two materials having different thermal expansion coefficients were also devised, probably as the result of scientific apparatus constructed to measure thermal expansion. These instruments were then called pyrometers (sometimes at a later date 'tasimeters') but to avoid confusion with current day usage of the term they are generally now referred to as dilatometers. Examples of early dilatometers are the first unit invented by Musschenbroek in 1731, and others reported by Chaldecott (1954), Brongniart's iron bar lying in a porcelain support at Sèvres, Deschanel (1891) pt. 2, and a late 18th-century artefact in the van Marum collection, Turner (1973). Ramsden, in 1785, used this principle in connection with an early geodetic baseline that was set-up on Hounslow Heath with an iron-bar length standard, Whipple (1972).

The thermometer soon found its way into closed-loop temperature regulation in the *c*1620 furnace of Drebbel, see Fig. 1.27, in its baro-thermoscope form. Reamur discussed designs (of Roma's) using thermal expansion of a bar to control heat flow and in America, William Henry constructed his *Sentinel Register* reporting it in 1771. This used a 'temperature feeler' based on the use of a sealed air thermometer that forced water to rise, moving an actuating float. It was responsive to the expansion of both the water and the air but this mattered little in a fixed setpoint application. The above accounts are taken from Mayr (1970) which provides an in-depth study of the early use of such controls with tight reference to the primary literature.

While on the subject of heat, mention must be made of the work carried-out on the measurement of specific heat. The first quantitative measurements appear to be those of Lavoisier and Laplace made with apparatus reported by them in 1784 (but dated 1780, see Chaldecott (1954)). The original is held in the Conservatoire des Arts et Métiers, Paris with copies existing elsewhere, Chaldecott (1954). They related their calorimetry measurements to the concept of determining how much ice was melted in the cooling of a heated substance. The van Marum collection, Turner (1973), has a later apparatus devised by Crawford that he used around 1780 for work on animal heat.

Pyrometry, in context as measurement of high temperature, has its workable origins in the 1782 device made by Wedgewood, the famous potter. Newton, earlier, had attempted to measure the temperature of a fire by observing the cooling time for a bar of iron heated to red heat but his publication of the results in 1701 appear to have had little influence.

Wedgewood's apparatus, Fig. 4.42, 'was based upon the permanent contraction of certain argillaceous bodies under the action of heat, the contraction which began at low red heat being thought to proceed uniformly with rise in temperature until the clay became vitrified', Chaldecott (1976). Wedgewood used his own scale of degrees. In use a piece of a selected clay was formed to just fit the wider gap of the tapering gauge bars. After heating the specimen was rapidly cooled and fitted into the bars at its tightest point to establish the temperature. It must be realised that Wedgewood was primarily concerned with contraction problems and his method, although unscientifically based, was probably more suitable than the next described.

A more direct method was Ferguson's method. This was based on mechanical amplification of the extension of a heated bar. He used several stages of levers to obtain a gain of 400, Chaldecott (1976).

Pressure

Berti's experiment, shown earlier as Fig. 2.45, was one of a number conducted in the 17th century to seek knowledge of what existed in space. Many held that space always must have a substance in it – the 'horror vacui'. Experiments were devised that eventually showed this to be incorrect. Knowles-Middleton (1964, 1969*a*) ascribes the invention of the barometer, not to Berti, but to Torricelli (1608-1647). His experiments showed that air-pressure could be used to support a column of mercury but it

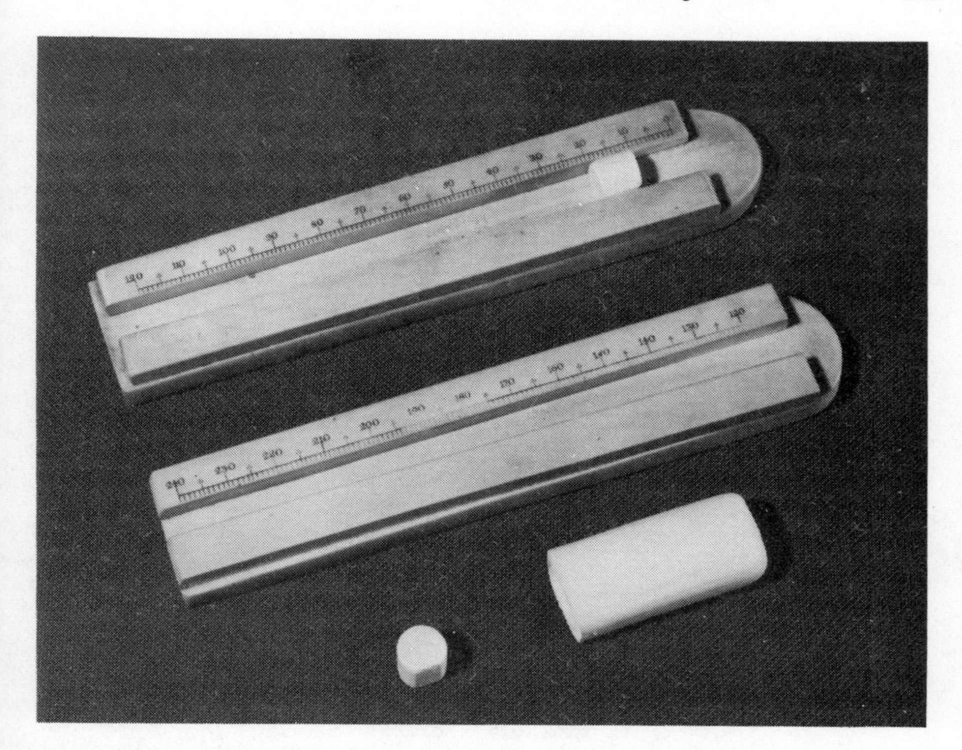

Fig. 4.42 *Wedgewood pyrometer introduced in 1782. Instrument made by Newman for the King George III collection*
[Photo. Science Museum, London]

is not known if he took the instrumental apparatus to the quantitative stage by adding a scale. Evidence specifically dates his device for measuring the 'changes of air' to 1644. Descartes added a scale in 1647.

The development of the barometer was based upon a considerable study of vacuum and knowledge gained in investigation of the reasons for the failure of early pumps to lift water from greater than a certain depth. Cajori (1962) and Jones (1971*b*) are brief introductions to this preceding work.

Knowles-Middleton (1969*a*) suggests that the next phase of development of the mercurial barometer, in the 17th century, was concerned with ways to increase the sensitivity of reading the comparitively small changes caused by atmospheric variations. Hooke expanded the scale by using a siphon form to actuate a rotating pointer. Replicas of this historic instrument, Fig. 4.43, have been made from Hooke's original etchings of it. Another method of obtaining scale expansion was to incline the tube; now commonplace in manometers. This was first done by both Morland and by Ramazzini in the mid-17th century.

A torrecellian barometer is known to have been used outside of a laboratory situation in 1648. It was taken up a mountain to observe pressure reduction with altitude. It was this kind of application that showed that air pressure varied for reasons other than height change.

Fig. 4.43 *Replica of wheel barometer according to an engraving by Hooke 1665* [Courtesy National Museum of History and Technology, Smithsonian Institution, Washington DC. Photo 61903]

Liquid filled barometers group into two basic types i.e. cistern and siphon forms. For each of these there exists numerous versions. Again, as with thermometers, the reader can obtain detail from such works as Middleton's several accounts, Pearsall (1974) and Frisinger (1977).

Mechanisms to enable the lower surface of the mercury to be zeroed and the upper to be read with precision came around 1770. Positive steps were implemented to define which part of the meniscus was used as the upper datum. The original Fortin design began with Nicholas Fortin around 1800, emerging as one of the two patterns now in general use.

Using the pressure of the air to bear upon a disk was well enunciated before 1800 (by Leibnitz, 1698; by Zeiher, 1763; by Conte, 1798) but it was not brought into practice until materials and processes of manufacture enabled it to be made as the aneroid barometer in 1843, Knowles-Middleton (1969*a*). The research of Knowles-Middleton on instruments of meteorology has been extensive. His works provide extensive detail of the numerous instruments involved. It is difficult to do justice to the available historical documentation on meteorological instruments in this brief space.

The barometer principle became important in early 17th century engineering as the principle used in pressure measurements associated with flow determination. With their introduction as part of the differential pressure flow meter, in the early 19th century, the manometer found extensive application for measuring small pressure-head differences. This application can be traced to measurement of pressure differences in the course of laboratory experiments on rough vacuum and water flow. The van Marum collection contains apparatus for the former, dated *c*1789.

Henri Pitot, in 1732, described a device for measuring flow velocity. It used, Burstall (1970), 'two parallel small-bore tubes mounted in a frame, one straight and the other with a short right-angle bend at its extremity'. Taps and a scale enabled the pressure-head difference to be stored within the device so that the velocity of the stream in which it had been placed could be gauged.This device is remarkably similar to an anemometer suggested around 1722 by P.K. Huet, Knowles-Middleton (1969*a*).

The work of D. Bernoulli (1700-1782), and later Venturi (1746-1822), required many set-ups for the detection of pressure differences. Fig. 4.44 shows extracts from Venturi's 1797 work *Recherches Expérimentales sur le Principe de la Communication*

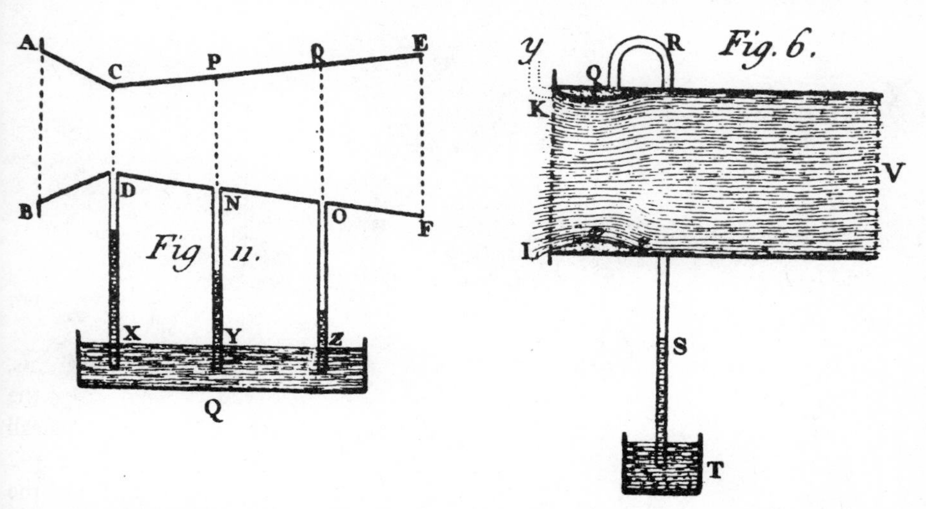

Fig. 4.44 *Extracts from Venturi's original diagrams showing use of water manometers to measure internal pressures in ducts*

[Kent (1912), p.16. From Venturi G.B. *Recherches Expérimentales sur le Principe de la communication Latérale du mouvment dans les Fluides appliqué a l'Explication de Différens Phenomènes Hydraulics,* 1797]

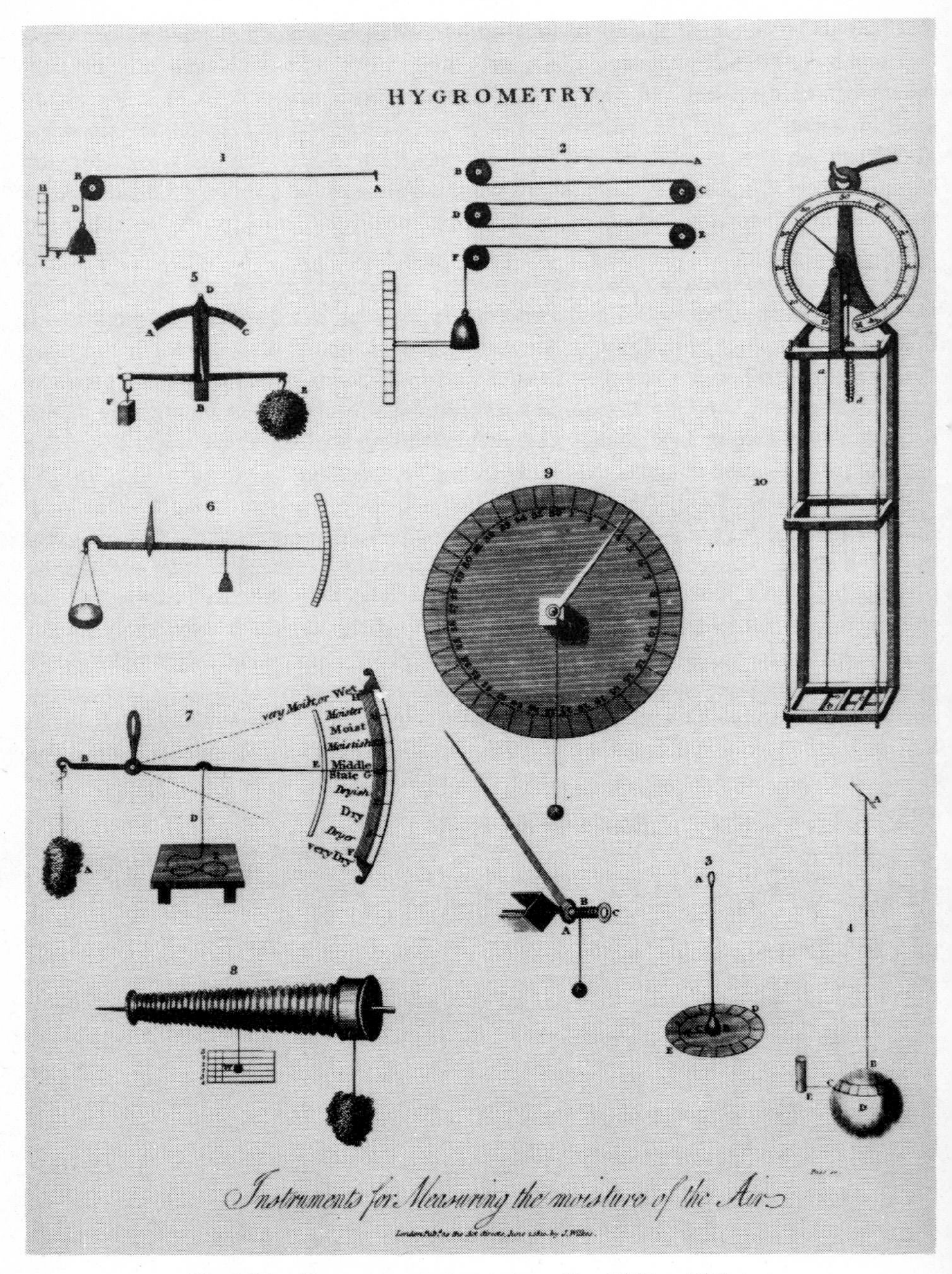

Fig. 4.45 *Hygrometer designs: released by J.Wilkes 1810*

Latérale du Mouvement dan les Fluides appliqué à l'Explication de Différens Phénomenès Hydrauliques. Venturi's pioneering work in flow metering has been briefly studied by Kent (1912).

Atmospheric humidity

Quantitative measurement of the water present in the air has been possible since the late 14th century (see last chapter). It was, however, subsequently found that the dimensional change of certain physical materials was also an easy measurement principle to use and this indirect method gradually supplanted direct weighing. Mizauld is the accredited – Cajori (1962), Knowles-Middleton (1969*a*) – inventor of the gut-string relative humidity hygrometer in 1554. This design became very common after that time, Fig. 4.45, taking numerous forms. The sensitivity of the oat-beard to relative humidity was known to children in the early 17th century. This phenomenon became established as an instrument through the later work of Hooke. He described a working oat-beard instrument in his *Micrographia.* It has since been reconstructed. The hair hygrometer was introduced by Saussure in 1783.

Condensation hygrometry was also a product of these early times for in 1655 Ferdinand II of Tuscany invented an apparatus that measured the amount of water of condensation that ran off from an ice-cooled surface. This appears to have been an isolated instance, having no subsequent links to the modern dew-point instrument. The course of development can be traced to le Roy in 1751. Again Knowles-Middleton's works are the definitive histories of the hardware as hygrometry forms a substantial part of the total gamut of meteorological measurements.

The realisation that a damp surface can be dried by moving air is ancient. It was first connected with thermometric measurement in 1740 when John Hutton was possibly the first to be able to measure the dryness of air by an aspirated thermometer – now called psychrometry – (this term also had another use, see Table 4.2). A theoretical basis for this, that would enable the observations to be given measurement units and a scale, was formulated early in the 19th century.

Miscellaneous

The above measurement methods, being reasonably well researched and documented might be thought to have been the main thrust of measuring instrument development prior to 1800. They are, however, but a few of the transduction effects put to work as measuring instruments in the period 1500-1800. Space does not permit further description of individual instruments. Table 4.2 is included to provide some measure of completeness for readers who may wish to gain entry into the literature. The wide range and varying nature of the application of measuring apparatus is evident from the table.

It must not be thought that the table is a complete list but it probably comes reasonably close to that for it was compiled by reference to more than seventy relevant works chosen across a wide range of subjects.

References cited may not be as useful as might be hoped but in the light of the inordinate amount of research that would be needed to fill out the data they must

Table 4.2 *Additional measuring instruments having foundation in the period 1500–1800* (see text for explanation and for other sensors of this period)

Named device	Associated date	Conversion process	Associated persons	Source of further reference
Actinometer	1777	Chemical effect of sun's radiation	W.H. Wollaston J.W. Ritter Scheele	Timbs (1860)
Anemometer	1667	Force, or speed of a wind or, later, air	L.B. Alberti R. Hooke	Pearsall (1974), Knowles-Middleton (1969*a*), Mayr (1970)
Atomgraph	*c*1620	Evaporation of water	C. Perrault J. Leslie	Knowles-Middleton (1969*a*)
Bathometer (also, sounding device)	*c*1662 onward	Depth of sea	R. Hooke A. van Stiprian Luiscius	Multhauf (1961*a*)
Centrifugal Governor (A late example is shown in Fig. 1.5)	*c*1788	Rotational velocity to linear displacement	M. Boulton J. Watt	Mayr (1970,1971*b*)
Chronoscope	1783	Brief time interval into observable variable	G. Atwood	Marmery (1895) Cajori (1962)
Echometer	1736	Duration of a sound		Pearsall (1974)
Engine indicator	1790	Internal pressure of (Steam) engine to record	J. Watt J. Southern	De Juhasz (1934) Burstall (1970) Lardner (1862) Vol. V
Eudiometer	1790	Oxygen content of air by combustion or absorption	A. Volta J.H. Magellan J. Priestley	Marmery (1895) Turner (1973)
Goniometer	1766	Angles of crystal structures and other refraction effects	G. Adams Whiston	Pearsall (1974) Turner (1973)
Heliometer	*c*1750	Angular separation of stars in telescope field	J. Dolland P. Bouguer	Pearsall (1974) Turner (1973)
Holometer	1696	Measure of angle		Pearsall (1974)
Hydrometer (Also called psychrometer, gravimeter plus others – see Pearsall)	1500 and before	Density, specific gravity of liquids (also salinity of water for shipping)	W. Nicholson R. Boyle	Marmery (1895) Gunther (1935) Pearsall (1974)

Named device	Associated date	Conversion process	Associated persons	Source of further reference
Otoscope	1600s?	Viewing inner ear		Pearsall (1974) Hamerneh (1964)
Photometer	*c*1760	Relative intensities of light sources	Count Rumford, J. Leslie	Marmery (1895) Pearsall (1974) Timbs (1860) Bray (1930)
Piezometer	*c*1800	Compressibility of liquids	H.C. Oersted	Marmery (1895)
Pluviometer	1791	Quantity of rainfall		Pearsall (1974) Knowles-Middleton (1969*a*)
Pressure sensor	*c*1680	Pressure to a displacement (for steam devices)	Numerous	Mayr (1970)
Saccharometer	1784	Sugar solution strength by polarised light effect		Pearsall (1974)
Seismoscope (An earlier example is shown in Fig. 3.47)	1703 (and before)	Detection of abnormal ground acceleration or tilt	Abbe de Haute Feuille	Matschoss (1925)
Spirit level	1666	Departure from horizontal line	Also ancient Greeks	Burstall (1970)
Tonometer	1725	Rate of vibrations of sound or effect of drugs and liquids on organs removed from dead body		Pearsall (1974)
Viscometer (antecedant)	*c*1750	Resistance of object to movement in a fluid	Robins	Burstall (1970)
Windvane	1578 (also *c*100 BC)	Direction of surface wind	E. Danti G. Geutmann G.F. Parrot	Knowles-Middleton (1969*a*) Mayr (1970)

suffice as a starting point for further study. The terminology used and a date are often valuable facts; they at least indicate that there is something to seek and in which time period.

All entries pertain to descriptions of instruments known to have been generated in this period. Dates given are sometimes those suggested by the writers themselves, often without the provision of supporting evidence.

The list gives only those transducer effects not previously covered in this chapter.

4.6 Recording and registration

Today we regard the availability of chart recorders, tape punches, storage oscilloscopes, and the like, as mandatory subgroups of instrument systems. They too, had to develop through scientific and technological paths. Their design is inextricably linked with measuring instruments for they often use similar principles and practice.

Recording and registrating devices are needed for a number of reasons. First, they can be used to produce a permanent record of measurement data – a method of storing experience for others to use – for later use. Secondly, they can release an operator from the task of watching the varying output of a measuring instrument this often saving expenditure but also, sometimes considerably improving the results by eliminating human errors. A third reason is for transforming a very rapid or slow set of data into a form that is more compatible with the time performance of the natural comprehension of man. The registration part of an instrument is an additional transduction stage that converts the primary signal form into a more convenient form. Some transducer stages require it to provide a suitable signal form – a clock does not but a resistance bridge does.

An instrument that produces a graphical form of output record will often have 'graph' on the end of its name, for example, the seismograph. A record so produced is correctly called, using that example, the seismogram.

When did this aspect of instrumentation begin? The short answer is not until the 19th century but there were some developments before that time that may have prepared the path.

The earliest method of recording data was to write values down on clay tablets or on papyrus. Singer *et al.* a (1954) presents illustrations of a late Babylonian clay tablet with cuneiform values of the motions of Jupiter upon it and a table of lunar months from the Carlsberg 9 papyrus. Breasted (1944) includes an illustration of a piece of clay tablet showing observations of the Moon from a series of observations that began in 747 BC.

A few, isolated, ancient devices could be considered as having in-built recording. The hodometer of Vitruvius *c*192 AD, shown in Fig. 3.24, recorded distance travelled by counting stones. Chang Hêng's seismoscope, Fig. 3.47, recorded the pressure of an acceleration exceeding some undefined amount by the position in which the ball dropped. Additionally, some forms of apparatus inherently retained the value of a determination e.g. the weights needed in the beam balance, for instance.

The basic design components needed for the manufacture of a mechanically-driven registration device, the first form, existed in Hellenic times. The Ktesibios water clock, Fig. 1.26, had a mechanism that moved a pointer across a drum. In that case the drum was a graduated scale, not a recording cylinder.

There existed considerable ability in devising ingenious mechanical mechanisms. There were also many methods for smoothly rotating a drum at any speed required. Suitable marking instruments also existed. Why then did the automatic registration method take so long to emerge? Perhaps it was primarily because there was little extensive need for automatic records until the 19th century.

It seems most unlikely that the ancient methods described above had influence upon registration devices built in the more recent time period. They would have been developed to suit needs that arose using intuition and available knowledge of mechanical skills.

In 1657, at the early age of 25, Christopher Wren, Crowther (1960), was appointed to the Chair of Astronomy at Gresham College, London. Before changing interests in his thirties to the architectural pursuits for which he is more generally known, he had established himself in the field of science. Weld, historian of the Royal Society, had this to say of him:

> his name had gone over Europe, and he was considered as one of that band of eminent men, whose discoveries were raising the fame of English science.

He devised and built many measuring instruments; his work, though little was reported, influenced many a scientist of the day. He was greatly responsible for the foundation of the Royal Society.

A design that interests us here was his 'weather-clock', Fig. 4.46, for it appears to be the first multiparameter automatically registering system. It was to be a century and more before similar weather stations were established. He devised this as part of his plan to continuously record that 'which concerns us as near to the breath of our nostrils'.

He left a long list of instrument innovations that the world may have had instead of buildings, if his interests had not changed.

In the middle of the 18th century registration became reality in relation to temperature measurements. The highest and lowest temperatures of the daily cycle were of interest but their periodicity was such that the detection of these two extremes would have been found to be rather tedious and inconvenient to observe by direct means. According to Knowles-Middleton (1969*a*) the first practical thermometers to record the maximum and the minimum values were made by Charles Cavendish in 1757. Black and Bernoulli are both credited with similar developments at this time.

It has also been suggested that in 1786, William Playfair the statistician, saw that a graphical display of data enabled the observer to visually process the set for such things as rates, trends, extremums, and unusual features.

Returning to the thermometer, Bernoulli in 1745 described maximum and minimum designs in his *Commercicum Philosophicum et Mathematicum*. Knowles-Middleton related how John Rutherford, in 1790, constructed a pair of thermometers

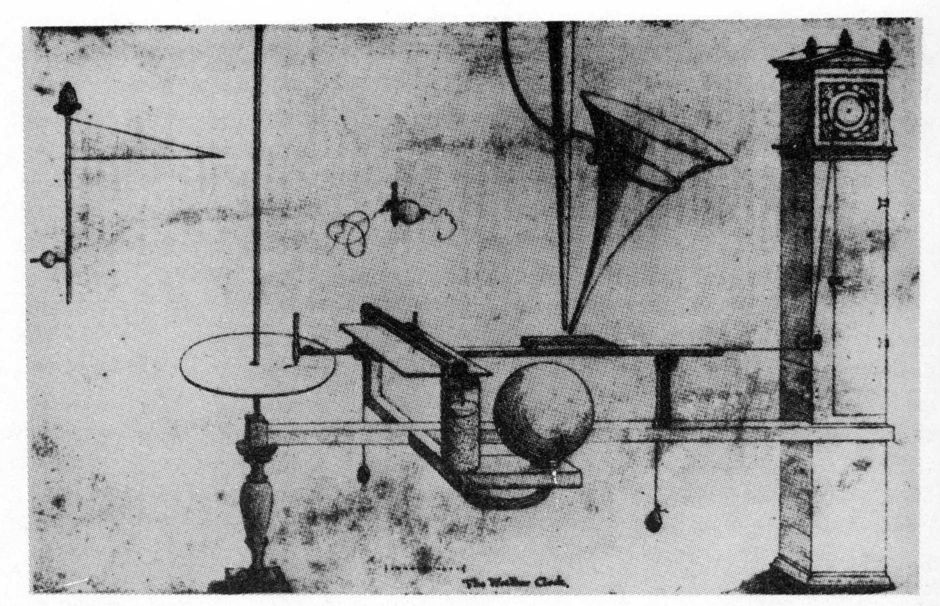

Fig. 4.46 *Christopher Wren's weather clock c1660. The first multiparameter, automatically recording, measurement system*
[Crowther (1960), plate V facing p.177. From 17th century etching]

for which the 'minimum' recording instrument design is basically that used today. His 'maximum' design did not survive. James Six devised a combined unit in 1782, Chaldecott (1976). The first successful maximum thermometer was that of John Phillips, invented in 1832.

The thermometer devices did not record a continuous signal. It appears that the first well documented use, De Juhasz (1934), of a continuous registrating mechanism is the Southern engine indicator of *c*1796. Watt *c*1790, had constructed an instrument that showed the internal pressure within the drive cylinder of a steam engine. It was used in conjuction with visual sighting of an appropriate engine part to monitor the pressure during a cycle. A collaborator of Watt, John Southern, then added the idea of running a pencil along a chart which was caused to move perpendicular to the pencil travel by means of a weight driven actuator mechanism. The indicator diagram thereby came into existence.

Together Watt and Southern further improved the concept to produce the engine indicator shown in Fig. 4.47. This device enabled the operation of a steam engine to be improved by observing the shape of the indicator diagram. Thus was born, not only the first chart recorder, but the concept of mapping measurements from the physical situation into shapes as well as numbers.

Little other evidence could be found of recording mechanisms prior to the 19th century. It will, however, be seen that shortly after this time there came about many registration devices, impetus arising from the need to record telegraph messages and then, later in the 19th century, scientific observations and industrial temperatures.

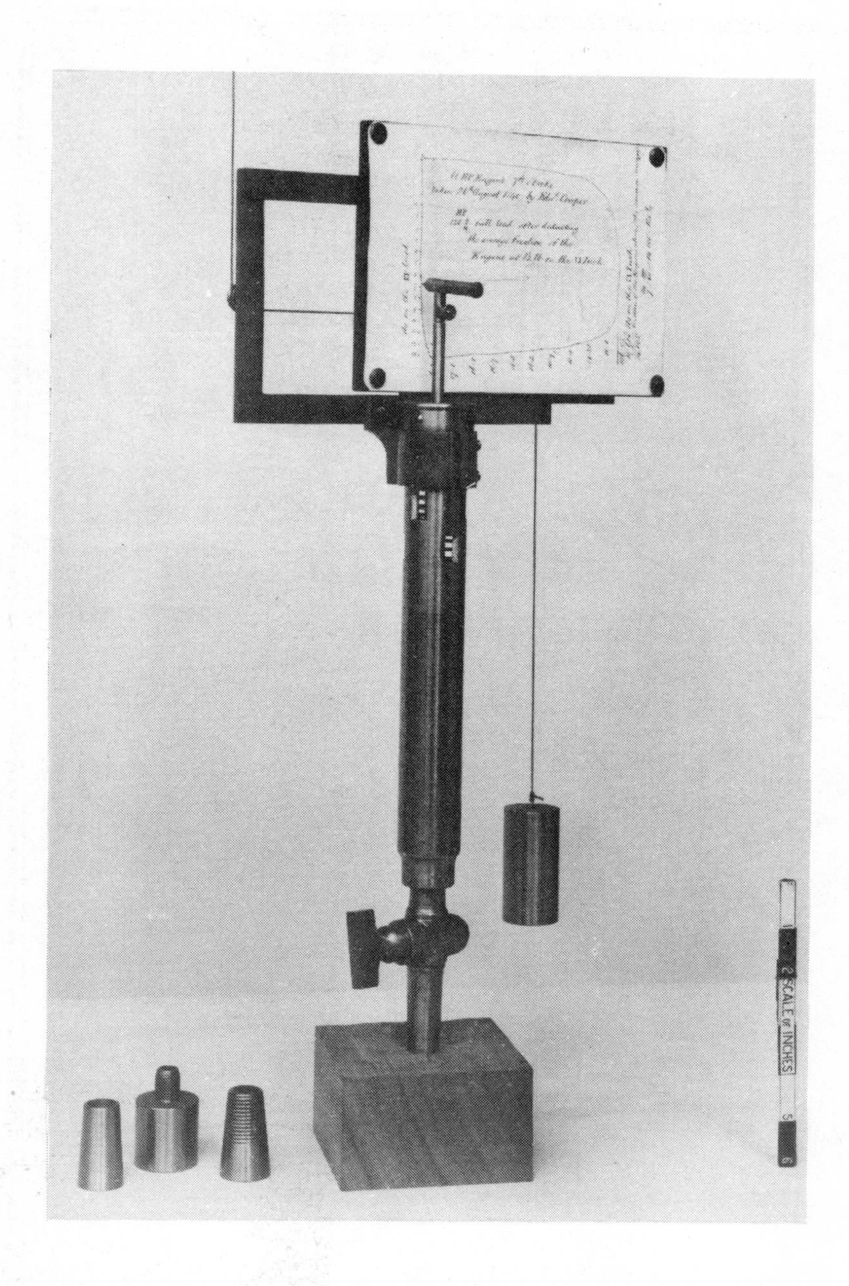

Fig. 4.47 *Watt-Southern recording, engine indicator c1800 – possibly the oldest existing automatic registration device for a continuous variable*
[Lent to Science Museum, London]

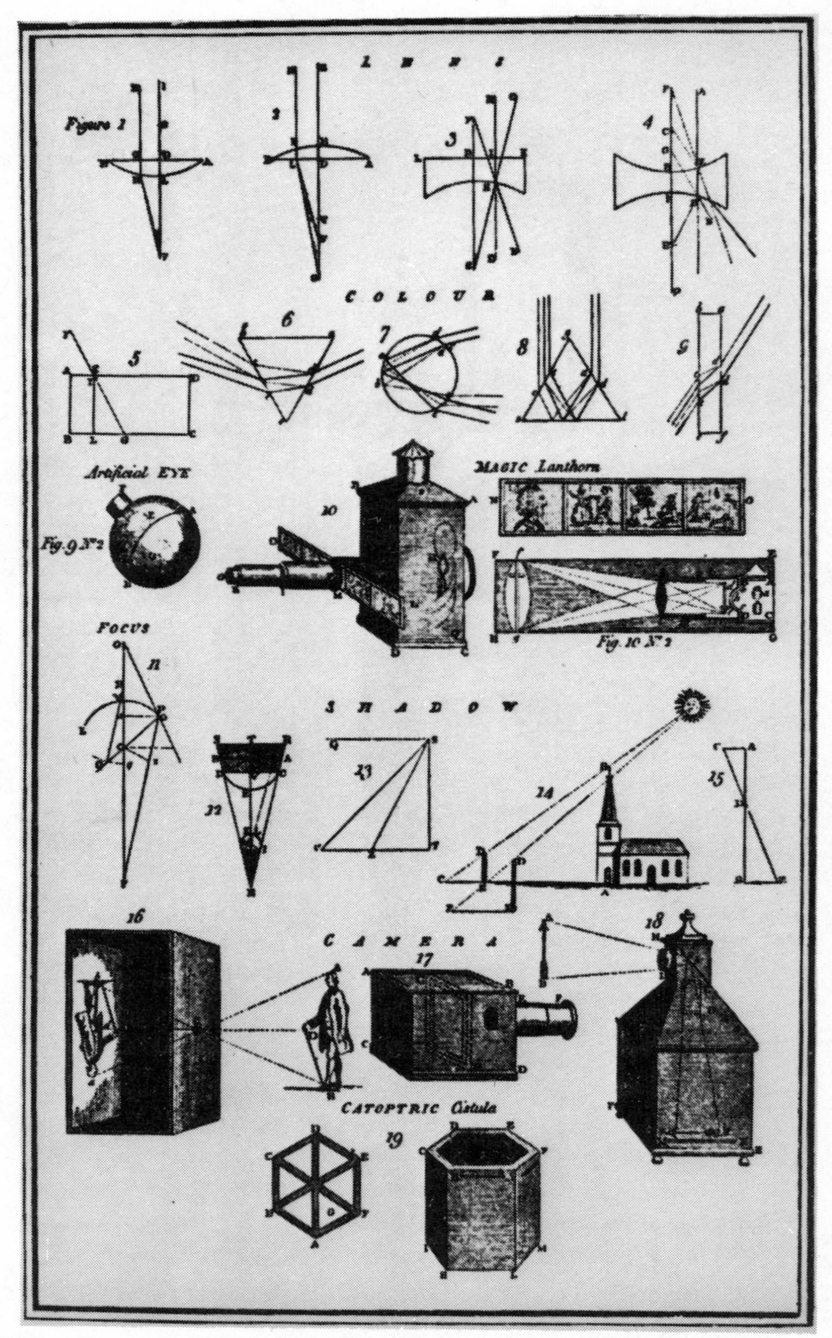

Fig. 4.48 *The camera, in the 'camera-obscura' form was invented in the 13th century*
The pin-hole principle is shown here in a page from an 18th century work
[Larson (1960), p.15, From 18th century work. Courtesy Hamlyn Publishing Group Ltd]

Although not having reached a stage of development that enabled records to be made until the mid-19th century, photography was, nevertheless, emerging by 1800 after a particularly long period of development, Gernsheim and Gernsheim (1969). The camera was known in Roger Bacon's time; he described, in 1267, a device for creating an image of a subject using the pinhole camera principle. This was named the camera-obscura. Accounts seem to vary somewhat but credit is given to Giambattista della Porta and others, who around 1550 constructed such apparatus. One writer suggests the lens was added in the 'pin-hole' by Zahn *c*1600 but some writers credit this improvement to Nièpce *c*1805. Fig. 4.48 shows that by the mid-18th century the camera-obscura was a commonly known optical device. Adding a lens was not an essential step toward photography but it was an important improvement.

The missing process needed was some way to record the image in a permanent manner. Although the ancient Chinese had methods of forming images on surfaces using sunlight intolerant dyes, this was not a factor in the development of photography. It is also known that the Arabs experimented with light sensitive substances. In 1727 J.H. Schulze observed that silver salts blackened when exposed to sunlight but this, too, had little impact. In 1777 K.W. Scheele and William Lewis advanced things a little further by conducting more research into silver salts. In doing so, Scheele discovered that red light had a lesser effect than light of the blue end of the spectrum. Lewis's notes here handed down to Joshiah Wedgewood's son, Thomas, who made contact prints of leaves. These negative prints were only transitory because the rest of the silver-salt treated area also blackened with time. Some method to fix the plate was needed. In 1835 Fox Talbot discovered that the plate could be roughly fixed by washing it in brine after exposure. This account has been derived from Timbs (1860), Grolier Society (1946) and New Caxton (1969). Larsen (1960) gives an account of early photography beginning with the work of Nicéphore Nièpce.

The ability by the 1850s to record instrument information by photographic means released designers from the limiting difficulties imposed by the friction in mechanical registration mechanisms thus opening the way to provision of continuous recording of the weaker transducer effects such as, for example, are found in magnetic field recording. It is, however, interesting to see (as will be shown in the following chapter) that the mechanical recording stage does not seem to have been a sequential stage toward the adoption of more sensitive methods.

Chapter 5

Growth of electrical method: the 19th century

5.1 Expansion of measurement technique

By 1800 the scene was set for the discovery of the laws of electricity which, in turn, allowed many useful and powerful transduction methods to be implemented as measuring instruments. As with previous periods the 19th century also was a time in which many new physical principles were discovered and quantified.

A most significant development for this period must surely be the growth of knowledge about the electromagnetic field. At no time in history had man discovered an effect that he could put to use about which he had no prior inkling as the result of first-hand experience. Previously measuring, and other observation enhancement devices, had mainly provided increased capability to observe what could already be partially witnessed with the natural sensors. The relationship between magnetics and electrics lead, through the work of Maxwell, to the realisation that, naturally unobservable, energy fields could be radiated from a source. Proof of their existence was provided by Hertz in 1888. It was not long after, that electromagnetic fields were no longer in the realm of fantasy. The modern sequel to this most significant course of events will, perhaps, be the eventual practical proof of existence of the much sought-after gravity waves. This weak energy field has been predicted from theoretical considerations but has, to date, gone undetected. Discovery of means to generate and detect these waves could have an impact on communications that surpasses the changes brought about by the application of radio waves.

The 19th century was a period wherein instrument users became aware that measured variables were dynamic phenomena and that steady state, or stationary values were not necessarily a correct indication of what happens during transient conditions. It was the failure to readily recognise that electromagnetic induction is a dynamic phenomenon that prevented Faraday (and the many others) from recognising the effect earlier. Ohm's law for electrical circuits was for unidirectional, steady state direct current flow. It provided the mathematical law for modelling many circuit conditions but could not explain what was happening in certain electric spark discharge circuits. At one time the spark magnetised an iron needle with one magnetic polarity, yet sometimes caused the other. Kelvin subsequently derived the laws of resonant inductive-capacitive circuits but his work of the 1850s had to wait a decade before a method could be developed that could record the cyclic waveform of the radio-frequency energy produced from the discharging arc, (by Fedderson *c*1857).

Discovery of the need to record signal events that happened faster than the eye could see or too slowly for patience sake, stimulated the development of recording methods, the dominant invention being Callendar's potentiometric chart recorder of the last decade.

Telegraphy and other later forms of teletechnique are important to the history of measurement for they provided a significant boost to the design of instruments in the 19th century. It grew out of the electromagnetic discoveries that provided the sought-after means to send fast, high data rate, messages over vast distances. One of its many requirements was the need for signal recording devices. It also heightened man's understanding of electrical circuitry and signalling techniques. Some idea of the impact it made and the extent of its application can be seen from Fig. 5.1.

Telegraphy was based upon digital signal methods using on-off devices such as electromagnetic relays, sending keys and discrete position indicators; it sent messages

Fig. 5.1 *The impact telecommunications had on the scenery can be gauged from this 1890 print of telegraph and telephone wires in Philadelphia*
[De Vries (1971), p.140, from De Natuur, 1898]

in a digital code form. This was followed, in the late 19th century, by the telephony method in which analogue or linear signal systems were put to use. Multiplexing in the time domain began in this period. Out of commercial message conveyance needs grew the means to control complex networks of railways and to conduct yet more scientific studies.

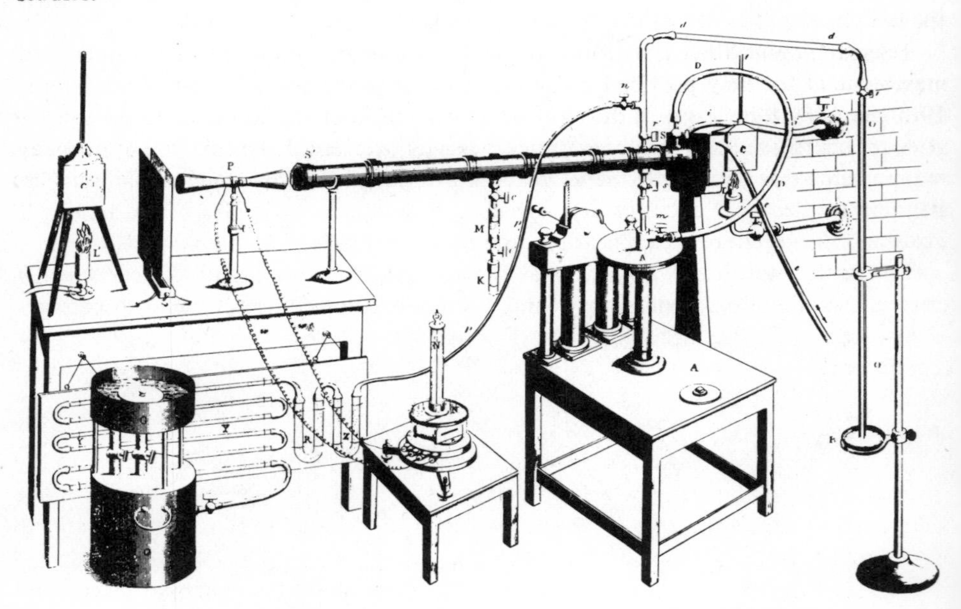

Fig. 5.2 *Thermoelectric pile and galvanometer used by Tyndall for investigating the nature of heat*
[Tyndall (1863)]

By 1900 the technical world of information machine design lacked three vital techniques. There was no adequate method for amplifying low-energy level signals, there was no adequate method for generating high-frequency signals and there was no adequate method for detection, that is, for rectifying alternating current signals. Considerable effort and ingenuity were expended on these problems during the closing decades of the 19th century.

Many transducer effects that came into man's comprehension in this time period provided only low-level signals, for example, thermoelectric junctions only generated a few millivolts. Those that provided electrical output forms could be investigated and put to use by coupling them in some manner to an indicating galvanometer. The weaker were the signals the greater was the sensitivity needed of the galvanometer. Improvement in galvanometer design (Fig. 5.2) went hand in hand with scientific investigation which yielded the discovery of more sensing effects. The galvanometer, however, converts electrical energy into mechanical output movement that was not capable of providing substantial energy. It was rarely feasible (there were some notable exceptions) for the galvanometer to be used to provide gain in the general purpose manner that we are used to today with the operational amplifier. Telephony and wireless, both in practical existence by 1900, needed some form of electric signal amplifier. It was, perhaps, an interesting coincidence that the knowledge and technology for the invention of the thermionic diode and triode were available at a time when the demand was being felt. Thermionic devices also answered the other two needs for they enabled rectification with the diode and signal generation with the triode.

Foundations of the electronics discipline can be traced back into the 19th century but general purpose amplification, detection and a.c. signal generation did not appear until the early 20th century, the subject of Chapter 6. Many attempts to provide for these needs in the late 19th century were ingenious indeed. They show that designers were aware of the kind of conceptual devices they needed, but were unable to realise at the time.

The 19th century was the first age of great expansion in the application of measuring apparatus. Toward the end of the century the demand for the new kinds of instrument, those based on some effect of electricity, for example, coupled with the introduction of better manufacturing methods and material availability, provided the conditions for the emergence of instrument industries in which larger numbers of similar instruments were manufactured together rather than being fashioned one by one.

The greater use of measurements and increased trade in ideas and products gave rise to the need for legal standards of physical units. The metric system gradually became adopted and late in the century several countries co-operated to start the international standards agreement, hosted by the BIPM in France.

Easier generation of measurement and other data allowed designers to place more of their attention onto conversion and generation of data with computational devices. Many machines were built; some, although sound in principle, were not successful because of the lack of available adequate engineering precision, Fig. 5.3. Some could generate mathematical tables and perform special computations, others converted data. It was in this period that both the analogue and the digital computer had their genesis. Although digital computing methods were not brought into widespread use until the 1940s, the fundamental bases for their construction can be traced to this time. Analogue computation, however, being constructable with electromechanical and mechanical principles to a readily understood philosophical basis was the first to emerge. By 1900 analogue computing units were able to calculate quite complex functions.

Feedback control also developed rapidly in application, if not in theory, during the 19th century but it lacked the general-purpose amplifier of signals and adequate theoretical understanding to be brought into rigorously designed application. Today we accept the feedback controlled amplifier without questioning its origins but many man-years of effort were expended in the 1930-40 period to realise it.

Emphasis with electrical signals must not be allowed to cloud the fact that there was also steady development in steady growth of the optical and mechanical regimes of instrumentation. The optical instrumentation highlight of the 19th century must surely be the spectrometer and the technique of spectroscopy. Photography, also, was

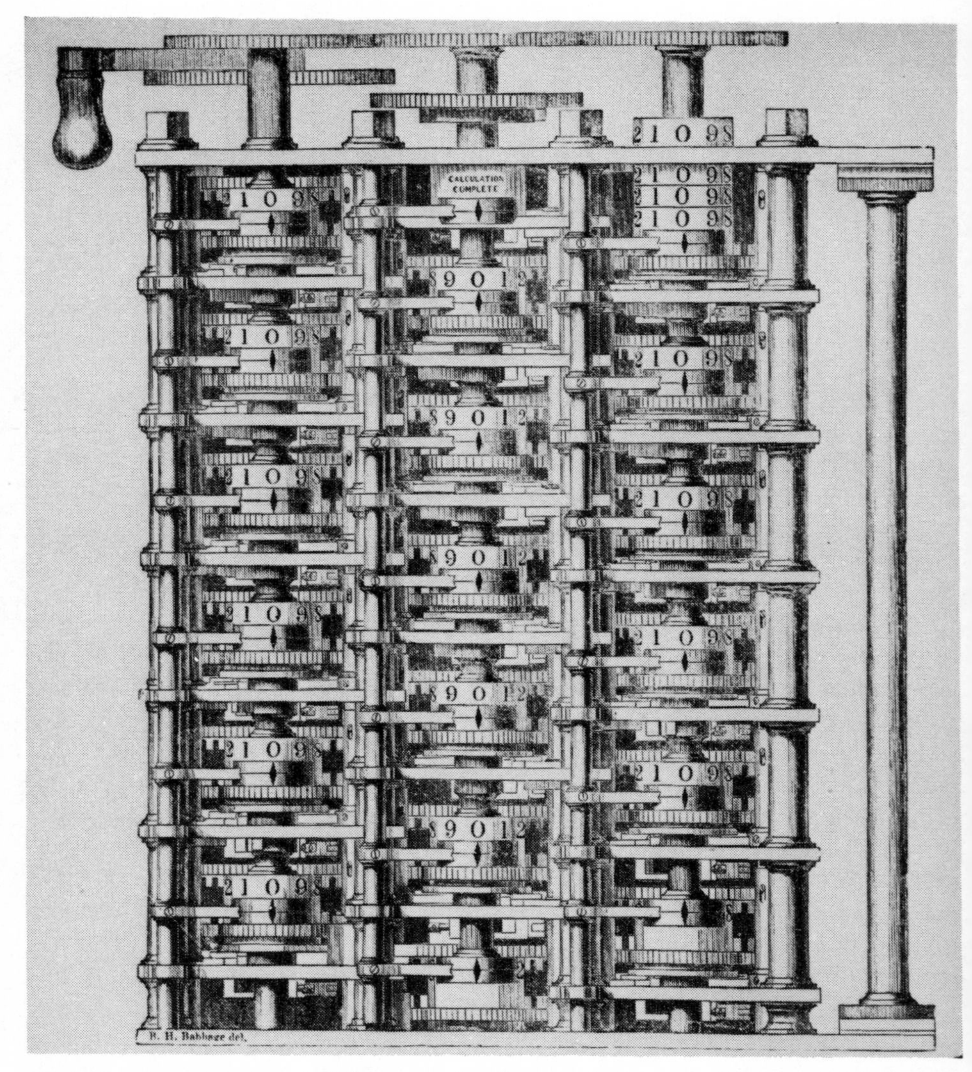

Fig. 5.3 *Part of Babbage's difference engine No.1. It could be conceived in principle, but not made with sufficient precision to work successfully*
[Timbs (1860), plate facing p.142]

important for it allowed the faster transient signals to be captured. In mechanical design it was realised that instruments have impossible-to-beat physical limits imposed upon them by Brownian motion. Clocks were further improved to a point where there was little to be done with mechanical forms of timekeeping. Combined with electrical method, clock design again gained a fresh impetus from the search for still better precision and accuracy of timekeeping.

It was in the 19th century that many, already witnessed, physical phenomena could be raised from the qualitative stage of knowledge to become quantitative measurements. Seismology developed from a descriptive stage to its instrumental form. Galton began to measure man in statistical terms, the life-sciences began to make use of more sophisticated instruments, physiology moved into its scientific era.

It has been said by many authors that the Victorian era (defined as the reign of Queen Victoria in Britain, 1837-1901) was one of great invention in the technical and scientific world. This theme has been taken up by De Vries (1971) who put together an extensive collection of reprinted Victorian etchings of scientific interest. It includes many illustrations of instruments. Brown and Jeffcott (1970), a reprint of the 1932 original, is another source showing the extensive ingenuity expended in those times – perhaps if often somewhat misguided for lack of scientific understanding of the principles of the inventor's subjects.

Other general works of 19th century invention and science include Byrn (1900) a definitive study of invention; Burstall (1970) who covers mechanical engineering history of this period over two chapters and the popular style contemporary works of Beckmann (1846), Lardner (1862), Timbs (1860) and that published by Nelson and Sons (1881). Marmery (1895), Cajori (1962) and Pledge (1966) provide accounts of scientifice progress in the 19th century. Harré (1969) provides seven biographies of 19th century British scientists which, seen collectively, portray a scenario of the times.

Works dealing with specific topics are introduced throughout the chapter. Chapter 2 deals with several topics that are also relevant to this period.

5.2 Electrical method

Measurement technique is integral with the general development of electricity; some material about electrical measurements is, therefore, to be found contained in general works. A considerable number of such sources exist. Works on natural philosophy such as Wells (1856), Pepper (1874) and Deschanel (1891) include sections on electricity. Glazebrook (1922), although published after the 19th century, devotes one volume to the subject, describing equipment and techniques that were developed at the end of the 19th century. Byrn (1900), Sherwood-Taylor (1945), De Vries (1971) and Brittain (1977) each devote space to the subject of electricity of this era. Mottelay (1922) provides information on workers in electricity and magnetism. In introducing their material, many contributors to the Institution of Radio Engineers' 50th anniversary issue gave information about 19th-century developments, IRE (1962) as did those to the series, IEEE (1976).

Published books on the specific subject of electricity began to appear at the end of the 19th century as the result of the discipline of electricity becoming recognised in its own right. Tunzelmann (1890) was among the first flush of general accounts of the practical use of electricity: he included a chapter on electrical measurements. Lodge (1892) is another contemporary statement. Frith and Rawson (1896) is a later account in the same vein. It also includes some photographically produced illustrations; it was at this time that such techniques came into use providing a more accurate printed representation of artefacts. In 1888 the first edition of *Electricity in the service of man* appeared as an adapted translation from the German by Urbanitzky. After re-

release under the later editorship of Dr. Wormell it was found that the material to be covered had expanded to the point where two volumes were needed. Volume I was released in 1910, Mullineux-Walmsley (1910), on the theme 'history and principles of electrical science'. The chapter on measurements was considerably expanded for this edition. Having over 1600 illustrations of high quality and intricate detail this work will stand as an important source book about late Victorian electrical apparatus.

Gray (1893) is concerned exclusively with measurements of electrical and magnetic quantities, being the revised version of work compiled a decade before. Hay (1907) is about a.c. electricity: it provides account of the state of this aspect at the end of the 19th century.

The Victorian era produced numerous new names for the many inventions proposed. Sloan's Standard Electrical Dictionary (1894) includes several terms that did not endure i.e. electrepeter, electromotograph, Joulad, the Mack unit, microtasimeter, paragreles, telephote and others. Another contemporary dictionary was published by Houston (1892): Hobart (1911) is another. In general, however, most of the words described in these dictionaries are those in use today showing that the basics of electricity had reached a high degree of stability by 1900. The same cannot be said of all of the names invented for 19th century instruments. Pearsall (1974) lists many that are virtually unknown today.

The history of electrical engineering has been covered by Dunsheath (1962) and to a lesser extent by Squier (1933), Morgan (1953) and Kemp (1950). Facets are covered by Molella (1976), and Reynolds and Bernstein (1976).

Historians have made much of the impact of the discovery of the laws of electricity placing emphasis on the energy and communication applications that emerged. Considering that neither of these could operate efficiently or compatibly without measuring technique and measuring instruments, it is surprising that more studies of the measurement aspect have not been carried out. There has certainly existed misunderstanding about the role of electricity and measurements in general: study of published works and courses shows the existence of emphasis on electrical parameter measurements by electrically trained persons with the other much larger group of 'transducer' devices that use electrical principles, being largely left for description and education by people with other training, such as mechanical engineers and physicists. Electrical technique has a very wide application in measurement applications, especially now that information processing can invariably best be done for the lowest cost, smallest space and least power consumption, by electrical techniques. It will become clear, as this Chapter proceeds, that electrical-indicating instruments and measurements of electrical quantities form only a part of the overall group of electrically-based measurements that were made.

Laws of electricity

Before looking into specific devices, and because of the overall importance of electrical technique in the later history of measurement, it is appropriate to discuss the growth of electrical theory throughout the 19th century.

The previous Chapter described how researchers of electricity were confined to work with high impedance, low current, very high voltage, statical electricity and how it enabled development of a certain amount of knowledge. In general, very little of it was of importance to the bulk of the uses to which electricity was subsequently put. The breakthrough came with the introduction of the primary electric cell for it enabled all manner of people to investigate the nature of electricity with simple to make and use equipment.

Somewhat surprisingly, and many writers have commented upon it, it was two decades before it was formally realised that electric current flowing in a wire produced magnetic field. This was expounded in a qualitative way by Oersted in 1819. A model of his apparatus is shown in Fig. 5.4. Ampere, shortly after this, enunciated the quantitative laws involved.

This release of the new fundamental principle was taken up by numerous workers and it was not long before the electromagnet was in both popular and in serious practical use. Sturgeon was awarded a medal for improving the electromagnet around 1825, with Henry building one of the first powerful units. Development of the electro-

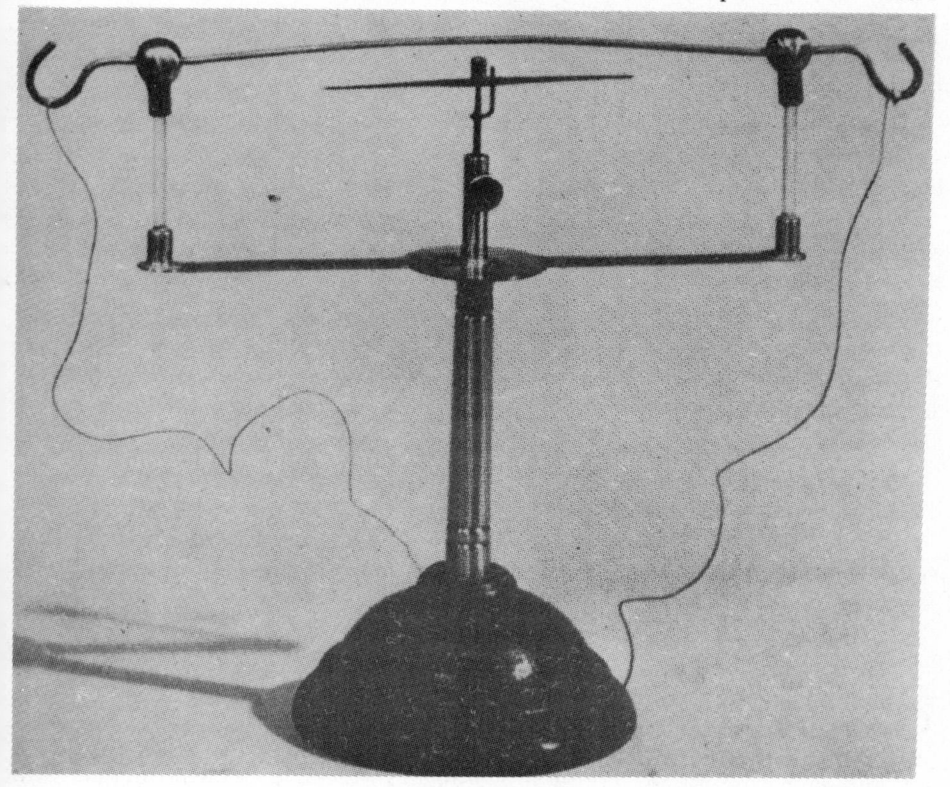

Fig. 5.4 *Model of Oersted's wire experiment in which a magnetic needle can be deflected by the passage of electric current*
[Courtesy National Museum of History and Technology, Smithsonian Institute, Washington DC. Photo 31504-A]

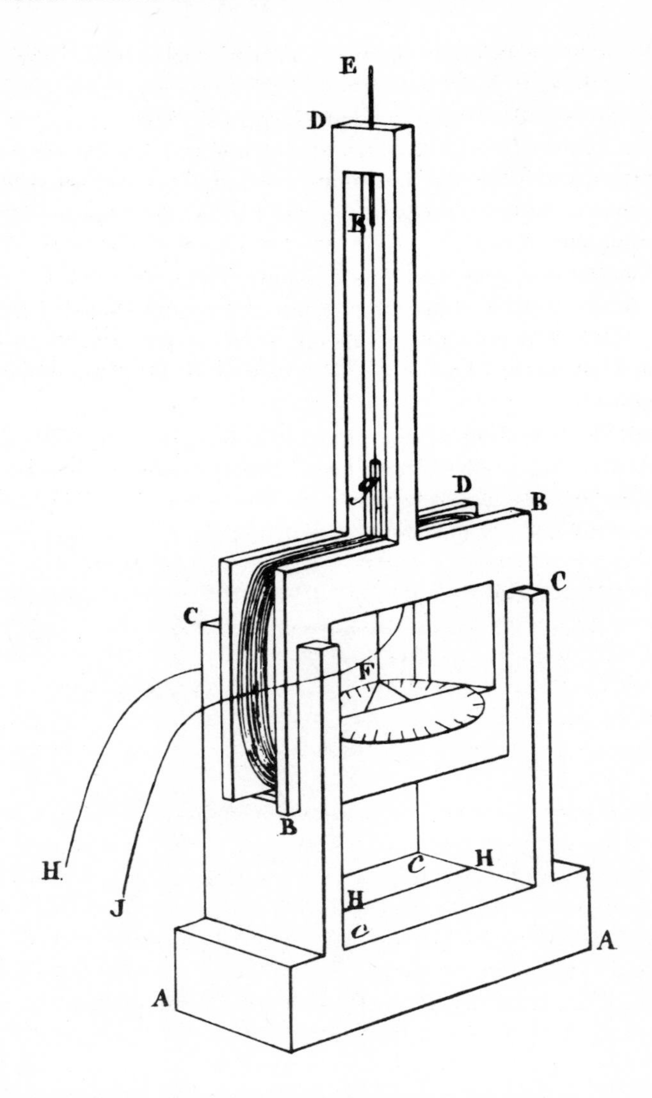

Fig. 5.5 *Oersted's 1823 adaption of the Schweigger multiplier principle*
[Chipman (1964), Fig. 7, p.135. From paper by Oersted, 1823]

magnet was considerably more complex than space allows it to be described; the early history of the primary electric cell and the electromagnet is published, James-King (1962*a*).

The electromagnet is an important device in the history of measurement for its contribution to the art of control as the dominant actuator used in the mid-19th century and because it assisted the discovery of the laws of induction. It also provided the basis for the construction of the, all-important, galvanometer which gained a significant boost when Schweigger, *c*1820, realised that a wound coil, Fig. 5.5, could multiply the deflection sensitivity. Schweigger, at the time, hailed Oersted's discovery as 'the most interesting to be presented in a thousand years of the history of magnetism'. This development has been studied by Chipman (1964).

Magnet coils, primary cells, insulated wires, compass needles, iron filings and primitive galvanometers formed the stock-in-trade of the electrical scientist of the 1820s. It is again surprising to learn that despite numerous people having the equipment and the idea that magnetism might be able to produce electric current as the reverse of the Oersted and Ampere effect, it was not until 1831 that Faraday realised the conditions to be those of changing magnetic field linkages that would generate electric current. Thus was born the 'generator' effect to go with the earlier 'motor' effect. Henry, independently, discovered the law of electromagnetic induction at the same time as Faraday.

This law provided the springboard, Larmor (1937), for many further realisations, the most significant being the theory of electromagnetic radiation that Clerk-Maxwell published in 1863. Whereas Maxwell's theory explained some already known observations, such as resonance of electrical inductive-capacitive circuits it also provided the realisation that energy could be radiated from a source without wires. This concept stimulated the Berlin Academy of Science to offer a prize for an experimental proof of certain aspects of the theory and this, in turn, encouraged Hertz to investigate the electromagnetic radiation. In 1888 Hertz managed to radiate, and detect, radio-frequency signals over the length of a room. Only six years were to pass before Marconi, in 1894, was able to claim the first transmission of a recorded wireless message. This line of development also brought about improved understanding of a.c. circuits.

Turning back to the path of the direct current d.c. circuit it was Ohm, in 1827, who formalised the relationship between current and voltage in a circuit, Fig. 5.6. Lenz's and Kirchoff's laws were added in the mid-19th century. Ohm's law provided the basis for scientifically devising magnets and signalling systems over wires i.e. telegraphy. The first really practical telegraphy systems were brought into operation around 1837. Morse constructed his celebrated relay in 1837 thereby adding amplification to the digital techniques then available. (There are other claims to its conception.)

Needle telegraphs used the Oersted, compass-needle, principle to display digitally-signalled code positions. It was problems with long lines, especially underwater cables – see Finn (1973) – such as the trans-Atlantic cable of 1858, that prompted work on the filtering effect brought about by the capacitance of cable. Black (1972) discusses cables in Victorian times. The telegraphy path of development overlapped the electromagnetic field developmental path in many ways. Inductive loading of cables was suggested by Heaviside in the 1880s.

The Wheatstone bridge, actually devised by Christie and originally called the 'differential resistance analyser' by Wheatstone when he popularised it in his 1843 Bakerian lecture to the Royal Society, Bowers (1975), was an important instrumental development for it enabled electrical quantities to be compared with high precision under conditions of good 'common-mode' noise rejection. It is interesting to note that Kelvin inadvertently re-invented it, reporting it in his Bakerian lecture of a decade

Fig. 5.6 *Reconstruction of apparatus used by Ohm c1830*
[Crown copyright. Science Museum, London]

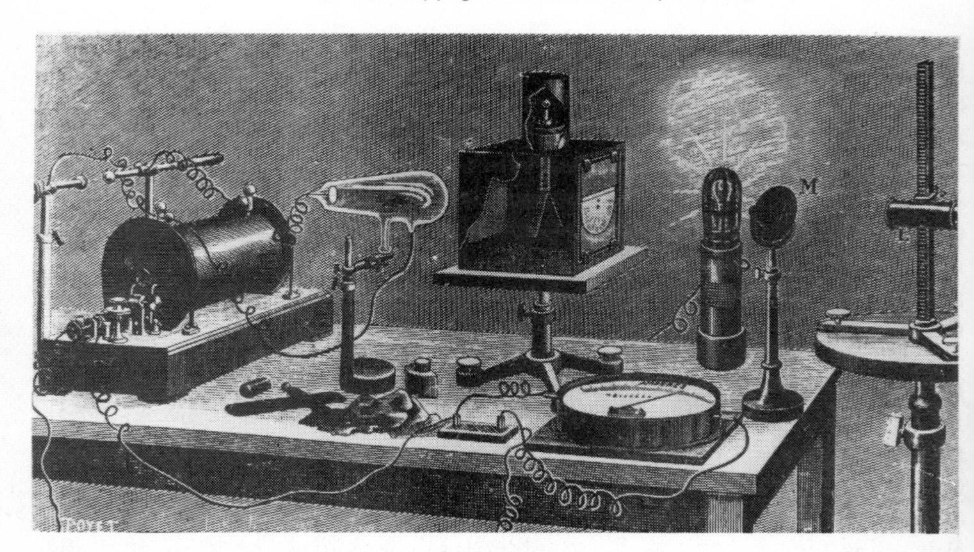

Fig. 5.7 *Electrical discharges were investigated with apparatus like this at the end of the nineteenth century*
[De Vries (1971), p.103, from *La Nature,* 1896]

later. Pepper (1874) provides an interesting account of its use in the Atlantic Submarine cable telegraphy system. Demonstration apparatus, as used in Pepper's time is shown in Fig. 5.7.

Kelvin's work on the resonant circuit was published, in 1853, in a paper entitled *Oscillatory discharge of a Leyden jar*. He provided theoretical grounds for the earlier attempt of Helmholtz who explained resonance in terms of the pendulum which has the same 2nd-order equation. Kelvin also provided theory relating to propagation of a current pulse along a long line. This was based on insights gained from previous work on heat flow and on some of Faraday's suggestions. These findings, at the time, went almost unrecognised. Watson (1969) has summarised Kelvin's work in this area. This theory provided the basis for design of radio-frequency oscillators such as was later used by Hertz. Coupled-tuned circuits were used by Braun in 1898 to improve the selectivity of wireless signals. The induction coil created much interest as the paper by Post (1976*b*) shows.

Quite unrecognised for the significance that it would later have, went Edison's chance discovery in 1883 that an electric lamp could be made to rectify current by adding an extra plate in the vacuum enclosure. This was called the 'Edison effect'. Thomas Alvan Edison's US Patent No.307031, 21 October 1884, included these words:

> If a conducting substance is interposed anywhere in the vacuous space within the globe of an incandescent electric lamp, and the said conducting substance is connected outside of the lamp with one terminal, preferably the positive one, of the incandescent conductor, a portion of the current will, when the lamp is in operation, pass through the shunt-circuit thus formed, which shunt-circuit includes a portion of the vacuous space within the lamp.

It was patented but was not applied to any purpose. In 1899 the effect was explained by J.J.Thomson who was investigating the fundamental nature of matter. In the 'Edison effect' and knowledge about it, lay the solution to the lack of gain, detection and signal generation that the teletechnique design community was seeking.

By 1900 electricity was firmly understood from both the practical and theoretical standpoints. Gradually, after that date, an emphasis developed toward more investigation and application of the weaker transducer effects, a main factor being steady improvement of the electronic amplifier that enabled previously impractically small transductions to be put to use. The preparatory stages for the emergence of electronic technique are worthy of special mention.

Electronic fundamentals and practice

By 1800 investigators of electrical phenomena possessed well developed experience with electrical charge, by then appreciating that it was of two polarities. They had experience of electric current flow in evacuated enclosures and were familiar with the glow discharges. There also existed an appreciation that matter existed of fundamental particles although they had little idea of their form, nor had anyone isolated them. Fig. 5.7 shows apparatus in use toward the end of the 19th century.

Work on electrolysis in the 1830s began an increase of activity that would culminate in the clear understanding that electrons did exist, this being the result of the, 1896, J.J. Thomson experiment for measuring the ratio of electric charge to proton mass. His work did not appear as an isolated event but was built upon the endeavours of many before him. Frisch (1977) is an account of this development seen from the point of view of discovery of fundamental particles. Morgan (1953) includes a small account of J.J. Thomson's work.

Significant terminology used in modern times arose out of Faraday's work on electrolysis. By 1834 he was in need of new names for his ideas of the entities involved. He distinguished the poles to which the elements were attracted, naming them the 'anode' and the 'cathode' (derived from the Greek for upward and downward). These were associated with the terms anion and cation. (Cathode was sometimes spelled kathode, a practice that persisted into the early decades of the 20th century and still prevails in other languages.) He termed the liquid that exhibited decomposition the 'electrolyte' and the process 'electrolysis'. He also coined the word 'ion', meaning the traveller. Faraday arrived at these terms in consultation with his polymath, the Reverend William Whewell. Accounts of Faraday's reasons for needing and generating these terms are to be found in his republished laboratory note books but are more easily read from Thompson (1898).

Apparently the first person to speak of the singular existence of the electron was G.J. Stoney who attempted an estimate of the unit value of the charge for 'one electron'. In 1874 he proposed a natural system of units based upon the speed of light, the constant of gravity and the charge of the electron (as he understood it to exist). He was ahead of his time – the proposal was unworkable then.

Merging from another direction were those investigators interested in the discharge tube. Mullineux-Walmsley (1910) includes a lengthy description of many of the experiments performed. One is shown in Fig. 5.8. The notable development in this field of interest came from a German glassblower, Heinrich Geissler (1814-1879), who worked to improve the vacuum pump to manufacture better toys. Geissler constructed high quality evacuated glass tubes for such scientists as Plücker, who called the tubes after their maker, and also for Hittorf. Plücker, in 1858, noticed that a beam of 'cathode rays' was produced from the high-vacuum tubes. Hittorf, in 1869, observed that as the vacuum was hardened the dark-space increased in width to finally fill the tube. He also recognised that these rays were deflected by a magnetic field, Fig. 5.8.

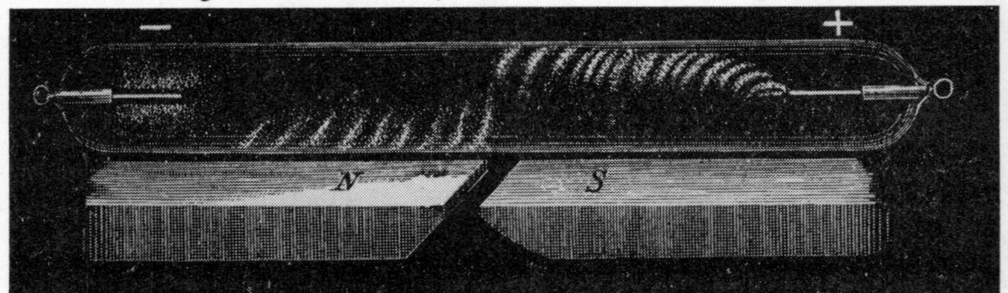

Fig. 5.8 *Effect of a magnet on the discharge produced within a vacuum tube . . . . one of the many experiments performed with Geissler tubes during the nineteenth century* [Mullineux-Walmsley (1910), Fig. 669, p.681. Courtesy Cassell Ltd.]

The seeds of the J.J. Thomson experiment, the mass spectrometer and the cathode-ray tube for television monitors and the oscilloscope were thereby sown ready for others to use later in time. Other workers made contributions to the gradually growing knowledge of the electron. Crookes studied their properties concluding, incorrectly, that they were negatively electrified molecules. This was not acceptable to other scientists and it was finally realised that the magnitude of the ratio of charge to mass would settle the argument. Along the way Hertz, Lenard, Schuster, Zeeman and others each added information about the nature and properties of the electron. After considerable delay in acceptance of the numerical value for the ratio e/m (it seemed impossibly large) J.J. Thomson finally provided satisfactory evidence in 1897. His apparatus, a clear forerunner of the cathode-ray tube for measurement purposes is shown in Fig. 5.9. It contains both electrostatic and electromagnetic deflection facilities.

Developing out of the interest in the nature of cathode rays sprung, with a great leap, the discovery of X-rays. Lenard noticed that 'cathode rays' passed through his hand but did not realise the significance of his observation. Crookes observed that photographic plates became fogged if left near his 'cathode ray' tube experiments. In 1895 it was Röentgen, Fig. 5.10, who realised that another form of ray existed.

Accounts of the discovery and early application of X-rays were published at a voluminous rate. Close to a 1000 papers appeared in the first year! It became a working tool in hospitals across the world within a year, see Fig. 5.11. Bowers (1970) provides an introduction to the history of X-rays. Byrn (1900) includes photographs of early hospital apparatus. De Vries (1971) includes some contemporary prints of c1900 apparatus.

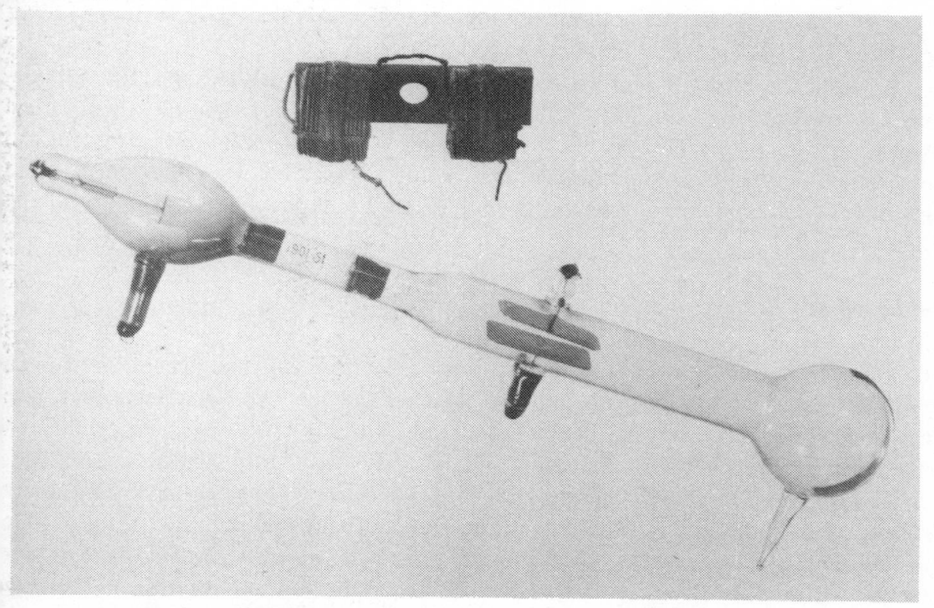

Fig. 5.9 *Original apparatus used by J.J. Thomson to determine the e/m ratio* [Lent to Science Museum, London]

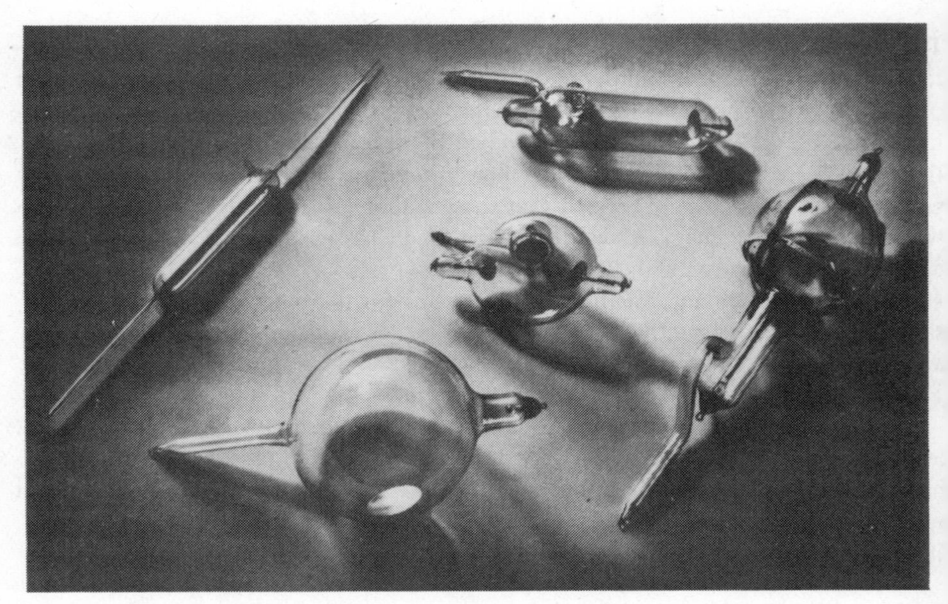

Fig. 5.10 *Wilhelm Roentgen used one of these discharge tubes when he discovered X-rays in 1895*
[Courtesy Deutsches Museum, Munich]

Nuclear radiation 'rays' were discovered by A.H.Becquerel but he thought them to be X-rays. It was E. Rutherford, in 1899, who showed that they were different and that the 'rays' emitted without energy input from certain materials, such as radium, could be grouped into alpha and beta particles.

By 1900 the Edison effect had been explained in fundamental terms, the cathode-ray tube had already been put to quantitative measurement of time-varying electrical signals by Braun in 1897, the electron was understood. The scene was ready for the emergence of the thermionic devices that would provide many of the long-awaited instrument requirements. Anderson (1964) is another study of the discovery of the electron.

Indication of electric variables

By the end of the 18th century many forms of electrometers had been devised and a few devices had been made to indicate some of the then known effects that gave electrical forms of output. By contrast, however, with the variety of electrical indicating instruments that existed a century later they formed an almost insignificant collection. As electrical circuit laws and more transducer effects became known, there steadily grew more and more varieties of instrument that could indicate the magnitude of the electrical variables voltage, current, resistance, power, frequency and power factor.

There developed for each of the variables, very many methods indeed. Variety arose in the provision of needs for a large range of variable magnitude, for extended frequency range, for numerous different applications and for a range of accuracy capabilities. Add to these reasons the need for manufacturers to employ a different principle to their marketing competitors for reasons of patent protection, and the result was an incredible variety of devices.

Electrical-indicating instruments certainly constitute an important part in the whole gamut of measurement devices, but to maintain a balanced presentation here, it is necessary to keep this subsection much shorter than the material would easily allow. It is not possible, in less than several volumes, to cover all indicating instruments that were devised in the 19th century. A general guide to the literature is now given, followed by accounts of dominant methods. It is worth remarking here that in approx-

Fig. 5.11 *One of the first X-ray equipments in use. It was devised by Edison to assist surgeons in 1896*
[*Scientific American*, 1896]

Table 5.1 *Classification of electrical phenomena and instruments*
[From Drysdale and Jolley (1924). Part I, Table I, p. 26. Courtesy Ernest Benn Ltd., London]

I.–ELECTRIC CURRENT

Phenomenon.	*Manifestation.*	*Discoverer.*	*Instruments.* LABORATORY	*Instruments.* INDICATING
A. Magnetic	(*a*) Deflects permanent magnet	Oersted, 1819	*Moving Needle Galvanometers:*	
			Schweigger's multiplier, 1820	Ayrton and Perry ammeter, 1879
			Nobili's astatic, 1825	..
			Pouillet sine and tangent, 1837	..
			Kelvin mirror, 1858	..
			Deprez magnetic control, 1879	..
	(*b*) Tends to move conductor in magnetic field	Faraday, 1821	*Moving Coil Galvanometers:*	
			Sturgeon, 1836	Weston voltmeter, 1888
			Kelvin siphon recorder, 1867	..
			d'Arsonval galvanometer, 1882	..
			Blondel oscillograph, 1891	..
			Duddell oscillograph, 1897	..
			Einthoven string galvanometer, 1901	..
	(*c*) Magnetises and attracts iron core	Arago and Davy, 1820	Ayrton and Perry ohm-meter, 1882 Blondel oscillograph, 1893	Ayrton and Perry magnifying spring, 1884 Kelvin ampere gauge, 1885 Schuckert attraction ammeter Nalder repulsion ammeter, 1885
	(*d*) Attracts or repels second current	Ampere, 1820	Weber dynamometer, 1843 Joule's current weigher	Ayrton and Perry wattmeter,1881 Kelvin balance, 1883 Siemen's dynamometer, 1883
	(*e*) Tends to reduce section of conductor	Bary, 1901	Northrup absolute current meter, 1907	
	(*f*) Induces current in neighbouring conductors	Faraday, 1830	Fleming cymometer Marconi wavemeter	Ferraris induction instruments ..

B. Thermal	(*g*) Heats circuit	..	Duddel thermo-galvanometer, 1904 Irwin oscillograph 1907	Cardew voltmeter, 1883 Hartmann and Braun hot-wire instruments
C Chemical	(*h*) Decomposes water	Carlisle and Nicholson 1800	Gas voltameter	Bastian supply meter
	(*i*) Deposits metals	Cruikshank, 1800	Copper and silver voltameter	Edison supply meter, 1879 Wright mercury meter
	(*j*) Produces standard polarisation E.M.F.	Volta, 1800	Latimer Clark cell, 1872 Weston cadmium cell, 1893	..

II.–POTENTIAL DIFFERENCE

				Voltmeters.
D. Kinetic	(*k*) Produces current pro portional to P.D. in fixed resistance	Cavendish, 1781 Ohm, 1827	Any galvanometer or ammeter of suitable constant resistance	..
E. Electro-staticforce	(*l*) Attracts or repels other charged bodies	Gilbert	*Electrometers:* Henley, 1772 Kelvin quadrant, 1851 Volta attracted disc	*Electrostatic voltmeters:* Kelvin, 1867 Ayrton and Mather Jona
	(*m*) Tends to increase capacity of system	.. ..	..	.. ..
F. Discharge	(*n*) Disruptive discharge	Hawksbee, 1707	Spark gauges	..
	(*o*) Ionization	Edison, Fleming, etc.	Valve amplifiers	Moullin voltmeter

imately a decade from the date of preparation of this work it will then be an appropriate time to compile the history of the mechanical form of the electrical indicator. It is clear that by 1990 indicators will almost exclusively be solid-state devices in which there are no moving mechanical parts. A long and important era in sensing history will have closed.

Accounts of electricity of the early 19th century gave little space to measuring devices used with the 'voltaic' current circuits because they were not generally known or in common use. Worthwhile explanations began to appear in the latter decades. Tunzelmann (1890) gives a short account of the reasons for measuring devices being made to give quantity to the electrical variables that had, by then, been isolated. It also relates how the unit of resistance gradually rose in importance as a means of standardising the work of electrical scientists.

Ohm's law expresses the ratio of voltage to current as the unit resistance. Pouliett, in 1837, expressed his measurements in terms of the resistance of a column of mercury of standardised length dimensions. Work of Gauss on the intensity of a magnetic pole led to Weber, in 1851, showing that Gauss's millimetre, second, milligram dimensional unit system was a sufficient basis for expressing the new electrical units in a traceable manner. In 1861 the British Association set up a study for the development of standards of electrical resistance. The unit of resistance was subsequently called the ohm, that of electromotive force, the volt, and flow of charge, the ampere. C.G.S. units came into some measure of general use around 1870. In 1881 an international meeting of electricians, held in Paris, argued that the c.g.s. system should be adopted worldwide. More detail on electrical standards work of the British Association is available in Howarth (1931) and in Gray (1893), the latter being a detailed account of the equipment and procedures used in the late 19th century.

Tyndall (1863) records the practical problems of galvanometer construction of the 1860s, a time when scientists usually had to design and manufacture their own apparatus, commercial sales of electrical equipment yet having to emerge. Various contemporary works providing information on electrical instruments over the period 1880 to 1920 are Gordon (1880), Kempe (1892), Houston (1893), Nichols (1894), Carhart and Patterson (1897), Reed (1903), Edgcumbe (1908) and Bushnell and Turnbull (1920). The catalogue of the University of Utrecht Museum (undated) includes several illustrations of early indicators but it has no commentary. Chipman (1964) traces the development of the electromagnetic moving-needle form of indicator covering from genesis to about 1830. Handscombe (1970) provides a concise account of the historical development of electrical measuring instruments. The main part of this text is useful for its explanation of the design principles of the many forms brought into use. Herbert (1923) provides detail of indicators in general use at that time, showing that most innovation took place in the 19th century.

The definitive work on electrical instruments was written in two parts by Drysdale and Jolley (1924). This extensive work contains a considerable quantity of information about the numerous types invented. It also includes valuable, still-current, design information that is highly relevant to the construction of fine mechanisms in general. The introductory chapter includes a table that classifies instruments according to their principle, who was associated with the principle and the form in which the principle was implemented as a measuring instrument. This table is reproduced as Table 5.1.

Another useful historical account of the development of instruments is that compiled by Whitehead (1951) on British designs. Glazebrook (1922) is another definitive work on the whole topic of electrical measurements. It is valuable for its detailed and scholarly accounts of the many precision instruments in use at the turn of the century in advanced institutions such as the National Physical Laboratory. Sloanes Dictionary (1894) is also a useful source of information.

Designers of outstanding significance in the commercial emergence of indicating devices were Kelvin, see Green and Lloyd (1970), Weston and Paul. Their biographies, see Appendix 1, include detail about the instruments they invented. A source of information about historical aspects of scientific instruments that may seem to be of value for electrical instrument information are those works written for the antique buyer market. These, in general, however, give little space to 19th-century electrical devices; Pearsall (1974), for instance, although being relevant to other forms of instrument has nothing on electrical apparatus. This presumably is because public interest in collecting electrical instruments, such as indicators, friction generators, electromedical generators, and the like, has only just begun to manifest itself in the antique market. It is made very clear, from perusal of works like Pearsall, that electrical instruments, although numerous in their shape and form, constituted only a small part of the whole range of measuring devices in use in the 19th century.

With exception of the electrostatic voltmeter instruments most electrical indicators respond in some way to the magnitude of current flowing through the structure. The current can be used to produce a force that is balanced by a restoring spring, this force being the result of interaction between the magnetic field produced by the current flowing in a wire and a magnetised needle, by field produced by a coil reacting against a stationary permanent magnet field or by two electromagnet coils. Alternatively, the current can be used to cause a length change in a wire by the heating effect. Power-measuring meters react to the nett force produced as the product of current and voltage source derived magnetic fields. Yet another group makes use of plating and other chemical decomposition effects of electric current.

Ohm's law provides the means for converting one form of the electric variable into another through the use of the appropriate conversion component, such as a resistor to convert current flow into a voltage drop. Galvanometers, therefore, although operating on current principles were often made with integral, internal, conversion facility or suitably matched windings to indicate voltage and resistance directly. Some forms are able to respond to both d.c. and a.c. quantities but some require the use of inbuilt or external conversion mechanisms.

Electrometers of the 18th century indicated potential either by inference from the force generated between two charged bodies, or from the point of breakdown of a known airgap. The electrostatic form of indicator continued to be developed for the measurement of voltages, especially for high values and in use in situations where current drain must be small, such as in charge measurements. They are free from magnetic hysteresis effects, have no heating error and read both direct and alternating voltages correctly.

Significant progress was made on the design of electrostatic instruments through the endeavours of Kelvin, who showed in 1853 that such an instrument could be made that would measure in terms of the absolute units. His first improvement on the 18th-century electrometer was to use the divided-ring instrument (Fig. 5.12*a*) in which a needle was brought into a central position by a restoring spring that provided a force to balance that exerted by the divided ring upon the needle. This was a relatively short-lived improvement for the principle used there was more efficiently implemented in the quadrant form, Fig. 5.12*b*. The instrument developed by Kelvin for standards work (Fig. 5.13) was very sophisticated and time consuming to use, Gray (1893). It was also very insensitive for low voltages i.e. those under, say, 100 V. For measuring low magnitude voltages Kelvin devised the multicellular instrument (Fig. 5.12*c*) in which a number of quadrant cells were placed on the same shaft to increase the torque generated for a given voltage applied to the terminals. Various forms of these came into use. The vertical electrometer (Fig. 5.12*d*) was designed to fill the need for a more practical instrument of lower cost. On application of a voltage, the capacity of the plates varies such that they generate a force equal to that of the gravitation force exerted on the vanes. Kelvin implemented these principles as working instruments and they were used in similar and adapted forms by many others. Until the advent of very high ($10^{12}\,\Omega$) input resistance thermionic valve and solid-state amplifying devices, the electrostatic voltmeter played an important role in the development and application of electrical method. The electrometer also found application in power measurement as the result of findings of Ayrton and Fitzgerald in 1882, Whitehead (1951).

Electromagnetic current detectors first took the form of the Oersted experiment shown previously in Fig. 5.4. It was not long before it was realised that more turns yielded greater sensitivity and the Schweigger multiplier, or 'multiplicator' as it was called in the 19th century, was universally adopted to form instruments like that shown earlier in Fig. 5.2. A later version, used extensively by the General Post Office GPO of Britain had two coils wound differentially around the magnetised needles. This was for the detection of differential currents, Herbert (1923).

An early improvement was the use of a second needle that was placed to balance out the torque exerted on the measuring needle by the earth's magnetic field. These designs were called astatic instruments. With one needle (Fig. 5.14) inside the loop and the other outside, and of the opposite magnetic polarity, the earth's field torque was cancelled while the effect of the field of the coils on the needles is doubled. Here was another early form of a differentially connected system in which common-mode signals are caused to cancel. (An earlier use of this concept was the use of dual thermometers by Herschel). Astatic designs like this were first made by Nobili in 1825. Kelvin virtually perfected the astatic moving-needle galvanometer, in 1858, by the use of four coils used in conjunction with a centrally mounted mirror for optical lever readout.

The above-mentioned needle galvanometers, although giving repeatable performance were not then capable of calculated calibration. Absolute calibration was possible through the use of calculable versions of the sine and tangent galvanometers, (Fig. 5.15) in which knowledge of the parameters of construction enabled the sensitivity to be

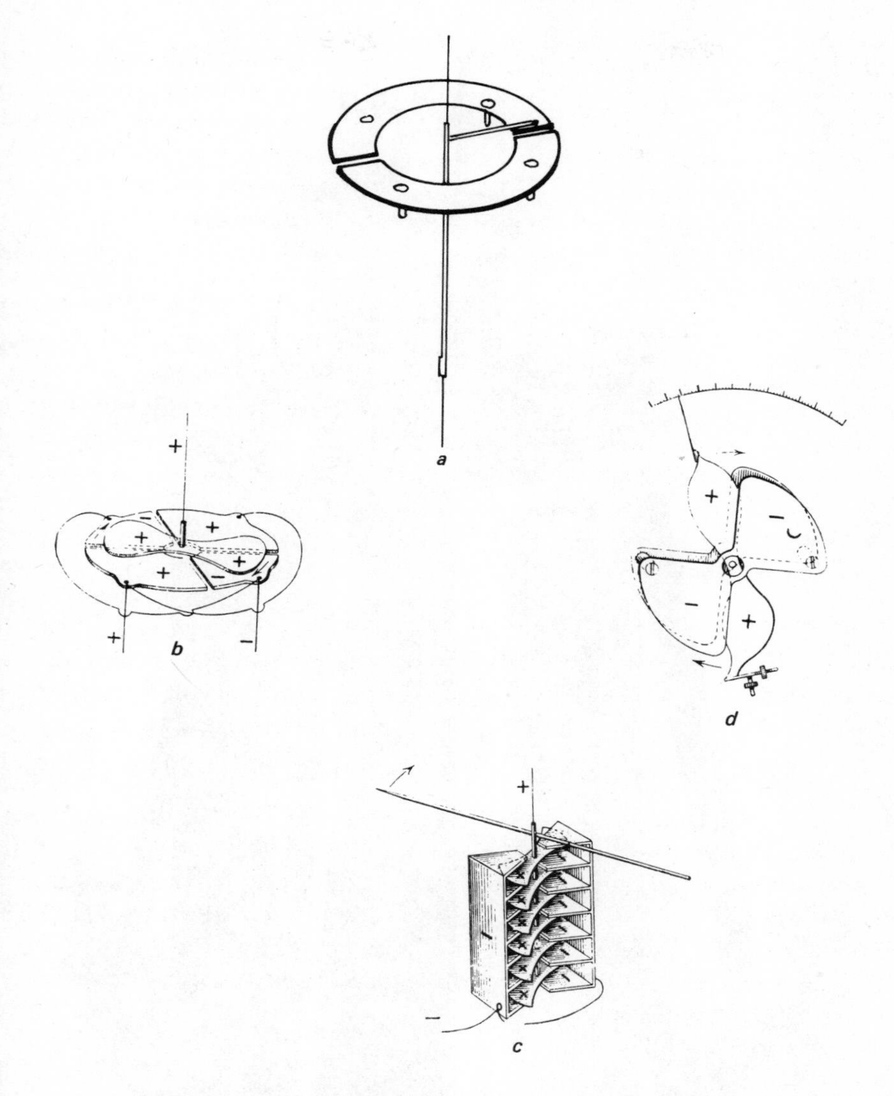

Fig. 5.12 *Development, during 1856 to 1890, of the electrostatic voltmeter by Kelvin*
a Divided ring electrometer
b Quadrant electrometer
c Multicellular voltmeter
d Vertical form
[(b), (c) and (d) Drysdale and Jolley (1924). Courtesy Ernest Benn Ltd.]

derived. These could also be used as instruments having known laws of deflection: for the tangent version the tangent of the angle of deflection is proportional to the current causing the deflection. For these laws to hold sufficiently well deflections had to be small and the instruments physically large. These were first devised by Poulliet in 1837, later being developed into accurate measuring instruments by Helmholtz.

Fig.5.13 *Absolute quadrant electrometer of Kelvin c1880*
[Gray (1893), Fig.19, p.116]

Fig. 5.14 *Astatic form of needle galvanometer* [Deschanel (1891), Fig.469, p.703]

Although moving, magnetised-needle, galvanometers were still in use in the 1930s they were, for more routine indication, gradually replaced by the moving-coil, permanent-magnet, type of instrument. As early as 1821 Faraday had performed a simple experiment showing that a current-carrying conductor would rotate around a fixed magnetic pole. In 1836 Sturgeon used a suspended coil held in a magnetic field as an indicator device. This was taken up by others: Highton used the principle as part of telegraphic receiving apparatus in *c*1856; Kelvin added the internal iron core to concentrate flux into the region of the rotating coil in his siphon recorder (see Fig. 5.45). Maxwell also built a similar instrument. It was d'Arsonval, in 1882, who brought the moving-coil meter into popular use, adding a mirror for optical lever use.

In 1888 Weston marketed the instrument as a commercial venture selling enormous numbers to the emerging general market. Weston meters held first place for many forms including the multimeter in which respective connections and additional components enabled the one basic meter movement to be used for measuring voltage, current and resistance. These instruments were made for the market that did not need

Fig. 5.15 *Standard tangent galvanometer*
[Mullineux-Walmsley (1910), Fig.730, p.731. Courtesy Cassell Ltd.]

great precision but did require simplicity of operation and high reliability. They were also made in smaller numbers at standards grade. The physical layout of the Weston meter, Fig. 5.16, was different to the galvanometers of prior times being pancake-shaped rather than a tall cylinder. Fig. 5.7 shows an ampere meter of this flat style. Drysdale and Jolley (1924) describes numerous variations of moving-coil meter arrangements. The 'Uni-pivot' form, in which only one suspension pivot was used was invented by R.W. Paul at the turn of the century, Mullineux-Walmsley (1910).

Most galvanometers are used as steady-state current indicators and are appropriately designed to be critically damped. Ballistic galvanometers, introduced prior to 1900,

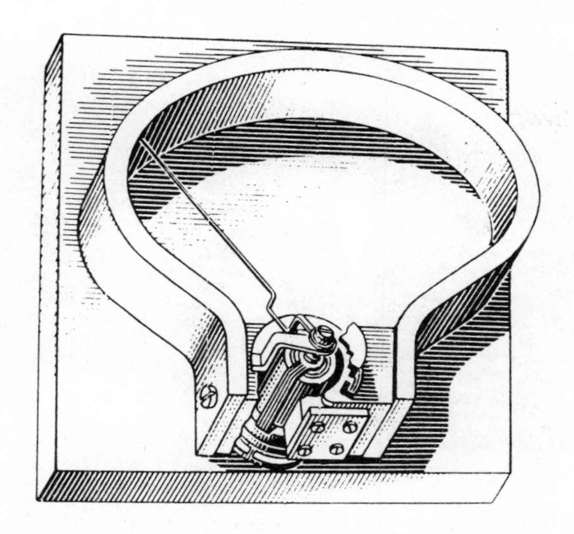

Fig. 5.16 *Sketch of early form of Weston moving-coil indicator*
[Drysdale and Jolley (1924). Courtesy Ernest Benn Ltd.]

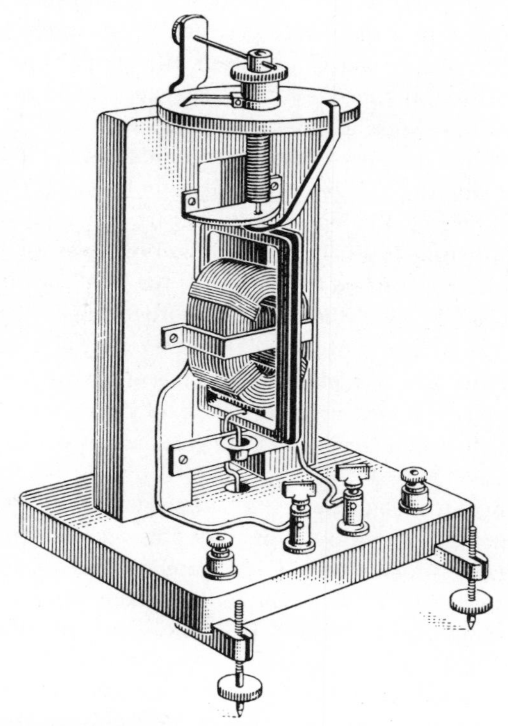

Fig. 5.17 *Early form of Siemen's wattmeter*
[Drysdale and Jolley (1924). Courtesy Ernest Benn Ltd.]

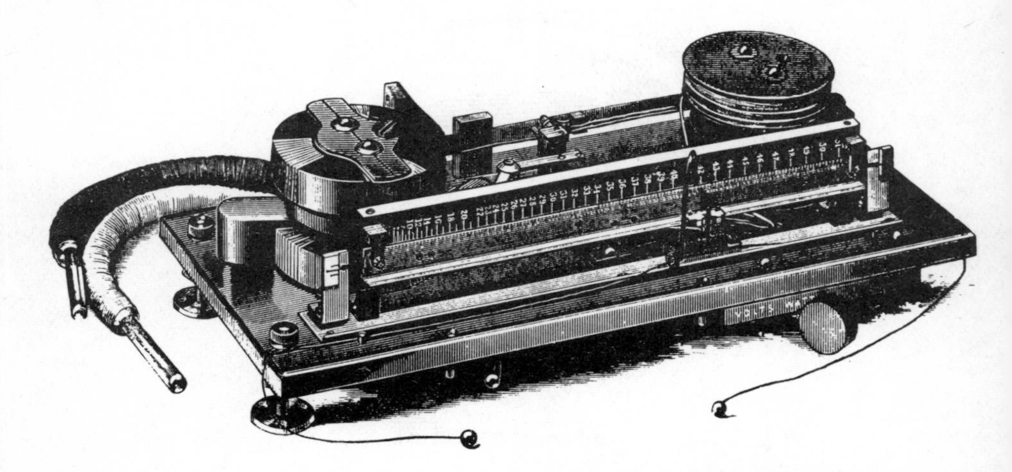

Fig. 5.18 *Composite form of the Kelvin current-balance in use around 1880 to measure current from 0·02 to 300 amperes or 100 to 25 000 watts, at 100 volts*
[Gray (1893), Fig.16, p.108]

were used to measure impulses of electrical energy. They differed from the more usual form of galvanometer in that they were designed to withstand high-impulse voltage levels, had small inherent mechanical and electrical damping and were shielded from external drafts. They also, in contrast to the usual need, had high armature rotational inertia. Amplitude of the swings was observed, this being used, by inertia calculation, to obtain a measure of the quantity of electricity contained in the impulse.

The resonance or vibrating galvanometer, Glazebrook (1922), was first introduced in 1891 by Wien. It was also independently constructed by many others, including Blondel and Rubens, before 1900. These were used for measuring alternating current; they were purposefully constructed to resonate at the frequency of the current supply that they were to measure. By virtue of the amplitude amplifying effect of the mechanically tuned resonant arrangement these exhibited greatly enhanced sensitivity. They were also very selective (mechanical tuned arrangements have a far greater quality factor than passive electrical element resonant circuits) and, therefore, rejected signals at frequencies other than that for which the instrument was tuned. Numerous forms were devised.

Another possibility for constructing a current-indicating device is to pass the current through two coils and measure the force exerted between them by weighing it or by balancing it against a torsion spring. Dynamometer instruments use this principle. The first to be built was that of Weber in 1843. Siemens later manufactured a design (Fig. 5.17) using spring restoration and, therefore, needing calibration. The current balance of Kelvin (Fig. 5.18) uses a sliding mass to bring about force-balance; it is capable of absolute current determination. The current-balance method is still in use today, in the Ayrton-Jones form, for defining the fundamental unit of the ampere, HMSO (1972). Dynamometer instruments are described in Gray (1893), Mullineux-Walmsley (1910), Glazebrook (1922) and Drysdale and Jolley (1924).

Another class of electromagnetic instrument is that in which a soft-iron moving armature is drawn into a coil as the result of induced electromagnetism. This effect was observed by Davy and by Arago in 1820 and many forms have been made. These became known as moving-iron meters but they were also called soft-iron or electromagnetic instruments. They have the advantage that they respond to alternating current as well as to direct current. The shape of the scale can also be tailored to obtain expanded scales by suitable adjustment to the shape of the iron vane. In general, this principle was best employed for measurement of relatively large current magnitudes.

Joule, in 1843, expounded the I^2R law of heating in a resistance. This effect has been used to build instruments of two major forms. The first made use of the expansion effect of wire that elongates as it is heated by a current passing through it. Cardew, in 1883, built the first practical indicator of this design (Fig. 5.19), using a long fine platinum wire. A later form (Fig. 5.20), by Hartmann and Braun (and others), used the magnification effect of the sag of a sag-wire to build a more compact hot-wire meter. This form is still used today, a significant feature being the ability to read power of alternating current accurately irrespective of the frequency and waveshape.

The second method of using the heating effect is to convert the heat into electrical energy via a thermocouple. These are called 'crossed thermojunction' instruments. Strictly speaking they are not indicators but converters for use with a suitable form of indicator. Sophisticated forms of these, in which the thermoelectric current was generated within the structure of the galvanometer suspension, were termed thermogalvanometers by D'Arsonval in 1886. They were used to detect minute levels of radiation by Boys in 1887, in a more advanced form. Descriptions of these instruments are available in Mullineux-Walmsley (1910), Drysdale and Jolley (1924). Chaldecott (1954) describes the thermogalvanometers.

The technique first used to indicate current flow did not survive in general use. It made use of the chemical action of current passing through an electrolyte – the 'voltameters'. They either decomposed water into calibrated gas collecting tubes or, alternatively, were used to plate-out metal, such as copper or silver. The silver voltmeter was adopted as the standard for current in the 19th century. Later Edison used the principle to build a public supply meter for his electricity distribution systems of the late 1870s. (The reverse process was used to define the standard of the volt in the Weston Standard cell of 1893 and others that preceded it, such as the Clark cell of 1872.) This form of supply meter was called a 'coulomb-meter'. It measured the current-time product which then had to be converted into power-consumed by multiplication with the mean voltage for the duration of the measurement. Voltameter devices were not very practicable and furthermore they were unable to give an indication of power consumed at any given time to the consumer. Voltameters and the coulomb-meter of Edison plus other designs are discussed in some detail in Mullineux-Walmsley (1910).

The need for a better wattmeter lead to the introduction of motor kinds in which the product of current and voltage cause a disk to rotate at a speed proportional to power consumption. Aron's meter of 1888, shown previously in Fig. 2.39, was the

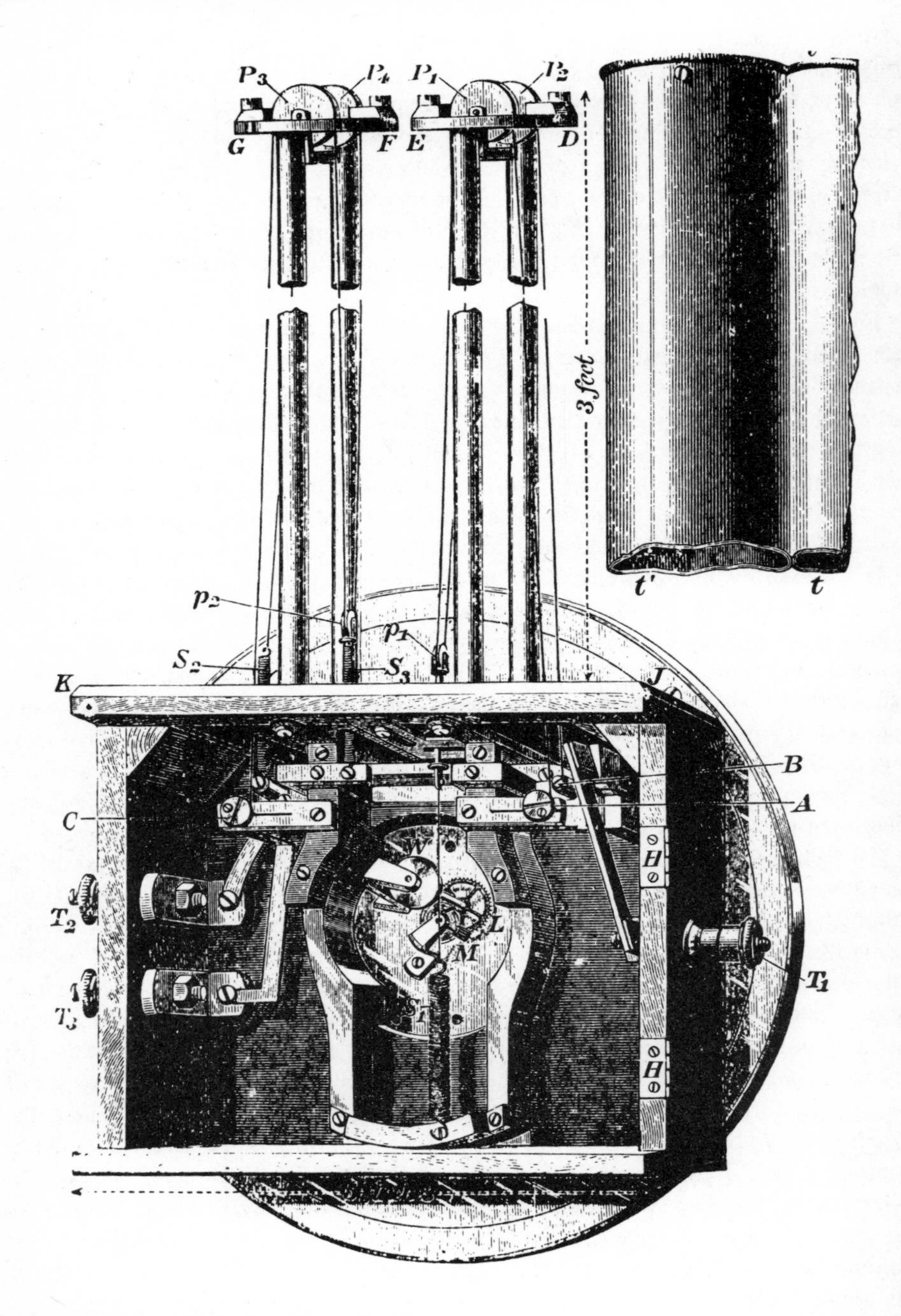

Fig. 5.19 *Cardew 'hot wire' voltmeter c1885*
[Mullineux-Walmsley (1910), Fig. 334, p.372. Courtesy Cassell Ltd.]

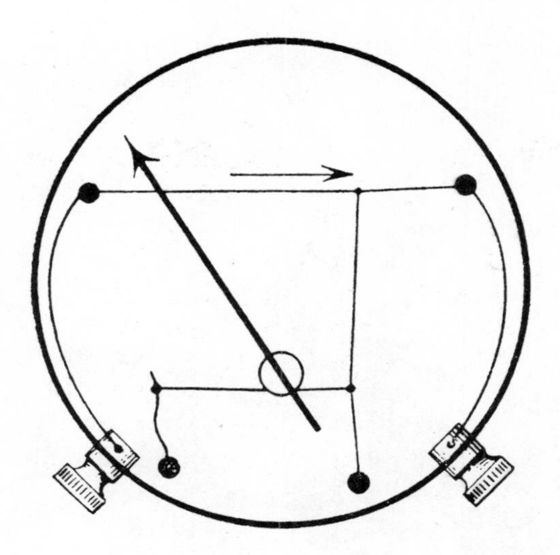

Fig. 5.20 *Schematic view of Hartman and Braun Hot-wire voltmeter*
[Drysdale and Jolley (1924). Courtesy Ernest Benn Ltd.]

first mechanical form of meter to be put into widespread use. In it a swinging permanent magnet, on the end of the pendulum, has a force exerted upon it that is related to the current flowing in the lower metering coil. The rate of the pendulum swing is a measure of the current consumption (for fixed voltage). This was recorded by the mechanical mechanism which detected the difference in swing periods against a pendulum clock. The Aron meter was unnecessarily complex and not particularly accurate. They were superseded by many varieties including the now familiar form of single-phase induction motor style, the Shallenberger meter of 1888 being among the first to find application. Table 5.2, taken from Drysdale and Jolley (1924) lists the various forms and the dates of their introduction. Lanphier (1925) traces the development of the wattmeter.

By 1900 instruments had also been invented to measure the new variables that arose through the use of alternating current in municipal supply. There were instruments to measure form-factor, frequency, phase angle, wave-shape, inductance, capacitance, power-factor and properties of magnetic materials. Some of these, those that used recording mechanisms, are covered below in Section 5.4. Space does not permit description of the others. The reader is referred to Glazebrook (1922), Mullineux-Walmsley (1910) and to Drysdale and Jolley (1924) for details. The use of measuring bridges had also reached a sophisticated stage by 1900. These took numerous forms, Mullineux-Walmsley (1910), Glazebrook (1922) each describe these techniques. Herbert (1923) gives information about procedures used in telephony at the turn of the century.

The galvanometer played a vital part in the development of 19th and early 20th century science. Fig. 5.21 shows two mirror galvanometers in use for calibrating Clark cells, *c*1874. For a century it provided means for men to 'see' minute electrical signals

Fig. 5.21 *Laboratory set up in 1874 for calibrating Clark standard voltage cells*
[James-King (1962*a*), Fig.37, p.255. From *Phil. Trans Roy. Soc. London*]

that they could not observe otherwise. Performance was gradually developed to the stage where, late in the 19th century, fundamental spontaneous fluctuations became the limiting factors on the sensitivities that were possible, they reached the order of 10^{-14} A/mm of deflection or 10^{-10} V/mm. The limiting factor was called Brownian motion after Robert Brown, who first noticed that suspensions of pollens were constanly in motion in his experiments of 1827. His discovery led to the eventual realisation by others that this was not a property only of pollens but of all materials in any phase that existed above absolute zero temperature. Delsaulx, in 1877, published a paper in which he put forward the correct explanation that it was a matter of energy relationships within a substance. Gouy, in 1888, prepared the way to a satisfactory explanation that was later provided in 1905 by Einstein and subsequent work by

Table 5.2 *Classification and dates of development of electricity supply meters*
[From Drysdale and Jolley (1924) Part 2, Table 1, p.22 (Courtesy Ernest Benn, London)]

Principle	Pioneer or Typical examples	
	Quantity meters	Energy meters
Electrolytic (d.c. only)–		
(*a*) Decomposition of metal salt	Edison (1879), M'Kenna (1892), Wright, Long, Schatner	None
(*b*) Decomposition of liquid	Bastian (1900), Holden, Thorpe	None
Thermal	Forbes windmill meter	None
Electromagnetic motor meters–		
(*a*) Permanent magnet field (d.c. only)	Ayrton & Perry (1882), Chamberlain-Hookham, Ferranti, Swinburne, O'Keenan, Busch, Siemens & Halske	None
(*b*) Electromagnetic field (d.c. or a.c.)	Ferranti (1889)	Thomson (1891) (rotating), A.E.G. (oscillating)
(*c*) Induction motors (a.c. only)	Schallenberger (1888), Ferranti, Wright (1890)	Ferranti, Westinghouse, A.E.G., B.T.H., etc.
Mechanical drive meters–		
Clock meters	Ayrton & Perry, Aron, Siemens & Halske	Ayrton & Perry, Aron
Intermittent integrators	Cauderay, Brille, Kelvin, Frager	Frager, Holden
Continuous integrators	..	Boys 'Mangle' meter (1882)

Smoluchowski and by Perrin.

Many papers have appeared on this limit to measurement. Ising (1926) seems to be the first rigorously to connect it with galvanometers. Barnes and Silverman (1934) and McCombie (1953) discuss it as a limit of measuring processes including some reference to galvanometers whereas Jones and McCombie (1952) provide a treatment of the phenomenon in terms of its effects on galvanometers and their associated amplifiers. Takahasi (1952) presents a generalised theory of thermal fluctuations. Wax (1954) is a collection of historic papers on noise. These papers can provide entry into the extensive theoretical literature. Until recent times the galvanometer was the most sensitive form of electrical signal detector. In some applications its inherent integrating characteristic and high stability makes it still a powerful instrument. The galvanometer also provided the sensing element for electrical signal recorders as will be seen in Section 5.4. In the early 20th century it was also used to derive derivative and integral feedback signals for early p.d.i. controllers.

The contribution of teletechnique

Telegraphy, telephony, television and radio, the teletechnique group of data transfer systems using electrical method, provided considerable assistance to the development of modern measurement systems. Everitt (1976) deals with the value, over time, of this service.

Although generally studied and reported from their social and technological aspects as communication systems they were, nevertheless, also the source of methods for transmission of measurement data.

From teletechnique use came signal transmission over wires, cables, waveguides and the electromagnetic medium. From it, designers learned of coupling problems, loading effects, termination difficulties, the theory of transmission lines and the effect of external noise sources. Telegraphy gave birth to the use of digital codes, of digital circuitry and of sensors and actuators in general. Telephony produced means for handling linear signals in the audio part of the frequency spectrum, spin-offs being an understanding of alternating current measurement techniques, of the composition of speech, of inductive sensing devices, the terminology of decibel, filter design using passive and, later, active electrical elements. Television took these concepts into the visual regime and led to appreciated understanding of optical and photo effects in combination with electricity. Visual messages, and other data, could be sent in visual form.

Radio provided another carrier medium for signal transfer, one that possesses extremely wide bandwidth capability.

Multiplexing techniques, in both the time and the frequency domain, were developed by communications equipment designers.

In short, the methods of signal communication used in telemetry and in circuit coupling came down to us from the considerable efforts of countless workers on teletechnique systems. It is, therefore, appropriate to provide a guide to the development of these areas. The published material is extensive but this brief guide should assist the reader to find where to seek more detail. This account concentrates on some of the specific areas that have obvious relevance to measurement systems, hopefully supplementing the normal descriptions given in works on teletechnique.

As we have seen in the previous Chapter telegraphy, the first of the tele methods, did not arise directly out of a chance discovery but evolved from the work of many people, over many decades.

Interest in frictional electricity, plus the growing need for faster communication than by a horseman, led to the emergence of many trial schemes. After several predictions and written suggestions about a practical telegraph system, the first to be built was that of Sömmering; in 1809 he built a system using electrolysis as the receiving-end detector of current flow. This is illustrated in Fig. 4.16.

The dominant difficulty until the 1900s was the lack of suitable detection methods for small electric power levels involved. It was just a year after Faraday's (and Henry's) 1831 discovery of the electromagnetic induction laws that Schilling devised a 5-wire telegraph system using electromagnetic sensing. From then on imagination, plus technological availability, gradually led to telephony, wireless telegraphy, broadcasting and television. Associated recording devices – tape, chart, wire, disk – arose as natural developments filling, in the main, auxiliary needs of each of these areas. Many developments can be traced to this availability, the pen recorder, for instance.

Historical knowledge of telegraphy, telephony and radio is now quite vast; of television, less so, and of the associated audio-signal recording, still lacking.

Telegraphy and telephony have fascinated both historians and contemporary writers since their inception. Thus we find that numerous 19th century technical books (from 1830 onward) have quite detailed accounts. The same applies for radio, where the general electrical texts (from around 1900 onward) that were written in the popular style more often than not have contemporary accounts of radio technique. General works later than about 1930 refer to television; the concept can be traced back to the 1890s but it was not until around 1925 that working systems were demonstrated. Television was rapidly accepted from then on. In 1928 the public was being advised how to build 'your own televisor' in secondary magazines.

Interest in the various aspects of electrical communication is such that the stage has been reached where numerous early texts are now available as partial and whole facsimile reprints. A recent work is a collection of relevant publications of the period 1857-1932, Halstead Press (1974).

It is not feasible to provide any more than a brief introduction into this area. The following comments are intended to be helpful to nonspecialists on telecommunication history and do not presume to be valuable to the many expert historians specialising in this development.

Numerous works, ranging from the popular to the learned, have included material about the general history of the telegraphy, telephony and wireless techniques. A selection is Angwin (1951), Byrn (1900), De Vries (1971), Frith and Rawson (1896), Gartmann (1960), Gibson (1915), Glazebrook (1922), Halstead Press (1974), Hawks (1929), Hibbert (1910), James-King (1962*c*), Kirkman (1904), Knox (1917), Lardner

(1862), Marland (1964), McKenzie (1944), Mullineux-Walmsley (1910), Squier (1933) and Tunzelmann (1890). Some of these sources provide quite detailed accounts of the instrumentation used, notably so in Byrn, Frith and Rawson, Glazebrook, Halstead Press, Lardner and Mullineux-Walmsley.

Works that are specifically about early telegraphy, or include only mention of telegraphic technique, and not the later forms of teletechnique include Arno Press (1974), Bright (1898), Deschanel (1891), Douglas (1877), Fahie (1884), Finn (1973), McNicol (1913), Pepper (1874), Reid (1879) and Willoughby (1891). Many of the earlier works are now reprinted in facsimile form by the Arno Press, New York and are, therefore, again easy to procure.

Telephony is covered in the works by Baldwin (1925), Bell (1908), Bell Telephone (1957), Bennett and Laws (1895, 1910), Du Moncel (1879), Finn (1966), Hanscom (1961), Herbert (1923), Preece and Maier (1889), Rhodes (1929), Schiavo (1958) and Watson (1913).

Historical aspects of the wireless, also called radio, method of transmitting telegraphy and telephony signals are covered in Archer (1938), Blake (1928), ETI (1975), Hancock (1950), Hawks (1927), Lodge (1900), MacLaurin (1949), Mazzotto (1906), McNicol (1946), Morgan (1953), Oatley (1932), Ratcliffe (1929), Risdon (1924), Sivowitch (1970) and Sydenham (1975).

Television became every day reality in the 1930s but its origins extend back into the 19th century. It is, therefore, appropriate to introduce the literature at this point. Histories of television are far less prolific than for the early tele methods but much can be gleaned from study of accounts contemporary with its commercial application from the late 1920s onward. A selected set of works might be Abramson (1955), Eckhardt (1936), Hawks (1929), Larsen (1960), Morgan (1953), Risdon (1935) and Shiers (1977).

Considerable detail is also available in biographies of workers in the above teletechnique fields (Appendix 1). Collections of apparatus of early telecommunication are extremely prolific and show the great variety of equipment that was used. Notable collections and presentations are at the Science Museum, London; in the Deutsches Museum, Munich; in the National Museum of History and Technology, Washington DC; in the Museum of Posts and Telegraphs, in Vienna and in the National Museum of Science and Technology, Stockholm each of which is listed in Appendix 2. There are, however, numerous other collections, existing in just about every local museum concerned with past lifestyle, such was the impact made by the teletechniques.

Attempts to construct electrical telegraphic systems began well before 1800, as was explained in the previous Chapter, but it became working reality when scientists discovered that a needle, (Fig. 5.4) could be deflected by a wire-carrying current, the work of Oersted in 1819. The earlier inventions of the simple voltaic power source, coupled with this law, enabled reliable and simple devices to be developed. From *c*1830 onward, due to the work and ideas of Schilling, Morse, Gauss and Weber, Cooke, Wheatstone, Steinheil, and many other persons, the simple needle galvanometer with a multiplicator winding evolved, see Fig. 5.22, into complex networks of intricate communication devices.

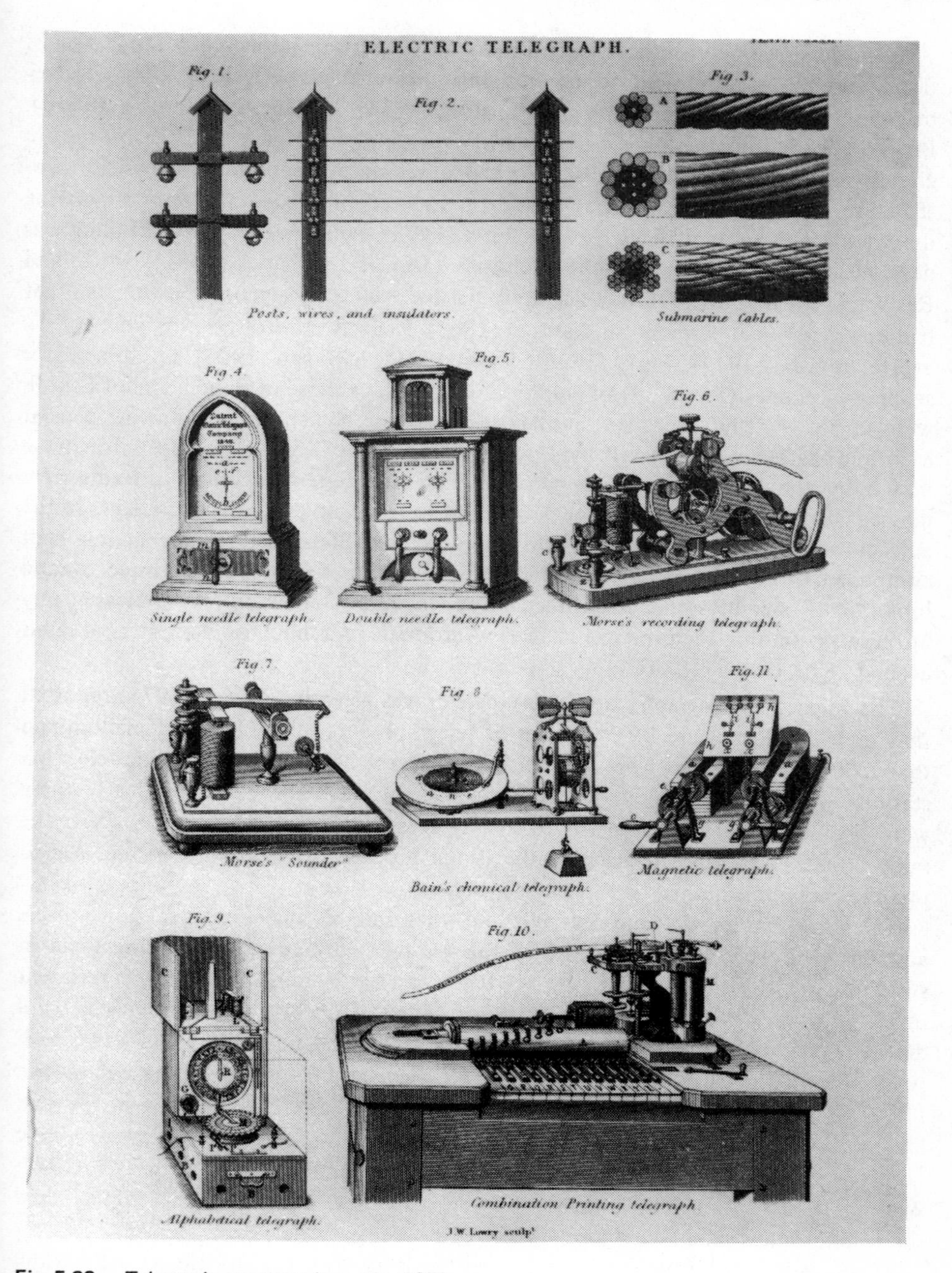

Fig. 5.22 *Telegraph apparatus in use by 1870*

Recorders, see Section 5.4, were developed to provide improved reading signals; faster and more convenient to use sending apparatus was developed. The electromagnetic relay was invented by Wheatstone *c*1837, Mullineux-Walmsley (1910), Bowers (1975) (but there are other claims) providing power gain to repeat or restore circuit states and to actuate 'alarums' and other output actuators. Open-wire construction was developed, Douglas (1877). Armoured multicore cables for submarine use were manufactured Black, (1972), Finn (1973). Synchronised signal sending was developed for multiplex systems, Lardner (1862), Deschanel (1891), Frith and Rawson (1896), Finn (1976*b*). Telegraphy assisted science to conduct experiments that had previously been impractical, Lardner (1862). Railway control systems were able to evolve, Bowers (1975), Burtt (1926), King (1921), Kirkman (1904). Equipment for sending facsimiles (Fig. 5.23) of graphical data, such as those of Bonelli and of Caselli, were in use by 1890 – called 'autographic' systems. Synchronously moving pens at each end station were devised by Cowper, Deschanel (1891). Tape perforators were in use by the 1860s along with perforated tape readers. Noncontact systems were devised in the 1880s for sending telegraph signals from moving railway carriages to the adjacent overhead wires, Byrn (1900). Henry, in the 1860s, used telegraph equipment components to telemeter river current meter data, Frazier (1964). The uniselector and bimotional, step-by-step, electromechanical circuit connectors were devised by Strowger, patenting the principle in 1891. Automatic switchboards were first patented around 1879, O'Dea (1939).

The impact of telegraphy on the lay person was gigantic. Bowers (1975) reported that in 1868 there were 90 000 miles of telegraph wire in the United Kingdom and that 6 000 000 messages were transmitted each year between 3 000 public telegraph stations. The visual impact has already been mentioned, see Fig. 5.1. Clearly there were also numerous other benefits to society at large that normally do not receive the same attention as those relating to the prime purpose of telegraphy systems e.g. to provide communication.

Telephony has its roots in telegraphy and when suitable analogue signal transmitters and receivers were invented much of the ground work needed for using them in systems was ready to use. Kelvin saw the telephone as 'the most interesting experiment in the history of science'. Although many had tried to devise some sort of electrical device that would 'pick-up' sound pressure variations and convert them into electrical signals the first person to virtually succeed was Reis in 1861. His apparatus is shown in Fig. 2.13. It was shown to the Physical Society of Frankfurt in that year. It was, however, not quite developed enough to be considered for reliable everyday use. Some others who added to the art were Yeates in 1865, Wright in 1865, Varley in 1877, Wray in 1876, Gray in 1874, van der Weyde, Pollard and Garnier. Many people joined the search for a workable design.

It was Bell, Finn (1966) in the United States of America and Hughes of Britain, Mullineux-Walmsley (1910) who succeeded in making workable devices, the work of these two finally complementing each other . . . one form of Bell's receivers and the Hughes' transmitter were the forms generally adopted. Fig. 5.24 shows 19th century etchings of the magnetoelectric system exhibited by Bell at Philadelphia in 1876.

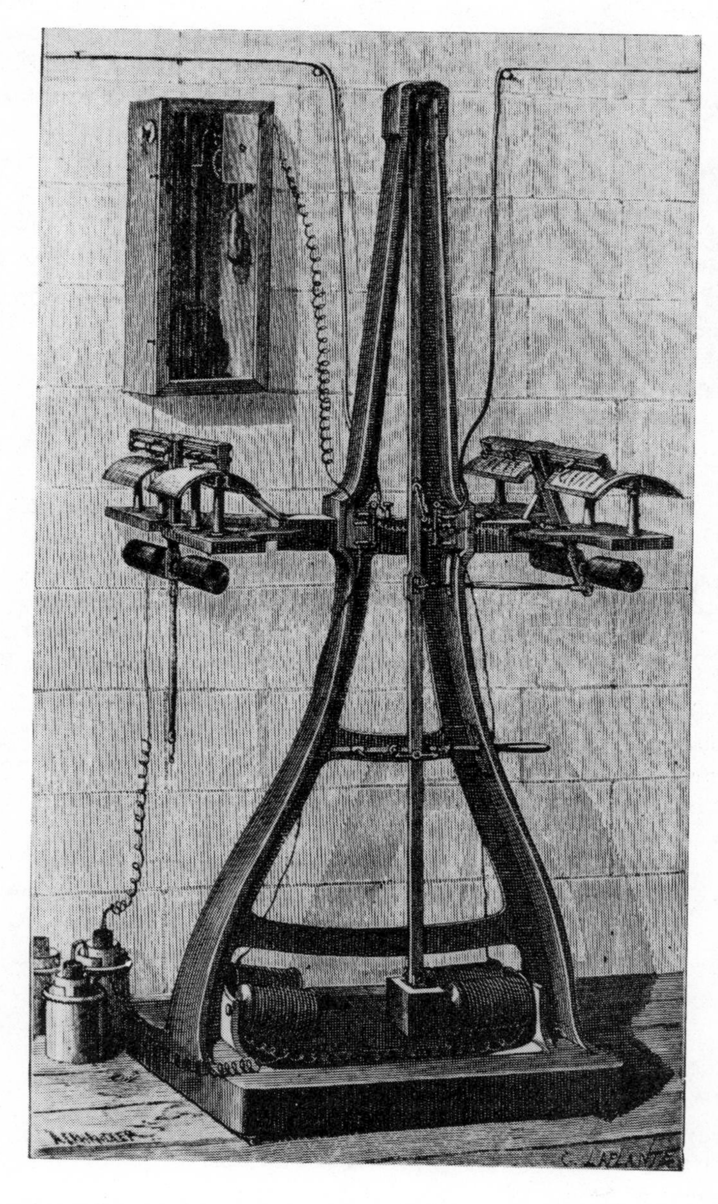

Fig. 5.23 *Caselli's pantelegraph, an autographic system for receiving graphical data transmitted over telegraph lines*
[Deschanel (1891), Fig.574, p.824]

a

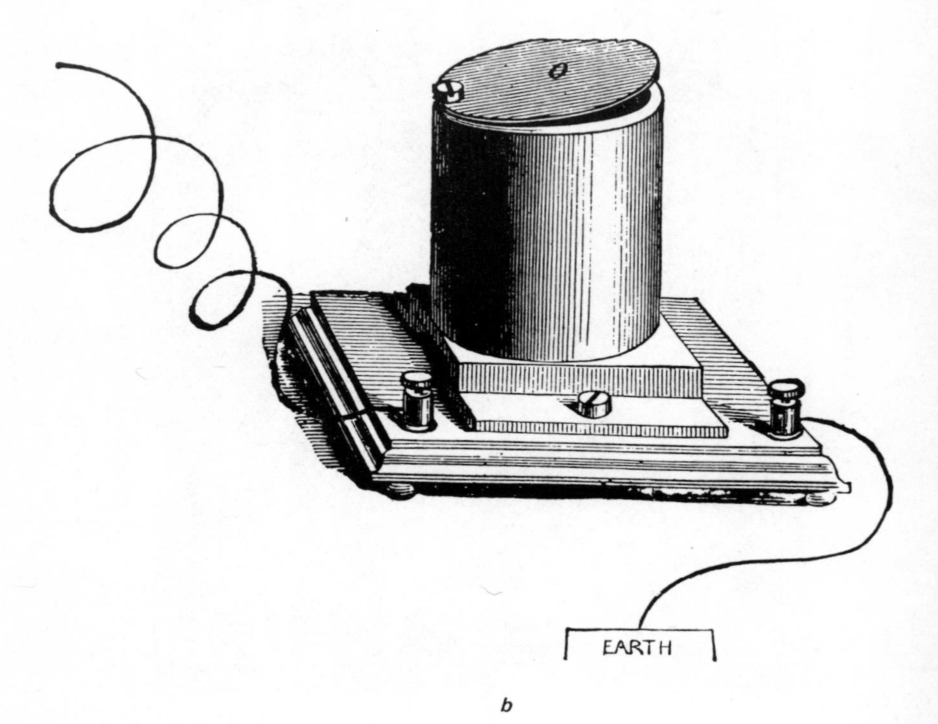

b

Fig. 5.24 *19th-century etching of telephone system exhibited by Bell at Philadelphia in 1876*

a Transmitter

b Receiver

[Frith and Rawson (1896), p.225-226]

a

b

Fig. 5.25 *'Photophone' of Bell and Tainter, 1881*
a Transmitter
b Receiver
[De Vries (1971), p.138-139, From *De Natuur,* 1881]

One of Bell's early systems was for transmission of signals from his electric harmonica. He constructed many forms of transmitter for speech. Finn (1966) reports tests run to assess them quantitatively. The first units (1875) made by Bell had an energy transfer of only 10^{-8} Watts compared with modern outputs of milliwatt level. Basically the problem in telephone construction was to establish an efficient method of modulating a standing direct current, or of producing a varying signal. It then had to be demodulated at the receiver. Bell's first commercial telephone was somewhat different to his other prototypes. (For illustrations of Bell's many prototypes see Bell Telephone (1957). It used a metal diaphragm to modulate the flux strength of a magnetic circuit containing a coil. Flood (1976) and Hounshell (1976) have also researched Bell's work. It was followed by numerous other designs see Byrn (1900), Frith and Rawson (1896), Mullineux-Walmsley (1910) for illustrations and descriptions. Schiavo (1958) makes the case for Meucci being the true inventor of the telephone.

Emergence of the telephone revealed to its users that the signal on one wire was heard in another; this led to studies of induction effects and to ways of reducing them by appropriate connections, such as line transpositions. It also considerably assisted early research into 'wireless' communication that was to follow before the end of that century. Another spin-off was the invention (Fig. 5.25) of the first optical speech link, the 'photophone' of Bell and Tainter in 1880, Preece and Maier (1889), De Vries (1971). This device used sound pressure waves to vibrate a mirror that reflected sunlight to the receiver. Radiation received by a large parabolic focusing mirror was detected by a grid of selenium connected to earpieces. It was reported that the photophone could work to distances of 200m and that over 50 designs were tried!

Telephony provided the instruments and techniques for building the Hughes induction balance, a method, Fig. 5.26, used to locate metal objects such as bullets. Apparatus was invented for the detection of underwater metal-encased objects, for cable break and wire finding. The telephone found its way into medical apparatus, the Ducretet stethoscope microphone and Boudet's microphone that was used for listening to arteries and muscles, were invented *c*1885, Preece and Maier (1889).

Speech analysis also came into being at this time. It assisted experimental proof of the vowel theory of Helmholtz by Du Bois in 1877. It found its way into physiological research, Kylstra (1977*a*). It was because of the desire to better understand speech composition, in order to possibly reduce the cost of transmission lines, that the study of speech arose. It led to the invention of the phonograph, Kylstra (1977*b*).

Wireless communication came into being in 1894 in the capable young hands of Marconi who packaged, see Fig. 1.28*d*, the then known, but only in relative isolation from each other, elements of a practical (and amplifier-less) wireless system. By 1900 wireless had taken up the coupled-tuned circuit, the forerunner of wire recorder was about to be applied in the magnetic detector device using Rutherford's principle of 1895. In 1900 a car radio was patented. In 1904 Christian Hülsmeyer reported a radar ship location device, named the 'telemobilskop'. Tesla had taken out a patent for radio control of the position of a ship in 1898! Wireless did not, like the other tele methods, appear overnight. Sydenham (1975) traces the historical events that preceded Marconi's

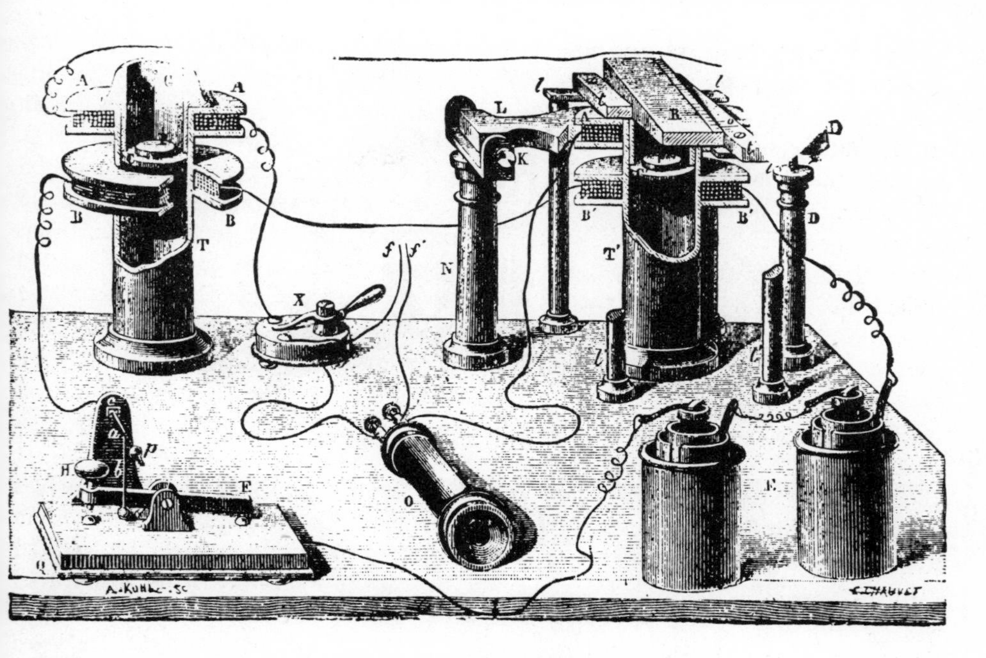

Fig. 5.26 *Induction balance of Hughes used for locating metal objects*
[Preece and Maier (1889), Fig. 286, p.469]

first transmission.

Television had its beginnings in discoveries that certain materials were light sensitive. In 1873 it was found that the electrical resistance of selenium decreases with increasing light level. In 1878 Paive suggested a method for a visual image system. Numerous means were suggested for scanning, for light modulation and for the necessary synchronous control of transmitter and receiver installations. One such attempt is shown in Fig. 5.27, that of Dussaud of 1898. Shiers (1977) has made a specific study of this period. Television, however, finally developed from ideas to working devices in the early 20th century and is barely relevant to this Chapter (see Chapter 6).

More electrical transducers

By 1900 a considerable number of transducer effects that operate in the electrical regime were in existence many, by then, being in routine use. Several have already been described e.g. the motor and generator effect, electromagnetic propagated fields, electrolysis, use of inductance and capacitance laws to construct sensing devices, use of Ohm's law to convert electrical quantities from one form to another, the Joule heating effect and so on.

Electrical methods of temperature measurement passed from the discovery stage to commercial instruments over a 70 year period.

First to appear was the thermoelectric method. Finn (1963) is an account of de-

a

b

Fig. 5.27 *Dussaud's 'telescope' of 1898 . . . one of the many early attempts to construct working television systems*
a Camera
b Monitor. Artist's licence has been used in portraying the output picture quality
[De Vries (1971), p.142, from *La Nature,* 1898]

velopments from 1850 to 1920, this beginning some three decades after thermoelectricity was discovered and covering to the time where the method was used routinely in industrial and scientific measurement.

In 1821, T.J. Seebeck (1780-1831) discovered that electric current could be generated by heating the junction of two dissimilar metals joined together. He reported this phenomenon preferring to call it a 'thermomagnetic' effect at the time. His work was carried out using junctions of bismuth and copper, antimony being used a little later. Knowing that the effect can only produce millivolt levels of voltage at quite low currents it can be seen that the simultaneous emergence of the crude but useful magnetic-needle galvanometer was of great importance to the discovery and investigation of this method. Seebeck's discovery, in turn, provided Ohm with stable enough currents for his investigation of the laws of d.c. circuits: the apparatus shown in Fig. 5.6 used junctions to produce steady current flow.

In 1826, Becquerel used the thermoelectric effect specifically for measuring quantitative temperatures having first calibrated thermopiles in terms of temperature, but the then poor reliability and performance of galvanometers did not allow much progress to be made. One method of increasing the sensitivity of the effect, first done by Nobili in 1829, was that of cascading several junctions to form a thermopile, see Fig. 5.28.

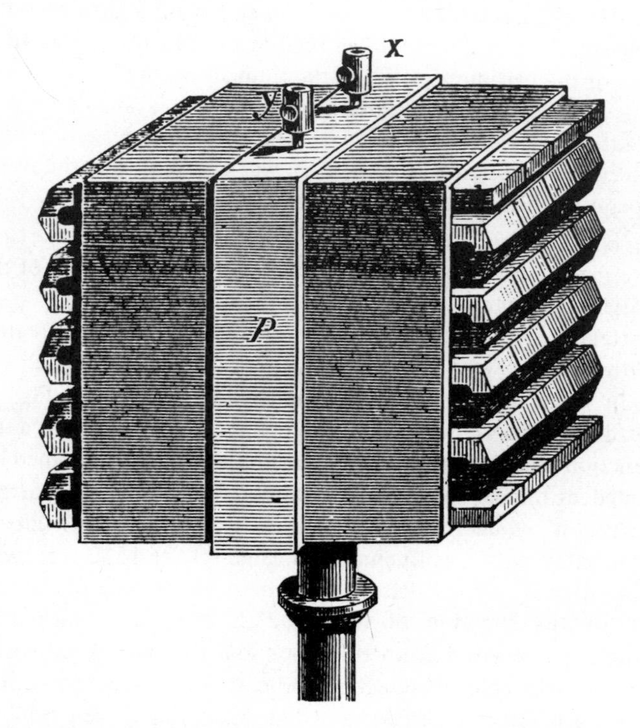

Fig. 5.28 *Construction of thermopile as used by Melloni from 1830 onwards* [Mullineux-Walmsley (1910), Fig.162, p.178. Courtesy Cassell Ltd.]

Although not directly concerned with measurement the reverse effect, now called the 'Peltier' effect, in which a temperature difference is produced between the two junctions by the passage of a current was discovered by Peltier in 1834. This has found its way into measuring apparatus as the means to control temperature by an electrical input signal, an example being its use to set the temperature of a surface in the closed-loop type of dew-point relative humidity detector.

Investigation of thermoelectric principles gradually revealed more of the laws involved. Thermoelectric inversion was discovered by Cumming in 1832. The Thomson law was enunciated a little later. By the 1890s it was well known that a piece of homogenous wire could become a thermoelectric generator couple by mechanical working or by surface contamination.

Melloni, around 1831, was one of the first persons to package the thermocouple into a workable measurement system; he used it for investigation of radiant heat. It was Melloni who coined the term 'diathermanous' for those bodies that easily transmit radiant heat, a term that persisted until much later in the century. Melloni was able to detect temperature differences as small as 0·0005°C, which was quite remarkable for the time. Thermopile sensors were widely known by the 1870s, Tyndall's apparatus shown in Fig. 5.2, being typical of that time.

The method of the thermocouples, however, was put on a firm everyday basis of use by Le Chatelier in France. He published accounts of practical application in 1886, reporting that use of the platinum/platinum-rhodium alloy couple gave reliable results. He combined this knowledge with the use of the then just available moving-coil, galvanometer of d'Arsonval, which had the advantage of reasonable sensitivity, reliability and a linear deflection scale. Barus, in America, was also active in the use of thermocouples for temperature measurement at that time.

Sensitivity was continuously improved upon over the latter half of the 19th century. Thermopiles were made with as many as 70 junctions. The sensitivity of the galvanometer was also raised significantly during this period.

In terms of attained sensitivity of the thermocouple device, the greatest achievements of the 19th century were those of Boys, see Allen and Maxwell (1944), and later Callendar. In 1887 C.V. Boys devised his 'radio-micrometer' Fig. 5.29 as an advancement of d'Arsonval's 'thermo-galvanometer'. In this design a small sized thermocouple junction made of antimony and bismuth was directly joined onto a loop of copper suspended, as the moving coil in a magnetic field, on a fine quartz fibre. This was highly sensitive: it would respond to a 'quantity of heat no greater than that which would be radiated onto a halfpenny by a candle flame 1530 feet away from it', Chaldecott (1954). It was, by calculation, shown to be capable of detecting a temperature change of the junction of 0·000002°C. Boys used such apparatus to investigate the radiation received from the moon and the stars. Used with a 16-inch focusing telescope he was able to detect a candle at three miles (5km) in 1889. His radiometer had a much faster time constant than would have been exhibited by the equivalent thermopile sensor.

Callendar pursued a similar line of interest a few years later in 1900, his work making use of the Peltier effect to bring about radiation balance adjustment of the through current.

The term radiometer is used today to describe any instrument for detecting and measuring radiation. In the late 19th century, however, it had first had the more specific meaning for devices that 'serve to measure the amount of radiation falling upon it by the velocity with which it revolves'. It was Crookes who coined the term in 1876, Chaldecott (1954). Otheoscopes were also in this class being devices to show the existence of radiation in other ways. The E.F. Nichols torsion radiometer, Allen and Maxwell (1944), used a torsion fibre to allow limited rotation of the mill vanes.

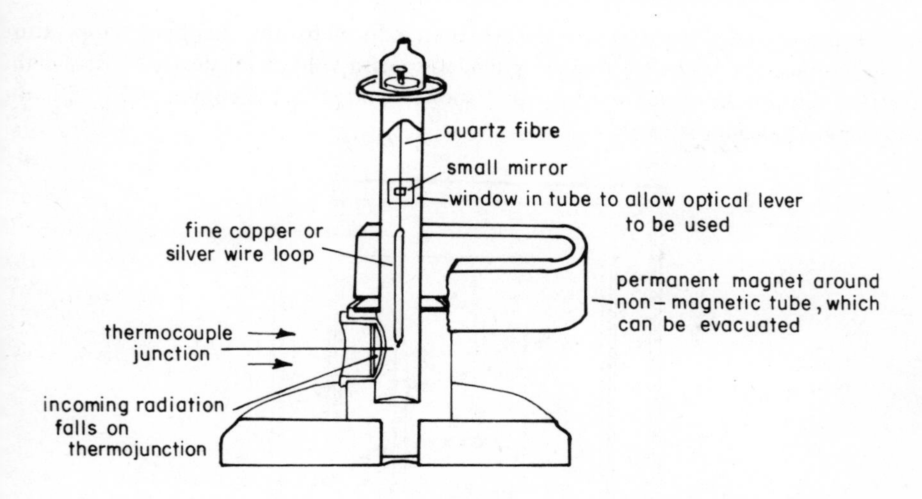

Fig. 5.29 *Schematic of Boys' radio micrometer c1887. It could detect temperature differences of 0·000 002 °C*
[Glazebrook (1922), Vol. 3, Fig. 16, p.707]

The thermocouple found its way into a wide variety of measurements across the whole field of science. In animal physiology, for instance, McKendrick in the late 1880s, used a thermopile (Fig. 5.30) to measure the heat generated by frog muscles, McKendrick (1894). Temperature rises experienced in his experiments were in the range from 0·001 - 0·018°C. Tyndall (1863) provided an extensive account of the use of thermopiles in the investigation of diathermancy. Jones (1972) is an account of the place that the thermocouple and the resistance thermometer played in the growth of knowledge about the infra-red and ultraviolet regions of radiation.

By 1900 the thermocouple was in industrial use as well as being an established tool of the scientist. It was one of the first of the new class, the electrical kind, of instrument to become commercial products. Darling (1911) is a contemporary account of the practice of higher temperature measurements at the turn of the century. Strong (1942) includes a chapter, with bibliography, on the manufacture and use of sensitive radiometers.

A considerable amount of information, including primary references, about thermocouple devices can be gleaned from the two catalogues by Chaldecott (1954, 1976). Deschanel (1891) has a little to say about thermoelectricity, as does Mullineux-Walmsley (1910). Popular works on electricity usually presented only very shallow accounts and are, therefore, of little value to the historian. Le Chatelier and Boudouard (1904) is about high-temperature measurements. Glazebrook (1922) is probably the best learned account; it includes references commencing around 1890. Glazebrook (1902) is a book of related teaching experiments dating from the practice of 1890.

The next electrical method to appear for measuring temperature was the use of resistance change with temperature rise. Although the effect was known in the early 19th century it was not until 1871 that the method achieved reality by the inspiration and endeavour of William Siemens. He devised a measuring method that used an ingenious electrical bridge to overcome the errors introduced by any change of temperature of the connecting leads. He originally made use of a voltameter device to detect the electrical output but that method was soon superseded by combination with the bridge and a galvanometer.

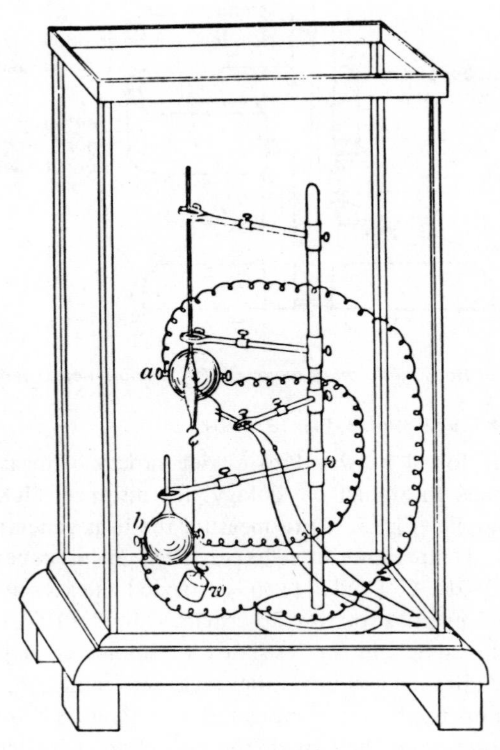

Fig. 5.30 *Apparatus used by McKendrick to investigate the temperature rise of frog muscl c1888. a and b are thermojunctions, a rests upon the muscle*
[McKendrick (1894), Fig. 59, p.117]

Siemens' apparatus, however, did not obtain the approval of a British Association Committee investigation held in 1872-3 and the method fell out of favour until its inadequacies could be overcome. The main problem was found to be that the iron tube used to contain the platinum resistance element became contaminated by reduction products at the high temperatures of use altering the resistance value with time Siemens overcame this by better construction, Fig. 5.31, using the platinum sheath over the element. Porcelain was also found to be satisfactory.

Other workers became interested, notably Callendar, Griffiths, Heycock and Neville; Callendar began his work on this topic in 1886. By the end of the 1890 decade the platinum resistance thermometer was the best thermometer available for the widest range of temperatures. The temporary decline of earlier interest that allowed the thermocouple alternative to flourish was over. Both methods became accepted for

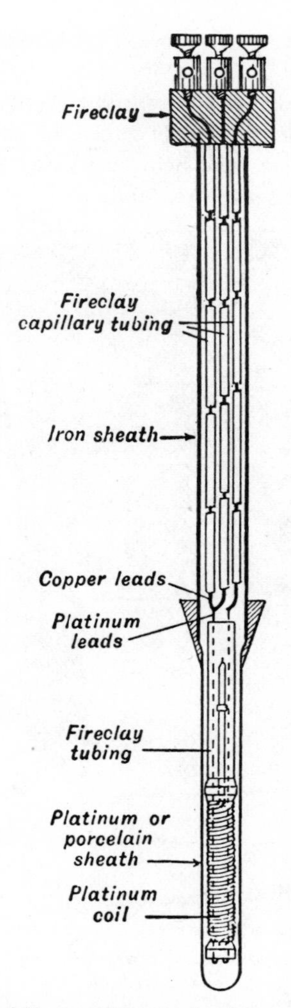

Fig. 5.31 *Cross-section of Siemens' improved resistance thermometer sensing element of c1885* [Glazebrook (1922), Vol. 1, Fig. 1, p.693]

scientific and industrial measurements in a routine manner. Applications widened: in 1887 Lea and Gaskell made a thermometer for measurements in a frogs heart.

In 1886 Callendar made a determination of the resistance of a platinum wire sample for temperatures up to 600°C. This revealed the characteristic equation for resistance against temperature.

To make best use of the sensing element numerous forms of measuring bridge were invented e.g. the Siemens three lead form, Callendar and Griffiths' version, the Smiths difference bridge and many more.

The self-heating effect was appreciated by 1900, some designs being arranged to circumvent the problem by arranging for the resistance sensing element to be operated under the same current conditions at all times. Fixed points were established for purposes of calibration by 1890.

Out of the availability of the reliable and sensitive thermometers evolved closed-loop control of temperature and the whole breed of recording devices . . . see Section 5.4. Resistance thermometers were being marketed by the 1890s. Fig. 5.32 shows the Whipple indicator marketed just after 1900.

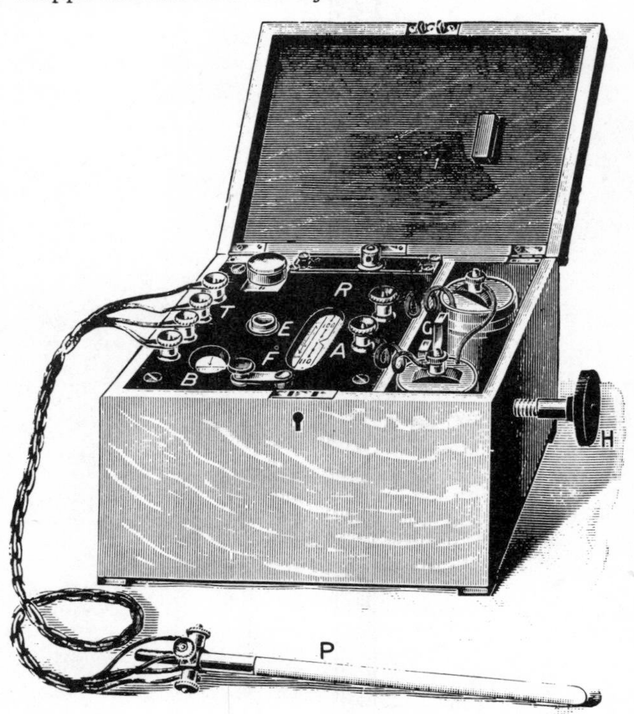

Fig. 5.32 *Whipple indicator for use with resistance thermometer element, as marketed by the Cambridge Scientific Instrument Company around 1905*
[Darling (1911), Fig.32, p.102]

As with the thermocouple method there were those scientists who pushed the detection limits of the resistance thermometer to extraordinary limits. The outstanding work was done by Langley. He invented the 'bolometer' or 'actinic balance' (which, in modern parlance, is now called a total radiation detector) that is based on resistance change of an element placed in a Wheatstone bridge arrangement. His balances, made in 1880, used a black grating of steel, platinum or palladium foil. These could detect 0·0001°C temperature differentials, its main advantage over thermopiles being the rapid response time. Callendar also experimented with similar forms of bolometer around 1900. Langley's, and later bolometers of Lummer and Kurlbaum, are described in Allen and Maxwell (1944).

Published sources of information for resistance thermometry are similar to those given above for thermoelectric means. The point needs to be made at this stage that the theoretical understanding of the processes was highly developed: this may not be obvious from this account.

To close this subsection on the influence of the emerging electrical method of the 19th century, it is worth mentioning that several other principles were discovered in this century but did not find application until much later. That the resistance of a longitudinally-strained wire increases as the wire is lengthened was known in the mid-19th century but this did find widespread use as the bonded strain gauge until 1938 under the simultaneous development of Simmons at the California Institute of Technology and Ruge at Massachussetts Institute of Technology. The electromagnetic flow meter, Shercliff (1962) was first described by Faraday who reported in 1832 that he had tried to measure the voltage induced by the flow of the River Thames using the earth's field as the magnetic field source. Wollaston, in 1851, succeeded in this experiment. Very little happened then until well into the 20th century when Smith and Slepian patented a method for measuring ship speed, relative to the sea rate, in 1917.

It is clear that, by 1900, electrical methods were having an enormous impact on the science of measurement, and it was only the beginning of what was to come, the foundation for numerous electrically based methods was laid in the 19th century. Much of 20th-century progress was to be more on improvement of application than on discovery of new laws to use as the basis of measuring instruments.

5.3 Control and computation

Measurement and control combine

Several isolated instances exist of closed-loop control systems being created in the times before the 19th century e.g. water flow regulators of the Greeks and the Arabs, incubator and windmill installations following the 17th century and temperature controls for furnaces of the 18th century. These have been discussed in previous chapters. Speed-control methods devised for windmills set the scene for the application of the centrifugal governor to control the shaft speed of the steam engine. Boulton and Watt, *c*1788, applied the centrifugal governor to their steam engines; Mayr (1970) describes the circumstances in some detail. They did their best to keep the details of this device secret but by 1800 it was more or less common knowledge. Thus began the general and widespread application of feedback control and a period when measurement became married to actuation as the essential partner of automatic control.

Previous to Watt's work, controllers were reported on an isolated basis. From his time, however, the principle of using a sensing device to automatically set the actuator of a process to suit the demanded energy became widely applied. During the 19th century it is possible to trace a growing use of feedback, with systems being built almost entirely devoid of fundamental understanding of the principles of tuning that are involved. A parallel development, one that rarely found application in the Victorian practitioner's work, was a gradual increase in the knowledge of the principles and mathematical approaches that could provide intelligent and economic solutions to difficulties such as instability.

Two areas of strong application for the use of feedback control in the 19th century were in prime mover engine speed-control systems and in arc lamp controllers. Rate control of clocks and the less dominant (of that time) applications in the potentiometric electrical recorder were other areas where feedback was applied.

The definitive works about feedback control of this period are those by Mayr (1970, 1971*a*, 1971*b*). Notes issued by Milner and Pengilley (1972) also dealt with the topic in considerable detail. The latter paper of Mayr specifically discussed speed regulation during the 19th century. Mayr traced the work of several scientists i.e. Airy, Siemens, Foucalt, Thomson, Maxwell and Gibbs who each investigated feedback systems. He drew a general conclusion that in that period the governors to find the greatest acceptance were not those built from theoretical understanding but those made simple and very practical by intuition alone. He also pointed out how in those times 'industrial success was not the rule but the exception, and such success obeyed laws that men of science rarely understood'. From 1836 to 1902, in the USA alone, over one thousand patents were granted for the design of governors! A patent model from that collection is shown in Fig. 1.5. Here was an example showing how technological development can proceed in a situation which almost totally ignored the pointers to scientific principles that were emerging through experience but that do not receive the organised effort needed to bring the many apparently isolated facts together into a general theory.

A theoretical basis of automatic control did not find practical application until the the 1940s, but, as with most developments, many individuals were developing the necessary knowledge well before that time. The major practical problem that has had to be faced in feedback design was how to reduce instability with its associated 'hunting' effect. Mayr (1971*b*), Burstall (1970) and Milner and Pengilley (1972) each attribute the first theoretical studies on control systems to Airy, who in around 1845, studied the friction-braked speed controllers used on equatorial telescope drives. He presented a mathematical investigation of governor instability viewing the problem from his astronomical background with orbiting bodies in space. He determined, in a rather tortuous manner, the conditions under which instability will occur and a means for deciding the frequency of oscillation. Associated work was that of C.W. Siemens who had devised a sophisticated governing apparatus, Fig. 5.33, for shaft speed using differential gearing in conjunction with a reference speed source to provide a controller having, in modern terms, integral compensation.

In 1868 Maxwell read a paper to the Royal Society, London entitled *On governors*. This had the purpose, he wrote, to 'direct the attention of engineers and mathematicians to the dynamical theory of governors'. In it he attempted, without the success he hoped for, to find a general criterion for high-order equations of motion to establish if the system in question was stable. That was later done successfully by Routh in 1877. Maxwell did, however, identify the general problem that needed to be solved and his method was able to handle up to 3rd-order expressions. Maxwell was apparently not as interested in the problem of stability as were the practitioners; his work was not presented in a way that technologists would have been able to apply, indeed some

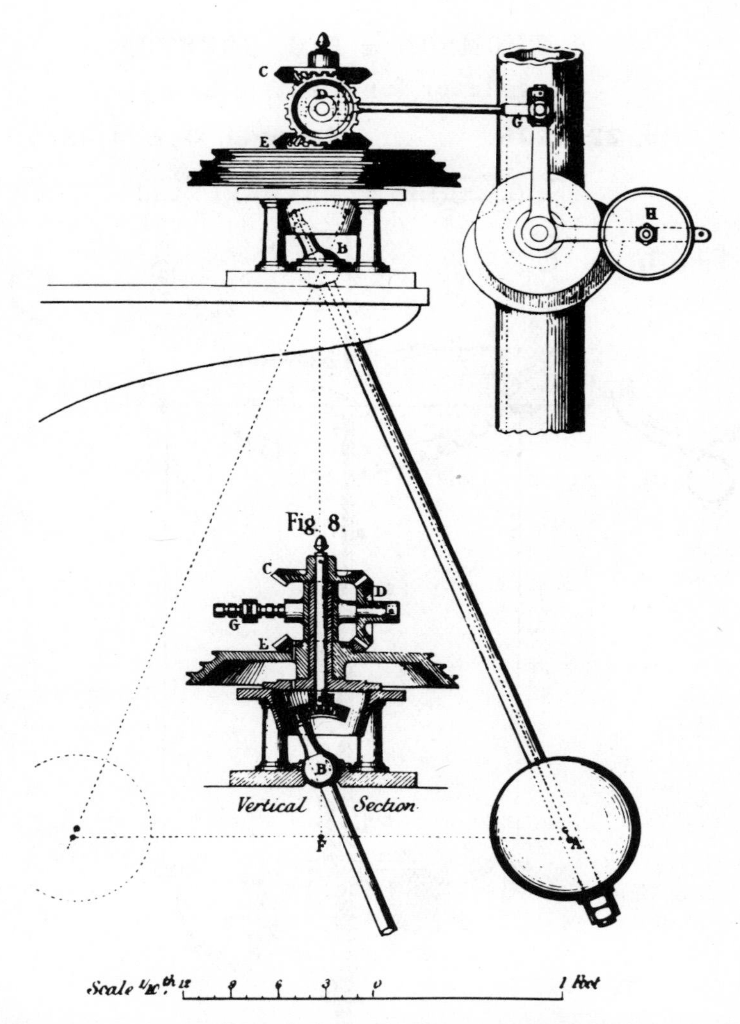

Fig. 5.33 *One version of the chronometric governor of C.W. and W. Siemens c1842. It produced an error signal related to the difference between the actual shaft speed and a reference shaft speed*

[Mayr (1971*b*), Fig.1, p.208. From *Proc. Instn. Mech. Engrs.*, 4, 1853]

of it turned out to be largely irrelevant. Even so, his contribution was a step in the right direction.

Maxwell's work was extended by Chebyshev in 1871 (work which has since found great usefulness in the design of electronic filters in recent times), by Vyshnegradskii during 1876 to 1878 and by others. Swinburne, in 1894, drew attention to the problems of backlash in linkages while other workers at that time published on the effects of coulomb friction. The unfortunately unpublished, extensively researched, notes by Milner and Pengilley (1972) provide a wealth of information for serious study of the rise of automatic control. It is worth noting that the three sources of reference

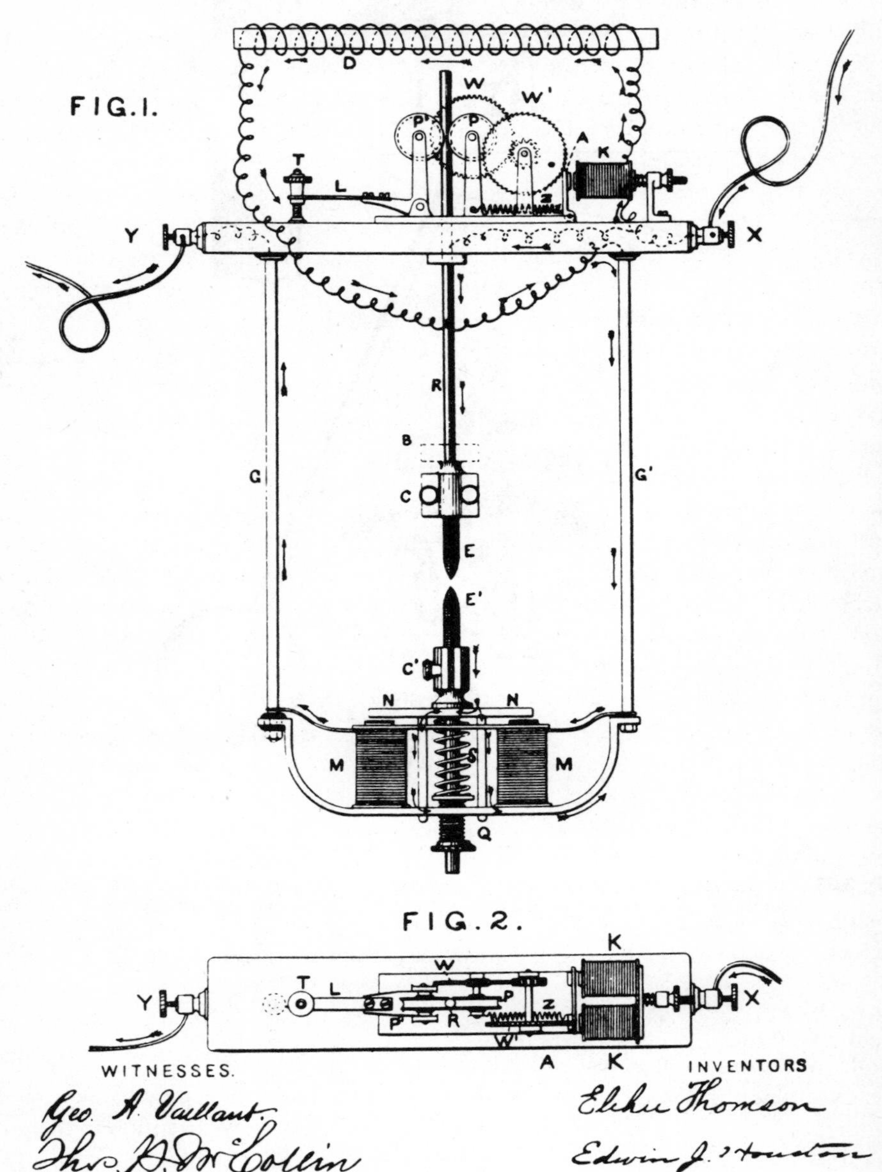

Fig. 5.34 *Patent drawing of advanced form of arc-lamp controller that could be used in series with other units*
[Mayr (1971*a*), Fig.128, p.113. From U.S.Patent No. 220508 of October 14, 1879]

mentioned as useful here are each the products of mechanical, not electrical engineers; a situation that largely applies for general interest in the science and technology of measurement and its instrumentation.

Directing attention back to the technological manifestations of automatic control, other applications for the centrifugal governor included shaft speed control of water turbines and later inventions, the steam turbines, tractors and gas and oil engines, Mayr (1971*a*). Burstall (1970) provides several examples with references to original papers. The friction kind of governor was applied to telescopes, at the beginning of the 19th century and found new uses in chronograph drum drives and in the various forms of phonograph that emerged at the end of the 19th century.

Automatic feedback control was applied widely for obtaining long-operating times of arc lamps, without having to constantly reset the arc gap, and for obtaining a steady light output. The first arc lamp was demonstrated by Davy in 1809 but his apparatus was far from a usable everyday source of illumination. Until 1848 no satisfactory practical devices had been constructed to hold the current constant as the carbon rods were consumed. According to Mayr (1971*a*) the invention of a workable controller is attributed simultaneously to both Staite in London and to Foucalt in France. The latter was subsequently improved in collaboration with Duboscq whose name was used in marketing the arc lamps.

From then on, until arc lamps declined in general purpose demand when the filament lamp appeared late in the 19th century, numerous designs were constructed. Most used the electrode current to produce a force, via an electromagnet coil, that was applied to alter the spacing of the carbons as was needed. Some used a solenoid driven ratchet drive, others used clutches to release the lamp allowing it to increment downward. Some supported the top electrode by the pull of the solenoidal coil. Mechanically acting, feedback-controlled, constant current transformers operating on similar principles appeared late in the century. Not all designs drew the actuation energy from the arc lamp current but used instead, clockwork drives. Later models, such as Fig. 5.34, also incorporated means to allow multiple operation of lamps from one source. Arc lamps are described briefly in Tunzelmann (1890), Byrn (1900) and in Mullineux-Walmsley (1910). Mayr (1971*a*) provides excellent illustrations of artefacts held by the National Museum of History and Technology in Washington DC. The definitive study is that by James-King (1962*b*) who traced the development of the arc lamp and its generator through the 19th century. Both Mayr and James-King are tied to primary sources.

Similar principles were also used by Siemens, in 1880, to control the arc of an electric arc furnace. That set-up made use of a series solenoid to position the electrode close to the surface of the melt. It is a particularly interesting set-up because the design incorporated fluid damping of the solenoid plunger magnet. None of the above-mentioned sources of reference discuss problems of instability nor is it clear that the designers saw their apparatus as having separable sense and actuate functions. Arc lamp controls did not have to respond rapidly and, in general, their design involved the inherently stable low-order equations of single position controllers without integral or derivative terms.

Arc lamp controllers embodied the kind of technology that was needed for the invention of the potentiometric recorder. It might be that Callendar was influenced by that use. During 1897 Callendar devised a method for automatically balancing a Wheatstone bridge making a permanent registration of value need for balance at the same time. In doing so he gave the world the first potentiometric chart recorder.

His original instrument of 1897 is shown in Fig. 5.35. A d'Arsonval galvanometer movement is used to sense imbalance of the electrical bridge. If this exceeds a given amount the coil of the galvanometer rotates to close a contact operating one of two electromagnets. This releases one of the clockwork carriage drives. The pen begins to

Fig. 5.35 *Callendar's original potentiometric recorder made in 1897*
[Crown copyright. Science Museum, London]

move across the linear carriage, which also carries a slider contacting the slide-wire resistance forming part of the bridge. The electromagnet is released when the galvanometer returns back to the zero position opening the contact. Error in the opposite direction produces a similar effect in the other direction of carriage movements. Provided the error is phased as negative feedback the carriage follows the resistance needed to bring about balance. Chart drive is also by clockwork. The then emerging theory of stability was most unlikely to have the influence on the design of this epic-making device as it would have today. Callender's recorders are reported in the journal literature, references for which are given in Chaldecott (1976) and in Glazebrook (1922), Volume 1, page 710. Darling (1911) and Williams (1973) each contain illustrations of relevance. Callendar reported his work in *Engineering* 1899, p.675.

The electromagnet was a valuable element for the design of many forms of instrumentation. It enabled the arc lamp to be controlled and Callendar's chart recorder to be made. In 1882 a solenoid operated steam valve was used to control the speed of a steam engine drive electrical generator set, Neville (1885). It also was used to punch indents in a circular paper chart from remote stations in an early form of the Bundy time clock. It enabled the telegraph to be implemented. It was used as the actuator of the step-by-step telephony selector switch. The lists of uses is extensive.

Controlling the movement of railways was another control engineering aspect that grew from infancy to hugely complex interlocked systems in which safety was a dominant requirement, Kirkman (1904), King (1921) and Burtt (1926). Analysis of the rise of railway installations, the accident rate, causes of accidents and other data are given in Lardner (1862) Volumes 1 and 2. More than half of railway accidents in the 1850's were caused by collision. Lardner went as far as listing safety rules for railway users!

Computation within measuring and related instruments

Calculating machines developed, as has been explained in the previous Chapter, Section 4.3, out of the mechanical digital adding mechanisms of Pascal, Leibniz and others of the 17th century. They were steadily improved, gradually becoming more sophisticated in construction and concept. The outstanding design was that of Babbage, see Fig. 5.3, whose calculating engines of the 1830s and the 1870s incorporated many of the features that have become the accepted systems arrangements for implementing the modern digital computer, Smiles (1879), Timbs (1860). The development of calculating systems, mainly of digital form, has been extensively expounded in pictorial form by Eames and Eames (1973) as a record of a history wall made for exhibition by IBM in New York. A similar history wall about data processing was subsequently designed and installed in the Vienna Museum of Work and Industry in 1973. A bibliography of the artefacts and illustrations has been prepared, Chroust (1974), for that display. Although on the same subject as the original wall of IBM this wall features different material complimenting the other. Other histories are those by Serrell *et al.* (1962), Goldstine (1972) and Randell (1973).

Stand-alone calculating instruments gradually developed as other forms that were built into measuring instruments. It is those forms that concern us here. Eames and Eames (1973), Chroust (1974), Trask (1971) and Pugh (1975) can each provide for interests in the broader field of general computation.

Digital computation has its hardware roots in the above mentioned calculating engines, in the 1880s electrical circuits proposed by C.S. Peirce for performing logical AND and OR gating functions and in the electrical machine of A. Marquand who proposed using switches to control electromagnets. Digital methods of performing arithmetical operations did not, however, have any widespread impact until well into the 20th century. Computing devices of the 19th century were almost entirely analogue operating machines. Thomson and Tait (1883) called them continuous calculating machines.

Analogue computing devices date from antiquity . . . the differential gearing that might have been used in the south-seeking chariot and the Antikythera mechanism that performed operations of continuous variables.

In the early 19th century mechanisms for computing more complex mathematical functions began to appear. Hermann, in 1814, was the first to invent an instrument for measuring the area bounded by an irregular curve; it was made in an improved form in 1817 and became known as a 'planimeter'. Pugh (1975) details the developments that followed, which were many. Notable makers were Amsler (who made over 12 000 up to 1884) and Coradi.

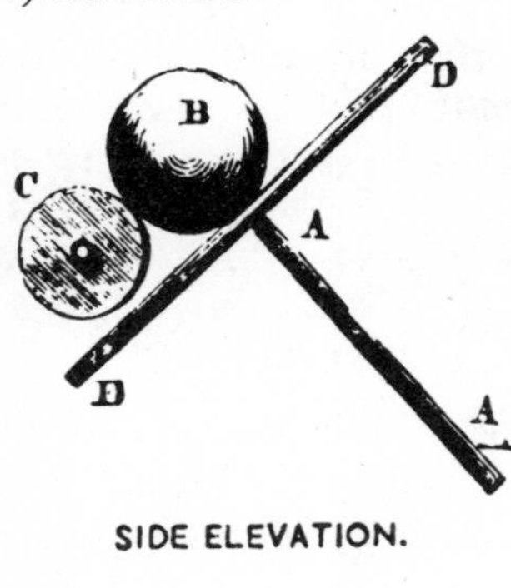

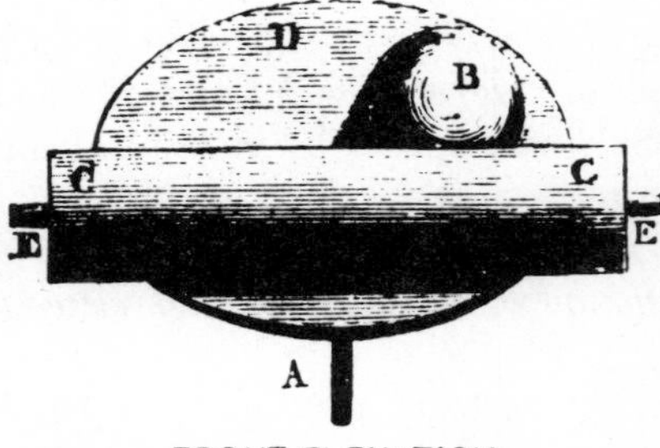

D, the Disk.
A, the Axle of the Disk.
C, the Cylinder.
E E, the Axle or the Journals of the Cylinder.
B, the Ball.

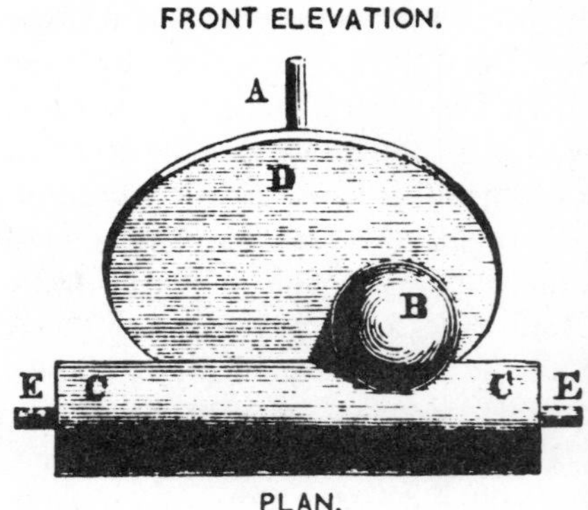

Fig. 5.36 *Disk, globe and cylinder integrating mechanism invented by James Thomson c1876* [Thomson and Tait (1883), Part 1, Appendix B, p.492]

Instruments of this form were devised during the 19th century to measure distances of a route traced over a map, areas bounded by irregular shapes, moments, moments of inertia, and for integration, in general, of a given curve by simply passing a pointer over it, the integral being traced simultaneously.

In 1855, Clerk-Maxwell designed a planimeter that used only pure rolling action but, (earlier versions were prone to error) it was not made. Kelvin's brother, James Thomson evolved the well known ball-and-disk integrator mechanism, Fig. 5.36, in 1876. This element became a key component in the evolution of many mechanical computing machines that were to follow. A study of the capabilities of mechanical computing equipment is given in Murray (1948).

Harmonic analysis of irregular signal traces and other sources was introduced by Fourier who showed early in the 19th century that any periodic function could be represented as the sum of a cosine and a sine series of harmonic terms. Initially analysis was performed by time consuming graphical and arithmetical methods. In 1876 Kelvin invented a mechanical instrument to do this operation. It used his brother's ball-and-disk integrator. A prime use of Fourier-series theory in the 1880s was for prediction of tide heights around the coasts and in the river estuaries. This, the reverse case of that just mentioned, requires means for generating and adding the harmonic components. Kelvin devised a machine to do this in 1872. An 1875 version is shown in Fig. 5.37. Several other machines were later constructed for harmonic function computational operations, International Hydrographic Bureau (1926). That of Stratton and Michelson, built in 1898 in the USA, could combine eighty components, Eames and Eames (1973). They used addition of the elastic forces of spiral springs. Pugh (1975) describes the operation of these devices. Thomson and Tait (1883) includes description of the principles of the tide predictor, a machine for solving simultaneous equations, the ball-and-disk integrator, calculation of the integral of the product of two given functions and integration of linear differential equations of second-order with variable coefficients.

Transformation of the amplitude of variables in simple linear ratio manner has been in use since the 17th century. It can be seen in the mechanical amplification pointers of the hygrometers shown in Fig. 4.45 and as the linear-to-rotary-to-linear amplification used by Hooke in his barometer shown in Fig. 4.43. Another later example of the 19th century was the use of cord-driven wheels, of different radii, to attenuate the original excursion from a water reservoir level sensor into a smaller ratio that suited the width of a recorder drum.

In the 18th century it became apparent to measuring instrument designers that systematic errors could sometimes be conveniently corrected by the addition of some special feature to the hardware arrangement. Harrison's grid-iron pendulum was an example. An alternative method of correction is to process the output signal with a computing stage that operates upon the signal to correct it for known deviation from the required law. This approach appears to have first been introduced late in the 19th century. No evidence could be found of a computing mechanism being used in cascade between the primary stage and the final output before 1897.

Fig. 5.38 shows a cross-section of the Kent 'Combined record' used to record integrated water flow with a venturi meter. This recorder was designed by John E. Grout

Fig. 5.37 *Kelvin's 1875 tide-predicting machine*
[Lent to Science Museum, London]

around 1896. The two floats rose to heights related to the pressure existing at each of the two measuring pressure points on the venturi tube. Shaped cams were incorporated between the floats and the recording drum to linearise the otherwise square-root relationship between flow and differential pressure head. The integrating arrangement comprised a constantly rotating drum at fixed speed with a cam profiled surface, the radius of which at any position decided the duration for which the counter mechanism was in a driven mode. The integrator rocking frame would rock in and out of gear

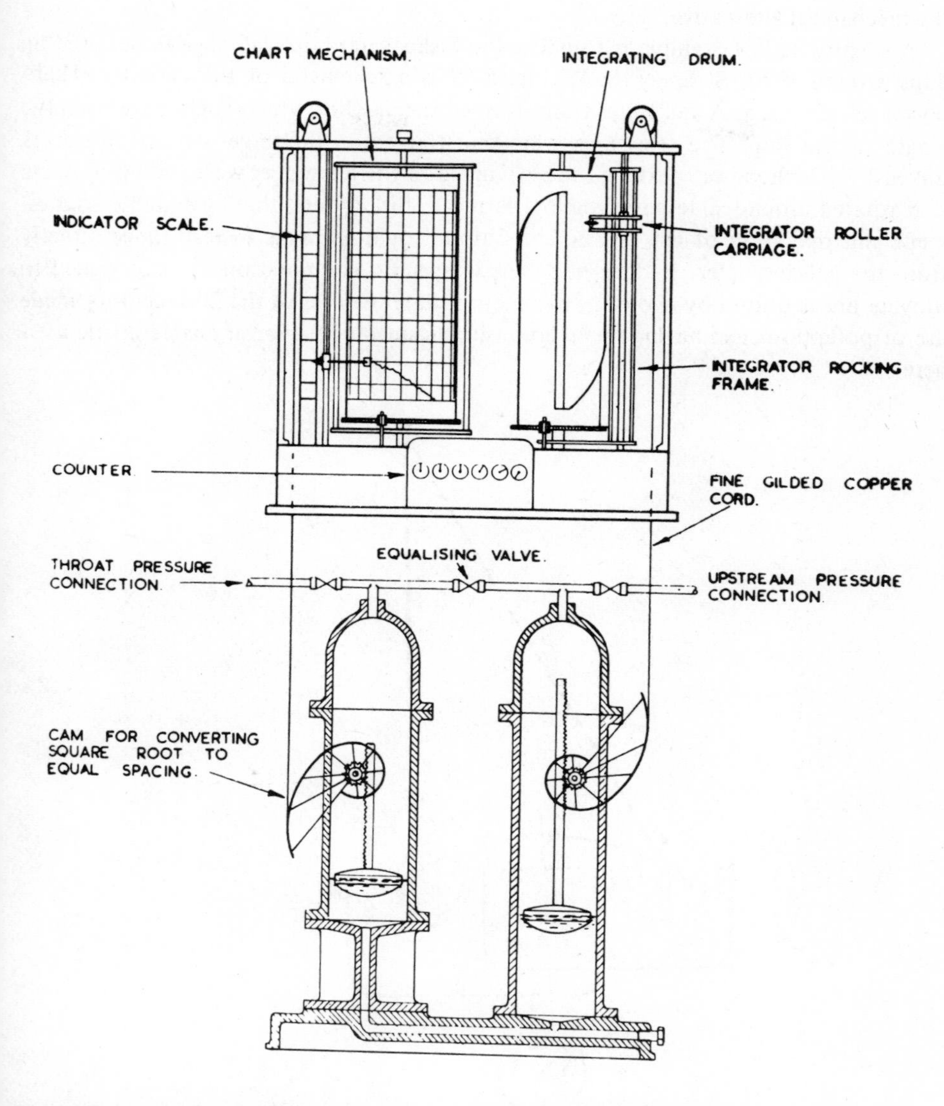

Fig. 5.38 *Combined recorder for use with venturi flowmeter; designed by Grout c1896* [Ardley (1953), Fig.1, p.83. Courtesy Kent Instruments, Pty. Ltd.]

accordingly, staying in gear to drive the counter longer when the float on the right is higher. It is interesting to read, Ardley (1953), that the buyers of this recording equipment were eventually persuaded that they could use the directly generated square-root relationship signal and the linearisation was dropped from later models, to be resurrected in more modern times with electronic computation.

The use of inbuilt mechanical computing mechanisms rose considerably after 1900 reaching a peak in the 1940s from which time electronic methods gradually superseded the mechanical alternative.

A slightly earlier example is found in the Fiske electro-optical range-finder used on ships around 1890, Sloane (1894). Fig. 5.39 is a schematic of the circuitry. High-power telescopes, at A and B positions, formed the baseline which might have been the length of the ship. The telescopes were kept trained on the target by two operators linked by telephone or speaking tubes. Coupled to the telescopes were sliding contacts that rotated around slide wires marked E and F. Fiske found that by suitable arrangement and operation of this device the output could be made to read range directly from the galvanometer. It does not include specific computation but is designed to provide linear output by choice of parameters. Later devices of the 20th century made use of potentiometers having sine relationships instead of the linear characteristic used here.

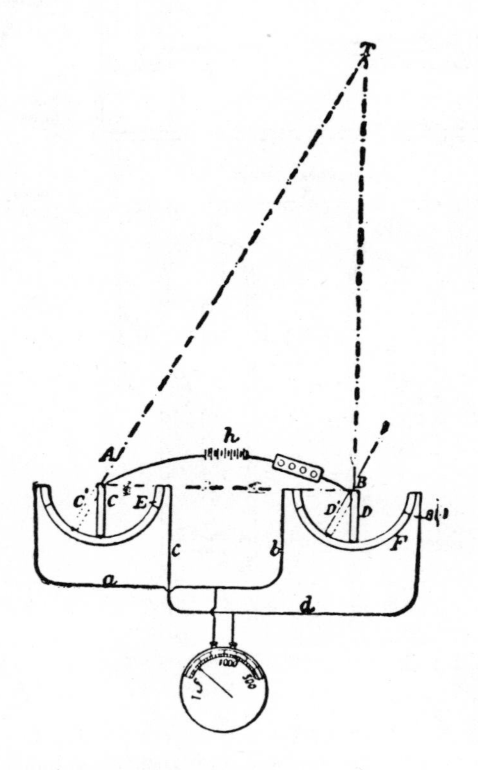

Fig. 5.39 *Fiske electro-optical rangefinder c1890*
[Sloane (1890), Fig.281, p.448]

Internal computation was also used in the dynamometer form of electrical indicating instruments. Energy is computed as the resultant of a current and a voltage source. Ayrton used this principle to design a direct reading ohmmeter prior to 1890.

5.4 Recording and registration

It is an historical fact that although men had occasionally written about self-registering instruments since the 17th century the concept only became everyday reality in the 19th century. Once it had been introduced there was no holding progress back. Development began slowly with the work of Watt, mentioned in Chapter 4, which probably had no influence of note on the progress that was to follow a few decades later.

Study of the dates at which information recording devices were first patented or reported shows that development occurred in three main areas of application. What was needed for devices to appear was a stimulus that would bring about construction of devices using already known principles. The first to emerge were integral mechanical registration parts of the meteorological instruments, these picking up the historical developments mentioned in the previous chapter but without having been the result of them.

About a decade later, in the 1830s, telegraphy needs for receiving terminal recorders gave rise to the subsequent invention of numerous methods for recording an electrical (rather than mechanical) digital signal coming in on a line. Later units recorded analog signals.

Coincidentally (or was there a link?) and another decade later, came the introduction of recording devices for physiological variables, such as the parameters of blood flow and later the responses of the nervous system. In 1847 Karl Ludwig reported using a rotating cylinder recorder. This appears to be the first physiological application for a recording device. According to Davis and Merzback (1972) the psychologists borrowed the techniques initially from the physiologists. Their dominant interest was in ways to record the occurrence of events happening over quite small time intervals, such as in measuring the time response of people's reactions to external stimuli. This course of events began in the latter quarter of the 19th century.

Although strong paths of application and improvement are evident in these three groups it must not be thought that progress was still at a standstill elsewhere. In the investigation of sound the same techniques as were adopted in the life sciences were in use. Various individual scientists developed equipment for recording rapid waveforms, for measuring the flight times of projectiles and, of course, for steam engine testing.

Almost all of the recording methods used today have their origins in the 19th century. This account begins with the telegraphy developments where the general requirement was to receive electrical signals of digital format.

The history of development and use of printing and copying devices in telegraphy can be traced from sections in Byrn (1900), Deschanel (1891), Lardner (1862), Mullineux-Walmsley (1910) and Tunzelmann (1890). The first initially crude telegraphs required the operator to be present to read the incoming code signals. It was soon apparent that a record of the message would enable faster sending, more accurate decoding, allow the operator to leave the terminal and provide a permanent record. The first recording device implemented was the pendulum instrument (Fig. 5.40) devised by Samuel Morse in 1837. The Figure shows the kind of output record made

Fig. 5.40 *Samuel Morse's experimental telegraph system 1837*
[Byrn (1900), Fig.5, p.19, from *Scientific American* 1837]

as the pendulum was deflected sideways across the clockwork regulated, weight driven, paper strip. A sending-code mechanism is featured as 2 in the Figure. Morse's original apparatus exists in the Museum of History and Technology, Washington DC.

Meanwhile other inventors were following different lines to record telegraph signals. Mullineux-Walmsley (1910) states that it was the American, Bail, who in 1837 invented the first printing telegraph followed closely by that of Wheatstone in 1841, Bowers (1975). These each printed, using type, alphanumeric formats instead of Morse code.

Bain's method of 1840 is particularly interesting to the history of recorders for it used electrochemical principles to produce the letters as bright blue marks on treated paper by passing the electric signal current through the paper. It was the forerunner of halftone facsimile recorders. The word to be sent was first assembled from metal letters. Over these were passed five sensing fingers, see Fig. 5.41. At the receiving end the words were again reconstructed from five contacts moving on paper treated with potassium iodine in starch. Others adopted his concept e.g. Stöhrer, Siemens, Gintl and others with the Caselli Pantelgraph (Fig. 5.23) coming later.

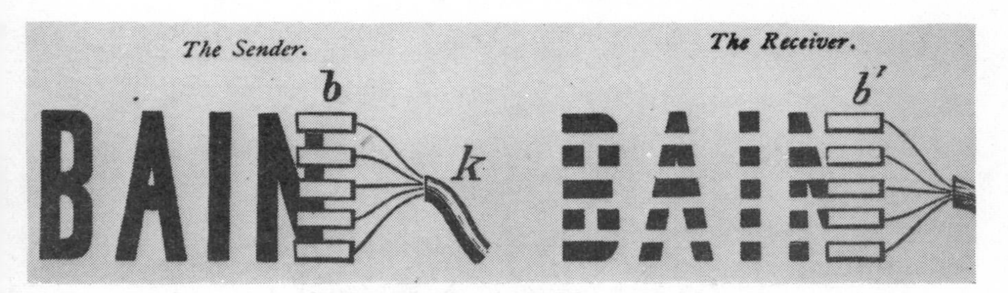

Fig. 5.41 *Principle of Bain's electrochemical telegraph c1840*
[Mullineux-Walmsley (1910), Fig.364, p.396. Courtesy Cassell Ltd.]

In 1844 Morse introduced the idea, Fig. 5.42, of indenting the paper rather than marking it with a pen or ink, Wells (1856), Lardner (1862). Versions on this theme either applied ink or perforated the paper, thereby originating the paper-tape punch. An improved later form of the Morse paper-tape register is shown in Fig. 5.22; the electromagnet caused the centrally pivoted point to move upwards indenting the paper as a dash or a dot. The paper drive was clockwork. Many equipments based on this principle were made. From 100 to 150 words per minute could be recorded with these registers.

Another recording device that was invented for telegraphy, and which later became the basis for other applications, was the writing recorder of Froment. As is seen from Fig. 5.43 the pencil stylus is deflected sideways by the signal magnet across its con-

Fig. 5.42 *Morse's early form of paper-tape register for recording morse-code messages*
[Wells (1856), Fig.135, p.301]

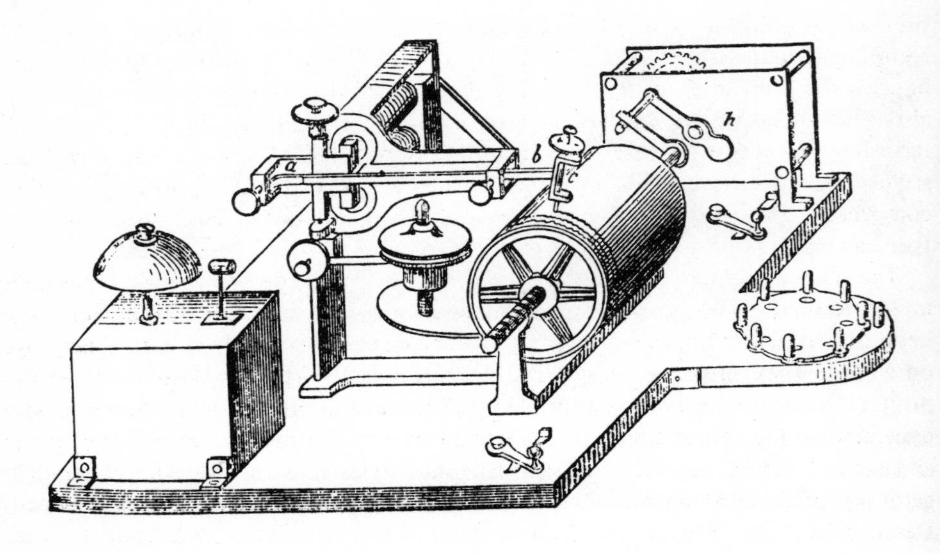

Fig. 5.43 *Writing telegraph by Froment*
[Lardner (1862), Vol.3, Fig.86, p.49]

Fig. 5.44 *Hughes' printing telegraph c1860*
[Deschanel (1891), Fig.570, p.820]

tinuous path drawn as the drum rotates. A central, coarse, lead-screw moves the recording drum axially, thereby, providing a greater record length. In this device can be seen the foundation of seismic and facsimile recorders now used extensively. It is also very similar to the mechanism first used by Edison to produce a phonograph recording, except in his case it was acoustically produced vibrations that moved the stylus not an electromagnet. Froment's recorder incorporated a mechanism that compensated for wear of the pencil lead. A radial, extended-record, mechanism by Bain can be seen in Fig. 5.22.

The teletype printer can be traced to the above and many other mechanisms built in the 1830 to 1880 period. Some systems provided the operator with a piano-type keyboard to send with, the received message being printed directly in normal language on a tape. They often were extremely sophisticated in design and manufacture. The printing telegraph made around 1853 by David Edwin Hughes – a Londoner who emigrated to the United States in 1838 to subsequently become a Professor of Music and then of Natural Science – was first taken up in France after he had been unable to get it accepted in Britain. A French origin etching of his terminal is shown in Fig. 5.44. Deschanel (1891) provides detail of the internal mechanism. Other keyboard instruments were made by Brett and House (see also that shown in Fig. 5.22). McNicol (1925) is about later printing telegraphs.

Ingenuity abounded e.g. punched paper tapes were read by two fingers sensing the two track holes in the Wheatstone systems *c*1867; many other, then novel, features were incorporated into contemporary designs. In general, however, they generally lacked receiver sensitivity, requiring the sent impulses to be amplified with relays along the route. The introduction of long length submarine cable telegraphy with the short-lived 1858 first Atlantic cable posed some severe difficulties, the most dominant being that the signal was distorted from clean digital pulses into attenuated lowpass filtered signals. It was also not amplified. Kelvin devised the mirror galvanometer to enable these low level signals to be detected and he also invented a recorder that would respond to them. It became known as the siphon recorder. As can be seen from Fig. 5.45, it used a glass tube which siphoned up ink from the well to mark the paper strip. Movement was produced by attaching the tube to the coil of a d'Arsonval galvanometer movement. Thus was born, in 1874, the first sensitive ink pen recorder in the form that is commonly used today in physiological and other similar pen recorders. Finn (1976*b*) includes a picture of an 1880s developed model.

Resolved action in *xy* co-ordinates, such as is now used in joy-stick controls, was first used by J.H. Robertson as the sending key of his facsimile letter sending apparatus, Tunzelmann (1890). He used two carbon disc stacks arranged at right angles to each other and against which a single stick control rod was held in contact. Movement of the top of the lever in the shape of a symbol produced two varying level currents which were transmitted. Upon receipt these signals were used to drive two electromagnets, arranged orthogonally, that each exerted a pull on a common armature on which was held a stylographic pen. The pen would resolve the various instantaneous positions of the symbol reproducing the shape sent. This was in use around 1880. The operator had a local receiving set that wrote on tape in front of him to give

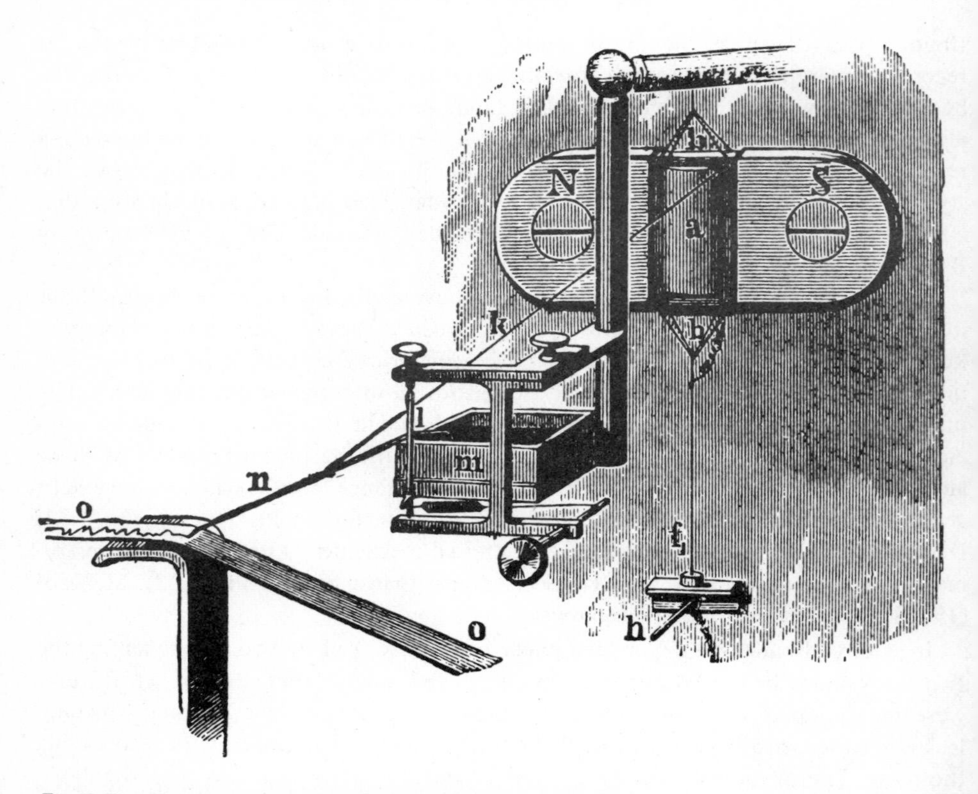

Fig. 5.45 *First moving-coil ink pen recorder by Kelvin 1874*
[Byrn (1900), Fig. 15, p.36]

him visual feedback of what was being written. It took about 20 min. of practice to become proficient at writing legible words.

In physiology the initial requirements for records were not for recording electrical signals but for recording small mechanical displacements formed directly or via a transduction stage from some other energy regime. In introducing the development of the use of graphic techniques in early psychology Davis and Merzbach (1972) published a brief introduction to the instruments used before the psychologists by physiologists. K. Ludwig, in 1847, published details of his work that used a recording cylinder to investigate the relationship between respiration and circulation of the blood. His apparatus used a vellum skin stretched tightly on a driven rotating cylinder against which rested the end of a feather pivot arm that made the registration. This kind of apparatus became known as a 'kymograph'. Life science workers also adopted the word-ending 'graph' for devices that produced a graphical record adding yet more bewildering terms to the already growing jargon of instrumentation.

The Marey sphygmograph, 1860, was a small portable device that was strapped to the forearm to record the pulsations of the radial artery on a small, spring-action, sliding recording platten. Marey devised the first simple instrument for this measure-

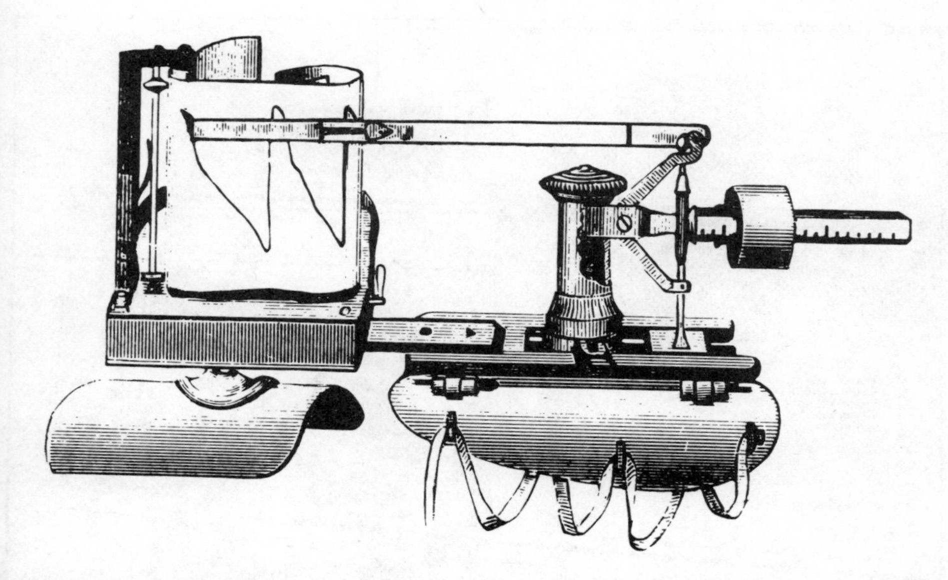

Fig. 5.46 *Verdin's sphygmometrograph used to record the action of the pulse* [Byrn (1900), Fig. 175, p.249]

ment creating a far less cumbersome unit than had been previously invented by Vierordt in 1853. A later version, due to Verdin, and called the sphygmometrograph, is shown in Fig. 5.46. Instruments of a relevant collection are described in a paper by Davis and Merzbach (1975). McKendrick (1894) provides detail of the uses of myographs (linear recorders) in muscle contraction experiments. He also used the glass tracings made on his 'railway myograph' as a projection slide enabling students present at his Royal Institution Christmas lectures to see greatly enlarged views of the traces as they were made during demonstrations. An etching of his complete apparatus is shown in Fig. 5.47.

An important accessory for use with the kymograph was a method for simultaneously adding precise time-interval data to the record. This device was the 'chronoscope' which, when combined with a chart recorder, was called a 'chronograph'. The chronoscope form, according to Ward (1966), was that timing device in which the indicating hand, running from a continuously moving gear train, could be thrown in or out of gear using an electromagnetic clutch. Wheatstone, in 1840, was probably the first person to use electrical control of a chronometer in this way.

Chronoscopes were needed to measure short intervals such as the reaction times of subjects in psychological tests, the transit movement of astronomical bodies and the falling time of masses. A well known unit from the 19th century was the Hipp chronoscope which could time intervals to within 0·001 seconds. Hipp, of Neuchatel, improved existing apparatus significantly in 1848. Ward (1966) contains an illustration of Hipp's apparatus. De Vries (1971) includes an illustration of the apparatus used by Rousselot to inscribe speech sounds on a drum in 1895.

One form of chronoscope made use of the relatively stable vibrations of a good

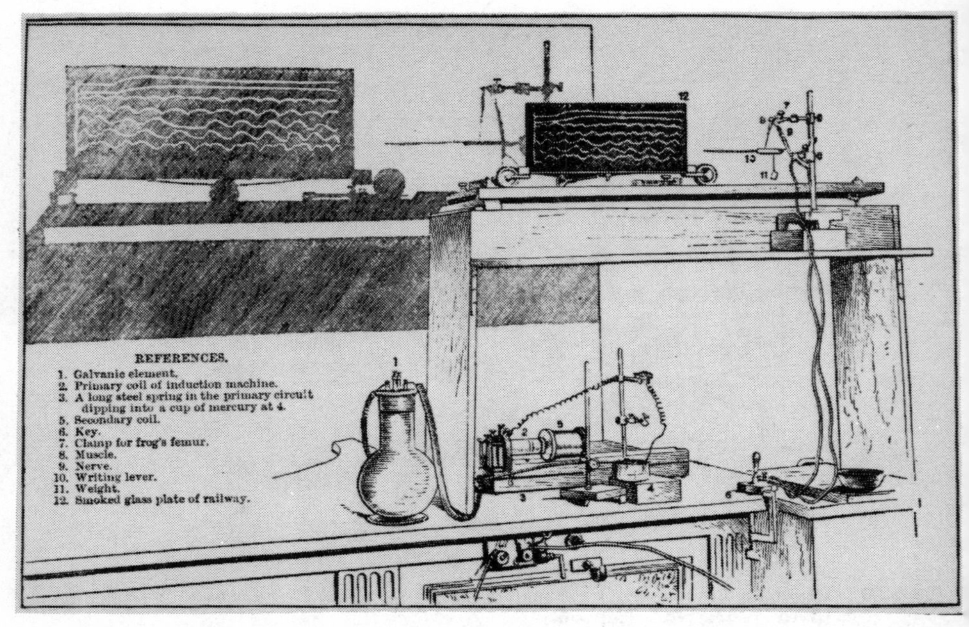

Fig. 5.47 *Railway myograph as set up for Royal Institution lectures c1892* [McKendrick (1894), Fig.29, p.61]

quality tuning fork. These were often directly applied to the kymograph drum forming a time-calibrating trace alongside the signal trace. Wheatstone's chronoscope had an air drive to maintain a tuned reed vibrating at 1000Hz. He used the apparatus for timing the free fall of objects and for measuring the speed of projectiles, Pepper (1874), Bowers (1975).

Chronoscopes were of great importance to the psychologists: these and other instruments are reported in Sokal *et al.* (1976), Davis and Merzbach (1975).

The third discipline where recording mechanisms were developed was in meteorology. An excellent, profusely illustrated, account of the introduction of self-registering techniques into recording of weather variables has been published, Multhauf (1961*b*). Knowles-Middleton (1969*a*) also contains pictures of original meteorological apparatus that featured recording mechanisms.

After Wren's 17th century weather clock proposal and some subsequent published accounts of Hookes' involvement with production of a working model some fifteen years later there occurred a 171-year gap in progress of a multivariable recording station. According to Multhauf (1961*b*) interest in the self-registering instruments came to the surface again in the period 1840-1850. Dolland, at the 1851 Great Exhibition held in London, exhibited an elaborate automatic weather station which recorded barometric pressure, atmospheric electricity, relative humidity, rail-fall, wind direction, evaporation and temperature. The seven records were made on a wide continuously driven paper roll using 'striking hammers' and 'ever pointed pencils'. Multhauf (1961*b*) provides illustrations. After Dolland's time numerous recorders were put into service to record the various weather variables.

Several atmospheric variables were easy to record for they inherently contained adequate energy to drive a mechanism and were easily coupled into, examples being, wind direction and wind force. For these, building mechanical drives would have given rise to few practical operating problems. Other variables, however, only gave precise and accurate traces if the recording mechanism did not load the transducer used. There lay the problem that designers had to circumvent. It must be remembered that they did not have readily available, high-input impedance, amplification devices at their disposal at that time.

The invention of photography, see Section 4.6, fortuitously enabled the limitation of friction in recorders to be overcome although, as will be seen below, the solution to many recording problems already was within the capability of designers in the form of simpler, more carefully thought out, purely mechanical, systems.

Within six years of Daguerre and Fox-Talbot discovering their two different photographic methods British investigators, F. Ronalds and C. Brooke, were constructing recorders in the 1840s using the new medium for the Kew and the Greenwich Observatories, respectively, Multhauf (1961*b*). Their apparatus recorded magnetic field, electric field, temperature and atmospheric pressure. Magnetometer installations using photographic recording were in use by 1847. A typical set-up for recording barometric pressure, as was used at the Stoneyhurst College Magnetic Observatory in England, is shown in Fig. 5.48, Pepper (1874). Multhauf (1961*b*) shows a similar commercial equipment. In operation the height of the mercury column F was used to vignette a collimated light-beam, produced by the gas light source H, so as to produce a varying height exposure strip on the photosensitised paper turning on drum D. Temperature compensation was built in. A reproduction of apparatus originally manufactured by Jean Adrien Deleuil of Paris exists in the National Museum of History and Technology, Washington, Knowles-Middleton (1969*a*). The Kew self-recording magnetometer installation recorded declination, horizontal force and vertical force.

Judging by the relative scarcity of extant 19th century photographic recording instruments and the generally little amount of discussion about them it appears that they were not widely used in meteorological instrument work except at major national observatories, such as at Kew, England.

As time passed other workers had reason to adopt the photographic method. In 1856 the frequency of the oscillatory electric arc was measured experimentally. A spinning mirror method was invented that scanned the illumination across a falling photographic plate to trace a time against amplitude curve of the wave form. Glazebrook (1922), Matschoss (1925) both give accounts of this method which was accomplished by the German, Feddersen. His original apparatus is in the Deutsches Museum. This method was still in use, Fig. 6.33, well into the 20th century for the precision measurement of the waveform and frequency of radio frequency signals. Around 1900 the Roberts-Austen pyrometer recorder appeared, Darling (1911). It used the optical lever principle to form a photographic trace of a light beam that was deflected according to the thermoelectric current present in the mirror galvanometer.

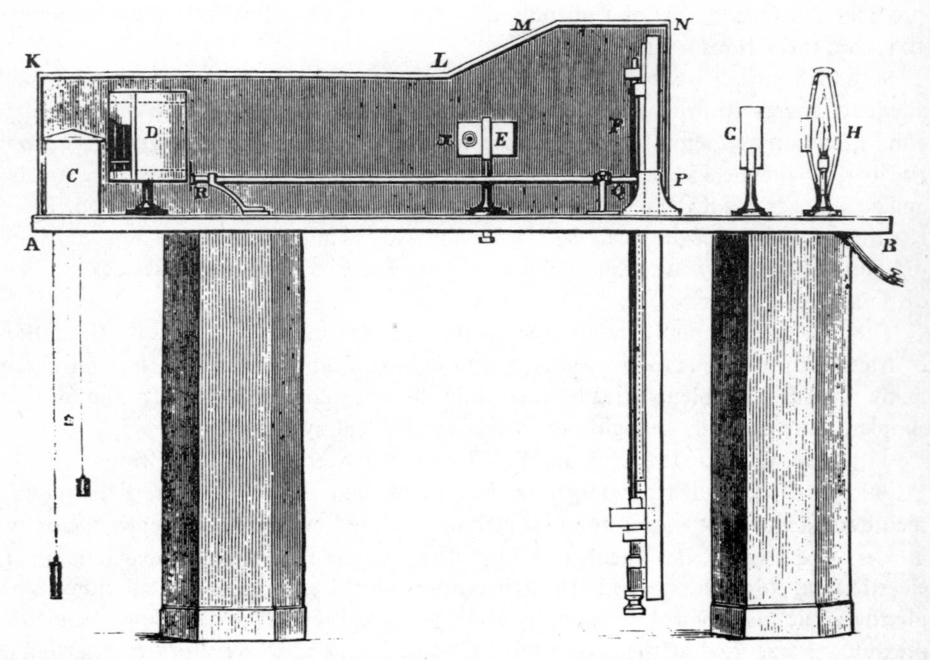

Fig. 5.48 *Photographically registering barometer used at the Stoneyhurst College Magnetic Observatory, England*
[Pepper (1874), Fig.416, p.453]

In its original form the sensitised plate was translated uniformly by driving it from a float falling as water was allowed to escape from a container.

In his 1858 patent, Nadar described the first aerial photography apparatus but it was J.W. Black who obtained the first satisfactory picture: Nadar's equipment was not entirely satisfactory. In 1877, Woodbury developed a more economical method than those used before him. Batut, a Belgian, used kites to take aerial photographs in 1887.

In 1893 Blondel proposed a photographic recorder design that could capture the rapid waveforms as were then being introduced in municipal a.c. mains electricity systems. Blondel realised that the resonant frequency of a stretched string could be made well above the main's frequencies. If the alternating current is passed through a tightly stretched string whilst it is suspended in a strong magnetic field it will vibrate in sympathy with the frequency of the alternating current provided its natural frequency was much higher. The idea was greatly improved upon by Duddell who built a working instrument around 1897, Glazebrook (1922). It was manufactured by the Cambridge Scientific Instrument Company. Light was passed through the strings, a method that became well established in the Einthoven string galvanometer. Arc lights were used to obtain the high light intensity needed for fast events.

If two strings are used so that they deflect in opposite directions a small mirror mounted across both will rotate, Fig. 5.49. An optical lever readout system can then be used to record on moving photographic paper or film. The introduction of continuous roll photographic film enabled these apparatus to be made versatile and compact. One model marketed around 1900 was able to project the traces so that they were made visible to a large audience. Mullineux-Walmsley (1910) states that these devices were more properly called 'oscillagraphs' but his efforts to stabilise the jargon were to no effect for they became known as 'oscillographs'.

The 'ondograph', Mullineux-Walmsley (1910), was a device that used a synchronised commutating switch to sample a fast waveform enabling its shape to be built up – the forerunner of today's sampling oscilloscopes. The 'rheotome' by H.Lenz 1868, was similar. It was used by Bernstein in 1868 to produce a time-amplitude plot of a nerve action potential, Dummer (1977).

It was possible, with vibrating string instruments, to record waveforms of 200Hz and higher, many of the strips being tensioned to resonate at around 40kHz. The high-speed response, ultraviolet (UV), recorder came later in the 20th century, a form that uses a high resonant frequency rotating moving-coil mirror galvanometer movement, not two strings. It is clear that the work of Blondel and Duddel paved the way toward the UV systems in vogue today.

Photographic recording enabled rapid events occurring in nature to be frozen. Before motion pictures had become reality, two experimenters, Marey of Paris and Muybridge of San Francisco, were exploring in 1882, means to capture the motion of

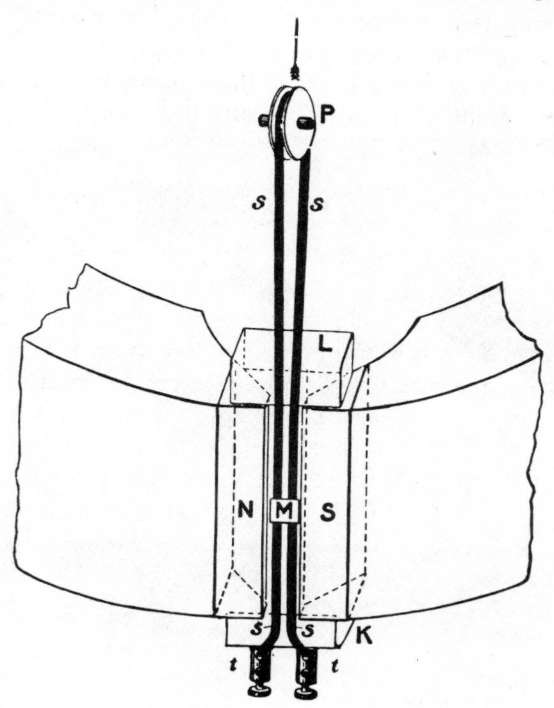

Fig. 5.49 *Schematic of the two-string 'oscillagraph' by Blondel and Duddell c1897*
[Mullineux-Walmsley (1910), Fig.759, p.773. Courtesy Cassell Ltd.]

birds in flight and galloping horses, respectively, De Vries (1971), Larsen (1960), Gernsheim and Gernsheim (1969).

Marey built a 'photographic rifle' in which 12 exposures of 0·7 ms were made at 0·4 ms intervals using silver bromide and gelatine plates brought into the focal plane by mechanical means. Pollock, 1867, and Jansenn, 1874, pioneered this field preparing it for Marey.

Muybridge's approach to the problem was to use 48 cameras triggered by electrical trip wires in the chronophotography system built at the Physiological Station in Paris. Later work used tuning fork exposure time control. He was able to achieve 0·2 ms exposures. He took over 100 000 exposures in his studies of the locomotion of man and beasts.

Shortly after this work, the use of celluloid roll film for photography continuous motion was patented by Friese-Green in 1889, 'roll photography' having been invented by Eastman and Goodwin two years prior to this. This development enabled photographic recorders to become more viable and easier to use in the hands of scientists and engineers who were generally untrained in the new photographic arts. The first automatically developed film recorder appeared in 1920 (Fig. 6.32).

The introduction of photographic methods did not deter other designers from seeking the elusive signal-driven mechanical self-registering instrument needed to record the low energy level, meteorological variables. Around 1880, Jules Richard, an instrument designer and builder of Paris, brought together simplicity of function in both the sensing and the recording stages of the common variables temperature, pressure and humidity. Multhauf (1961*b*) suggests that this development came from

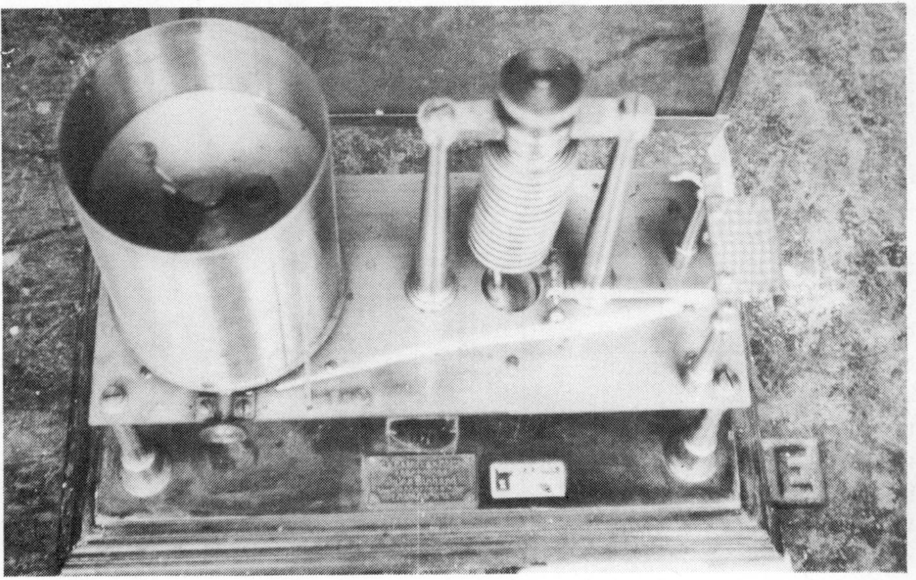

Fig. 5.50 *Late 19th century Jules Richard barograph*
[Object in Bureau of Mineral Resources, Geophysical Instrument collection, Oaklands, Australia]

stream gauging technology devised by Bourdon in the form of the 'Bourdon' tube used to measure pressure differences. Richard used such a tube to drive a well designed, comparatively low friction ink-pen that marked a record on a clockwork driven drum. Fig. 5.50 shows an early Richard design. This style has since been made in the millions. It is interesting to ponder on why it was not made earlier than 1880 for the principles and the practice required were available at least a century beforehand and the demand certainly existed that stimulated designers to invent in this area.

In many applications direct mechanical recording was quite satisfactory. It was easy to maintain, easy to comprehend, cheap to devise and required no additional power source. This method has remained viable ever since. Examples of 19th century use were Kelvin's tide predictor shown in Fig. 5.37 and the Kent flow recorder shown in Fig. 5.38. In botany, MacDougal (1908), mechanical means adapted from the Richard design of registration module were used to record growth (called an auxanometer), weight loss or gain and, of course, atmospheric variables.

Of recorders invented in the 19th century the most ingeneous was that devised by Professor H.L. Callendar around 1897, who, at that time, was at the McGill University, Montreal, Canada. It is described in Section 5.3 and is pictured in Fig. 5.35.

As the next developments in recorders occurred only a little into the 20th century it is appropriate to mention them here for completeness. They are further described in the following Chapter. The Callendar potentiometric recorder was very effective but its complexity gave rise to a high-cost product. In 1905, H. Darwin (of the Cambridge Scientific Instruments Company Ltd.) invented the string recorder. In this design an inked thread is held under the needle of a galvanometer just clearing the needle and a chart roll moving under it. A chopper bar, driven from a source that periodically lowers the bar onto the needle, presses the thread onto the paper making a dot. During the dotting time, the needle is constrained but it is then freed allowing it to take up a new position. This was the forerunner of the many forms of dotting recorder that are used extensively today. Early dotting recorders are described in Darling (1911), Glazebrook (1922), Chaldecott (1976) and Cambridge Instrument Co. (1945, 1955).

Just prior to 1911 Morris E. Leeds (of Leeds & Northrup) devised an improved form of Callendar recorder in which the electrical contact was not needed to be made by the galvanometer movement. His version enabled all sensing and drive functions to be performed by mechanical means. The galvanometer needle is clamped periodically by a driven clamping mechanism. The position of the needle is then sensed by mechanical feelers that transmit the position to the drives of the crossfeed. It then reduces the positional error in steps which were proportional to the size of the error. The needle is then unclamped allowing it to take up its new equilibrium position. His original recorder used clockwork drives but later units used a single electric drive motor. Fig. 5.51 shows the marketed form of this Leeds & Northrup recorder. It is described in Williams (1973) and Griffiths (1943). A similar instrument was still being marketed in the 1950s as the 'Multielect' by an English firm.

The beginning of the end for mechanically sensed potentiometric recorders came in 1929 when Williams of Leeds & Northrup, invented as the result of an exploratory

Fig. 5.51 *Leeds & Northrup mechanically-sensing potentiometric recorder, first developed in 1911*
[Reproduced by permission of the publishers, Charles Griffin & Co. Ltd. of London and High Wycombe, from Griffiths THERMOSTATS, 1943]

exercise, Williams (1973), the electronically balanced equivalent using chopper amplifiers, thyratron controls and tachogenerator feedback stabilisation. It is, however, quite impressive that with purely mechanical means and no electronic amplifiers, mechanical recorders were able to record and control with submillivolt level signals, being a very robust and reliable device that could be employed in industrial locations.

One last course of development that must also be included here is that of electric signal recording on wires, tapes and disks. In July of 1877 Edison built the first working method, Chew (1967), that could record acoustic sounds. He was actually investigating ways to build a telegraph signal repeater using a diaphragm to emboss sound vibrations on a travelling strip of paraffined paper which then was used to reproduce very crude sounds with a similar head. This led him to build improved versions that were specifically designed to record analog speech signals. His tinfoil phonograph was successfully operated in November 1877 with the recording and playback of the historic passage 'Mary had a little lamb'.

The concept, however, was anticipated by the Frenchman C. Cros who had lodged proposals for another device earlier that year. His description was of what we now know as the disk recorder; he also used acoustico-mechanical operation of the stylus. Much of the interest in recording machines has been researched for their entertainment uses. It was, however, due to Berliner, of Washington, that we now use sideways modulated disks and disk reproduction methods. The above methods did not record electrical signals.

Of interest here is the use of this idea for recording instrumentation signals. The magnetic recorder has its practical origins in the designs of Valdemar Poulsen of Denmark, the idea having been published by O. Smith in 1888. In 1898 Poulsen constructed a drum device that recorded sound on a magnetic surface using electromagnetic heads to record and playback sounds. It exists in the Deutsche Museum. He called the device a 'telegraphone'. Interest in 1903, for a disk version, was to enable records of telephone messages to be posted. He incorporated in his designs most of the features found today in the basic tape recorder. It had an electric motor drive and used bias on the recording signals, Jewkes *et al*. (1958). It would seem that there may be a link between the Rutherford bias method devised in 1895 (later used in the Marconi 1902 magnetic r.f. signal detector) and that used by Poulsen for he had a deep involvement in radio technique.

Poulsen's work on magnetic recorders was in advance of its time for without electronic amplifiers the output had to be listened to using largely inadequate state-of-the-art earpieces. Magnetic-coated tape was introduced much later in 1927 by Pfleumer. Poulsen's work was mostly with wire but his patents of 1898 clearly suggest the use of all three media, including the use of coatings.

Thus it was that almost every method that is now used for recording data was invented and developed to become working devices in the Victorian era. It was matters of detail and performance that have been added to over the past eighty years. With the advent of solid-state memories, bubble memories and optical recording cells, future recording devices may well pass into another era.

5.5 Optical instrumentation

In the 19th century optical instrumentation expanded out of the simple image amplification era into one in which sophisticated principles, learned from scientific research into the properties of light, were put to use to learn more about nature and into technological invention of tools of utility and measurement.

Gaussian optics, that aspect concerned with optical component design by ray tracing, continued to be improved as did the quality of the optical glasses, mirror surfacing, optical working processes and the ability to handle increasingly larger size optical elements.

A significant development was contributed by Josef Max Petzval who, in 1840, designed the first lens using a mathematical procedure. He produced a fast, *f*/3·4 portrait lens which is today held by the Technical Museum, Vienna. The design was used in a Voigtlander daguerreotype camera, Ostroff (1977). Thus began rigorous means by which the abberations of a lens can be assessed without trial and error methods to obtain a good lens. Rohr (1899) is an historical account of photographic lenses. The range of optical instruments increased over the century. Liébert, of France, devised the first enlarger in 1864: it made use of sunlight to provide enough illumination. Microphotography, Scharffenberg and Wendel (1976), began in the work of the Englishman, Dancer c1839, when he produced the image of a 500mm page at only 3mm size. During the Franco-Prussian war of 1870 Dagron, a Frenchman, was responsible for microphotography that sent 2 500 000 messages by carrier pigeon. A pigeon could carry 50 000 microphotographic messages at a time.

Colour pictures were first made by Maxwell in 1861. Fresnel invented the so-named stepped lenses that are commonly used in light-houses and overhead projector systems. He also devised the rhomb element for turning plane light into the circular polarised form. The polarising prism became available in 1828 as the work of Nichols. Books appeared on optical instruments, an example being Brewster (1813). Kitchiner (1824) was concerned with aspects of opthalmic optical instruments.

In 1893 Professor Abbe of Germany patented the stereoscopic binocular in which the object glasses are mounted wider apart than the eyes giving a better sense of vision. His design is that commonly used today. It reduced the size of the instrument, flattened the field of view and reduced the focusing adjustment needs.

Surveying instruments became smaller whilst being given increased accuracy scales, many taking the forms and shapes still used today, Stanley (1901). Navigational instruments also stabilised in design during the 19th century, Pearsall (1974).

These above developments are but a few of the many improvements made in this era. They have been selected from Marmery (1895), Byrn (1900) and Ostroff (1977).

The most spectacular Gaussian optical devices of the 19th century were the giant telescopes that grew, and grew, in size over the period. The great Rosse telescope, the 'Leviathan of Parsontown', Ireland, was built in 1845 and remained the largest until 1919 when the 3·9m Mount Wilson unit exceeded it in size. It had a speculum mirror of 2·8m diameter and focal length of 25m. It's limited mounting arrangement restricted its use but it was powerful enough to show that some celestial objects thought to be nebulae were, in fact, star clusters. A model of it exists in the Science Museum, London, Thoday (1971). It was described contemporaneously by Timbs (1860).

Every advanced country was engaged in making large telescopes. Foucault's reflector erected in Paris in 1862 had a mirror diameter of 1·24m; Newall's Gateshead telescope of 1870 had an object glass of 0·98m. The 1873 United States Observatory, Washington, unit was a 1·20m refractor, and so on. Byrn (1900) and De Vries (1971) include summarised details and illustrations of many of these spectacular instruments. Neal (1958) is another general account.

The greatest telescope project of the 19th century was surely the 'Grande Lunette' built at the Paris Exposition of 1900. The object glass was 1·89m in diameter and the focal length was 95m. This gigantic structure was laid out in an entirely novel manner

Fig. 5.52 *Grande Lunette: gigantic telescope made for the 1900 Paris Exposition*
[Byrn (1900), Fig.195, p.288]

to overcome structural problems, Fig. 5.52, brought with its immense size. A unit, called a 'siderostat' by Foucault, moved so as to maintain a fixed image in the viewing field whilst the sky moves by. The components were mounted on cast iron pedestals; the siderostat weighed around 40 tonnes! The mirror, which weighed about 3 tonnes was floated in a mercury pool allowing it to be moved to follow the field of sky chosen for study. A contemporary statement said it took eight months to cast, grind and polish the mirror, that it had a magnification of 10 000 and that it could 'permit an observer to follow the manoeuvres of an army group or transatlantic steamer on the moon'.

Sheer size does not necessarily make a good telescope. The Great Melbourne telescope, made in Ireland for the Royal Society, London, was shipped to Melbourne for installation in 1869. It was the last large telescope to utilise metal mirrors. In cleaning the protective varnish from one of the two mirrors the surface was damaged. That was the start of many unfortunate occurrences and it never performed to the hopes held for it. In 1949, it was taken out of service at Melbourne, and from there, went to the Mount Stromlo Observatory, Canberra, where it was extensively rebuilt for further useful service until 1975, Watson and Watson (1977). In contrast the Alvan Clark and Sons refractor, Warner (1968), at the Lick Observatory, Carlifornia was held in high repute until it was surpassed in performance by the Yerke's unit exhibited at the 1893 World Exposition held in Chicago. That, too, was improved upon with time. The work of Holcomb, Fitz and Peate, three American telescope makers of this period, has been studied by Multhauf (1962).

That natural light can be broken up or dispersed into its various wavelengths was well known at the end of the 18th century. In 1802 Thomas Young tried to interpret the yellow line that was so predominant in the light dispersed from a candle flame. Thomas Melvill, 1752, was the first to report the presence of the sodium lines by dispersing the flame's radiation with a prism.

The 19th century saw the development of spectroscopy from a minor observation to a major discipline, McGucken (1969), one that would enable much to be learned about the nature of light and matter and provide a tool for analysis of chemicals existing on earth and in outer space.

Young determined, for the first time, the wavelength of the sodium line in 1802. In that year Wollaston observed what became known as the Fraunhofer lines, the absorption lines of the atmosphere. These scientists each made a contribution to the emerging discipline. Fraunhofer stands out above all others. In *c*1813 he conducted a detailed examination of the sun's spectrum and in doing so devised a spectroscope that had, by the then standards, extraordinary resolving power. He found 'an almost countless number of strong and weak vertical lines'. He mapped 700 of them. In developing fine slits he came to study diffraction effects from sharp edges with greater care than had been done before him. He also explored the use of gratings made with fine silver wires, the finest being 19 lines/mm. In 1823 he constructed apparatus to make the first ruled glass transmission gratings; with 300 lines/mm density. The Deutsches Museum has an extensive exhibit of his original apparatus.

The history of spectroscopy can be gleaned from Byrn (1900), Cajori (1962), Nelson (1881), Pepper (1874) and Sawyer (1945).

The spectroscope, as an instrument system, passed from the crude method due to Newton in which a beam of uncollimated light was passed through a prism onto a screen the whole generally having little structural support or measurable features, to a portable quantitative measuring tool in the 19th century. Wollaston, in 1802, used a slit instead of a round hole to form the beam and, thereby, found the black lines. Simms, in 1830, used a lens to focus the beam onto the slit forming the collimating design. Kirchoff, using these concepts, with a theodolite telescope as his collimator and a precision goniometer, was able to do better than those before him because of improved apparatus. Multiple prisms were later used to increase the dispersive power of the instrument. Fig. 5.53 shows a *c*1870s unit by Gassiot. It used eleven, bisulphide of carbon filled, hollow prisms in one version.

Spectroscopes opened the world of light and its applications yet further, as had the invention of the telescope and the microscope centuries before. They enabled the composition of materials to be investigated through spectrum analysis showing that

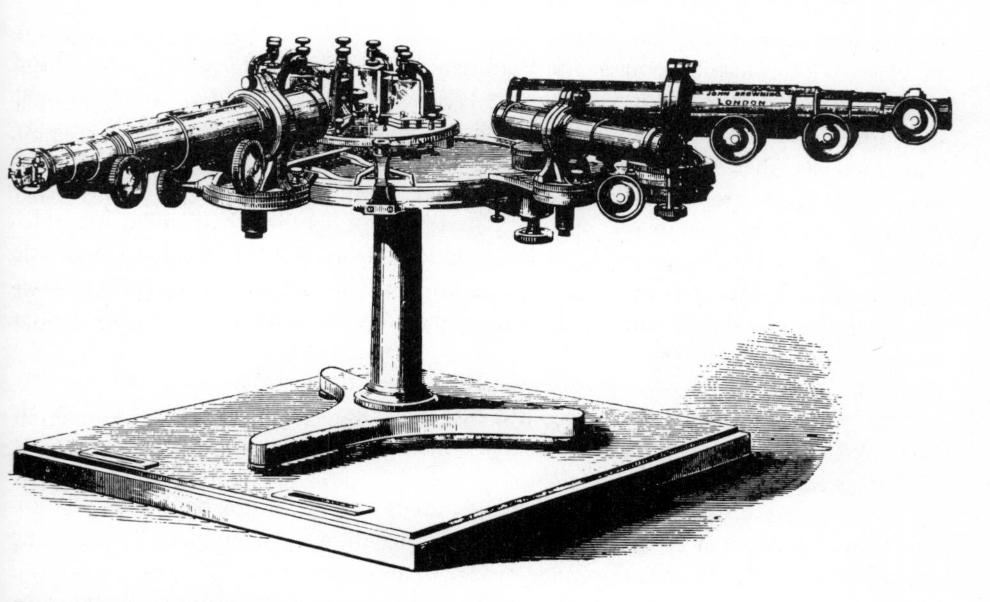

Fig. 5.53 *Gassiot, multiple prism, spectroscope c1870*
[Pepper (1874), Fig. 104, p.96]

yet undiscovered elements existed. They enabled the composition of the distant stars to be investigated with telescopes. Scientific telescope installations invariably have spectroscope attachments provided. Caesium, rubidium, thallium and indium were discovered this way. Lockyer, in 1868, while investigating the yellow line near the D-line in the solar spectrum, came to the conclusion that there must exist an element near to hydrogen. He called it helium. Just how he came to this conclusion using spectroscopy is described in Lockyer and Lockyer (1928). In astronomical use a telescope so equipped was called a 'telespectroscope'. The spectroscope also made metallurgical control possible in steel manufacture.

The first photographs of spectra were made in 1842 by Draper of New York. In 1843 Joseph Saxton made Draper a glass diffraction grating which Draper used to record the spectrum, again by photographic means. To enhance the dispersion, and, hence, sensitivity, many people embarked on their own grating ruling programmes. Friedrich Nobert had the reputation for producing the best gratings until around 1880 when that honour passed over to the United States owing to the machines of Lewis Rutherfurd. Rowland of Johns Hopkins University then came to the fore, building several excellent mechanically-controlled, ruling engines. To give an idea of the mechanical care and accuracy needed the 220mm long drive-screw was made to have no greater than half a wavelength pitch error. His gratings had densities of 4800 lines/mm. Compare that with the first made by Fraunhofer. Historical aspects of grating ruling engines are to be found in Candler (1951), Coogan (1966), Davies and Stiff (1969).

This brief account has not mentioned the work of Bunsen, Kirchoff, Angstrom, Kayser, Runge, Zeeman and others. They each made significant contributions to the complicated subject of spectroscopy.

Electro-optics has its roots in the Victorian period. The 19th century was a period of discovery and gestation in which many relationships between magnetism and light and, therefore, also electricity were established and, in the main, explained. Faraday was convinced that there existed a relationship between these and beginning in 1845 he systematically created experiments to observe it. In that year, he established that magnetic field could rotate the plane of polarisation of light i.e. the Faraday magneto-optic effect. In 1862 he returned to this field looking then at the effect of a magnetic field on light from a sodium flame. He was correct in his assumptions but his measuring apparatus was not capable of showing the line broadening that occurs. It was later observed by Zeeman in 1896.

It was Lord Rayleigh who, in 1885, quantified the Faraday effect using an extremely crude, 'string and sealing wax' apparatus that is in the Science Museum, London. He measured the constant of rotation of light in carbon bisulphide, reporting it in the *Philosophical transactions* of that year.

Kerr discovered that electric field also can induce rotation of polarised light (electrostatic birefringence) in 1875 and a year later that rotation can be induced by reflecting plane polarised light from the polished surface of an electromagnet.

Zeeman's work, Zeeman (1913), was taken up by A.A. Michelson of Chicago who showed that line broadening was a complex process. Improved spectroscopes showed that the lines actually split into triplets. Such work formed part of the growing theoretical knowledge of the electron. It was many years later before it found practical application in measuring instruments for other than basic scientific work. Current stabilised-frequency laser sources, those that may become the next length standard and the alkali-vapour magnetometers used in military detection and geophysical prospecting are based on the scientific work of the period of this Chapter, work that advanced as new principles were converted into new instruments. Another path of electro-optical development that was to have profound influence on the later development of instrument systems was that of photoelectric sensors.

Willoughby-Smith, in 1865, was conducting tests on cables using a high-value resistor made from bars of crystalline selenium. He discovered, by accident, that these resistors were highest in value when he closed the box that they were held in. After experimenting to establish what caused this it was realised that it was a light-induced phenomenon. Shiers (1977) gives background to this sequence of events in his notes on the formative years of television. Selenium had been discovered much earlier by Berzelius in 1817.

The comparatively high sensitivity of the resistance of selenium to light variation made it an obvious choice for early electro-optic systems. The slow response, however, restricted the chances of success of many proposals. The Bell and Tainter photophone, Fig. 5.25, was one example of a working system (of a kind!). According to Shier's research there were at least 24 proposals for television systems made before 1899. Much of the basic conceptual features, such as scanners, synchronous control, picture definition, colour reproduction, voice channels, image projection, transmission by radio and actual practical detail were included in those early claims. The selenium cell featured in most and was a major reason for the burst of inventive effort. Carey's 1880 camera unit used a mosaic of parallel-outputted selenium cells, but he had been anticipated by Redmond in the previous year. Shiers (1977) contains a chronology of events and a chart showing the methods and the people involved in television development up to 1900.

Proposals were well engineered to the conceptual stage. Fig. 5.27 shows a contemporary picture of the 'telescope' system proposed by Dussaud in 1898. It used the Nipkow disc scanner (of 1883 origin) to modulate a segmental strip selenium cell module. The received signals were then to be used to modulate, via a sensitive telephone unit, the radiation of a powerful arc lamp. The flicker rate would have placed spots on the screen at ten per second. There is no evidence to prove that this and most of the methods proposed were ever built, but it is clear that many would-be makers realised that they could not meet realistic data-rate requirements. The artist of the Dussaud apparatus has exercised his licence for the system could not possibly have produced such fine quality reproduction as shown.

Each designer used a different name, teleoscope, telectroscope, telephotograph, electric telescope and others. The term 'television' was eventually coined by Perskyi in 1900.

In Shier's opinion of all the systems proposed in those years the one that was feasible was the 'telephane' by Sutton, of Ballarat, Australia (reported in 1890 but claimed to have been devised in 1885). His combination, Fig. 5.54, of the then existing technological devices included use of electro-optic modulation with a Kerr cell, phonic wheels, an induction coil, and the Nipkow disc scanner. Matrix television systems have been discussed by Sydenham (1974). Hawks (1929) includes relevant comment.

Another electro-optic sensor was also in its early stages of development. The photoelectric cell, that emerged to find extensive use by the 1930s, has its origins in the discoveries of Hertz and Hallwachs around 1888. They found that light falling on bodies gave rise to the generation of negative electricity. This became the subject of more theoretical and practical studies by Lenard, Einstein, Millikan, Blake, Duane, de Broglie and others. Instrumentation applications emerged as the vacuum and gass-filled photoelectric 'magic eyes' of the early 20th century.

In 1839 Edmond Becquerel discovered the photovoltaic effect. In the 1880s A.C. Becquerel observed that certain substances used as electrolytes in a primary cell generated differing voltages if the two plates were exposed to different light levels. Around the same time Svante Arrhenius found that the resistance of silver halides increased with increasing light level.

Another significant optical principle that passed from physical principle to practical application in measurement during the 19th century was optical interference. This was a consequence of the wave theory of light revived by Thomas Young in 1801 through a paper read to the Royal Society which included material on the principle of interference, . . . 'When two undulations, from different origins, coincide either perfectly or very nearly in direction, their joint effect is a combination of the motions, belonging to each'. He used this to explain the colours of thin plates and scratched surfaces, the forerunners of diffraction gratings.

Thirteen years later Fresnel independently discovered interference not knowing of Young's work. Fresnel used two plane mirrors to produce two sources which set up fringes thus avoiding the problems of diffraction around sharp edges existing in Young's double slit experiment. As the century progressed proven evidence and theoretical understanding increased as more scientists became convinced and carried out further work.

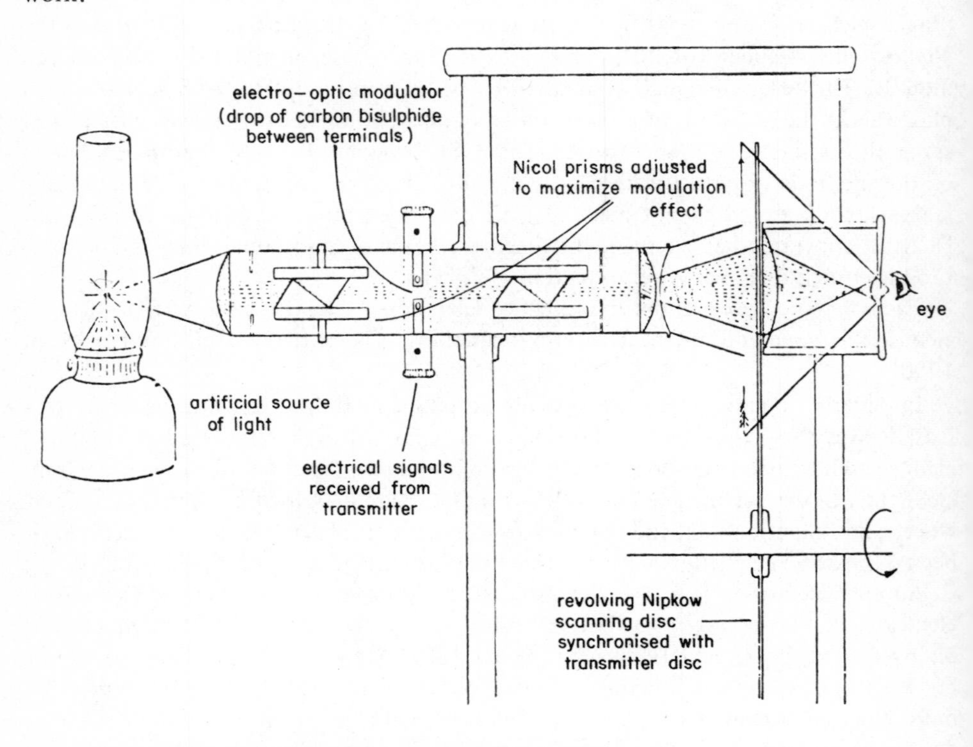

Fig. 5.54 *Schematic of Sutton's 'Telephane', devised 1885*
[Sutton (1890), Fig. 4]

This new knowledge became a measuring tool in the pursuit of further knowledge in the famous Michelson and Morley experiment completed in 1887, Jaffe (1961), that used the interferometer principle to investigate aspects of the speed of light. Michelson used the interferometer to look for the ether drag effect and also to weigh the earth by a new version of the Cavendish attracted-masses method. Fig. 5.55 shows the apparatus used in Michelson and Morley's interferometer. The top square block floated in a trough of mercury so that it could be rotated. This method was able to detect displacements of fractional wavelength order in metre distances, or the corresponding effects of changes to the refractive index of the path, to the wavelength, or to the speed of light. It was an important event in the history of measurement technique.

Interferometry has now enabled the versatile laser interferometer to be used for engineering and scientific length metrology, has assisted the invention of holography, has provided means to test optical elements and a way to measure and calibrate gauge blocks. It has also provided the basis for many electromagnetic distance EDM measuring equipments now used in surveying; and the means to measure star diameters. Michelson carried out work at the International Bureau of Weights and Measures that

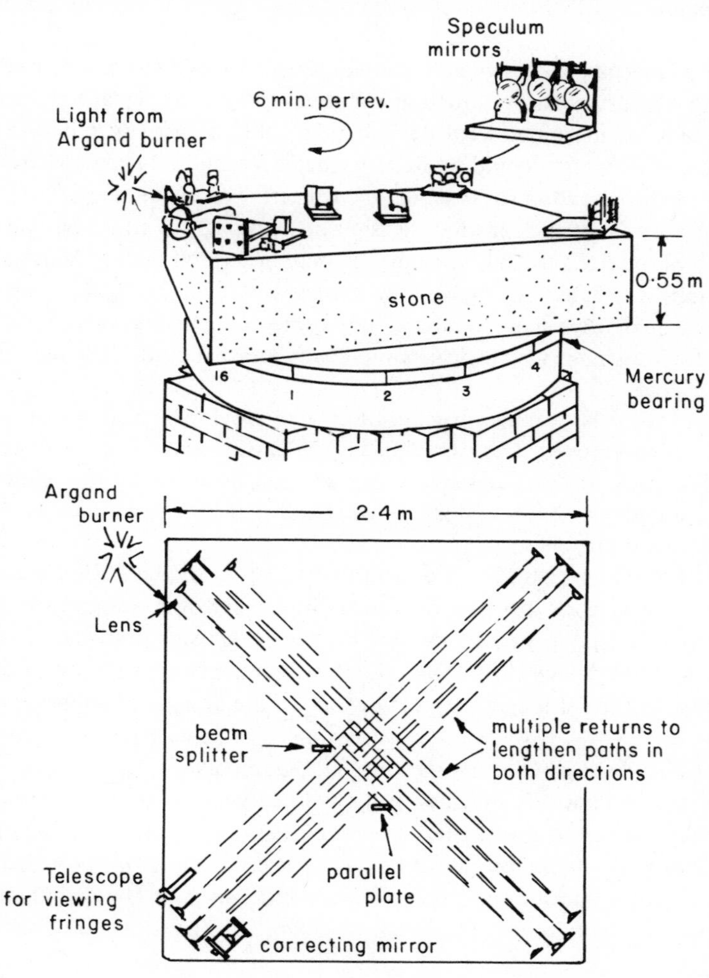

Fig. 5.55 *Interferometer set up for the Michelson and Morley study of ether 1887* [Michelson (1928)]

pioneered means to define the length standard using the cadmium red line in an interferometer. He later published a book on the subject, Michelson (1928). His work is covered in Jaffe (1961). Applications of interferometry have been covered over time in Williams (1930), Candler (1951) and Dyson (1970).

By 1900 optical methods of measurement had expanded from relatively simple lens and mirror devices to include new transducers that could convert light into electricity, and vice versa via the incandescent, arc and discharge lamps. The new principles of spectroscopy and interferometry opened up many previously impossible measuring techniques.

5.6 Mechanics

The quality of mechanical design and manufacture continued to be improved throughout the 19th century. Mass production eventually began to replace the early hand-crafted products as machines were devised that could be used to replicate the same part tolerances better than could be done by hand methods. Texts, such as Carpenter (1892), provided foundation for systematic engineering of experiments.

Improved metrology assisted the precision of manufacture and eventually allowed the concept of interchangeability to be phased into production. Materials testing became established. Table 2.2 shows how knowledge of materials and their strengths passed from a qualitative stage to possess applicable rigour as the various theories were established for designing mechanical components and structures. Burstall (1970) has traced this change.

Ferguson (1962) shows how mechanism aspects developed from a craft to a theoretical skill which enjoyed high standing in academic studies. The emergence of the subject of kinematic analysis and synthesis was brought about by the realisation that mechanical components are not perfect entities and that they cannot always be treated by static systems approaches.

Much of the fine mechanism used in instrumentation of the 20th century was invented and refined in the 19th century. Straight-line linkages were a favourite study after Watt used a simple parallel motion linkage in his steam engine. The Russian mathemetician Chebyshev, 1831-1894, produced mathematical theory of these linkages that later, in the 20th century, was applied to the design of high-order electronic filters.

Devices to produce parallel translation, pantograph action, various shaft couplings, such as the Oldham and the draglink forms and many mechanical computing linkages, for example, the scotch yoke, the Geneva mechanism, Ackermann steering, quick-return, ball-and-disk integrator and others were devised. Thus was provided the theoretical and practical background later to be used extensively in the manufacture of the various mechanical computers that began to emerge toward the end of the Victorian period.

Illustrations of equipment given in this Chapter show the high level of mechanical ability of these times.

In timekeeping clocks and chronometers continued to be improved. According to Ward (1970), the error in seconds per day for the best of these was reduced by an order of ten, to be within less than 10ms per day over this century. He lists the introduction of barometric compensation by Robinson of Armagh Observatory in 1831 as the reason for the first noticeable increase in rate keeping. At the end of the century it was the introduction of the alloy Invar, by Guillaume of Paris, and later in 1899, his special alloy that had varying thermal expansion coefficient as the temperature changed, that enabled better temperature compensated balance mechanisms to be used.

Other improvements were the invention of means by which a pendulum could be driven with effectively smaller loading by the driving source. The free pendulum clock, as these were called, began to emerge through the ideas of Riefler whose clocks had a daily variation of the order of 10ms. Rudd's approach was quite different. In 1898 he introduced the concept of the slave clock that provided 'kicking' impulses to the main pendulum each minute. The slave was adjusted to keep pace; it did not need to be highly accurate for the theory of the simple harmonic system shows that the resonant system will keep good time despite small changes in the actual time at which it is given the impulse to restore the energy stored in it. Other free pendulum clocks were built. It was that of Shortt 1921 that made a significant change to the rate of improvement of the accuracy of clocks.

Electricity also found its way into time-keeping in this period, Alexander Bain is said to have invented the first electric clock in 1840. Wheatstone claimed that place also but it is Bain that usually gets the credit, Hackmann (1973), De Carle (1965), Ward (1970). It is, however, more correct to attribute the place as first maker of an electrically driven pendulum clock to Bain for several other electric clocks were made before his time. Aked (1971) reports allusions to electric clocks having been made by Cumming in a late 17th century work. Ferguson's 1775 treatise *An introduction to electricity* gave details of an electrostatically driven clock and orrery. Its failing was, however, that the user of the clock had to crank a generator so the exercise was rather futile as a practical clock.

De Luc built a pendulum clock-like device driven by electric charge in *c*1809. It was for studying atmospheric electricity. He used high-voltage batteries, now called Zamboni piles, to drive it. In 1815 Zeller of Munich displayed for sale an electrostatically-driven clock made by Professor Ramis. It contained two voltaic piles in which 2000 cells were used. Apparently there exists a pendulum of this kind in the Clarendon Laboratory that has been swinging for over 130 years.

Around 1835 Steinheil, at Munich University, devised what seems to be the first electromagnetic clock system. He used a De Luc idea in which a normal pendulum clock provides electrical impulses that drive remote slave clocks. It was, therefore, not strictly driven by electricity. History then passes over to the work of Bain and his trials and tribulations with Wheatstone. Aked (1971) is a scholarly account of the use of electricity in clocks. Ward (1966) contains illustrations.

The synchronous clock was first proposed in 1895 by Hope-Jones but did not become popular until two decades later. Factors mitigating against it, to begin with, were the unreliable nature of a.c. mains supplies, the lack of good mains frequency control in public authority supply systems and the lack of a suitably small synchronous motor. In 1916 synchronous electric clocks appeared on the US market.

The subject of clocks is extensively researched and the literature published is voluminous. This sketch will, however, serve to illustrate the development of skills of mechanical design and construction.

Another area where improvement in mechanical design skills is evident is in weighing apparatus. Stock (1969) has many illustrations of the changes that were introduced and gives a useful bibliography. Oertling, the founder of the British firm, lived in this period. The well known Oertling long-beam design originated in 1847. The rider system of fine balance adjustment may have originated with Oertling for he was awarded a medal for a product incorporating it at the 1851 Great Exhibition held in London.

Vacuum balances were in use by the 1880s. Stock attributes the introduction of the so-called modern balance (those with short beams for rapid speed of reading) to Bunge in 1866. Some balances exhibited at the 1851 event had optical projection reading systems. Aluminium beams were used in late 19th century balances.

Operation of the finest and most sensitive mechanical devices that could be made in the 19th century e.g. galvanometers and microbalances, made users aware that there were limits to the maximum sensitivity that could be utilised. These limits were initially the degree to which atmospheric variables of temperature and pressure could be controlled. Compensation mechanisms and ambient control were devised to reduce these second-order errors. With these perturbations apparently in control experimenters still found that their apparatus was not absolutely stable and that output indicators, such as the position of the light spot in a mirror galvanometer system, would constantly deviate around a given point with unpredictable variation in amplitude and frequency. They had reached the Brownian motion limitations (see Section 5.2).

5.7 Still more measurements

Earth sciences

Until the 19th century meteorology was the dominant branch of the observational aspects of earth sciences, with magnetic and gravity observation occasionally reaching prominence through special expeditions such as that by Bouguer to Peru and Equador, in 1735, to measure the meridian arc and Cook's voyages, Rienits and Rienits (1968), to the southern continents to observe the transit of Venus and take magnetic measurements.

Earth sciences gained experimental and observational impetus in the 19th century. Oceanography became firmly established owing to the Challenger expedition of 1872 onward, Linklater (1972). An interior view of that ships' zoological laboratory is given in Fig. 5.56. Applied geophysics, that for the discovery of mineralisation, became established practice; there had been isolated instances beforehand, such as magnetic survey for iron-ore carried out in Sweden in the mid 17th century. In 1830 Fox used voltage measurements to detect sulphide-ore veins in England. Von Wrede, 1843, suggested that a magnetic theodolite could be used to look for magnetic-ore bodies; the idea was taken up in the 1870s. Seismology gained its name in 1858, then coined to describe the study of earthquakes. It became an applied geophysics tool in 1899 after Knott developed an explanatory theory of acoustic wave travel in the earth.

Scientific aspects of geophysics began with investigation of the earth's magnetic

Fig. 5.56 *Zoological laboratory on the main deck of HMS Challenger* [Photo. Science Museum, London]

field in the 16th century. Gravity investigations, for the shape of the earth soon followed. This has been researched and reported in detail by Lenzen and Multhauf (1965) who reported gravity measurement by pendulums. Hugill (1978) provides information on gravity meters. Walker's Adam's Prize essay of 1865 was concerned with the magnetic regime, Walker (1866).

It seems that no detailed history has yet been prepared on the general history of geophysics. Sydenham (1978*b*) provides a useful bibliography of this field of historical study. Table 5.3, a chronology of dates in geophysics, has been prepared from those sources and lists many of the key dates. A history of seismology has been published, Wartnaby (1957). Illustrations of early seismic apparatus are to be found in McConnell (1977).

Gravimetry in the Australian continent from the first measurement made there in 1819 has been studied by Dooley and Barlow (1976). These determinations were part of various world networks set up by many visiting overseas scientists. The first, spring-mass gravity meter, Fig. 5.57, was made in Australia in 1898, Threlfall and Pollock (1899), making use of skills developed by C.V. Boys with quartz fibres that were taken up by Threlfall while at Sydney University, Threlfall (1898). Scientific interests in gravity determinations and the various corrections led to the use, in the early 20th century, of the method for locating anomolies which might indicate a valuable geo-physical deposit. Poynting (1894) is relevant to gravity measurement.

Table 5.3 *Historical chronology of geophysical measurements*

Magnetic inclination independently reported by Robert Norman and Gerhard Mercator	1576
Magnetic surveys made to locate iron-ore deposits in Sweden	1640
Magnetic declination measurements made by Tasman in Australasian region	1642
Mersenne made first determination of length of seconds pendulum, not in use at that time to measure gravity	1644
Bouguer, La Condamine, Godin and Ullo measured meridian arc in Equador Maupertuis and Clairaut made similar expedition to Lapland, both groups proving earth to be the oblate spheroid as predicted by Newton	1735
Bouguer's original work with 'invariable' pendulum yielded methods for determining internal structure of the earth	1737
First comprehensive measurement of magnetic field in Tasmania by de Rossel	1792
Captain Henry Kater designed, built and used compound pendulum for first absolute determination of gravity	1817
Self-potential or spontaneous polarisation SP invented by Robert Fox in England who used it to detect sulphide veins	1830
Gauss built, and using oscillation method, measured earth's field with a torsion magnetometer	1832
Lieutenant J.H. Kay became Director of Rossbank Magnetic Observatory in Tasmania	1840
Von Wrede suggested magnetic theodolite can be used to discover magnetic ore bodies, but idea not taken up until 1879s	1843
Mallet experimented with artificial 'earthquakes' in attempts to measure seismic velocities	1845
Barlow, in course of British telegraph system studies, established existence of telluric earth currents	1847
Weber made first absolute earth's field measurement using an earth inductor coil	1853
Robert Mallet introduced word 'seismology' to describe earthquake phenomena and theories on such	1858
'American (Swedish) mining compass' introduced with two axis of freedom for its needle	1860
Lt. Gen. J.J. Baeyer of Prussian Geodetic Survey began action to standardise gravity measurements in Central Europe	1861
Thalen and Tiberg constructed magnetometer system based on laboratory sine and tangent methods for use in field	1870
Charles S. Peirce took charge of pendulum gravity measurements in the USA	1872
Faye suggested swinging two gravity pendulums in opposition – critics were divided on its value	1877
Thalen's book *On the examination of iron ore deposits by magnetic methods* prompted magnetic prospecting	1879

(Table 5.3 continued)

R. Von Sterneck, Austria, introduced 1/4 metre non-reversible pendulums reducing size of apparatus considerably – used for fundamental geophysical measurements	1887
Milne built 1m, iron-bar, earth strain meter	1888
Development of first spring-mass gravity meter (in world?) by Threlfall and Pollock in Sydney	1888
Baron Roland von Eotvos of Hungary perfected a gravity balance	*c*1890
Milne described first seismograph capable of detecting distant earthquakes using principles from 1832 work	1895
Milne inaugurate first Seismological service on Isle of Wight, England. He implemented the first epicentre estimation procedure	1895
Becquerel discovered radioactivity	1896
Knott developed theory of seismic reflection and refraction at interfaces	1899
Radioactivity logging in boreholes recorded	*c*1900s
Oldham applied theory of P and S waves to epicentre estimation	1900
Potsdam–based international gravity ties begun	1900
Oil domes discovered by geophysical method in Texas (Six in 1901 - 1924, forty in 1924 - 1927)	1901
Oddone improved Milne's earth strainmeter	1901
Eotvos used his gravity balance in geological mapping	1902
Galitzin applied electromagnetic transducer to seismic instruments beginning trend to the modern electro-mechanical designs	1904
Zoeppritz and Wichert published seismic wave theory	1907
Captain Scott and Sir Douglas Mawson operated simultaneous magneto-graphs on opposite sides of the South Magnetic Pole	1912
Schlumberger developed practical resistivity and equipotential-line methods	1912 - 1914
S. P. rediscovered and applied by Schlumberger	1913
R. Fessenden's work that led to seismic reflection method in 1921	*c*1913
Adolf Schmidt built his precision vertical magnetic-field balance	1915
Hugo de Boeckh suggested gravity torsion balance should be able to locate domes	*c*1915
Egbell used Eotvos gravity balance to make survey over Czechoslovakia oilfield	1915 - 1916
Work of groups, throughout the world, on sound ranging of big guns led to seismic prospecting in ground	*c*1917
Dr. Mintrop, Hannover, Germany built his first seismograph	1917
Lundberg and Nathorst resistivity method introduced, mainly in Scandinavia	1918
Lundberg developed inductive techniques using high-frequency signals. (Made airborne in 1950s)	1920s

(Table 5.3 continued)

Lundberg conducted aerial magnetic survey in a captive balloon in Sweden	1921
Reflection seismic method validated as tool in oil research by Karcher, Haseman, Perrine and Kite in Oklahoma City, USA	1921
N.S.W. Government supported expedition to Goodiwindi, Queensland to observe solar eclipse to help verify Einstein effect	1922
Golyer sponsored first refraction surveys for oil in Mexico	1922
First geophysical exploration for oil in US using Eotvos torsion balance under Golyer	1922
Orchard Dome discovered by seismic refraction using mechanical recording	*c*1924
Fan shooting seismic refraction method exhibited spectacular success at locating salt domes	1924
New Milne-Shaw seismograph released with increased gain and electromagnetic damping (optical lever and photographic record)	1924
Electromagnetic methods perfected with work in Sweden, Australia and America	1925 - 1940
Karcher built one of the first moving coil geophones	1925
Maud field, Oklahoma, – reflection seismic survey used vacuum tube amplifier	1927
Schlumberger brothers made first electrical well-logging measurements (resistivity)	1928
Fluxgate principle patented, development prompted by need for airborne magnetometer	1931
Schlumberger Brothers conducted first S.P. well logging	1931
L.J.B. La Coste Jr. published design of new vertical seismograph on which were soon based several gravity meters	1934
Poulter used air shots in seismic survey of Ross shelf Ice, Antarctica in Byrd expedition	1934
Benioff obtained solid-earth tidal records with 30m, steel-bar, strainmeter	1935
Logachev put earth inductor in airplane but it's 1000 gamma sensitivity was not adequate for prospecting use	1936
Geophysics Journal began	1936
Rieber published idea of processing seismic data using variable density record and photocells	1936
Turam electromagnetic method devised by Hedstrom	1937
Boliden gravimeter developed in Sweden; incorporated electronic detection and balancing	1938
First underwater gravity measurements	c1939
Howell and Frosch published first account of continuous borehole radioactivity logging	1939
Gamma-ray log first used in oil wells	1939
Instrumental prospecting for uranium began by detecting gamma radiation	1940 - 1944
Magnetometer (fluxgate) became airborne in blimp for location of submarines	1940
Laboratory radioactive detector began to be modified for field work	1944 - 1950

(Table 5.3 continued)

First airborne magnetic survey using fluxgate magnetometer	1944
NMR discovered, leading to proton precession magnetometer by 1954	1946
Original Worden gravimeter appeared	1948
First 3-component spinning-coil magnetometer	1948
Induction well log first used	1948
Renewed work on induced polarisation IP method began	1948
Multiple shot and multiple geophones became popular	1950
Some of the earliest scintillation counter surveys made in Saskatchewan	1950
Back scatter gamma radiation logger introduced for formation density measurement in boreholes	1950
Continuous velocity logging introduced to assist interpretation of seismic methods	1951
MIT study began to use information theory and computer processing to extract information from seismic signals	1953
Geophysical Exploration Journal began	1953
Nuclear resonance magnetometer developed by Varian and Packard	1954
Nuclear resonance (proton) magnetometers first used in airborne prospecting operations	1955
Enhanced sensitivity magnetic recording enabled seismic survey using dropped weight excitation	1955
Magnetic recording entered instrumentation	1955
Optical pumping first observed leading to alkali-vapour magnetometer (Bell and Bloom)	1957
International Geophysical Year	1957-1958
Induced polarisation method came into accepted use	*c*1958
Slingram and Ronka EM became established	1958
Ward introduced audio-frequency magnetic field AFMAG method	1959
Digital recording introduced in seismic work	1960s

Radioactivity was discovered in 1896 and by the first decade of the 1900s bore holes were being logged to determine RA intensity with depth.

As has already been discussed earlier, meteorological measurements began prior to the 19th century but it was in this period that concerted programmes were set up to gather reliable data in an attempt to fathom out, in a systematic way, what causes the behaviour of the climate. Multhauf (1961*b*) discusses the period when the US Army Signal Corps took over the burden of official meteorology for that country in 1870. As well as the instruments being given self-registration they also often incorporated telemetry systems for transmission of the data to some other location. An example is the 'tele-thermograph' by Jules Richard, Knowles-Middleton (1969*a*).

Public utility control of water resources led to the need for hydrological measurements to be made. The design of bridges, dams and culverts needed stream data. Litigations have been in existence over the rights to use natural water since the earliest times.

Aspects of the history of stream-gauging and the equipment used, are to be found in papers, Frazier (1964, 1967), on William Gunn Price and Daniel Farrand Henry. The current meters used by them were mainly of the cup, or propellor, design in which the rotations of the stream driven spindle were transmitted to shore or the boat as electrical impulses that were counted to integrate the flow. One design, Fig. 5.58, used an acoustic listening tube that provided a sharp sound every ten revolutions enabling the operator to time the rate. To give better resolution at low stream flows later Price models incorporated a second electrical penta-contact that made a circuit five times per revolution. Price was critical of the existing Herschel designs which were not practical enough for typical stream conditions of working. The problem with stream gauging instrument design, and still is, not that of devising a suitable principle but a case of implementing it in a manner that suits silting, weed fouling and inaccuracies arising due to bearing wear and slippage.

Henry's design grew out of the first really successful wind gauge – the Robinson 1841 anemometer in which a small windmill was turned by the wind so as to move at the same speed as the wind, thereby, giving a direct form of measurement. Not being satisfied with the instrumental methods of his time, Henry built his own version of a stream flow meter that was largely based on the Robinson design. It was made, in fact, by combining parts of an actual Robinson anemometer instrument with a Morse telegraph sounder, another instance of the influence that the availability of telegraph instruments had on the fields of measurement.

According to Frazier the first person to use the electrical contact method in a flow meter was Charles Ritter of Paris who used a galvanometer, the needle of which flickered each time the contact was made. That was in 1859. His apparatus worked well in the laboratory but failed to be useful in the field, producing more noise flickers than signal events.

Clemens Herschel, graduate of Harvard University in 1860, was a notable water engineer in the United States. He invented the venturi flow meter reporting it in 1888, Kent (1912). This development arose in 1879 when, at Holyoke he needed to monitor flow in half-metre diameter and larger pipelines. He first used an orifice-plate method

Fig. 5.57 *Threlfall gravity meter built in Sydney, Australia 1898*
[Lent to Science Museum, London]

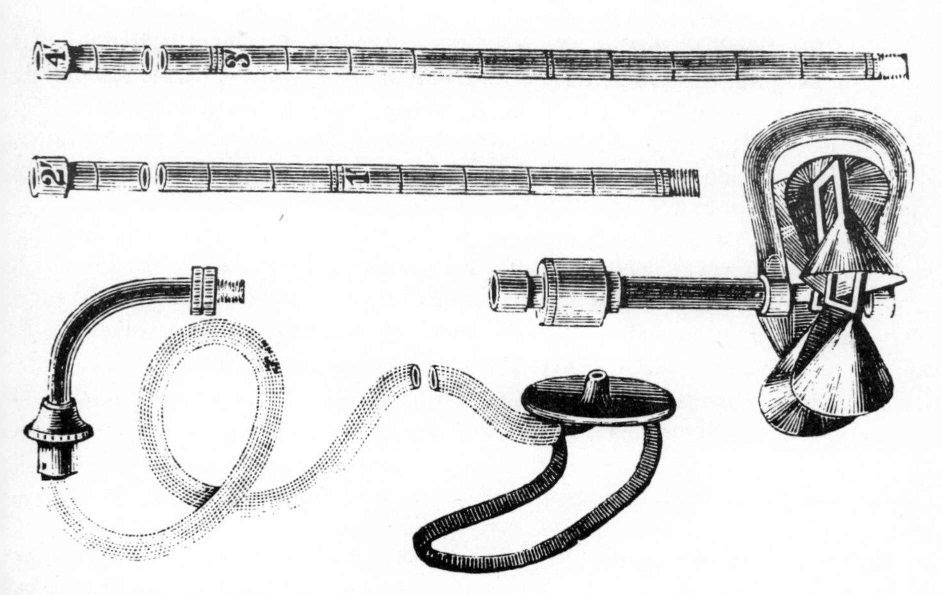

Fig. 5.58 *Price acoustic water current meter c1880*
[Kent (1895), p.62]

but later added a venturi section in his experimental test facility. He wrote 'the meter was by me named the 'Venturi' meter out of respect for Venturi's experiments of 1797'. Figures available for the period to 1912 state that the largest venturi meters then made were the two installed at Divi, India. They had nominally 3m bores and were 22m long. The venturi meter was introduced into Britain by the George Kent firm being included in the 1895 catalogue of that firm, Kent (1895).

Transducer effects

The discovery of electrical laws and the relationship with other energy regimes brought with its main paths of development numerous other discoveries of cross relationships that would be eventually turned into measuring tools or components to form subcomponents in subsequent measurement systems.

In 1835 Roschenschold discovered that certain solids only conducted electricity in one direction. Faraday discovered, in 1830, the magnetohydrodynamic effect upon which electromagnetic flow-meters could be modelled. Magnetostriction, the effect in which the length of suitable materials such as nickel shorten in the presence of magnetic field, was first reported by Joule in *Philosophical magazine*, 1847. It is a reversable effect in that mechanical strain will induce field strength changes which in turn can generate electricity.

In 1850 Faraday discovered that there existed – he used silver sulphide – a material that has a large negative temperature coefficient of resistance. Others were found afterwards. He had discovered the 'thermistor' but that component did not become practical reality until the second decade of the 20th century, Dummer (1977).

Kelvin first described the use of a resistance as a sensor of strain in 1856 but it was not put in use until the 1930s. Hall, in England, in 1879, was the person responsible for enunciating the Hall effect in which a voltage that is a linear function of field strength, is generated by a steady magnetic field in a device in which current flows. This has become a most useful measuring tool for the investigation of magnetic field in generator and motor devices.

Another valuable transducer principle is the piezoelectric effect. This was discovered by the Curies in 1880. It was further studied by Rieke and by Voigt in 1892. Early research of that time was able to establish the link between this and the previously discovered pyroelectric effect exhibited by tourmaline crystals. Later work came to the conclusion that the thermal contribution was due to mechanical deformation caused by thermal expansion, Glazebrook (1922). Piezoelectric crystals had become routine transducer elements by 1919.

Measurement in the inexact sciences

During the last decades of the 19th century it became apparent to a number of researchers working in the so-called 'inexact' sciences – those that were still working more in a qualitative stage of understanding – that the adoption of numerical techniques could provide a more satisfactory understanding of their field.

In 1877 Galton constructed a demonstration device that showed how a large number of random individual events will fall into a well ordered form of distribution, the 'normal distribution' or as he called it his 'binomial apparatus'. The Birmingham Museum, England, has a demonstration model of Galton apparatus, which is described in their Museograph series of pamphlets.

Galton made a significant contribution in bringing about quantification in the life-sciences through the anthropometric laboratory he established, Fig. 5.59, at the South Kensington Museum as part of the 1884 International Health Exhibition held there.

Fig. 5.59 *Part of Galton's anthropometric laboratory, South Kensington, London 1890*

[Eames and Eames (1973), p.28. From plates of Vol.11 The life, letters and labours of Francis Galton K. Pearson. Courtesy Cambridge University Press, 1930]

The laboratory was installed to capitalise on the presence of large numbers of people who visited the scientific and other collections. For a small charge (four pence to be registered) people could be measured for their physical characteristics and personal capabilities. Typical parameters measured were eyesight, judgement of the eye, hearing, breathing power, swiftness, strength and physical dimensions. Over four years 9000 people were 'quantified', an incentive for them to be assessed being that this process might assist them establish if they were abnormal in some way so that remedial treatment could be carried out. Galton studied the results for 10 years, his analysis methods being corner stones of statistics, Eames and Eames (1973), Howarth (1931).

It has been shown, in the study of registration devices given in Section 5.4, that the psychologists were another group to adopt more exact scientific methods in this period, Boring (1929). The education researchers were also trending in this direction by 1900: the French psychologist Binet, with Simon, established a numerical scale of intelligence in the first decade of the 20th century, Ross (1941). Many other events were happening that were then apparently unrelated to measuring instrumentation. In 1847 George Boole published his book *An investigation into the laws of thought*. In it he expounded a system giving rigorous foundation for a method of symbolising logical thought processes. This work, plus that of de Morgan, Jevons, Peirce (who wrote down the first known suggestion on how to use series and parallel switches to form AND and OR gates with circuitry), Russell and Whitehead and others gradually set up the background conditions enabling the emergence of the electrical digital computer in the 1930s.

The 19th century, then, was indeed a period of great inventiveness and utilisation of devices that would help man to learn and to control. To that period is owed a debt to the tremendous amount of pioneering work conducted, upon which rests a considerable number of the transducers, controllers and instrumentation that are currently used.

Chapter 6

The first half of the 20th century: 1900-1950

6.1 Data galore

At the opening of the 20th century the scene was set for dramatic increase in the application of measuring instrumentation. As has been shown in previous chapters, numerous measuring devices had been developed out of discovered physical principles. Electrical method had a firm foundation by 1900. Recording devices having high data-capture rates were in existence and so on.

But there were three vital missing elements. The first was the lack of some kind of convenient method to add gain to feeble signal levels available from the sensor effects. There was also need to be able to convert a.c. electrical signals into the d.c. form by rectification, this initially being needed primarily in the detection of radio-frequency signals. The third missing element was improved means to generate a.c. signals, especially at the higher frequencies as were then being exploited for radio transmission.

The first half of the 20th century grew into one of tremendous application and development of Victorian era and earlier findings. It was also one of still more discovery and invention of ideas that would become the foundation of yet more measuring devices. The account of the Bell Laboratories, Mabon (1975), provides a reasonably short and well illustrated example of the quickening of pace and expansion of ideas.

Disciplines became complex in their structure and knowledge content. The specialist developed as the person who had sufficient in-depth knowledge to advance a particular study further. The significant discipline to emerge was that of electronics for it was the invention of the thermionic 'valve' device that provided the technological solution to the needs for gain, detection and signal generation occurring virtually all at one time. The applications impact of electronic technique firmly established it as the means that designers would strive to use for the generally low energy level (there were exceptions) information or control situation. Chestnut (1962) is a study of the interaction of control and electronics in the period of this Chapter. Bennett (1979) is a companion volume to this work dealing with the history of control engineering which reached theoretical maturity in this period.

Ready availability of general purpose gain in the form of the 'valve' encouraged emphasis on the use of electrical forms of measuring and data-processing devices. Sensors, previously too unproductive in output power level, became useable by the addition of an amplifier. Signals could be more easily processed. Greater demands for commercial forms of apparatus brought greatly reduced costs which, in turn, boosted the market supply situation. Application became easier and more widespread for these reasons. Electronics also caught on as an amateur pursuit in areas of radio communication and television.

Complexity of the principles involved rose markedly to the point where even the experts needed to be careful about assessment of the true value of reports. The classic tale is that of the discovery of the so-called *N*-rays. This was the term coined in 1903 by a French professor who reported the existence of rays akin to X-rays but having different penetrative powers. For a year these rays were the subject of fierce scientific debate and profuse experimentation, Firth (1969). Eventually they were laid to rest as being nonexistent (in that first defined form) owing to carefully devised measurements – which also included a certain degree of surreptitious manipulation of the apparatus, unbeknown to their would be mentor.

The initial development of thermionic 'valve' devices can be firmly established as the result of commercial needs of the teletechnique industry, especially the sectors of long distance working, submarine cable work and radio. Without these financial incentives it is possible that electronics would have developed more slowly. Application aspects spurred development of the discipline of electronics. Stripped of input and output interfacing devices – the transducers of measurement and actuation – electronics has little useful purpose for the techniques inherently involved do not relate to the existent pragmatic world.

Computing, and other mathematical operation machines, gradually improved in data-processing power and speed as bandwidth was increased. Internal electronic bandwidth capability demanded improved response from peripherals used to feed in and out of the mathematical equipment. Data rates rose dramatically to rival some aspects of the physiological brain by the end of the half century.

No longer was a simple static understanding of a system of measurement satisfactory for many instruments rarely settled to the more easily handled steady-state condition. This increased demand for knowledge of the transient state, the dynamic regime of systems.

Measurement systems grew to such an extent that they produced data, as numbers and graphs, at rates not dreamed of previously. But this ability to capture information was not matched by a corresponding improvement in the understanding of what happens during the act of measuring; very little new knowledge was added about the meaning aspects of measuring. Information theory, signal theory, information science and other subjects were developed to assist designers create efficient and satisfactory data transfer channels but very little of consequence was added in the area of measurement meaning. Isolated bursts of work arose spasmodically with papers and books of significance appearing in the 1930s, the 1950s and again in current times.

In the mean-time gigantic quantities of data have been taken and recorded about almost every conceivable topic. But all too much comprises merely meaningless numbers and squiggles.

Contribution to this important factor of measurement has gradually appeared in those fringe-science disciplines where its practitioners could not simply plug-in a transducer, thereby easily obtaining the often dangerous appearance of having hard 'truthful' data. These workers had to think out what they were doing and in the process came to think hard about the meaning of data obtained.

This was the period in which purely mechanical instrument solutions gradually gave way to electromechanical alternatives which then succumbed to electronic methods wherever possible. Ability with fine mechanisms reached a peak of manufacturing accuracy in the form of mechanical computers such as those devices used to solve equations in the laboratory and those mechanisms used in conjunction with gunnery control systems of the 1940s.

The first swing to electronic method was based on the use of analogue signal methods but as linear signal methods were developing so were the practical building blocks and philosophies of a digital alternative that was to burst forth to replace many analogue processes. In the pursuit of application of electronic method it was too often the case that the alternatives were overlooked.

Easy availability of relatively cheap and reliable instrument system modules and the widening applications for measuring and controlling equipment brought about the concept of extensive systems design and management. Instrumentation steadily was applied to an ever widening range of applications. Theoretical methods were needed to gain a proper scientific understanding of how to manage and tune such expansive set-ups. Fig. 6.1 shows a comparison between two aircraft instrumentation systems just two decades apart. Control theory enabled feedback systems to be designed by scientific procedures.

Instrument manufacturing firms emerged at the end of the Victorian era as highly competitive groups that began to play a key part in the application of instruments. They greatly decided, by their sales decisions, which instruments would be marketed and, therefore, would be available. They also established the tradition of contributing a considerable degree of know-how as part of sales service. Today it is still accepted that instrument firms must supply a consulting service as part of any negotiation for a sale.

As more commercial products were marketed the demand for one-offs production of commonly used measuring instruments declined, except in the areas of research and development.. Standardisation did not arise easily and today there still exists considerable inefficiency in the instrument industry because of the existence of too many options and too little proper training of persons specifying purchases and designing instrument systems. The range of available different transducers now numbers in the millions, many being for the same variable and having, in reality, very small differences in specification description.

Electronic aspect of instruments brought incredible cost reductions in terms of data handling power. By 1950 the other regimes of an instrument, notably the mechanical components, had become and were later to become still further, the significant proportion of the total cost. Such trends were just becoming noticeable in the 1950s as electronic assembly methods were improved and mechanised. It seems, however, that few people were able to predict the great advances in the technology that would arise

Fig. 6.1 *Flight-deck instrumentation*
a 1920s light aircraft
[Hewitt (1934), Plate 1, opposite p.48. Courtesy Smiths Industries]

from the invention of the transistor and its subsequent integrated circuit form. Sir George Thomson, in his 1955 book, attempted to see ahead from the period at which this study is terminated . . . the 1950s. About the germanium transistor he said, Thomson (1955):

> A variety of minor advances in communication may be expected quite soon as the result of the discovery of the germanium transistor . . . Its advantage lies in the great reduction in size and weight that it will make possible in electronic gear . . . It is possible that a short range 'walkie-talkie', light enough to be carried regularly, may in the course of a few decades replace a good deal of telephony over wires.

Fig. 6.2, reproduced from Dummer (1977), shows schematically the rise in application of the subsystem blocks used in instrument systems.

For the first time in man's history the whole world went to war two times in this 50-year period. The significance to this study was that both wars were also the first at such large scale and the first to be organised, in a significant way, on scientific principles using technology to attempt to overcome the other side. The effect of these two wars on the application of science and technology following each period is a complex issue but one in which it is clear that the wars had considerable influence over the adoption of scientific ways. This subject was the theme of a one-day discussion meeting held by the Royal Society, London in 1974, Jones (1975). Another series of

Fig. 6.1 *Flight-deck instrumentation*

b Flight Engineer's station on 1950s four-engine transport; the pilot and co-pilot have still more instruments

[Ward Lock (1955), p.125. Courtesy Lockheed Corporation]

documents related to this question are the reports of the National Bureau of Standards on the NBS work of each war period, Dept. of Commerce (1921), Briggs (1949), Cochrane (1966). Trewman (1949) puts the case that the applications of electronics in the Second World War showed, and provided the reasons for using electronics more widely, as was the case in post 1945 years.

As would be expected from the study of all previous periods in history this half-century was also one in which still more transducer effects were discovered as scientists and engineers were able to 'enter new domains of existence' because of the availability of new knowledge about measuring devices and systems and new hardware.

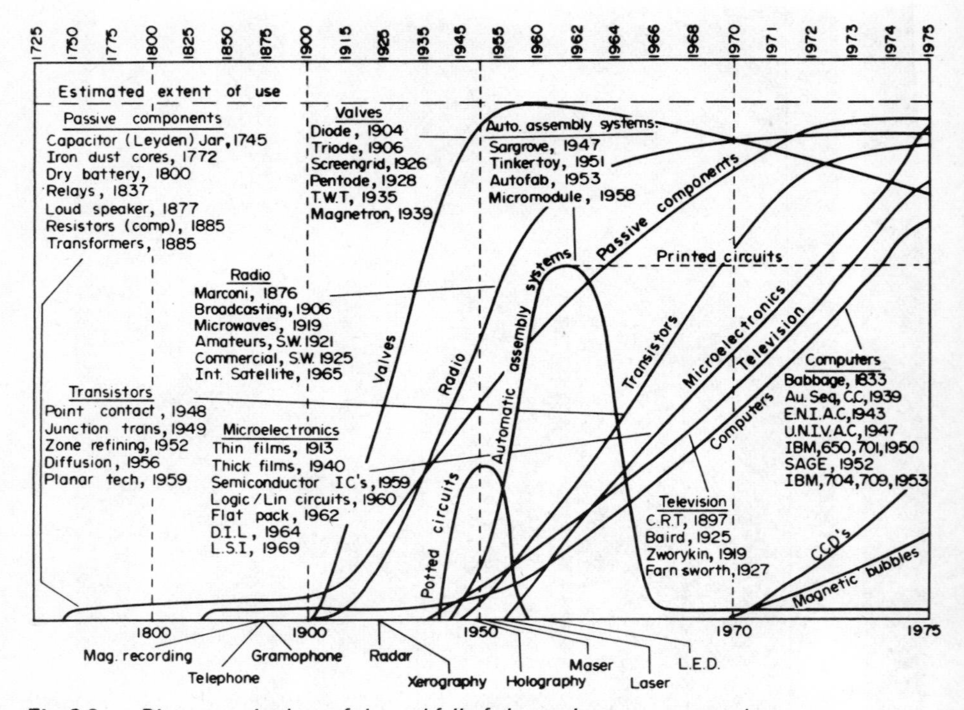

Fig. 6.2 *Diagrammatic chart of rise and fall of electronic components and systems.* [Dummer (1977), Fig. 6, p.8. Courtesy Pergamon Press]

Optical methods continued to be important the significant new discovery being the concept of holography that had to wait until the 1960s for a suitable radiation source by which it could be demonstrated. For a short period optical data processing was the advantageous method for processing optical images but these generally gave way to the use of digital computing methods by the 1960s. Military interests in radiation detection gave a boost to the use of infra-red and other wavelength sensing. I-R found application, Jones (1972), in military needs during the First World War as work of the Germans in detecting British torpedo boats. This stimulated other nations to enter the chase to gain greater detection capability than the potential and real enemies had. Military science has become greatly dependent on this theme from that time.

In all, the first half of the 20th century is now just about old enough to be commented upon as history. It was one in which it is thought the workers of previous centuries had been paving the way for the emergence of vast deployment of instruments for data capture, but one for which man was not quite ready owing to a lack of proper understanding about the philosophical nature of meaning in measurements.

Several facets mentioned above are now considered.

6.2 Electronic technique

Originally, the discipline of electronics was associated with the thermionic electron

Table 6.1 *Devices, components and principles of electronic technique*

Dates given generally relate to the date at which introduction of the topic was clearly evidenced. Cited references usually will give further sources wherein more detail may be found. Some dates are indicative only; for those further research appears necessary to provide a more positive date. Tables 2.1, 2.2 and 2.4 and appropriate biographies; see **Appendix 1**, may also prove useful as sources of dates.

Before 1900

The following were each available at the turn of the century but, in general, they continued to be improved in performance, cost, specification and understanding.

Electromechanical relay
Phase-sensitive detection
Capacitor (condensor)
Inductor
Cathode-ray tube
Filament lamp
Evacuated tube, diode effect
Battery
Multiplexing
Tuned circuit
Transformer
Resistor
N.T.C. resistance material
Vacuum photocell
Amplitude modulation
Loudspeaker
Thin films by sputtering
Iron-dust cores

Chronology of the first half of the 20th century

Device, component principle	Persons associated	Date	Further detail
Secondary emission discovered		1900	Summer (1957)
Vacuum photoelectric cell	P. Lenard	1902	Summer (1957)
Frequency modulation (invented)	C. D. Ehret	1902	Tucker (1970)
Electromechanical repeater	Shreeves	1904	
Oxide-coated cathode	Wehnelt	1904	Summer (1957)
Radar (demonstrated)	Hueslmeyer	1904	**Panofsky (1970)**
Thermionic vacuum diode	J.A. Fleming, R. Lieben (independently)	1904	Dummer (1977) Larsen (1960)
Triode as a.c. amplifier	L. de Forest	1906	Dummer (1977)
Ultraviolet photodetector	Elster and Geitel	1906	(Consult Deutsche Museum staff)
Crystal detectors came into use	Dunwoody; Pickard	1906	Knox (1917)
C.R.O. tube as image reconstruction device	B. Rosing	1907	Dummer (1977)
Alternating current thermionic amplifier (as detector)	de Forest patent filed	1907	Kloeffler (1949)
A.C. amplifiers in common use		1908	
Bandpass filter for A.M.	G.A. Campbell	1909	Heising (1962)
High-frequency 'valve' generator	Bronck	1911	
Heterodyne detection	H.M. Fessenden, E.H. Armstrong	1912	Dummer (1977)

(Table 6.1 continued)

Regenerative radio circuit	L. de Forest, E.H. Armstrong, I. Langmuir and A. Meissner	1912	Dummer (1977)
Improved triode	H.D. Arnold	1912	Swinyard (1962)
Thin-film techniques	W.F.G. Swann	1913	Dummer (1977)
High gain a.c. amplifier (20 000: Mercury vapour triode)	Lieben-Reisz	1913	
Triode oscillator	L. de Forest	1914	Herold (1962)
Thyratron	I. Langmuir	1914	Dummer (1977)
Repeater valves in use		1915	Herold (1962)
Crystal rectifier peak voltmeter		1916	Sinclair (1962)
Crystal pulling	J. Czochralski	1917	Dummer (1977)
Multivibrator	H. Abraham and E. Bloch	1918	Dummer (1977)
Shot-effect noise	W. Schottky	1918	Dummer (1977)
Flip-flop (bistable)	Eccles and Jordan	1919	Dummer (1977)
Thallium sulphide, light-dependent-resistor	T. W. Case	1920	Summer (1957)
Electron multiplier principle	Slepian	1920	Summer (1957)
Copper-oxide rectifier	Grondahl	1920	Kloeffler (1949)
Crystal detectors fall out of use owing to existence of thermionic forms		1920	Kloeffler (1949)
Quartz crystal oscillator	W.G. Cady	1921	Dummer (1977)
Resonator principle of Magnetron	A.W. Hull	1921	Dummer (1977)
Telephony thermionic repeater	van Kesteren	*c*1921	Herbert (1923)
Probable first vacuum tube voltmeter	Moullin (Cambridge University)	1922	
Battery eliminator rectifier pack	P.D. Lowell and F.W. Dunmore	1922	Cochrane (1966) Risdon (1924)
Theory of F.M. side bands published	J.R. Carson	1922	Heising (1962)
Iconoscope (t.v. camera tube)	V.K. Zworykin	1923	Dummer (1977)
Radar (first accepted)	E. Appleton, G. Briet, R.A. Watson-Watt	1924	Dummer (1977) Page (1962) Rowland (1963)
Sawtooth time-base	R. Anson	1924	Dummer (1977)
Johnson noise	J.B. Johnson	1925	Dummer (1977)
Insulated gate, field-effect, transistor (proposal)	J.E. Lilienfeld	1926	Durrant (1970)
Screen grid in valve	H.J. Round	1926	Dummer (1977)
Automatic gain control (volume)	H.A. Wheeler	1926	Dummer (1977)
Integrated, multistage, active device (thermionic)	Loewe	1926	
Negative feedback (electronic amplifiers)	H.S. Black	1927	Dummer (1977)
Pentode	Telegen and Holst	1928	Dummer (1977)
Geiger-Muller tube	H. Geiger and W. Muller	1928	Summer (1957)
Standard signal generators		1928	Sinclair (1962)
D.C. amplification (chopped input)	A.J. Williams	1929	Williams (1973)
Crystal rectifiers (research resumed)		1930s	Kloeffler (1949)
Scaling circuit (nuclear counting)		*c*1930	Dummer (1977)
Ignitron, mercury-arc, rectifier	Westinghouse, USA	1933	Dummer (1977)
Frequency modulation (promotion)	E.H. Armstrong	1933	Dummer (1977)

(Table 6.1 continued)

Image dissector tube	P.T. Farnsworth	1934	Summer (1957)
Liquid crystals	J. Dreyer	1934	Dummer (1977)
Field-effect transistor patented	O. Heil	1935	Dummer (1977)
Photomultiplier tube	V.K. Zworykin	1935	Dummer (1977) Summer (1957)
Cold cathode trigger tube	Bell Laboratories	1936	Dummer (1977)
Close-spaced valves introduced		1936	Herold (1962)
Long-tailed pair	A.D. Blumlein	1936	Dummer (1977)
Vacuum tube voltmeter	S. Ballantine	1938	Sinclair (1962)
Magnetron	J.T. Randall and H.A.H. Boot	1939	Dummer (1977)
Klystron	R.H. and S.F. Varian	1939	Dummer (1977)
Miniaturisation of active components (thermionic)	NBS and others	1940s	Cochrane (1966)
Thick-film circuits	Centralab.	1940	Dummer (1977)
Amplidyne amplifier	E.F.W. Alexanderson, M.A. Edwards and K.K. Bowman	*c*1940	Kloeffler (1949)
Miller integrator circuit	A.D. Blumlein	1942	Dummer (1977)
Printed wiring	P. Eisler	1943	Dummer (1977)
Potted circuits	in UK and USA	1945–1950	Dummer (1977)
B.C.D. counting stage with binary display	J.T. Potter	1944	Oliver (1962)
B.C.D. counting stage with decimal display	I.E. Grosdoff	1946	Dummer (1977)
Permeability tuning replaced capacitative method		*c*1946	Swinyard (1962)
Silicon, light-sensitive, transistor	G.K. Teal, J.R. Fisher and A.W. Treptow	1946	Summer (1957)
Automatic manufacture of electronic circuit system	J.A. Sargrove	1947	Dummer (1977) Cochrane (1966)
High performance a.c. amplifier	D.T.N. Williamson	1947	Dummer (1977)
Field effect transistor (practical proof)	W. Shockley and G.L. Pearson	1948	Durrant (1970)
Point contact transistor	J. Bardeen, W.H. Brattain and W. Shockley	1948	Dummer (1977) Kelly (1953) Brotherton (1972)
Mercury delay line (in SEAC)	NBS	*c*1949	Mason (1977)
Junction transistor	Bell Laboratories	1949	Dummer (1977)
Cold cathode stepping tube	Remington-Rand	1949	Dummer (1977)
Commercial digital counter timer marketed		1950	Oliver (1962)
Integrated active circuitry, solid state (Concept)	G.W.A. Dummer	1952	Dummer (1977)
Chopper stabilised feedback op-amp	E.A. Goldberg	1950	Goldberg (1950)
End of valve era in computers		*c*1958	Bell (1978)
Commercial digital voltmeter marketed		1954	Oliver (1962)

valve, or tube, with which a vast range of information and power handling capability developed. Since then the boundaries of electronics, as such, have become very blurred indeed.

The valve era started in the 1910s and began to wane in the middle fifties with the successful introduction of solid-state circuitry alternatives that offered greatly superior benefits to construct the awaited systems that the existence of valve technology had shown to be feasible and highly useful yet not quite economic.

In some ways the arrival of the thermionic devices delayed the emergence of solid-state devices by several decades. The existence of valve diodes clearly altered the path of those interested in the operation of the point-contact, 'cats-whisker' diodes, Fig.6.3, used in the early stages as the detector of crystal sets. These were the forerunner of solid-state diodes, as they became known later, when adopted in a permanent 'contact' form, Torrey and Whitmer (1948).

Valves, however, did give practitioners means to implement and gradually develop many of the basic building blocks that are today used to implement electronic systems.

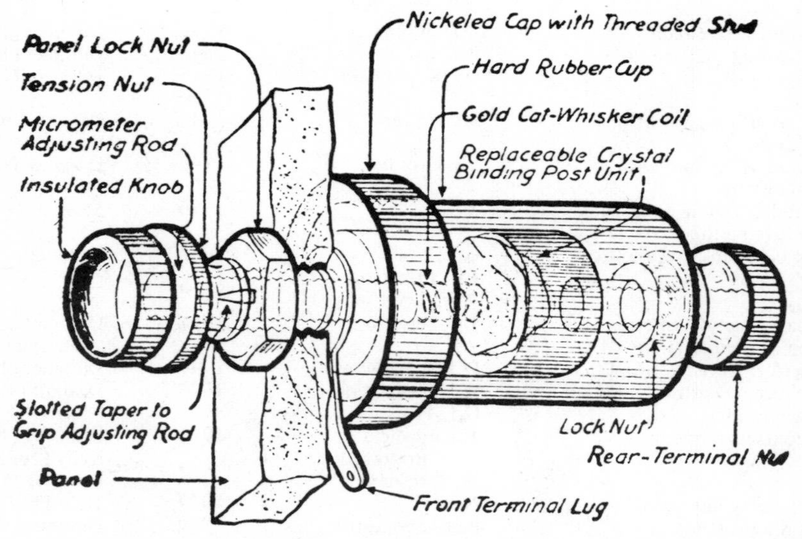

Fig. 6.3 *Construction of 1920s cats-whisker crystal detector. This unit used Foote-brand galena or pyrite crystals contacted with 14 points on a gold 'whisker'* [*Wireless Weekly*, December 5, 1924 p.7]

Too many individual developments occurred for each to be discussed in detail here. Table 6.1 has been prepared to portray the situation and show the trends over the 50 years concerned. It has been, in the main, compiled from the chronological work published by Dummer (1977), with additions from Summer (1957) plus others from works cited. It is quite likely that some dates will be proven incorrect as more research is expended.

The introduction of the two-element thermionic device (diode) by Fleming, in 1904, and simultaneously in a different form by Leiben, began the electronic age.

Their work, however, was of course based on that of many before them. Electronics did not appear suddenly as a group of concepts, devices and practices but emerged from a gestation period to provide, in clearly manifested ways, the badly needed gain, detection and signal generation that had been sought to assist the maturity of radio and other teletechnique applications. These three subjects are given special attention later. The traceable prior invention, although not the one that had impact, was Edison's work, Fig. 6.4, on the blackening of his electric lamps. Out of this he patented the 'Edison-effect' in 1883.

Fig. 6.4 *An Edison electric lamp 1880*
[De Natuur, 1882, p.49]

General histories of the emergence of the electronics discipline and industries can be found in the special issue of the Institution of Radio and Electrical Engineers, IRE (1962). It was concerned with communications and electronics of the period from 1912-1962. In that series of papers, White (1962) provides an account of the early times of industrial electronics that emerged as another aspect to communications from around 1930. Another special issue, IEEE (1976), also has some useful papers about this period. Papers of relevance to this study from those sources are individually referenced here but there may be others not mentioned that are relevant, depending upon the historical aspect of interest. HMSO (1977) includes a chronology of relevance.

The first texts to be published were concerned with the thermionic devices, examples being Lodge (1910) on electrons and Appleton (1932) specifically on valves. A chapter in Glazebrook (1922), Volume II, includes a key bibliography of early papers. Herbert (1923), on telephony, includes a final chapter about the thermionic triode; it was written around contemporary papers reporting Post Office use of triodes. It is clear from the preface of that work that the valve was seen by that time as a most important device in the future of telephony.

Gibson (1915), however, apparently saw no reason to add comment about valves in the 1915 revision of his work originally published in 1907. Hibbert (1910) also has no mention of valves. Knox (1917), a reprint of the 1914 edition, mentions the crystal detector but an illustration of 'a ship's wireless cabin, latest type' shows a passive component radio apparatus. It is apparent that the introduction of valves in radio equipment had not generally filtered through to the knowledge of the popular technical writers until the 1930 period.

The considerably high cost of the valves was a severe drawback to widespread amateur application in the initial years. Fleming, when introducing a work on wireless published in 1924, Risdon (1924), mentioned his work on the two-element diode of 1904 (several units are shown in Fig. 6.5) and how it had been adapted to become the triode that enabled signal carrier to be generated and very sensitive radio receivers to be built. Fig. 6.6 is of a triode used by the Post Office in England in the 1910 decade. He included a picture of a bank of Marconi transmitting valves having 80kW capability. Risdon included chapters on the 'Fleming thermionic valve' on crystal receivers and the detectors themselves, on valve circuits, on how valves were made and many illustrations and accounts of various applications of valves. The 1920s saw the introduction of the popular magazines on wireless. These provided many designs for home construction using valves and they were an important factor in helping to develop the 'amateur radio' following that had grown out of such factors as training given to military personnel in the 1914-1918 war. HMSO (1977) provides information on the development of the radio receiver.

The history of the vacuum tube has now been well researched. The Institution of Electrical Engineers published a series of papers on these devices, IEE (1955). More recently has appeared the work of Tyne (1977). Herold (1962) considers the effect that valves had on receiver design. A vast collection of valves is held in the Studieverzameling Elektrotechniek at the University in Delft, Netherlands. The chronology of HMSO (1977) lists some key events in valve development.

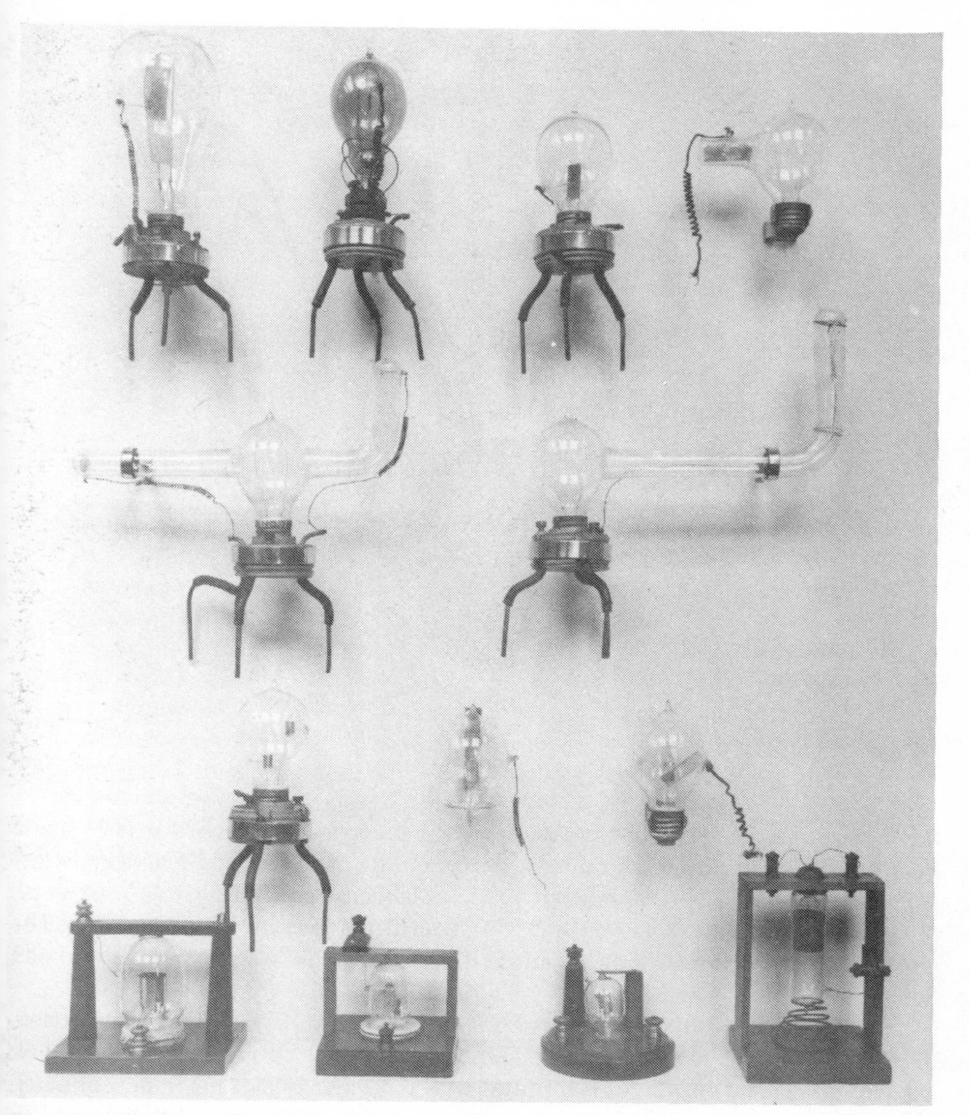

Fig. 6.5 *Valves constructed for Fleming's experiments, c1904. Construction is based on incandescent lamps of the day*
[Crown copyright. Science Museum, London]

Electronics gradually found its way into other than communication uses. No longer were thermionic devices to be called 'radio-tubes'. By the 1930s measurement and control applications were in common use under the general name of industrial electronics. Golding (1932) in an extensive account of the general use of electronics in instrumentation in its first decade of this kind of application. Included are chapters on the photoelectric eye, on the automatic age, automatic traffic control, medical electrical uses and measuring and controlling instruments. White (1962) traces

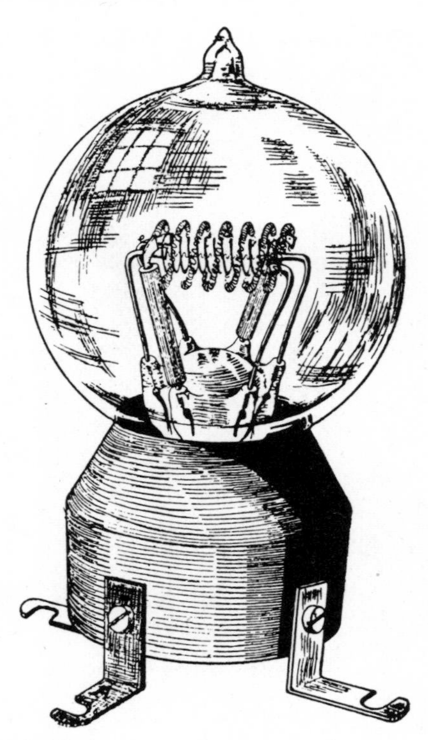

Fig. 6.6 *Triode valve, c1915*
[Herbert (1923), Fig.604, p.819]

this change. Chaffee (1933), McArthur (1936), Morecroft (1936), Koller (1937) and Fink (1938) each provide an impression of the state of the art and practice of electronics in the 1930s. Books relevant to the following decade are Greenwood *et al.* (1948), Kloeffler (1949), Trewman (1949), Benedict (1951) and Mitchell (1951). The growth of the electronics industry in the USA is the subject of Norberg (1976) and Susskind (1976).

The development of the necessary components to go with the active devices to form electronic subsystems are also reasonably well covered in the literature. Darnell (1958) and Brothers (1962) deal with the general field of component parts. Marsten (1962) and Bennett (1964) discuss resistors whilst Podolsky (1962) deals with capacitors. Circuit theory history is the subject of a paper by Belevitch (1962). Heising (1962) deals with modulation methods. Further detail about specific histories can generally be found by use of the sources mentioned in the citations given in Table 6.1. Braun and MacDonald (1978) trace the emergence of semiconductor electronics.

Judging by the number of popular accounts and text-books published it appears that the marriage of electronic gain to the photoelectric cell, in the late 1920s, was the first of the noncommunication applications of the valve to emerge in widespread use. Allen (1925) provides an early summary of knowledge about the theoretical aspect of photocells. Three works that followed shortly after were Campbell and Ritchie (1929),

Walker and Lance (1933) and then Zworykin and Wilson (1934). These books contain considerable account of practical experience and reported the use of photoelectric cells in such uses as newspaper counting, one-way counting of objects, race track timing, speeds of shafts, operator safety guard beams, gas detection, smoke detection, neon-tube modulating advertising equipment, proximity sensors, light-controlled model vehicles, talking-film sound tracks, facsimile transmission (also called photo-telegraphy), television, microphotometer, ultraviolet dosage meter, colour matching, pyrometry, an infra-red direction finder and more. The electronic amplifiers used were generally very simple using only one stage of amplification. Golding (1932) gives many illustrations of contemporary photoelectric installations. Fig. 6.7 is of the 1930s facsimile equipment used by the 'Daily Mirror' Newspaper, in London.

Schlesinger and Ramberg (1962) published a review of photoelectric devices includ-

Fig. 6.7 *Bartlane picture transmission system. The electronics can be seen on the vertical panel shown in the top centre*
[Walker and Lance (1933), Fig.75, p.127. Courtesy *Daily Mirror*]

ing an extensive bibliography. Most general works on electronics from around 1920 contained explanations of photoelectric devices.

Electron optics also became an established electronic subject in this period, the first transmission electron microscope being developed in Germany in 1931 and marketed in 1934 by AEG. The scanning form was born in 1935 as the idea of Knoll and von Ardenne. The fundamental concept, that of electrons being used instead of light rays, was realised by H. Busch in 1926 who subsequently devised an electron lens. Knoll and Ruska used magnetic lenses in the first transmission microscope mentioned above. The history of the development of this important measuring and observing instrument has been published by Freundlich (1963). Dummer (1977) also contains some primary references to pioneering publications. By the late 1930s Klemperer (1939) was able to compile a text on electron optics. MacGregor-Morris and Henley (1936) contains material on the electron optical design of cathode ray systems.

Teletechnique, in the form of radio and television, was the subject of many books in the period of interest here. Several histories have been compiled about these early times. On radio of this period there are Cadell and O'Kane (1951), which is about air-radio progress, and Batcher (1962) which covers the influence of product design on radio progress. The latter contains an illustration of the huge electromechanical generators that were used to generate carrier (of 1MW) for Poulsen transmitters in 1915. Fig. 6.8 is a cross-section of an Alexanderson alternator, *c*1920, that produced 200kW

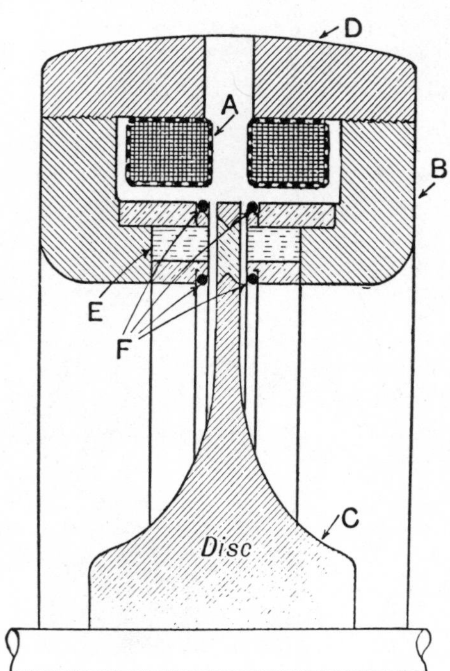

Fig. 6.8 *Generation, in the pre-valve era, of radio-frequency carrier required alternators having exacting mechanical design*
[Glazebrook (1922), Vol.2, Fig. 5, p.1052]

at 30 000 rev/min. A is the exciting winding, F the output windings. It had 300 slots, Glazebrook (1922), Hawks (1929).

On television history there are Bingley (1962), which gives an extensive list of references, Fink (1976) which deals with the political aspects of USA television control and Herold (1976) where colour television displays are studied.

In 1924 the Bureau of Standards in Washington DC, now the NBS, published what became a classic text on radio measurements, Dept. of Commerce (1924). Alternating current was sufficiently important to rate specific texts about its measurements, Owen (1937) being an example.

Important facets were the new availability of active electronic devices, the revival and extension of digital methods and the realisation of fundamental limits in electronic equipment.

Active elements

Physical systems comprise passive and active elements. The former can transform input energy but never into a greater energy level than appears at the input. Active elements are able to increase that signal level to one greater than flows at the device's input, doing this by allowing an additional power source to flow into the output under the control of the lower level signal input. Active elements are able to provide amplification of either the 'through' or 'across' variable, or their product i.e. power. In electrical terms this means that an appropriate electrical form of active element can be used to increase the level of a voltage (across variable), a current (through variable), or the product of the two which is power. Without active amplification properties a system will dissipate its input energy, the signals getting smaller the more they are processed. If an input signal is required to do more work than the energy available some means of supplying additional energy is needed, that is where the active element assists.

Amplification enables many other additional systems effects to be realised. An active element can be used to create oscillating signals, to match impedances whilst providing gain, to produce highly stable low-gain, feedback amplifiers, to form digital function blocks and many more useful effects.

Telegraphy was able to procure the necessary gain with the electromagnetic relay. Weakening electrical digital signals could be restored to desired levels by inserting a relay repeater stage in the signal path. Early radio was just able to function by the correct choice of circuit parameters and signal strengths. Resonant tuned arrangements helped transform variables into more favourable combinations. In these gain was obtained by altering the product of the through and across variables so as to increase the amplitude of one at the expense of the other, the energy product remaining somewhat the same.

In mechanical systems the simple lever also provides gain by a transformation action. Fig. 6.9 is of an 'electrical micrometer' said to have been able to detect movements of 5×10^{-10} m in tests of the magnetostriction properties of metals, Mullineux-Walmsley (1910). Although gain is obtained this is not an active device but an arrange-

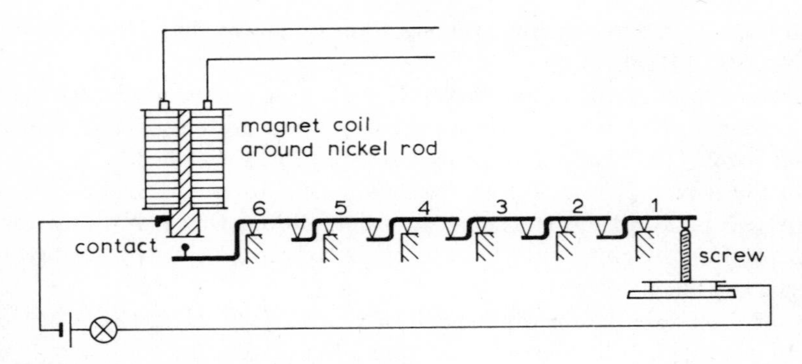

Fig. 6.9 *Electrical micrometer c1890*
[Mullineux-Walmsley (1910), Fig. 272, p.310. Courtesy Cassell Ltd.]

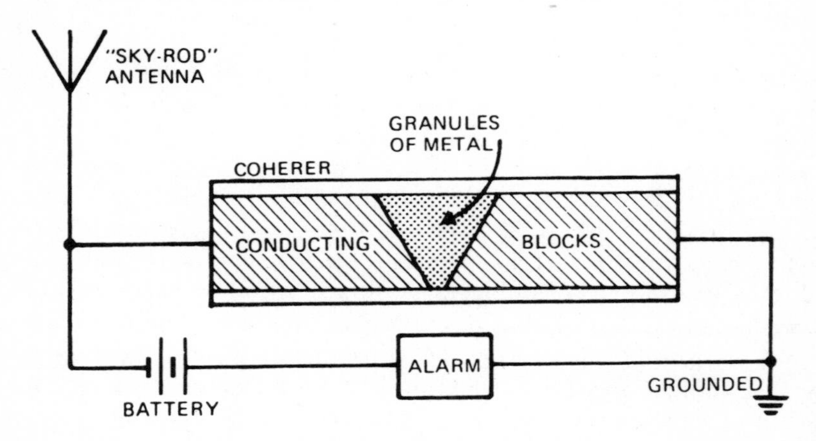

Fig. 6.10 *Typical coherer radio detector circuit. RF energy reduces the resistance of the coherer granules to a low value enabling current to flow in the alarm or recorder. A de-coherer tapper was used to reset the coherer*

ment of passive elements providing transformation.

Many devices were invented in the pursuit of better wireless telegraphy detection, the reason being that the incoming energy level was barely adequate to allow much inefficiency in the receiver. Transmitted code bursts of carrier needed to be detected by some arrangement that would reproduce the digital signal at a considerably greater power level than was available from a reasonable length of antenna wire (microvolts per metre of low impedance wire).

The first successful radio-frequency detector was the coherer, Hibbert (1910), Mullineux-Walmsley (1910), Hawks (1929), Sydenham (1975). These were commonly used, from inception in the 1890s for some two decades more, to form the detection stage of the first radio receivers which were built entirely of passive elements. Coherers were devices whose electrical resistance reduced substantially when an alternating, high-frequency current flowed through them (from megohms to tens of ohms). They came in many forms but were basically all similar. Passage of r.f. current through the

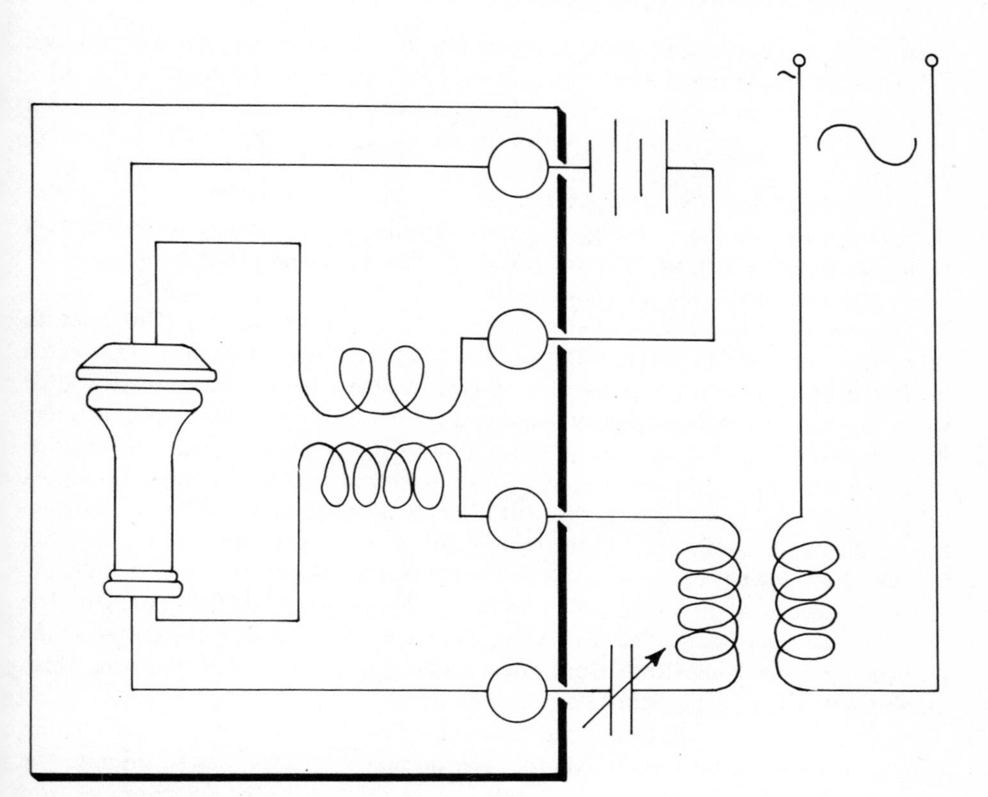

Fig. 6.11 *Kumagen audio-frequency oscillator c1910*
[R.W. Paul Cat. 1914]

element formed a low resistance component that was used to allow greater power to flow into the output sound or recording device. This is shown diagrammatically in Fig. 6.10. Once they had cohered they usually had to be decohered to recover the high-resistance state. The Hughes design, based on a telephone hand set, incorporated spontaneous decohering action. Others often had a small electro-magnetic tapper, an anticoherer, that gently knocked the coherer body shaking the particles again when the resistance went to a low state.

Coherers were active amplifying devices but they were neither sensitive nor uniform in signal level change; nor could they handle analogue, continuous, forms of signal. They were of very limited usefulness, being in a similar class to the electromechanical relay but suited to high-frequency signals instead. Before the invention of the coherer, detection was by visually observing a small spark-gap. Electrolytic cells were also used in a similar way to the solid contact coherers mentioned above.

In telephony, repeaters were made, prior to the introduction of valves, from a telephone receiver unit driving a transmitter element fixed in close proximity. Well known units were the regenerator of S.G. Brown devised around 1899 and the Shreeve units used later from 1904 onward (British Patent 9605 9606/65). The latter used a carbon microphone to obtain power gain. This arrangement of telephone parts was also used to generate audio frequency alternating current from direct current sources. Fig. 6.11, from a 1914 component catalogue of R.W. Paul of London and New York, is of a 'Kumagen' source. It provided a distorted sinewave, adjustable in the range 800-2000 Hz.

Electromechanical repeaters were apparently reasonably satisfactory in performance for they were used on long telephone lines in the USA and by the British Post Office, Herbert (1923). In 1914 New York to San Francisco became connected by direct telephone lines using triode repeaters.

Another detector used was the magnetic detector, Hawks (1929). This came to prominence through its adoption by Marconi in 1902. It was based on 1895 research by Rutherford wherein he found that an electromagnet became more sensitive if it were exposed to radio-frequency energy. Fig. 6.12 shows a developed Marconi detector. Risdon (1924) includes an illustration of the original Marconi experimental unit. The continuous band of iron wires was driven as an endless belt by a clockwork drive. Two horse-shoe shaped permanent magnets provided steady remanent magnetism to the recirculating wire. At the same point as the magnets are two coaxial coils surrounding the wire; one was connected to the antenna the other to the headphones. Under no-signal conditions the wire experienced no change in field strength. The steady unchanging value induced no voltage in the headphones. If r.f. appeared on the receiving coil it, by the Rutherford effect, altered the flux level in the wire which induced a signal in the headphone driving coil.

Magnetic detectors had earlier been investigated by Wilson and Evans in an attempt to trigger torpedos by remote control. The magnetic detector was superior to the coherer but it also did not provide a satisfactory solution to the detection needs of radio. Operators could comprehend Morse code at 150 letters per minute with this improved detector.

Fig. 6.12 *Marconi magnetic detector, 1902*
[Lent to Science Museum, London]

It was also discovered that the addition of bias current to the 'cats whisker' crystal detector made it more sensitive, the threshold of conduction to incoming signals was lowered allowing more signal energy to pass into the headphone stage. Despite this technique and the detectors mentioned above none was adequate enough. The scene was set for some other method to be devised. It was for purposes of better detection that Fleming, in 1904, devised the 'oscillation valve', a thermionic diode, Fig. 6.5. With this device it became possible to rectify alternating, high-frequency current converting it into unsmoothed direct current that could then be integrated over enough carrier cycles to produce reasonably adequate, audio-frequency, signals. Later it was found that it could also be used as an oscillator.

In 1907 the triode was invented by the addition of the control grid between the cathode and the anode (plate) of the diode. Fig. 6.6 is an early example. With this grid it was possible to control the flow of electrons between cathode and anode as a reasonably linear function of the applied grid voltage. Little current was drawn by the grid so the device provided comparatively good active-gain characteristics. De Forest called his triode valves 'Audions'. His 1907 patent document is shown in Fig. 6.13. Fleming (1919) is an account of development and patent history of the diode and triode.

Until 1912 triodes were workable devices but they were far from a consistent product. The vacuum was poor and they were notoriously unreliable. Brothers (1962) states that a de Forest triode had a typical life of 25h and the filament would last only 50h at the most. In 1912 Arnold improved the triode to make it a more acceptable component. He improved the vacuum and introduced the use of the oxide coated, lower temperature, cathode. Further improvements resulted in subsequent years owing to the introduction of better vacuum pumps.

Early amplifier circuits used one triode and the elements were often direct-coupled. The input to the grid was modulated by the signal received from the antenna or from a device such as a photocell. It was soon seen that stages could not satisfactorily be cascaded without decoupling the d.c. levels of each from the other stages. Amplifiers first became developed for a.c. work as that was the dominant need of the early times prior to 1930. Their use in radio apparatus is reported in the chronology given in HMSO (1977).

By the 1930s the three classes of a.c. amplification had become ordered in the text-books. Multiple element and many special-purpose valves were in use – see Fig. 6.14. The problem of direct-current amplification, however, was rarely mentioned.

The d.c. instrumentation amplifier, now known as the operational amplifier or simply op-amp, seems to have begun to emerge around 1920. The earliest reference that could be found in this study was that of Turner who published details of his regenerative, resistance-coupled, two triode amplifier in 1920 . . . see Chaffee (1933). It was unable to provide high gain as it became unstable. He called it a 'kallirotron'. Chaffee (1933) also described work of his own on a d.c. amplifier of two stages that he and other workers used to measure electropotentials from the retina. It was reported in 1923. Stages were connected to the following grid with a separate high tension B battery placed in series with the plate of the previous stage. Each stage was adjusted

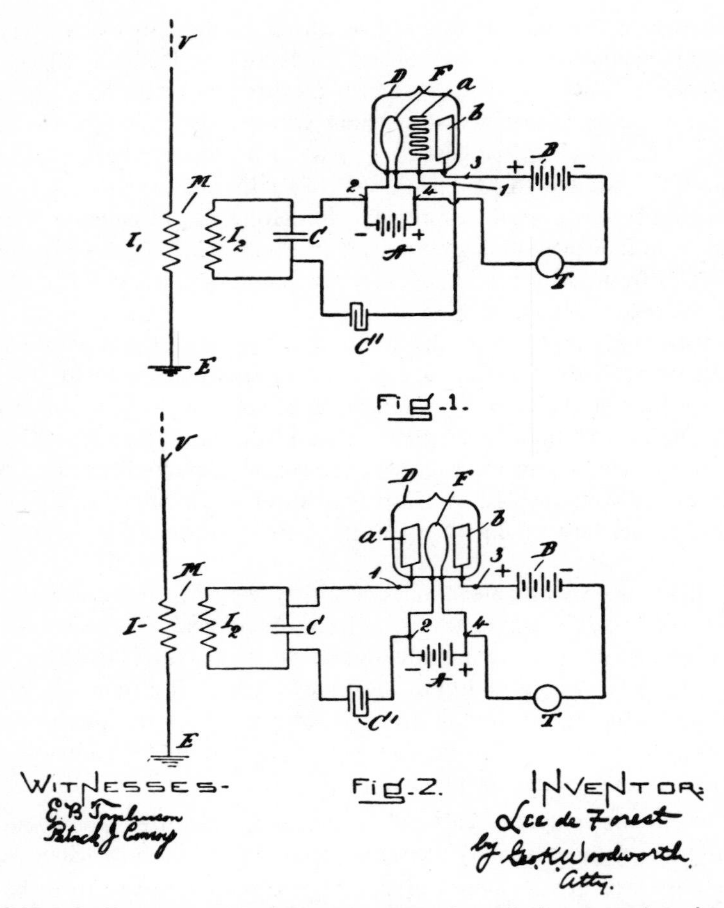

Fig. 6.13 *De Forest patent using 'Audion' triodes for radio frequency detection, 1907* [Kloeffler (1949), frontispiece. Courtesy John Wiley & Sons Ltd.]

to obtain the required output level by a tapping from the battery. Chaffee (1933) discusses the upper frequency limit imposed on d.c. amplifiers by valve input admittance values.

Williams (1973) reviewed his 1929 work in which he devised, Fig. 6.15, a chopper style of d.c. signal amplifying arrangement needed to drive the control motor of the Leeds & Northrup electronic potentiometric recorder that he was developing. Williams also devised an automatic zero stabilising arrangement making use of the fact that the gain was known.

Another form of cascaded d.c. amplifier, that was used by Professor Morecroft, was devised around 1930 to drive a Duddell oscillograph. It used one valve to drive four more connected in parallel as an output stage. The reported gain was 31 at 5 kHz dropping to 19 at 10 kHz. It was able to drive a 500 ohm load giving around 60dB power gain. The design was credited to Lofton and White who reported their circuitry in 1930.

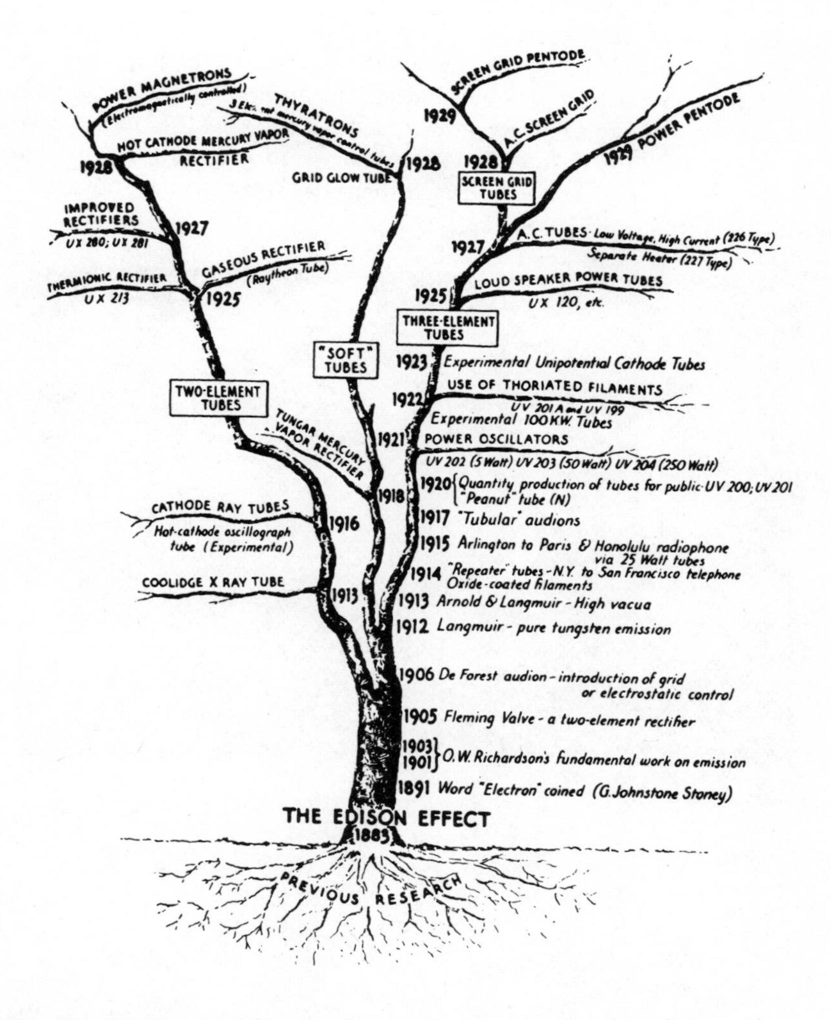

Fig. 6.14 *Family tree of thermionic valves*
[Morecroft (1936), frontispiece. Courtesy *Electronics* of McGraw-Hill Inc.]

Each of the above accounts about d.c. amplifiers mentions the inherent difficulties of drift, adjustment of gain and offset. None of the designs used the long-tail pair differential arrangement. According to Dummer (1977) that was the invention of A.D. Blumlein who constructed it, in 1936, not to obtain d.c. amplification, but to obtain common-mode rejection (as it is now called) for use with video signal cabling for early television in London. Cabling in those times was very susceptable to stray induced noise. A decade later the long-tailed pair was appearing in texts as a form of balanced d.c. amplifier along with a degenerative form in which one half is formed by a resistor divider chain instead of a second identical valve arm of the bridge, so formed. The direct-coupled amplifier continued to be developed, being one of the design problem areas that lingered in the growth of electronics over a considerable period. This slow progress may be due to the fact that d.c. amplifiers were not required to the same extent as were a.c. designs. The many designs used can be gleaned from the references to original papers given in Langford-Smith (1955) who covered the 1935 to 1950 period, to Ginzton (1944) for reviews, to Artzt (1945), to Valley and Wallman (1948) and to Dickinson (1950). Strong (1942) reviews their use as photocell amplifiers.

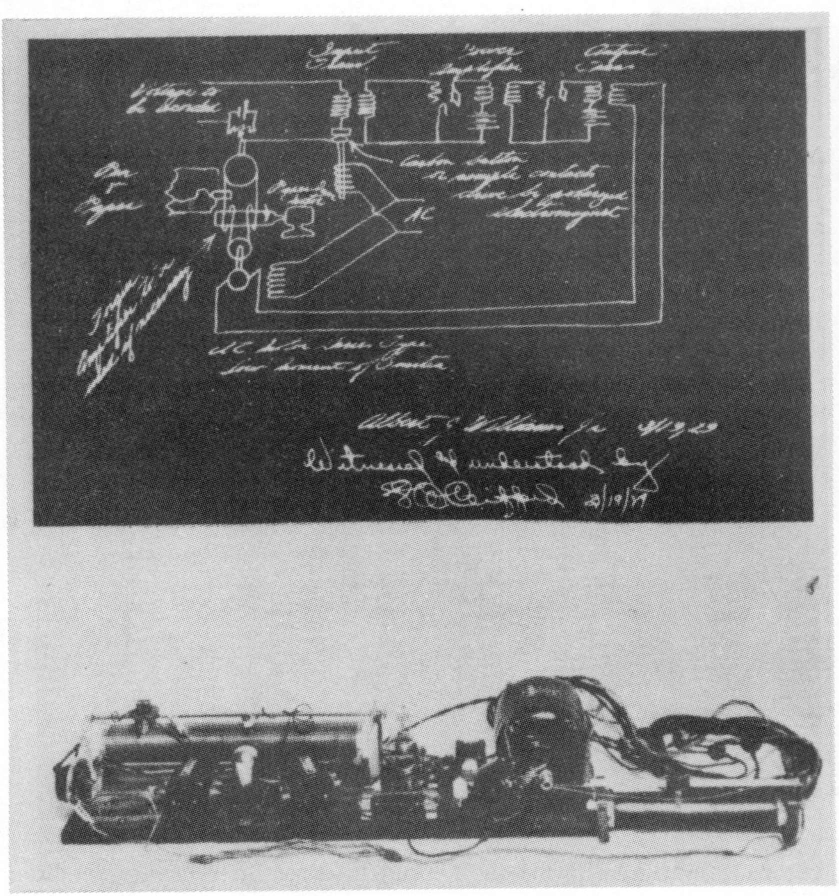

Fig. 6.15 *A.J. William's 1929 record and apparatus for a d.c. electronic potentiometric recorder marketed in 1933*

[Williams (1973), Fig. 3, p.13. Courtesy American Society of Mechanical Engineers]

Progress was gradually made in obtaining improved stability against temperature, power supply and stray signal variations but in 1935 when Penick (1935) compared the designs of six contemporary authors he was unable to state that any were able to cope with changes in overall gain resulting from the inherent gain variations of the valve itself. Progress on that problem had started many years earlier but was yet to be used to help the op-amp needs.

After several years of work on the problem of how to reduce the distortion of telephone line amplifiers, H.S. Black in 1927, realised that this was done very effectively and simply by feeding back some of the output signal into the input. The negative feedback amplifier was thus born: he also suggested the use of feedforward at that time. He gave this concept mathematical rigour in papers that became cornerstones of control theory, Black (1934). Amplifiers using his techniques were first used as coaxial cable repeaters in 1936. The use of negative feedback in this way, provided the amplifier has high gain (over 1000 for instance) can reduce the effect of drift in gain of the basic amplifier element and components by many thousand times that of the open-loop system, in fact, given high-enough stage gain the gain of the open-loop amplifier does not, in any first-order way, decide the gain of the closed-loop arrangement. As well as improving stability with time the connection also provides improved linearity of amplification over the dynamic operating range of the unit. It was a most important invention, as any person using modern linear ICs now appreciates.

Another significant improvement to the directly coupled amplifier was to follow in 1950. That was the year that a paper was published, Goldberg (1950), laying out the method known as 'chopper stabilization'. This, Fig. 6.16, was an extension of the idea used by Williams in 1929, and by others, in which the incoming signal is chopped to

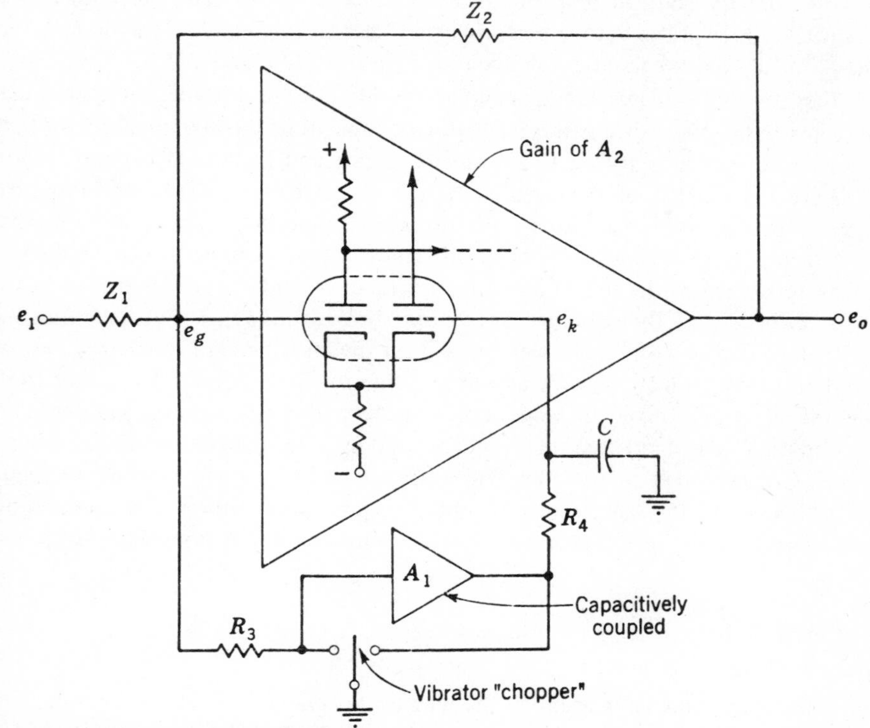

Fig. 6.16 *Goldberg, chopper-stabilized, d.c. op-amp 1950*

[Scott (1960), Fig. 4.13, p.135. Courtesy McGraw-Hill Inc.]

form an a.c. equivalent used to drive an output device. Goldberg used d.c. restoration, after the chopping action and a.c. amplification, to recover an amplified d.c. signal representing the input d.c. level. Goldberg also employed negative feedback around the unit to form a high quality, highly stable, directly coupled operational amplifier.

The quest for the general purpose d.c. operational amplifier continued well into the 1960s when it was still a topic that was often the subject of a Ph.D. in electronic engineering. Analogue computers needed them, as did many special kinds of sensing apparatus. Dickinson (1950) is a text about electrophysiological measurements. Scott (1960) is a book written out of notes delivered from 1950 onward on analogue computing. It is another example of the use of op-amps in the 1950s and contains a chapter on the circuitry of amplifiers used at that time.

Demand, however, was still too low to entice the very large electronic device manufacturers to enter the field. Then, at the start of the integrated circuit revolution that began in the late 1950s, it was realised by the IC makers that a cheap directly coupled amplifier would be able to be marketed in their millions provided the design was sufficiently versatile. It had to be able to cope with either polarity signals, be stable by the use of feedback, have reasonably wide bandwidth and be abuse tolerant, all of this being at low cost. Today we now use, as could not have been predicted in the 1930s, directly coupled op-amps instead of a.c. units that can be purchased at undreamed of low prices with a performance that exceeds many of the designs that people put many man-years of development into in the 1930s and 1940s.

The first use of electronic directly coupled amplifiers in the arithmetical operational mode seems to be in the automatic calculation equipment of the M-9 gun director that used radar to direct guns against flying targets. Eames and Eames (1973) and Mabon (1975) have illustrations of this system. It was the first unit to carry out extensive analogue computation by electronic rather than electromechanical computing means. The spectacular performance figures for the gunnery system must have helped overcome any resistance to the use of electronic techniques. In October 1944, in one engagement with the V-1 flying bombs, the system shot down 4672 out of 4883 missiles engaged. Only two out of the 91 were missed on the last day. This was a real test of electronic technique under very hazardous and rigorous field conditions. In the 1950 decade the electronic analogue computer became a common tool, texts began to appear and units were marketed that could be used by other than electronically trained personnel. Greenwood *et al.* (1948) includes an excellent review of electronic analogue computer hardware. It is interesting to note, however, that this review includes no specific mention of the feedback use of a general purpose directly coupled op-amp.

Revival and extension of digital techniques

In the telegraphy, and later again in the wireless-telegraphy pioneering periods, new methods and techniques were developed to cope with the digital signals that were used. Telephony, and then wireless telephony, emphasised the linear, analogue signal aspects of circuit design.

Apart from telegraphy and a few simple on-off control applications the demand for digital devices prior to 1940 was small. Progress in their development was gradual as new circuits and devices appeared. But by the time that the digital computing field began to emerge most of the necessary subsystem building blocks were in existence, see Table 6.1, as thermionic valve arrangements. The underlying operating philosophy of binary logic had been pioneered in the previous century and had been steadily extended to the point where logic could be set up entirely with electrical devices. The multivibrator, square-wave source (commonly called the clock source in modern computers), was devised by two Frenchmen in 1918, Dummer (1977). The bistable (or flip-flop) came into being in 1919 to become an important element in the counting circuits used a decade or so later in nuclear radiation counters. The monostable (confusingly called a flip-flop, or gate-circuit, in Mitchell (1951)) was known in the 1940s as a means of providing set time delays or as a way for restoring the shape of pulses. Coincidence circuits and other forms of digital gating arrangements were known.

Digital systems, as such, began to appear after the time release of the reports about the first digital machine, ENIAC, which used 18 000 valves in its 1945 construction. Examples of these reports are Hartree (1946) and Electronics (1946). Eames and Eames (1973) includes several illustrations of the subsystem equipment of ENIAC and other early digital machines.

Counting stages that use decimal coding, rather than the inherent binary number system, were devised in 1944. Digital instrument systems had begun to emerge. In 1950 the first commercial digital counter-timer was marketed to begin the new, still current, phase in which an ever increasing number of traditionally analogue instruments are progressively designed with a greater proportion of their circuits being replaced by superior, cheaper, digital equivalents.

The digital trend demanded devices not previously required. Counters needed displays. Potter, in 1944, built a decimal-coded counter using on-off indicator lamps to show the state of each flip-flop of the decade units. Grosdoff, in 1946, made use of the so-called columnar display in which one out of ten lamps was energised to illuminate the appropriate number in the 0-9 group. Work carried out previously on cold cathode trigger lamps in the late 1930s paved the way for the multicathode stepping tube development by Remington-Rand. These displays were best known as the 'Dekatron' tube. They were better than the columnar display method but considerable improvement was still possible. It came in the form of the so-called 'Nixie tube' (a trade name) in which the cathodes were formed in the shape of the numerals, or other symbols, and in which each was stacked on top of the other so that when any one was energised it appeared as the only symbol in a decade digit position. These tubes were first developed by Hagelbarger of Bell Laboratories Oliver (1962). Summer (1957) provides information on the Potter counter.

Mass assembly of numerous similar digital and analogue subsystem blocks exerted strong influence on the development of automatic means to manufacture and assemble circuits. Miniaturisation was also desirable, if not vital, in some applications of the 1950s. From this line of approach came hybrids, printed circuit boards, thin and thick film circuits and integrated circuits. The shape and form of electronic circuit assemblies changed from open, large structures, Fig. 6.17, to the tiny high-density packages used today.

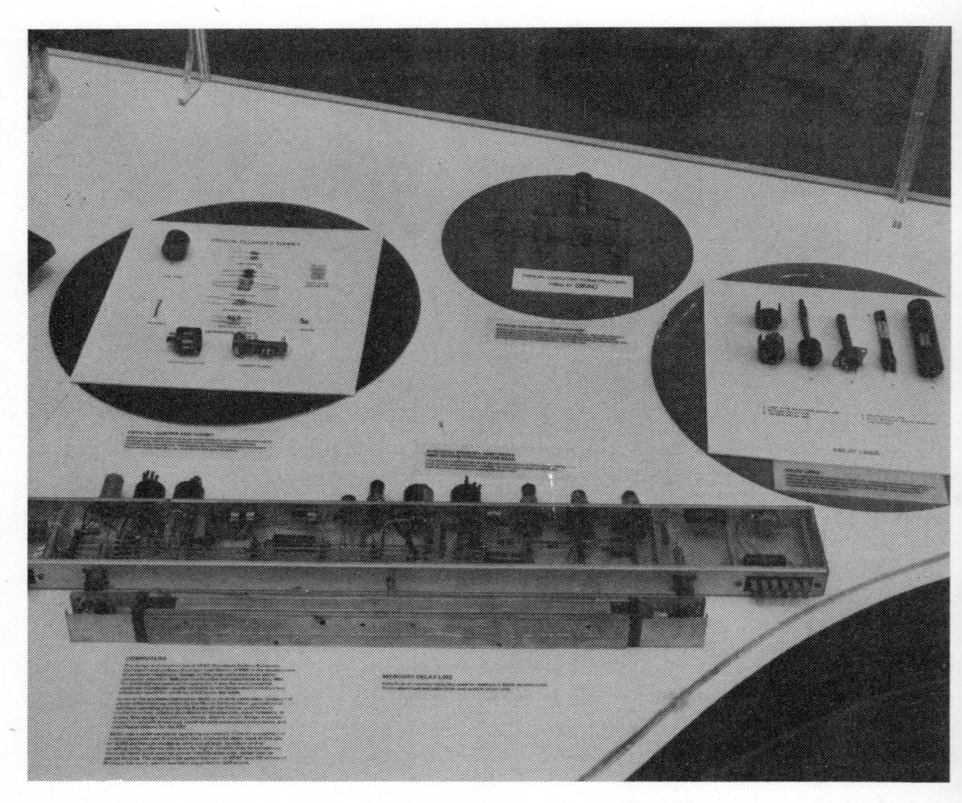

Fig. 6.17 *Mercury delay line unit from Standards Eastern Automatic computer SEAC, c1949* [Mason (1977), M538 (part).Courtesy,US Government, National Bureau of Standards]

Fundamental limits

Unlike mechanical systems design where fundamental physical limits on sensitivity are hard to reach, electronic devices exhibit various noise phenomena relatively easily and thus noise limits were realised early in the adoption of thermionic devices. Appropriate basic theory was already developed when experimenters began to find their circuitry beset with troublesome and unwanted noise. According to Dummer (1977), it was attempts to build very high-gain amplifiers during World War I that revealed the existence of a limit to the number of stages that could be cascaded before the background noise became excessively troublesome. Schottky predicted the reason for this in a paper of 1918 giving the theory of what is now termed 'shot-effect' noise, suggesting its source to be due to the arrival on the plate of discrete electrons. This was confirmed experimentally by Hull and Williams, by Hartmann, by Johnson and by Williams and Vincent over the period 1922 to 1926, see Chaffee (1933). Schottky's theory, however, applied only for the case where the current is not limited by space charge, which is not usually the case in practical amplifiers. In practice the effect is smaller than the Schottky expression suggests. Llwellyn, in 1930, published an expression that allowed for space charge conditions. A spacecharge-limited valve was shown to exhibit virtually no shot effect noise but experimenters still found that noise existed in such cases suggesting that there were other sources to be explained.

A second major predictable source of noise became known as Johnson noise (also called thermal or resistance noise) after the work of J.B. Johnson who tied this type of noise down to it being caused by the random movement of thermally agitated molecules of any material, gas, liquid or solid, forming effective electrical resistance in a circuit. He showed Johnson (1928) that it is related to bandwidth of the signal present, to the temperature at which it is operating and to the magnitude of the resistance concerned. It is the electronic parallel of Brownian motion seen in very sensitive mechanical systems, a fact pointed out by Johnson in his paper.

A third noise source showed-up in low-frequency applications for it becomes larger in magnitude as the frequency falls, being generally observable below about 400Hz. This is the 'flicker' noise; it was observed in the 1950s but is still not explained satisfactorily even today.

Although noise was observable in certain instances the bulk of electronic instrument applications generally was, and still is, not troubled by it. Where it was troublesome means were devised by which added electronic equipment added minimal additional noise (noise figure), or where noise was present in a signal it was effectively reduced to enhance the signal to noise ratio. Radar needs provided significant research effort on such matters in the post 1940 period. Contemporary texts on radar include Hallows (1946), CSIR (1947) and Sonnenberg (1951). Radar signal theory papers were prolific in the primary journals in the 1950 and 1960 period.

At the end of the period of this chapter came the first transistor, Fig. 6.18, for which contribution Bardeen, Brattain and Shockley shared a Nobel Prize in recent times, Mabon (1975). Transistors were generally available in the 1950s becoming subjects in teaching programmes in the late 1950s as an established, if not still rather expensive and still rather poorly performing, electronic alternative to the valve. They were so similar to valves in their operation that much of what had been learned in the 1910-1950 valve era was directly transposable into solid-state equivalent circuits. Transistor technique was, at first, taught in terms of valves. Today that approach has just about gone; valves are rarely designed into new equipment or lectured about in courses. Electronic technique of today owes a great deal to thermionic devices for they allowed men to forge ahead with the implementation of ideas, if not at the grand system scale now made feasible with integrated solid-state circuitry.

Development of electronic technique enabled measuring devices to be implemented with greater information sensing capability. More transducers effects could be put to use because of improved signal amplifiers. Data processing enabled imperfect sensors to be utilised. Costs were reduced enormously and a coherent approach to the design of instrument systems became possible in which relatively poorly trained persons could construct systems of great internal complexity and trained persons could devise and execute information and control systems of vast complexity and power.

But with all these great improvements the electronic thermionic decades of 1910-1950 would seem to be only part of the steady continuous development of man's ability to extract information and put it to use.

Fig. 6.18 *First laboratory transistor constructed at Bell Laboratories in 1947* [Brotherton (1972), p.1, upper picture. Courtesy of Bell Laboratories]

6.3 Accurate mechanism and mathematics

The design of measuring and energy machines of the 1900s and on, needed continued improvement in mechanical design and manufacturing skills. Measuring instruments of many kinds demanded better precision of geometric form than did 19th century apparatus; there were, of course, exceptions such as telescope mirrors. Embellishment came to be seen as less important to functional mechanical accuracy, given a fixed level of expenditure.

Precision manufacture was needed for the new breeds of recorders, like those shown in Fig. 5.35 and 5.51, that were marketed as commercial, reasonably large-scale, production lines from the early decades of this century. Oscillographs, Irwin (1925), MacGregor-Morris and Henley (1936) and other galvanometer-based devices needed careful design and construction to yield their theoretical detection potential. Gravity balances, Fig. 6.19, involved very fine mechanical construction. Seismometers became routine instruments, Walker (1913). Gyroscopes needed high-speed, well balanced, rotating mechanisms. Astronomy called for fine guidance mechanisms, as did meteorological instruments, Knowles-Middleton and Sphilhaus (1953), and so on.

This demand resulted in many excellent books on mechanical design of fine and accurate mechanical mechanisms. Drysdale and Jolley (1924) summarised what was

Fig. 6.19 *Gravity balance by Oertling c1920*
[Object in Bureau of Mineral Resources Geophysics Instrument Collection, Oaklands, Australia]

known about the static and dynamic regimes of mechanism used in the, by then, very wide range of electrical indicating instruments. They included a considerable amount of experimental data on such parameters as air damping and material properties. Rolt (1929) was one of the first books to appear on gauging aspects of mechanical components. His work included much about the, by-then, established tool-room metrology and its equipment. In 1933 Whitehead (1954) was first published. It was a text specifically on the general aspects of fine mechanism. This was not a catalogue of mechanisms as had appeared in the centuries before but a study of the underlying principles needed for ordered mechanical design of instruments. Catalogue style works, such as Barber (1906), gradually declined in number in this period.

Optical measuring instruments also needed accurate fine mechanisms for successful operation. Martin (1924) and Jacobs (1943) deal with this aspect of mechanical design needs. Simple devices like bells and alarms, Allsop (1902), and teletechnique components required accurate fine mechanisms as did many other electromechanical devices.

The other branch of fine mechanism that was developing in parallel was the use of mechanical devices to perform mathematical operations. Development prior to 1900 enabled a certain amount of mathematical manipulation of data and generation from functions to be performed with mechanical machines, Kelvin's tide predictor (shown in Fig. 5.37) and the many marketed calculating machines (Eames and Eames (1973) contains many illustrations) being representative of the skills and design then available.

The Michelson-Stratton harmonic analyser of 1898 was an improved and extended version of Kelvin's harmonic series generator. It could sum 80 elements of a complex function, plotting the total as a waveform train.

Electromechanical components were added to the purely mechanical systems of computation, e.g. drive motors, relays, contacts, potentiometers and others gradually replaced mechanical components. Notable calculating engines were the Torres 1920 'electromechanical Arithmometer', the 'Great Brass Brain' tide predictor of the US Coast and Geodetic Survey built around 1914, German tide predictors used for assisting submarine navigation in shallow waters during the First World War, the 1970s 'product integraph' of Vannevar Bush, the Dodd mechanical correlator of 1926, the Wilbur simultaneous equation solver that was able to handle the computations for an economic model which needed half a million multiplications, the Bush differential analysers of the late 1920s, the Hartree 'Meccano' analyser of 1933 used for studying problems in atomic theory, to bomb and gun aiming equipment of the Second World War. Many of these equipments are illustrated in Eames and Eames (1973).

Gradually electrical components were used where-ever they would perform better than their mechanical analogue. In 1935 Bush started construction of a general-purpose computing machine using electrical components. Electricity power authorities began to install network analysers which provided a model of the network that could be studied in the laboratory. As new power generators, major loads and line changes were made to the real system appropriate alterations were made to the analyser. Westinghouse installed a network analyser in 1930. Mortlock and Davies (1952) contains a chapter on the types of network analysers in use at that time.

It is now clear that the mechanical computer was, in the 1940s, already being replaced by an electrical equivalent, just as the design was reaching maturity. Several books on the design of mechanism appeared at that time Murray (1948), Svoboda (1948), Beggs (1955) and Mabie and Ocvirk (1958) being examples. Crank (1947) is a report about the Manchester University differential analyser. Murray, in his foreword, pointed out that the art of mechanical computing machine design was then rapidly approaching the intellectual maturity corresponding to the research level. The contents include both digital and analogue methods. Greenwood *et al.* (1948) includes a little on mechanical mechanism but their book concentrates predominantly on the electronic analogue computer showing the transition was just about complete by the middle 1940s. Electronic directly coupled amplifiers, the operational-amplifier feedback connection, special function generators and mathematical operations with electronic apparatus had almost completely replaced mechanical methods by that time. Several units defied conversion for a longer period, multiplication and synchronous generator functions being some examples that continued on as electromechanical units for some years owing to the lack of an adequate electronic alternative. Fig. 6.20 is of an electromechanical computing arrangement designed to assist flight in a straight line when using radio location signals.

The 1940 decade saw the arrival of the digital computer which very rapidly began to replace much, but not by any means all, of the mathematical work being performed on the new breeds of analogue machine. It was the beginning of a period of intense interest in control in which people would seriously expect to soon be making intelligent machines and self-organising devices that would outperform the human being. That age did not come as fast as expected; progress is still edging toward it as sensors, data processing, actuators and systems design gains the necessary ingredients

Fig. 6.20 *Electromechanical straight-line flight indicator made by CSIRO, Australia c1950* [Object in Bureau of Mineral Resources Geophysics Instrument Collection, Oaklands, Australia]

of technology and understanding. Berkeley (1949) hints at the expected future at that time of initial computer expansion. The collected papers, Scientific American (1955), is a general account of what was expected from automatic control.

Control in this period is generally associated with the spectacular computing achievements but that was not all that was being developed. A very important field, if not such a fascinating one to the general populace, is that of railway signalling and control.

According to White (1962) it was the pioneering work of the Union Switch and Signal Company of the US that helped pave the way for the acceptance of electronic methods of signalling when it began to implement a signal relay method for cabin displays in 1916. To overcome problems of poor visibility work was commenced to relay the track signal data into the cabin via an inductive transfer system from coils laid near the track. It took several years to develop sufficiently robust amplifiers but the system went into operation in 1924. Many texts were published on the subject of railway signals and safety measures, Sperry (1913), Prior (1919), Burtt (1926), Union Switch (1928). Much was learned about the practice of sequential switching from this field. Cock *et al*. (1951) provides an overview of the history of railway signalling.

6.4 Instrument systems emerge

Until the 1900 period measuring instruments and allied equipment were relatively simple arrangements. The level of sophistication, by comparison with what was to come from around 1920 onwards, was very basic. The most complex information systems then in existence were the telegraphy and telephony networks. Although they were not specifically concerned with measurement, it was in that experience that many of the subsystems concepts developed that later were to be of value to instrumentation systems.

Measurement systems of the 19th century were becoming complex, but the lack of easily built, altered, compatible units and the lack of many of what today are regarded as vital units (such as, amplifiers, recorders and other peripherals) inherently kept the degree of complexity at a low level. Figs. 5.7, 5.21 and 5.47 show 19th century measurement systems. They were generally characterized by the *ad hoc* assembly that grew as need-be and could be allowed.

The most complex measuring instrument systems appear to have been the meteorological stations and other earth science observatory complexes that came into being late in the 1800s.

The arrival of electronics, coupled with the emergence of recorders and the existence of data transmission systems, provided the means for more extensive measuring and controlling systems to be developed. Initially they grew 'like topsy' being assembled from whatever proprietry units and components could be assembled to do the job. This phase was short-lived for it was realised that methods for handling large-scale systems were needed. Standards of practice were set by mutual agreement. Companies began to market regular lines to reasonable well adhered to specifications.

By the 1930s many extensive instrument systems were in use. Several examples of dominant fields of use are now discussed.

Aircraft and ship instrumentation

In 1900 the first experimental Zeppelin airship was flown. In 1903 Orville Wright made the first controlled, power-driven aeroplane flight. Powered flight rapidly developed from the primitive aeroplanes and dirigibles, a development requiring an ever increasing amount of instrumentation, both in research and design and in flight management.

Fig. 6.1*a* shows the cockpit of a light plane of early origin. Initially, the instrumentation needed indication of a few engine variables such as oil pressure, temperature and engine speed as would be found today in a high-performance motor car. Additionally, the pilot was soon given basic flight direction indicators via the magnetic compass and, also from quite early, the gyroscope. He also had altitude readout and a crude altimeter made from a liquid manometer. The Wright Brothers aircraft used a turbine drive dial indicator for wind speed.

As time passed this simple set of instruments, that had been tucked into the cockpit wherever there was a suitable space, grew into orderly panel systems in which were included airspeed, maximum safe airspeed, artificial horizon, radio-location, radio communications, radar, fuel level, many more temperature and oil-pressure gauges, undercarriage position indicators and so on. Fig. 6.1*b* shows how, in the larger aeroplane of a few decades later, many of the flight engineering condition instruments had been separated out into a panel for the flight-engineer, these being variables that were not directly related to the task of flying the machine. Military devices added still more instrumentation to the instrumentation system of a plane. Aircraft instrumentation became the most complex and extensive arrangement of data indicators to be placed in a given space.

Illustrations of cockpits at various eras are to be found in Hewitt (1934), Ward Lock (1955), Tupholme (1942) and Schaedal (1973). They cover the period from 1920 to 1950. Ward Lock (1955) contains a well illustrated chapter on instruments in use in the 1940s, giving sectional drawings for many of them. Hewitt (1934), Royal Air Force (1934) and Air Ministry (1939), provide details of how such instruments were used in practice. Rathburn (1929) includes some detail of instrumentation.

Electronic technique was gradually adopted beginning as radio communications and later moving into the instrumentation area. Weihe (1962) and Davies (1976) consider growth of the use of electronics in aircraft. Still (1957) is concerned with air-conditioning equipment for use in aircraft.

The gyroscope, invented by Foucalt in 1852, entered transport machines as part of roll-stabiliser proposals for ships in the late 19th century. Horner (1905) states that it was in use in his time for torpedo and submarine course guidance. Oberg and Jones (1917), a decade later, give a longer account of its uses including mention of an

Admiral Fleuriais using it as an artificial horizon at sea; for keeping torpedoes on course as the result of original work by Obrey; of Brennan's use of it in a monorail; Schlick's application for damping the roll of ships; Russel Thayer's proposal for stabilising dirigibles and the work of Sperry, who by then, had made the gyrocompass. By the 1910s it was in routine service in ships and aircraft. Golding (1932) contains a chapter on the application to ships in the late 1920s. Its use in aeroplanes of the 1940s is explained in Ward Lock (1955). Eames and Eames (1973) includes several illustrations of early gyroscopes. Its introduction was somewhat slow to take on, perhaps because of the high level of mechanical and electrical skills required in its design and maintenance.

According to Eames and Eames (1973) the first practical solution to ship guidance, when the ship contained vast amounts of steel that distorted magnetic compass readings (as was the case with warships of that period), was the gyrocompass patented by the German, Anschutz-Kaempfe in 1908.

In America it was Elmer A. Sperry and his sons who became associated with the application of the gyroscope to ships and planes. Yost (1962) gives a short biography of this family endeavour. Elmer Sperry and Hannibal Ford were the first to instal a gyrostabiliser in an aeroplane. Sperry discovered that the dynamics of flight control were more complex than for the previously experienced torpedo control. He overcame one condition of instability by making use of the wind speed indicator data to correct the gyroscope data thus beginning the use of interconnected instrument systems in a computational, interactive, way. Eames and Eames (1973) gives several pictures, Fig. 6.21*a* is one of Sperry's early plane stabiliser systems. Spectacular demonstration of his improved designs, at a French competition in 1914, greatly assisted gyrostabilisers to become accepted practice.

The Sperry arrangement made use of mechanical linkages from the four gyro units to the servomotors used. Eames and Eames (1973) contains a reproduction, Fig. 6.21*b*, of a drawing of a 1910 scheme by Paul Regnard which is interesting as it appears to make use of electromechanical actuators. Naturally the gyro units became more complex as times passed, as can be seen by reference to manuals issued to military personnel over the years, such as Admiralty Compass Department (1941). The ship navigation manual, HMSO (1938) also gives some detail and lists the various publications concerned.

In Britain the name associated with gyro systems in the early 1900s was that of S.G. Brown. His second gyrocompass of 1916 is held by the Science Museum in London. He began his work in 1914.

Stabilised aeroplane flight was of great international interest at that time. For example, the Austin Automatic Balancing Airship Syndicate issued shares in Melbourne, Australia from 1912 on. In 1914 it was over subscribed. Fig. 6.22 shows the aircraft involved with that enterprise.

Repeaters were subsequently added to ship gyrocompass systems so that the indication could be where the officers needed it, not where the gyroscope was required to be in the most stable, lower deck, position.

The first National Bureau of Standards tests in the US, of a radio guidance system

a

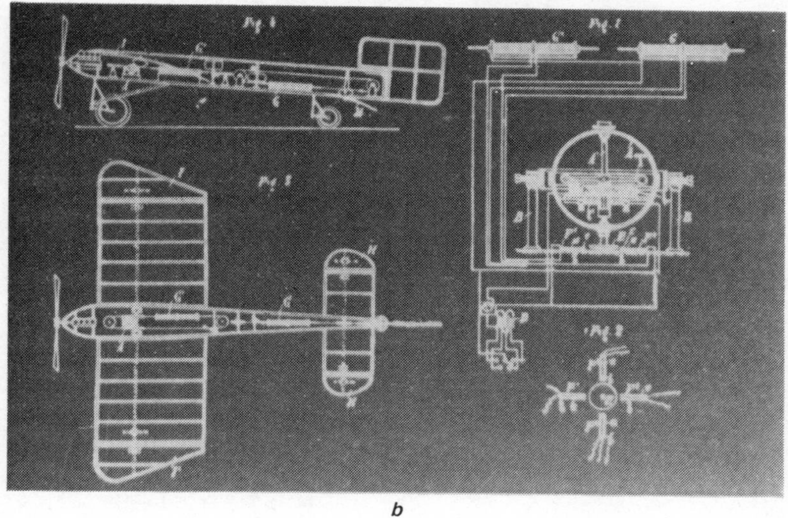

b

Fig. 6.21 *a* Sperry and Ford gyroscopic stabiliser 1909
b Regnard's 1910 proposal for aircraft stabilisation
[Eames and Eames (1973), p 61. Courtesy (a) Sperry Rand Corporation (b) *Scientific American,* 103, p.228-9, 1910]

that would guide an aeroplane were conducted in 1921, Cochrane (1966). The pilot flew a course set by two transmitting coils on the ground. This work was then dropped for several years and resumed in 1926. Work of H. Diamond, who joined the NBS in 1927, led to a successful visual indicating system that enabled guided flight. It, see Fig. 6.23, used two tuned reeds (resonating at 65 Hz and 86·7 Hz) that were excited by two radio beacons, Mason (1977). Visual appearance of the vibrating reeds indi-

Fig. 6.22 *Aeroplane associated with a 1912 project to obtain stabilised flight in Australia*
[Courtesy of Mr. N.Harwood, Museum of Applied Arts and Sciences, Sydney]

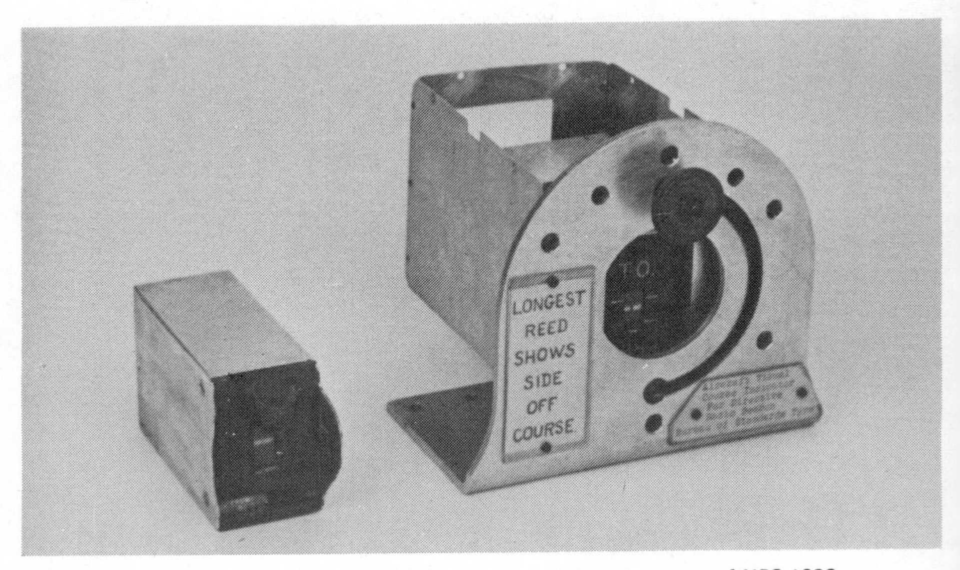

Fig. 6.23 *Two-reed, flight course indicator from radio guidance system of NBS 1928*
[Mason (1977) M87. Courtesy US Government, National Bureau of Standards]

cated course error. The unit had to be up-ended for flight away from the source. Associated persons with Diamond were Dunmore, and Dellinger. Air-traffic control began in the 1920s getting underway in the next decade. Hudson (1931) describes a radio network to assist talkdowns.

This work set the situation for a blind landing system. For this was required the addition of information about the remaining degrees of freedom in space. Diamond proved an arrangement of his design in satisfactory tests of 1931, Cochrane (1966). In 1933 the system was passed to the Department of Commerce as a workable system for use on aeroplanes. In the 1930s airport control towers equipped with radio were established along with a teletype weather-map service.

Aviation instrumentation systems provided large amounts of research and development expenditure which greatly assisted the longer-term development of instrument systems in general.

Another aspect of instrument systems applications that was emerging from a simple systems stage to become very complex by modern times was the field of electromedical apparatus.

Electromedical systems

The foundations of the use of electricity in diagnosis and treatment in medical fields began in Ancient times. Statical and inductive electricity sources were in extensive use in the 19th century for therapeutic treatment. By 1900 a discipline of medical electricity was well established as seen by the titles of Lewis-Jones (1900) and Cumberbatch (1921). Magill (1916) includes the teaching syllabus for the medical electricity examination of the Society of Trained Masseuses.

In 1900 medical electricity was confined to the use of various forms of electric current for shock and 'ionisation' treatments, to the use of X-rays for measurement and some treatment and to the use of very basic devices like the 'panelectroscope', a form of electric light endoscope, introduced by Leiter of Vienna, see Lewis-Jones (1900) for this and other turn of the century electromedical apparatus. Cumberbatch (1921) is a similar work published a decade later (but which was originally written in 1905). Fig. 6.24 is a page of advertisement from that book that shows the range of electromedical apparatus in use at that time.

More general accounts of medical uses of electricity are to be found in Knox (1917), Hawks (1929) and Golding (1932). These show that very little change occurred up to the 1930s in the general aspects of hardware, especially when those accounts are contrasted with that of a chapter given in Trewman (1949).

Dowse (1889) includes two lectures on electricity in medicine giving another person's experience of early 20th century medical electricity. The practice of use and locally made equipment, as it was in Australia in the 1930s, is to found in Dark (1934). These also show little change in the use of sophisticated instrumentation in the years before 1935. Photographs of various exhibits of the Division of Medical Sciences at the Smithsonian Institution, Hamarneh (1964), and the bibliography given there are relevant to the history of medical apparatus. At the time of writing a history of medi-

Fig. 6.24 *Electrotherapeutic apparatus on sale in 1920*
[Cumberbatch (1921), advertisement at rear]

cine and its technology was in preparation, Davis (1979). The historical role of technology in medicine has been studied by Reiser (1978).

X-ray methods were established as routine equipment very rapidly after their discovery just prior to 1900. An example apparatus is shown in Fig. 5.11. Subsequently, the period of interest in this Chapter was one in which the generation tubes were improved to give higher energy levels and greater quantities of energy into a more specific radiation solid-angle. The safe practices of X-ray use by operators were gradually improved and monitoring apparatus developed. These developments can be studied from Mullineux-Walmsley (1910) and Bowers (1970). The engineering use of X-rays was written up in Terrill and Ulrey (1930) at the time when electronic amplifiers, rectifiers (kenotrons) and indicators were just coming into use in X-ray plant. It includes a small chapter on what the writer said was a new field – that of radiography for general testing operations. Hawks (1929) includes a chapter on the non-medical uses of X-rays. This new subject had become quite ordered by the 1940s, as is shown by the account in Crowther (1944) on industrial radiology. Glasser *et al.* (1947) is concerned with medical uses of X-rays and nuclear radiation at that time.

During the First World War period, X-ray high-tension rose from 50 000 to 200 000 volts and they were beginning to be used for cancer therapy, Cochrane (1966). Standards of dosage were vitally needed for reasons of safety. An international congress, held in London in 1925, raised this issue and by 1928 standards had been instituted. Cochrane described the work of the National Bureau of Standards in this area. Trewman (1949) discusses the X-ray technology used in medicine in the 1940s.

Research into animal electricity and electrical potentials in the physiological body that was conducted prior to 1900 created the knowledge and technology for the invention of the electrocardiograph. The first apparatus to record the low-level electrical signals taken from electrodes placed on the body was made by Professor Einthoven in 1903. He used his now well-known string galvanometer as the sensor-indicator of currents produced by electrodes placed on the body. His apparatus design was marketed by the Cambridge Scientific Instrument Company in England from 1909 onward. It was, by modern standards, very bulky, elaborate and difficult to use. Fig. 2.40 is of such apparatus. It was developed in co-operation with Sir Thomas Lewis, see Cambridge Instrument Co. (1945).

Around 1936, in Adelaide, South Australia, the firm of E.T. Both designed, and commenced marketing, a cardiograph that gave instant records and which was portable, Fig. 6.25. This apparatus made use of an electronic amplifier to drive the moving coil assembly of a permanent-magnet loudspeaker unit which, through a stylus, formed a trace on a clockwork driven rotating smoked disk. The disk was viewed with a small microscope fitted to the panel. Two of these instruments are preserved at the Institute of Medical and Veterinary Science, Adelaide.

The Cambridge cardiograph was also reduced in size as the years passed. By 1935 it was of similar small-suitcase size to the Both unit, Cambridge Instrument Company (1955). This company diversified into related products, such as direct writing recorders, amplifiers, transducers and other electrical measuring apparatus for electromedical use.

Fig. 6.25 *Both portable electrocardiograph, designed c1935*
[Courtesy South Australian Division, Institute of Instrumentation and Control, Australia]

Electronic amplification enabled other medical apparatus to be devised that gave the practitioner greater observability and detection, stethoscopes with a vacuum-tube amplifier being an example.

Hearing aids began as primitive devices in which as many as three microphones, Fig. 6.26, were wired together in a chest slung group, these feeding the earpieces. They were far from un-noticeable, the assembly also needing two 'small' dry-cells. Miniature valves were developed by the 1940s and gradually the size of hearing aids were made more tolerable and the performance more reasonable. Electronic aids of the 1930s were far from small. A design is given in Simer (1936).

The practice and sophistication of electrophysiological measurements and experiments had grown considerably by 1950: the simple systems of 1900 having been replaced by racks of gear such as is illustrated in Fig. 6.27, which is taken from Dickinson (1950). Following somewhat behind was the study of psychology which was beginning to deploy a greater amount of recording and sensing apparatus after 1900, Myers (1928), Davis and Merzbach (1972, 1975), as was the study of plant physiology, MacDougal (1908).

Hospitals gradually acquired sufficient technology of advanced kinds to warrant the training and employment of specialists in such areas as radiology, electronics and instrument maintenance.

Fig. 6.26 *Hearing aid, c1910, that does not contain an electronic amplifier*
[Object in Birdwood Mill Museum, South Australia]

From manual to automatic control

Industrialisation demanded improved means to manipulate materials into products. Initially, in the 19th century, this generally meant the invention of an improved method that produced the product more speedily and of a better quality by the use of a better principle.

The availability of measuring and recording devices that could monitor processes, which started most noticeably from around 1900, began a new phase in manufacturing and processing. Variables of a process could be monitored for their variation and magnitude. Better control occurs when there are more variables being monitored so that more information is available for decision making. Thus began, around 1900, an era in which progressively more industrial variables were measured as part of the advance of better management methods of inanimate processes, this inherently leading to loop-closures using hardware.

It is quite striking that of the many books consulted to seek evidence of closed-loop control on a wide spread scale in the first fifty years of this century only those from around 1950 onward give specific mention of what is now so commonly known as automatic control. Donald (1925), an account of industry at that time lists and illustrates many processes in action without mentioning the use of automatic control through the use of measurement data feeding closed loops. Similarly so for Odhams Press (1948). Kemp (1950), a three volume treatise (compiled in the 1940s) on what

Fig. 6.27 *Contemporary electrophysiological research equipment c1950*
[Dickinson (1950), frontispiece. Courtesy *Electronic Engineering*, London]

an electrical engineer might need to know also has nothing on control systems. Trewman, in his book reviewing electronics in the factory in 1949 at least included a chapter showing that the experience with control loops gained in the war years had begun to filter through to practical use in industry. An interesting review of how the mathematics of control was developing is to be found in the appendix of Griffiths (1943).

In the 1950s literature there is extensive evidence of the acceptance of automatic control as a subject. Thus it would appear that the period prior to *c*1950 can be regarded as one in which manual control, via the use of indicated variables, developed to the stage where it was clear that closed-loop operation was the next evolutionary stage to pass into to improve productivity. Knowledge of a processes variables certainly assists its tuning but a point is reached where the characteristics of the human operator are no longer adequate; the loop must be closed via a 'fixed meaning coded' information hardware link. In such cases the human then acts in a supervisory role watching trends and safety levels.

Deficiences of human operators began to be aided from the time when man first used measuring instruments and observing apparatus to enhance his limited, naturally given, senses. This use of aids was particularly noticeable in the industrial field from 1900.

As an example, consider temperature measurement and control. Electrical methods of sensing temperature mainly originated in the 19th century. Continuous measurement of the temperature of furnaces, ambient environment for meteorology, incubators and the like, was not economic using the manual method of registration. This particular need was a strong factor in the development and marketing of electric recorders in the early 1900s. An additional factor was that furnace operators were not always reliable and consistent stokers so means were required to check their performance, especially in the small hours of the morning.

The author of the history, Cambridge Instrument Company (1955) stated that temperature measuring equipment was the fastest advancing field in those early decades. Thermocouples came on the market as regular items from 1902, albeit in a primitive form. Thermocouple resistances and voltages were standardised in 1905. The thermopile, high temperature, Féry mirror pyrometer, Fig. 6.28, was sold from about 1904 onward being, apparently, the only satisfactory means to measure the high range when it was first introduced. Cambridge Instrument Company first marketed a temperature controller in 1909.

Other developments followed that made use of the thermocouple, resistance-element, optical pyrometer, closed-tube expansion thermometers and other sensors, Cambridge Instrument Company (1945).

Many texts have been published over the first half century on temperature measurements. A sample is Le Chatelier and Boudouard (1904), Darling (1911), Day and Sosman (1911), Burgess and Le Chatelier (1912), Weber (1939), Wood and Cork (1941) and Griffiths (1947). A work specifically concerned with methods for temperature regulation application is Griffiths (1943). Glazebrook (1922) contains several sections on aspects of temperature measurement.

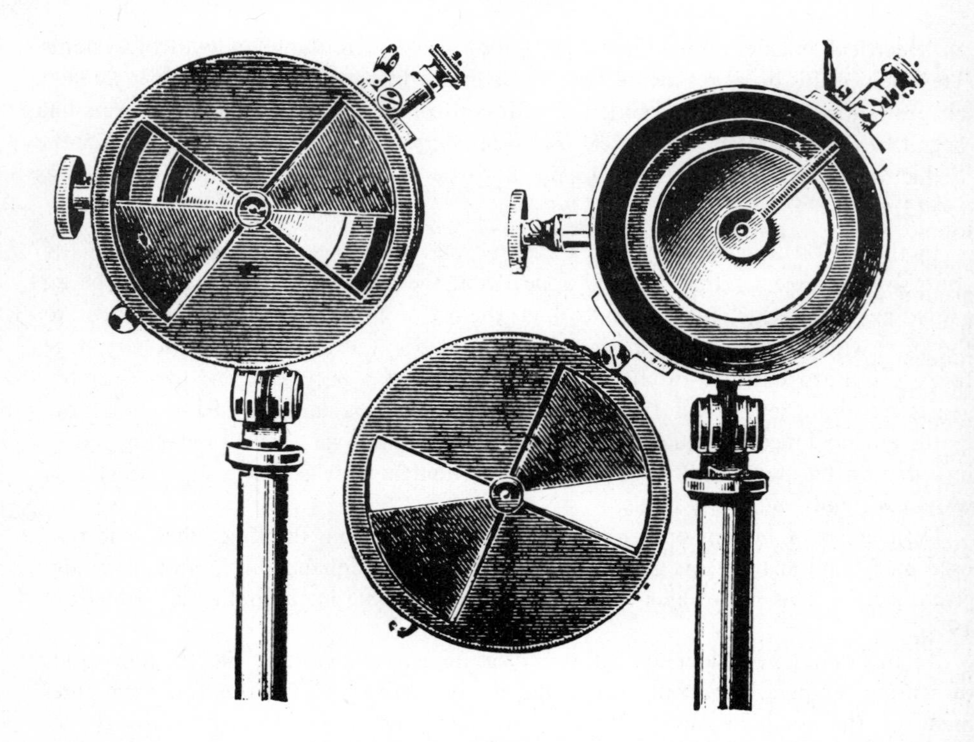

Fig. 6.28 *Féry mirror pyrometer; view looking into mirror that focused radiation onto a thermopile c1910*
[Darling (1911), Fig. 41, p.127]

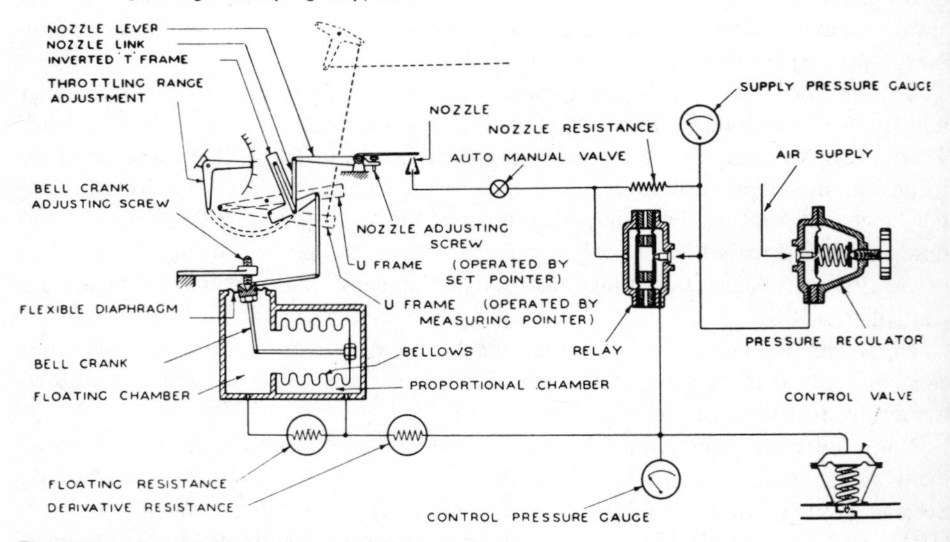

Fig. 6.29 *Kent Mark 20 pneumatic control unit*
[Kent (1950), p.10. Courtesy Kent Instruments Pty. Ltd.]

Additionally to the needs of industrial production were the measurement and control requirements of large engineering enteıprises such as ships, buildings, lifts, railways and many more. There was also the need to provide better control of public installations such as water treatment and town gas supply, sewage treatment and disposal, road traffic control and so on.

Gradually there came into use more indicating sensors followed by more control loops. As an example, by 1950 steam generation used automatic control for gas, steam, air and water flow, furnace pressure and stoking operations. A survey made around 1936, according to Trewman (1949), showed that of several hundred firms, having a wide variety of activity, only 20% were then making use of electronic techniques, their uses being mainly for measurement and counting. Trewman suggested that at that time there existed a deadlock preventing advancement. Marketing firms would not enter significant development of electronic equipment because the market was suspicious of electronic apparatus due to its lack of experience with it. Trewman stated that it was, in his opinion, the war effort that followed which demonstrated that electronic methods were powerful tools in measurement and control. The war experience also provided the knowhow and methods with which industry could forge ahead, from 1945 onward, with its use of advanced instrumentation.

Initially signal transmission and control systems for industry developed using pneumatic energy, but by the middle of this century electrical and electronic systems had just begun to challenge the firmly established pneumatic equipment, Instrument Engineer (1954). Fig 6.29 is a schematic diagram of the *c*1950 Kent Mark 20 pneumatic control unit which incorporated porportional, integral and fast derivative action. A mechanical linkage feeds signal from the sensor type used with it.

6.5 Recording data

In 1878 *Punch* published the visionary cartoon shown in Fig. 6.30. The caption with it was:

> By the telephone, sound is converted into electricity, and then by completing the circuit, back into sound again. Jones converts all the pretty music he hears during the season into electricity, bottles it, and puts it away into bins for his winter parties. All he has to do, when his guests arrive, is to select, uncork, and then complete the circuit; and there you are!

Just exactly what the artist thought about the concept of recording is debatable but it may have been that, having heard about the conversion process of changing sound into electricity and that electricity is stored in Leyden jars, he may have come to the conclusion shown. It is most unlikely that he would have seriously thought that everyone would do just this many decades later.

In 1878 electric recorders were being used to record digital telegraph signals but audio, analogue, recording was still no more than a gleam to a few. In that year Edison's phonograph was released to the public but it did not make use of electrical transduction. The concept shown in the cartoon was to come into common use in

Fig. 6.30 *Cartoon suggesting how sounds can be bottled for later use — see text* [*Punch*, 1878]

the 20th century after the occurrence of several more inventions and as the result of much development. This account about recording methods continues on from the pioneering contributions described in the previous chapter to reach the stage where marketed electrical signal recording devices were in routine use.

As had been said in Section 6.4 considerable interest existed around 1900 in means to capture data variables as permanent records. The use of in-built mechanical recording mechanisms was continued, as it still is today, where-ever that method is the most suitable. A catalogue, Kent (1930), contains illustrations of many such devices in use for flow reading. Multhauf (1961*b*) and Knowles-Middleton (1969*a*) also include brief mention of 20th century recording mechanism of this form. Knowles-Middleton and Sphilhaus (1953) discuss their use in meteorological devices.

Some measuring instruments incorporated the mechanical recorder as a secondary function of some other primary mechanism. An example of this was the Whipple indicator (Fig. 5.32), Darling (1911), in which the position of the slidewire bridge is read out, at balance, as degrees of temperature.

Recorders of analogue signals generally were initially in demand for recording and replaying sounds, the Poulsen 'Telegraphone' being devised to record telephone messages. Magnetic recorders were first put into regular use as a journalism aid during the Second World War. The various forms of phonograph were developed primarily for the consumer market.

Industry and science had also seen the value of recorders and their use steadily grew over the years. One report, Cambridge Instrument Company (1945), states that there were over 40 of the Callendar potentiometric units (Fig. 5.35) in use for controlling a bank of annealing furnaces in a large steel-works in South Wales, this being the first large-scale use of them after their release for sale after 1900. The trade publication, Kent (1930), shows a handsome wooden, side-board-like, cabinet containing the electrically transmitted, recorded traces for eight venturi meters, five water level gauges and a tide gauge. But this order of demand was not sufficient to bring about the great development and lowering of costs that has been experienced in modern times since the consumer decided that every house wants a tape recorder.

In those early times the factors that were of importance were very much concerned with the speed of response, reliability, and cost. Some former electrical, self-balancing, recorders required up to 20s to catch up with a full-scale-deflection demand. The Callendar units were 'remarkably fast' taking only around 1-2s for a full span movement. For many industrial purposes this degree of frequency response was adequate. Attention is now directed toward many kinds of recorders that developed to be accepted into routine from 1900 to 1950.

Potentiometric-recorders

Callendar's potentiometric, self-balancing, recorder was the first in a long line of designs that were based on his principle of operation if not exactly so in their execution. His design has already been covered in the previous chapter being born in the 19th century. It need not be detailed here.

The Callendar instrument was marketed from around 1900 and held sales until the 1920s. Some instruments were still in active operation in the 1950s.

In 1911 the Leeds & Northrup self-balancing recorder was introduced to hold a strong sales place until it was replaced by their electronic version of 1930 origin. Their first design also was mechanically sensed. It, too, has been described in the previous chapter. The electronic version is covered in Section 6.2 inthe discussion on electronic d.c. amplification. Their 1940s, model G, is discussed in Vogel (1949).

The line of mechanically-sensed potentiometric recorders was held in the market by such lines as the Kent 'Multielect' which was reaching the end of its sales life as a valid marketable design by the early 1950s. These recorders were, by today's standards, so over built that they were more likely to be withdrawn from service because of the lack of servicing ability or lack of performance speed than because of wear. Mechanically sensed units were comparatively slow, a 6-channel dotting recorder taking around a minute or two to cover each channel, but this was not always a serious limitation.

The self-balancing mechanism itself was also a useful device as the heart of many controllers, Griffiths (1943).

Chopper-bar or dotting recorders

The high cost of the Callendar recorder restricted its use for many recording needs. A simpler recorder was needed that could plot electrical signal data with a reasonably fast, order of seconds, response at a cost that would entice less convinced people to adopt them. The chopper bar recorder provided the solution, first appearing around 1905. Many designs appeared over the decade following.

Horace Darwin devised his 'thread recorder' in 1905 Darling (1911), Cambridge Instrument Company (1945, 1955). In this a continuous inked thread is held between the chart and the, free to move, needle of a sensitive galvanometer movement. A drive arrangement periodically lowers a chopper bar onto the needle causing it to mark the paper at that point with the thread. Between chops the needle finds its new position in an unconstrained manner. Typical chop periods for this form are from 10-30s between depressions. If the chart speed is slow the trace appears as a smooth line.

R.W. Paul also built and sold a similar instrument making use of his 'Uni-pivot' suspension to allow the needle to be depressed without damage, Fig. 6.31. It was first sold in 1914 being featured in the 1914, R.W. Paul catalogue. One version had a two-colour typewriter ribbon as the 'thread', the colours being used alternatively to mark the values of two separate time-multiplexed data channels. These instruments were the forerunners of the modern single and multipoint dotting recorders, as they are termed today.

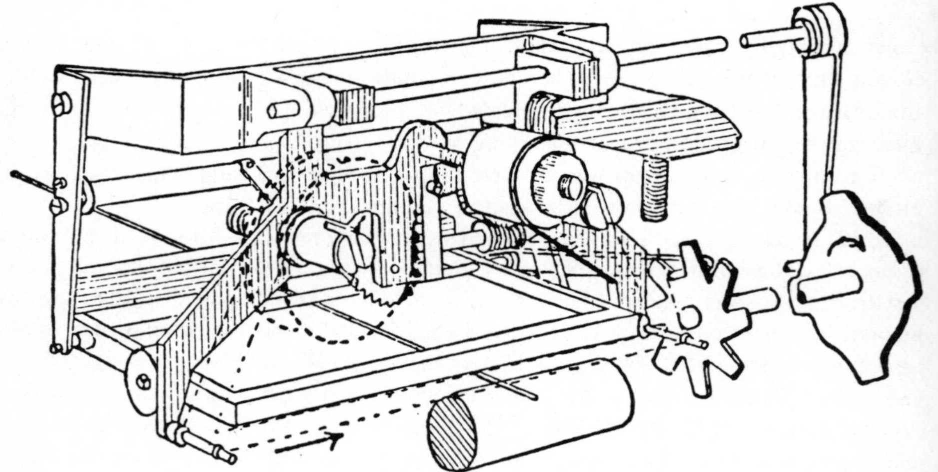

Fig. 6.31 *Mechanism used in R.W. Paul chopper-bar recorder c1914*
[From catalogue of R.W.Paul, 1914, Section M, p.889]

Drive for the charts of recorders in the early decades was usually by clockwork spring or, if electric, by a motor that was governed to provide a fixed speed. Synchronous motors were not viable until the 1930s for before that time, as is shown by the list given in Lewis Jones (1900), existing alternating current mains supplies were not by any means at the same nominal frequency or held to reasonably fixed values. Furthermore not all supplies were of the alternating kind. Darling (1911) gives details of several dotting recorders which were used at that time for temperature work. He discussed a Siemens-Halske version in which the needle stylus forces the paper down onto an inked ribbon running under transparent paper. A similar concept was used in the 1920s Brown, of the US, circular chart recorders. They had a carbon-paper sheet in between the recording paper and the stylus.

Darling also described constant-contact pen recorders which, not being feedback controlled, were prone to errors caused by friction when used to record small energy input signals. He suggested (that being in 1911) that they must be regarded as less accurate and reliable than the intermittent chopper-bar design.

Like so many instrument designs they have not yet been replaced by any obviously better unit for certain purposes and chopper-bar units have continued to be used. The chopper-bar mechanism has also been applied to form a controller unit, Griffiths (1943).

A more recent version of the 1930s, called the 'Canadian radiosonde' version is described in Knowles-Middleton and Spilhaus (1953). It makes use of the concept of time-interval recording. In this design a small endless chain carries four styli arranged at equal intervals along the chain to run perpendicularly across the moving chart. They are caused to make a mark, wherever they might be, by slightly moving the whole chain drive from on-off input signals so as to cause a small portion of a line to be drawn.

Optical recording

Photographic methods of recording data continued to be advanced, a noteable development being the first self-contained developing unit being that designed by the Cambridge Instrument Company for the gunnery ranging system in the 1915 era. Records made of 'telegeophone' sensor output signals were formed on moving strip-film as traces of light transmitted through aperture wires of a multichannel Einthoven galvanometer. It was imperative that the traces be examined speedily so the film was passed directly through an automatic development unit that provided ready-to-view film in one minute. These systems are described in Glazebrook (1922), Vol. IV, and in Cambridge Instrument Company (1955). Fig. 6.32 is of the camera unit of such a system that was later taken to Australia for investigations of the seismic prospecting method. Mason (1977) and Department of Commerce (1921) deal with USA work on this type of equipment.

Photographic recording came into common use once celluloid strip film was introduced. The Minthrop field seismograph of 1919 used it. Walker and Lance (1933) includes illustration of a contemporary recording microphotometer with this form of recording. R.W. Paul began to market a photographic recorder from 1914. Examples are numerous.

Fig. 6.32 *Camera unit of Cambridge Instrument Company sound-ranging system. Certain chemical containers are missing from automatic development sections*

[Object in Bureau of Mineral Resources Geophysics Instrument Collection, Oaklands, Australia]

A prime advantage of the photographic method was its ability to capture fast events. High speed methods were generally called oscillographs (the 'a' spelling suggested by Mullineux-Walmsley not having been taken up).

The falling plate method, used so successfully in 1861 by Feddersen in Germany to record the characteristics of spark radiation, found application in the commercial falling-plate camera shown in Fig. 6.33. This camera was used to record the frequency of sparks by obtaining a record of the number of waves on a photographic plate falling under controlled conditions of speed. The rotating mirror that scans the spark ran at around 5000 r/min. This apparatus was used by the National Physical Laboratory, England in research of 1910 vintage. Glazebrook (1922) Vol. II, and the 1914 trade catalogue of R.W. Paul products both provide extensive descriptions of this and similar recorders.

Other oscillographic equipment of the early 20th century includes the hot-wire recorder to the design of I.J. Irwin, marketed by R.W. Paul around 1912 (see the 1914 catalogue of that company), and the Duddell and Blondel forms already mentioned in the previous chapter. Mullineux-Walmsley (1910) and Glazebrook (1922) have excellent working drawings and etchings of this kind of unit. Hay (1907) and Pohl (1930) discuss oscillographs, the latter giving a record of a very simple kind in which the length of the glow in a small glow tube is photographed on a moving filmstrip to reveal the waveshape of the signal applied to the tube.

The cathode ray form of oscilloscope began to be developed in the 19th century: the work of many people resulted in the measuring device as used today. It is clear, however, that early ideas of who invented the device generally attribute this event to Ferdinand Braun whose original tubes, from the years 1897 and 1898, exist in the Deutsches Museum, Munich. Further information on these developments is available in Schlesinger and Ramberg (1962), Dummer (1977) and Glazebrook (1922) Vol. II, which contains many contemporary references. Both magnetic and electrostatic deflection were incorporated in various pioneering Braun tubes.

It has not been possible to establish just when cathode ray tubes were given graduated screens and calibrated deflection characteristics needed to make them quantitative measuring instruments but they were certainly in routine use in this way by the early 1930s. A graduated unit, dated 1930, exists in the Deutsches Museum. Glazebrook (1922), in the section on radio frequency measurements, gives a list of references that shows that they were in use as form-displaying devices from as early as 1912, perhaps earlier. Mabon (1975), in the biography of J.B. Johnson, accredits the first commercial cathode ray oscilloscope to Johnson's work of 1922.

The cathode-ray tube found considerable interest as a component in television systems from Campbell-Swinton's 1908 use of it for this purpose. Walker and Lance (1933) gives a description of this application. Scroggie (1952) provides an introduction to the historical developments.

Recording cathode ray oscillographs have been described by MacGregor-Morris and Henley (1936) who stated that their book was the first review work on the subject. According to them Dufour built a high-speed oscilloscope in 1913, this being only a few years after the invention of the triode amplifier. A relevant work of slightly earlier origin is Irwin (1925). Cathode ray oscilloscope technology of the 1940s is covered in Greenwood *et al.* (1948). Scroggie (1945) includes a list of contemporary works on cathode ray equipment.

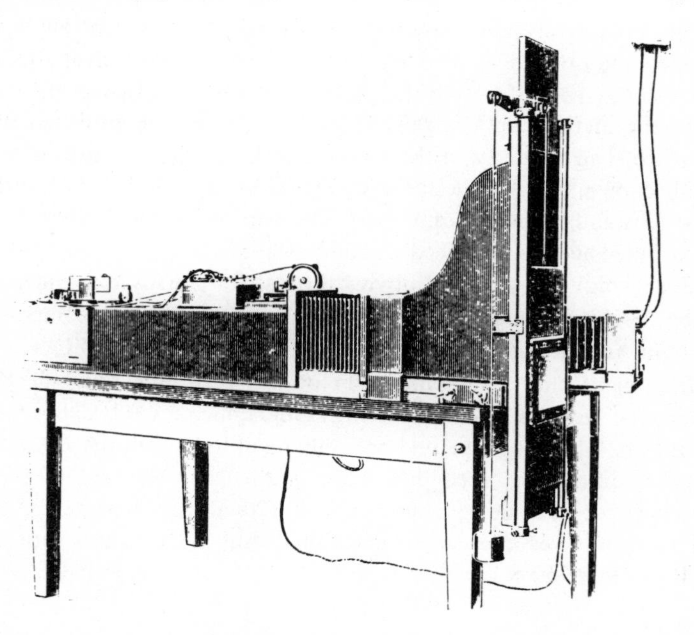

Fig. 6.33 *Radio-telegraphic spark counter made by R.W. Paul*
[From catalogue of R.W.Paul, 1914, Section H, p.703]

Another form of optical recording, that became highly developed as the result of the need for sound synchronised with visual effects in the new 'talking pictures' industry, was the optical sound track method in which sound data was recorded alongside the photographic frames. In 1924 T.W. Case devised his 'Aeo' recording light used by Fox Movietone News, this being about the year that de Forest 'Photrons' came into general knowledge. Photronic cells are described in considerable detail in Morecroft (1936). They were apparently marketed under many trade names, such as the Rayfoto cell and the Weston Photoronic cell. In 1929 E.C. Wente reported his work (originating in 1922) on a variable density light valve, the paper being in the *Bell System Technical Journal*. It used an electromagnetic optical shutter made from a fine metal ribbon. The RCA used a cathode ray oscilloscope to modulate the film sound track.

Walker and Lance (1933) has a chapter on sound recording for the talking films; Batsell (1962) has reviewed the history. Fig. 6.34 shows the general arrangement for recording sound on film. Note that this was made possible with the aid of quality electronic amplification that provided linear gain with wide bandwidth. The Walker and Lance work gives detail about the performance of these systems in terms of dynamic and noise aspects of both film and electronic components. The account also explains how the frequency response of the photocell could be measured. It is of interest to state that the first talking-picture apparatus was not optical but mechanical, Mabon (1975).

The electroreprography method of optical recording, more generally known as the 'Xerox' process and originally called 'electrophotography' by its inventor C.F. Carlson, was first devised in 1938. In 1944 the *Electronics World* magazine, then called *Radio News*, wrote the method up as a general interest article. This provided the publicity that the inventor had needed and the method, at last, began to find commercial interest. A small company in Rochester, New York State, called The Haloid Company, took up the idea and it soon became part Xerox in ownership. The massive development that has since been experienced was underway.

Although generally known as a copying process for already existing material it has since, in the 1960s, become the output stage for a long-distance facsimile system. Accounts about Xerography are to be found in Dummer (1977) and Celent (1965).

Two other significant developments in optical recording arose in the first half century. The first was the invention of the holographic method, Gabor (1972), which was proposed in 1947 but that could not fully be utilised until the 1960s for lack of a coherent light source. The second was the gradual development of the high-speed ultraviolet chart recorder using a moving coil, rather than dual string, galvanometer movement to form a recorder somewhat similar, but not identical with, the Duddell oscillographs of the 1900 era.

Magnetic recording

The idea of using electromagnetic principles to build an electric-signal recorder can be first attributed to the Poulsen and Pedersen 'Telegraphone' that has already been

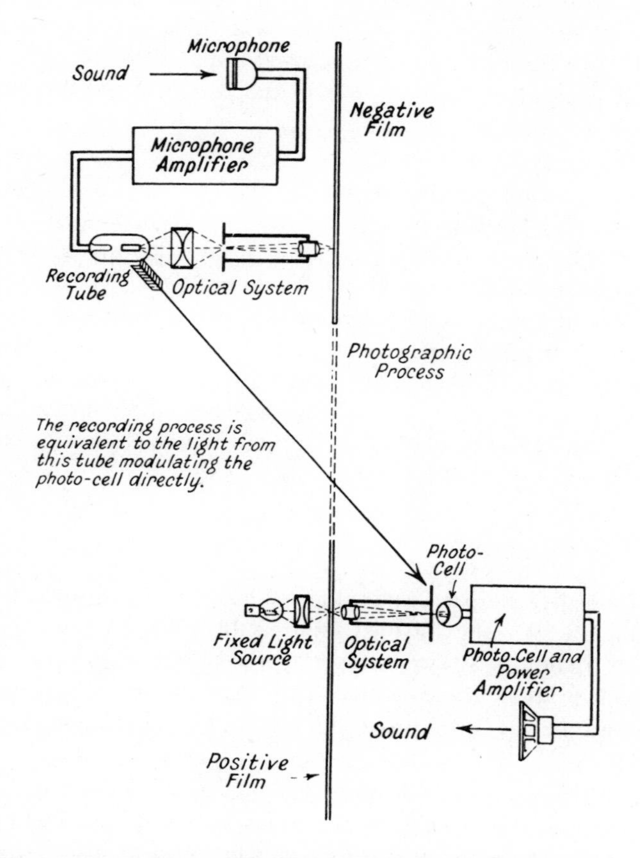

Fig. 6.34 ***Recording and playing a sound-track with talking-picture film***
[Walker and Lance (1933), Fig.47, p.86. Courtesy of Pitman Publishing Ltd.]

mentioned as a late development of the 19th century. They were not the only persons interested in this method; Berliner had used an electromagnetic head in 1882. The concept was patented in several forms and at that time the use of the media of disk, tape and wire were each considered.

In many ways the 'Telegraphone' was ahead of its time for although the design incorporated electric motor drive, electrical input signal form and d.c. tape bias, contemporary workers did not have satisfactory amplifiers to obtain worthwhile performance from the low-level electrical signals that are inherently involved with apparatus using this principle.

Several years were to pass before the next significant development. In 1928 F. Pfleumer, in German Patent No 500 900, described the use of paper tape that is coated with a magnetic material. BASF introduced such a tape around 1931 at the time when electronic technique and electrical sound recording had become topics of general interest with a reasonably capable level of known technology.

The ring-type recorder head was suggested by E. Schuller in 1933. The AEG organisation produced its 'Magnetophon' apparatus in 1935 using Fe_3O_4 coating on a cellulose tape. The modern form of tape recorder had begun to emerge. They were still of a very specialised piece of equipment prior to the Second World War. During those hostilities journalists began to use wire recorders and their experience showed the value of the electrical recorder. Non-electric forms, of course, had been in general office work as the Dictaphone for many years beforehand. P.V.C. tape was introduced in 1944, with polyester tape coming into use 14 years after that.

By 1950 the magnetic-tape recorder was becoming a reliable and versatile method. The well known example of that period is the Wirek recorder.

General histories of this form of recording are to be found in Dummer (1977), who gives references to the papers of value, and in Bachman *et al.* (1962). Hilliard (1962) contains an extensive bibliography. An unpublished lecture, reported briefly *Engineer* (1949), adds a little to the material available on the history of magnetic recording. Ruiter and Murphy (1961) shows how rapidly magnetic recording developed after 1950.

And more recorders

The majority of designs have been covered above but there were others that appeared from time to time. In 1920 staff of the Cambridge Instrument Company noticed that the coefficient of friction of a fine stylus being moved at high speed across celluloid film was considerably lower than expected. This was a chance outcome of work using tuning forks to measure the torsion in a ship's propellor shaft, Cambridge Instrument Company (1945).

This principle was put to use in a recorder marketed as the 'Vibrograph'. The plastic deformation of the clear film caused by the stylus was easily read due to its optical magnifying properties. The principle was used in many sensitive instruments including vibration recorders, a microindicator for engine indicator work and in a stress recorder. Illustrations of some of these appear in Cambridge Instrument Company (1955). This form of recorder was still being manufactured in the middle 1940s.

Glazebrook (1922), in the Vol. 1 discussion of thermocouples, makes mention of a method whereby the tip of a galvanometer, which is not quite in contact with the paper, is used to discharge an electric spark into the paper at each half minute interval. The error due to spark wander is noted but said to be not significant.

Finally, on recorders, mention is needed of the development of digital signal systems. These, as had been shown in the previous Chapter, grew out of telegraphic indenters and punches. Paper-tape punches and readers were well developed by 1900 and continued to be used for many applications of teletechnique. But that was not the only use to which they were put.

Study of the many illustrations given in Eames and Eames (1973) shows that punched tapes and cards played an important part in the 20th century development of statistical data processing equipment. They were used in census and insurance data processing, for example, and formed the input-output peripherals of many calculating devices.

Walker and Lance (1933) describes several systems using punched tape for picture transmission. The code, across a row of five holes, was used to decide the grey level of a dot transmitted. Initially only five levels were coded, one for each hole, but it was extended over time to make better use of the coding range possible with 5 holes.

Recording digital electrical signal information as normal alpha-numeric symbols is performed in the teletype. The modern form of teletype is accredited to C.L. Krum in 1900 but it was in existence in various forms well before that – refer to the previous Chapter. The electric typewriter came into being as the 'Electromatic' made by IBM in 1933, this offering service as a recording peripheral as its keys were operated by electrical impulses.

6.6 Further growth in sensing instrumentation

Over time the types and forms of measuring instruments have increased enormously. In discussion of the methods of Ancient times it was possible to list the few classes of variable that were measured and then treat their subset components. This was done with little fear that some other significant set had been overlapped. In those times measuring instruments of any particular type had few applications in different fields.

Continuous development, modification and new introduction, has today yielded such a vast number of instruments that it seems impossible to satisfactorily classify all transducers by a simplistic flat diagrammatic network of branches. The number of interconnections, the variety of forms of device and the number of basic principles invoked are truly vast. If restrictions are placed on the classification arrangement then it becomes a little more manageable. Such restrictions occur naturally when workers are not informed of the use of similar techniques in other fields or when they consciously decide boundaries to enable them to think more clearly. Whichever the reason, the fact is that a large amount of measurement instrumentation that has been applied over the last fifty years, has come as reinvention. So often one hears of a new idea that, in reality, has already been reported elsewhere.

Another interesting observation is that people tend to regard an idea that apparently could have been made in a much earlier period, but has not been taken up, as already having been tried and not used for the reason that it would now exist if it had been attempted. This philosophy is very far from sound. Many of the ideas that have been applied to great effect often may have been possible in the past but, for a variety of reasons, often do not come to fruition until much later. An important reason is that there may not have appeared a suitable person to develop the concept to a point where it gains acceptance. It is people that make measuring instruments and apply them, not some mystic force. Lack of suitably placed enthusiasm can retard development as equally as can lack of suitable technology.

The period from 1900 to 1950, and thereafter, was one of still greater application of basic sensing principles. Amplification and signal handling methods, added to the stock of measuring tools as electronics during that period, allowed a wider range of people to apply the wider range of available techniques. Increased industrialisation, two major technological wars and increased international technological competition provided incentives for nations to divert resources into this form of activity.

The greatest use of measurement occurred in the 20th century but this study has shown that this is but part of a continuous trend. It is part of the nature of man. He seems destined to build a world to his own design using the resources of that world given him in another form.

New levels in technical achievement and new laws demanded new measuring instruments. Pollution and contamination, for example, gradually became a significant field of measurement. Its beginnings can be traced to the decades before, evidence being seen in such works as Austin (1883). Haldane and Graham (1935) is concerned with the quality of air in such places as mines. It covers poisonous gases, particulate levels, lack of oxygen and other variables that are obviously important to health and survival. Public Health Acts introduced to control pollution required improved means to test levels. In the early times such vague criterion as 'shall not emit smoke' were used. These have gradually, but by no means entirely, given way to more objective measurable criteria which, in turn, required new forms of measuring instruments. An extensive study on pollution, illustrating this point, is Meetham (1952).

Methods of measurement have generally developed from an initial application that spread to other users who adopt them for the same, or different use. Commercial manufacture of instrumentation has altered that trend. Manufacturers, having made a product, seek other fields for sales and so have helped education and transfer of technique across traditional boundaries. In very recent times there has emerged national institutions, such as research laboratories and academic groups who view the art of measuring on the hardware as a starting point.

The field of testing is today an identifiable activity, although there appears to be little difference between testing and measuring as both require the quantification of attributes to given standards. Batson and Hyde (1931*a, b*) is a two-volume work on mechanical variables. Sweeney (1953) is a later account of the same field. Testing of metals became very important in technology with its roots going back several centuries. In the 20th century it became very scientific as can be seen from study of O'Neill (1934), Chalmers (1939), Smith (1940) and Chalmers and Quarrell (1941). Astbury (1952) is on a more specialised aspect, that of magnetic testing.

Gauging metrology, the measurement of mechanical parts, came of age in this period, Rolt (1929) being the accepted classic, but not the only published work, on toolroom metrology to be published at that time. Interferometry found extensive use in this field of metrology, Candler (1951). Glazebrook (1922) provides an extensive coverage.

Mechanical measurement requirements brought about the need for other forms of dimensional measuring devices. The new breed of electrical transducers for converting position, length and angle into electrical signals can be traced to such devices as the 1920 'ultra-micrometer' of R. Whiddington. This device made use of the principle that variation in the capacitance of a grid-tuned thermionic valve could be used to vary the frequency of oscillation. It was shown that the principle could be used to detect extremely small displacements, Dummer (1977). Rolt (1929), Vol. 2, contains a chapter on the electrical devices then being introduced into the testing room as alternatives to the mechanical comparitors. They include frequency-varying, valve micrometer circuits, such as the Dowling and Thomas forms of Whiddington's arrangement. He also described several forms of on-off electrical methods for sensing positional contact. A dual carbon-disk stack, Wheatstone-bridge form of extensometer, Fig. 6.35 is also discussed. The bibliography given is valuable. It is noteable that some of the methods reported performed tolerably well without the use of electronic technique and could have been implemented in the previous century.

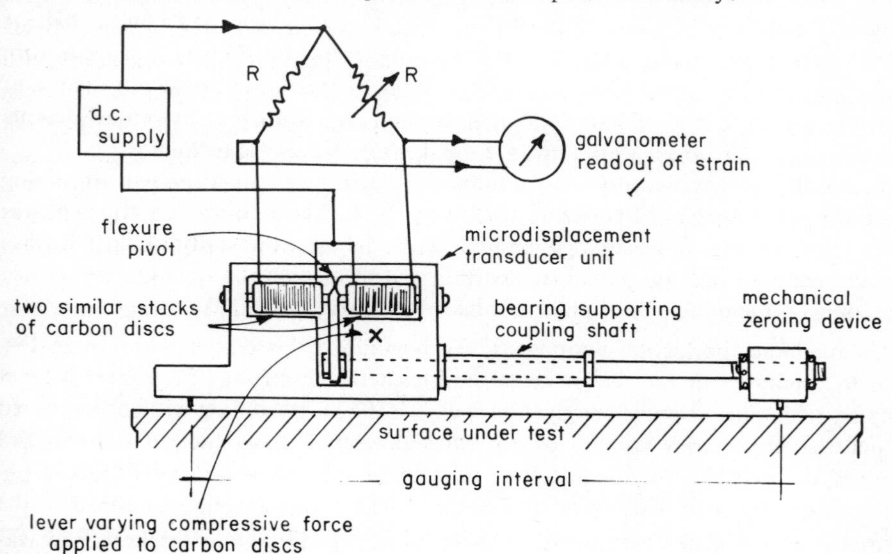

Fig. 6.35 *Schematic arrangement of carbon-disc stack extensometer c1920* [Rolt (1929), Fig. 276, p.256 and Fig. 277, p.257]

The inductive transducer, in the differential transformer connection, was in use by 1938. There exists, on display, an original commercial system made by Bauer and Schaurte-AEG in the Deutsche Museum. It incorporates a carrier modulation detection arrangement giving readout to 0·01 micrometre resolution.

The strain gauge, invented in principle by Kelvin (in 1856) around a century beforehand, began to be used in practice in the 1930s. Kelvin's ideas were confirmed by Bridgman in 1923. It's immediate forerunner was the carbon stack gauge mentioned above, which would have developed out of telephony apparatus. The carbon resistance sensing element was replaced by the more predictable metal resistance wire appearing first in the unbonded form of gauge.

The bonded resistance strain gauge was brought to a satisfactory state of performance as a measuring instrument in 1938. According to Perry and Lissner (1955) this was the result of work from two independent places; through Simmons at the Californian Institute of Technology and Ruge at the Massachusetts Institute of Technology. An initial barrier was lack of a suitable means with which to adhere the wire to a surface so that surface strains were transmitted uniformly to the wire. Their work was recognised by the use of the initials of their names in a classification of these as SR-4 gauges. In 1955 the Baldwin-Lima-Hamilton Corporation held all patents to these gauges and their applications. The metal foil gauge came from the British in the 1950s.

The strain gauge, once introduced as a workable tool, was taken up with enthusiasm; F.G. Tatnall is acknowledged as having greatly assisted its adoption. Gigantic measurement tasks were undertaken with them in the years from 1938. During the Second World War they were used on a very large scale throughout the hulls of some of the ill-fated 'liberty' ships to investigate why the ships were cracking in half. A central telephone exchange was used to enable observers to dial up each gauge in turn to take readings. Perry and Lissner (1955) provides an excellent overview of the use of the strain gauges in the 1950s. Stein (1967) provides insight into the improvements that were introduced from 1938 onward to make it such a useful technique.

Possibly the earliest sensitive use of inductive electromagnetic displacement sensing was in the seismometers of Galitzin, devised in 1904. Prince Galitzin, a Russian, was one of the pioneers of seismology credited with 'dragging the study of earth-quakes from the region of ignorance and superstition and in making it a quantitative science proceeding on the principles of physical philosophy', Walker (1913).

An aperiodic, horizontal-component, seismograph of his design is shown in Fig. 6.36. Incorporated in the design are electromagnetic sensing and permanent-magnet damping of copper. Details are given in Walker (1913). Recording was accomplished by photographic means using a mirror galvanometer. One wonders if Galitzin had heard of the existence of oscillographs for they could have been used to record signals with great effect. This form of sensing provided greater amplification than the traditional mechanical instruments, such as those of Milne and Wiechert, and was devoid of a frictional registration error. It is recorded that they were 'by no means difficult to handle'.

Earth sciences provided demand for a wide variety of measuring apparatus not previously existing. Already mentioned are the gravity balance and the seismograph. The sound-ranging equipment originally developed to locate the position of the big guns deployed in the First World War (mentioned in the previous Section) was later put to peaceful purposes as a contribution in the development of the seismic exploration method. Sydenham (1978*b*) provides a starting point into the literature of those early days of applied geophysics. Chapman (1936) is a work concerned with the magnetic regime of the Earth. Knowles-Middleton and Spilhaus (1953) is a good account of meteorological instrumentation.

It seems that in whichever activity one seeks evidence of measurement hardware, there are developments to be found. The short-lived airship era posed certain important measurment problems, one being the rate of diffusion of gas through the material coverings. As early as 1831 it had been shown that materials were not absolute barriers to gas flow. Rough measurements of diffusion were made by Mitchell in that year. Balloon losses provided a very real need for practical instruments. Many tests were devised, see Glazebrook (1922), Vol. V.

Fig. 6.36 *Galitzin seismograph having electromagnetic output c1910*
[Walker (1913), plate 3]

Fig. 6.37 shows equipment developed to the designs of G.A. Shakespear of Birmingham University. He was commissioned, in 1905, by the British Admiralty to develop an instrument for measuring the permeability of balloon fabrics to hydrogen gas. The instrument was called a 'katharometer'. It was based upon the principle that gases exhibit different thermal conductivity with changing concentration. Here is seen the early foundation of the sensing stage now used in gas chromatographs and other gas analysers. The katharometer is also described in Cambridge Instrument Company (1945, 1955). Gas leak detection was studied in the US, Department of Commerce (1921), for balloon, submarine and aircraft situations.

Oxygen analysers began through the theoretical work of such people as Senftleben whose work of 1930 eventually led to appreciation of the magnetic wind effect that can be set up due to the paramagnetic properties of oxygen. In 1941 Lehrer devised an instrument to make use of this principle as an oxygen analyser. The development of oxygen analysers is covered in Medlock (1952).

Sound recording and replay for films, records and, later, domestic tape decks and the needs for devices to replay recorded and transmitted sound received via broadcast radio programmes was another area in which considerable development occurred and many transducers and actuators for converting sound pressure into electrical signals, and vice versa, came into being.

Dummer (1977) credits the first diaphragm type of microphone to Reis' of 1860, that shown earlier in Fig. 2.13. Movement of the membrane due to sound pressure variation was coupled to a variable metal-to-metal contact that modulated an electrical energy source. It was Edison, in his 1877 telephone apparatus, who built the first carbon style of microphone. He made use of carbon granules, details are given in Dummer (1977), to modulate the supply. This technique was used in many forms, see Preece and Maier (1899) and Herbert (1923). The condensor form was the in-

Fig. 6.37 *Hydrogen permeability apparatus after Shakespear and Daynes c1920* [Cambridge Instrument Co. (1955), Fig.25, p.13. Courtesy The Cambridge Instrument Co. Ltd.]

vention of E.C. Wente in 1917 who designed a 'uniformly sensitive instrument' by trading sensitivity for flatness of frequency response making up the level with active amplification, Dummer (1977).

At the output end of an audio system is an actuator, the loudspeaker. Benefiting from Dummer's research we find that he is able to accredit the first loudspeaker design to E.W. Siemens who, in 1877, devised a 'motor mechanism consisting of a circular coil located in a radial magnetic field'. He state that Lodge, Pridham, Jenson and others further improved the concept of this first moving-coil loudspeaker.

Apparently little improvement then occurred until a more theoretical understanding had been developed in which the important parameters of an acoustic system were recognised. In 1925 Rice and Kellogg laid down three basic design requirements concerning sound-output power, dynamic impedance laws and resonant frequency behaviour.

The electrostatic form of loudspeaker came into being around 1925. Dummer (1977) declines to list an individual to which this design can be accredited. Instead he lists many patents and, through a short explanation, leaves it to the reader to interpret the course of events that led to its emergence as a working, practical device.

Hunt (1954) provides a short history of loudspeaker engineering; Olsen (1962) also is relevant. Microphones are studied in Bauer (1962). General aspects of electro-acoustics are to be found in Beranek (1962). The use of the piezoelectric effect in electrical communications is the subject of a review by Bottom (1962). Transducers used in record players of around the 1950s period were often referred to as the 'transducer'. Lay understanding of the word was that a transducer was a specific term for the pick-up element of such a replay head.

From the study of acoustics in the 1920-1930 period emerged the concept of electromechanical analogies. This was a significant contribution to the systems approach that became dominant in later decades. Olson (1943) is a key systematic work on electromechanical-acoustical analogies.

The nature of this work does not permit extended coverage of the many transducer principles and measurement techniques that emerged in the first half of the 20th century. They occurred in agriculture, Addyman (1904); in paper testing, Jahans (1931); in illumination engineering, Stine (1900) and in many more fields of endeavour.

Electrical engineering continued to need an ever-increasing range of measurement apparatus. Many of the references given in the previous chapter are relevant to this period as thay span the 1900 point. Others of specific relevance to the early 20th century are Parr (1903), Griffiths (1920), Westinghouse (1920, 1921), Golding (1940) and Edison Electric Institute (1940) which provide accounts of electrical instruments.

6.7 Measurement in the empirical sciences

The first 50 years of the 20th century was a time when the numerous inventions and discoveries of the Victorian era came to maturity, finding out which were viable and which were not, through the test of time. To close this general history of the development of measuring instruments and allied instrumentation the account finishes with an appraisal of what are the beginnings of a most significant development in measurement, the realisation that it may be possible to conduct measurements and design the necessary hardware using rigorous and systematic rules.

Such an understanding is developing very much as the result of work in areas of knowledge seeking where the variables are not always obvious, where they are often so interconnected that they cannot be isolated for individual study and where measurement variables are known to exist but are often not accessible.

These academic studies generally go under such general descriptors as the soft sciences, the empirical sciences, the inexact sciences or the fringe sciences. These terms have been adopted in an attempt to say that these kinds of pursuit seek to apply the rigorous scientific method but with less success as has, say, the discipline of physics.

Ellis (1966), when introducing the reader to his book on the basic concepts of measurement commented upon the lack of interest in the philosophy concerned:

> Measurement is the link between mathematics and science. The nature of measurement should therefore be a central concern of the philosophy of science. Yet, strangely, it has attracted little attention. If it is discussed at all in works on the philosophy of science, if is usually dismissed in a fairly short and standard chapter. There are notable exceptions; but for the most part the logic of measurement has been treated as though it were neither interesting nor important. Why this should be so, I do not know. Its importance can hardly be questioned: to understand how mathematics is applied is to understand the most significant feature of modern science.
>
> The pioneer work in the field was undoubtedly Campbell's *Physics; The Elements*, which appeared in 1920. This was followed by his *Account of the Principles of Measurement and Calculation* in 1928. But apart from Bridgman's *Dimensional Analysis* (1931), there have been no major works of a primarily philosophical nature dealing with measurement since that time – although there have been many individual papers in scientific and philosophical journals, and many technical books concerning measurement.

He then continued with a discussion of the prevailing state of affairs. In Chapter 1 this subject has been covered from the instrument designer's viewpoint, rather than as a philosopher, the same conclusion being expressed. Chapter 2 has further reflected the situation in discussion of the past lack of organised instrument design using basic principles that extend continuously down from the specific, well developed, hardware level of instrumentation to the philosophical concepts level upon which an endeavour hopes to base its prime decision-making processes.

There is very little evidence before this century, of any measurement practitioner giving serious, prolonged, thought to the philosophy of a measurement. Measurements were simply allowed to happen as an extension of what had already been achieved. Development, shown in the preceding chapters occurred, nevertheless and has been able to advance enormously without practitioners needing to be concerned with what happens in the process by which knowledge is gained.

In many instances of the late 19th century and continuing well into this century, where discussion of measurement was a key subject in the work concerned such as Kollrausch (1886), Thomson and Tait (1883) and Glazebrook (1922), it is as though the writers believed that their readers were already totally familiar with the fundamental nature and characteristics of a measurement. Philosophers and people working close to philosophy began to question the process; periodically over the last fifty years there has arisen circumstances where a small effort sprang up forming the nucleus of what could have become a critical-mass of effort that might have continued the study with some inertia. These have, however, generally faded away after providing a small contribution to progress.

Writers in this aspect of measurement have not usually been practising engineers or physical scientists, for people engaged in those pursuits generally have been content with the fact that they have been able to devise what ever hardware they needed to measure the variables that have arisen.

In certain disciplines, however, such as economics, education, psychology, human geography, certain aspects of physiology and many more, the variables are not as easy to identify and isolate. Indeed in many instances debate still centres on what are the important variables.

Difficulties that these disciplines have had to face in this way and the inability to measure very much at all with conventional hardware measuring apparatus has led these areas to seriously question what a measurement is and how they might use the process to gain knowledge in their particular situations. It has been more from such sources that measurement philosophy is emerging than from the areas of the exact sciences. It is only in quite recent times that both extremes have come to jointly recognise the situation prevailing.

Brakel (1977) has compiled a review study of measurement in the empirical sciences, seeing it from the standpoint of a fundamental physicist. His extensive bibliography (which is currently being extended into a major bibliographic project) provides evidence of the many comparatively isolated contributions that have been made in this century. Out of numerous, generally nonconnected, studies is emerging the realisation that all disciplines are attempting to find means to handle the same kinds of measurement requirement. It is their applications, the situations, that are different, not the fundamental issues involved.

Whereas it seems that the exact scientists were not formerly interested in the nature of a measurement beyond simplistic mechanistic errors handling levels, the psychologists certainly were. Various references and statements about the measurement in McLellan and Dewey (1907), which were compiled in 1895, show that they were considering the fundamental issues involved rather than the mechanics of implementing them. Mention has already been made in the previous Chapter to the work of Galton in the field of anthropometrics. Fig. 5.59 shows how he made use of measurement instrumentation, albeit of simple kind, in his quantification of key attributes appropriate to early understanding of that subject.

In education there arose a strong interest in quantifying the performance of students. Binet's intelligence tests, first introduced as the Binet-Simon scale of the 'mental age' of children, came into being in 1908. The tests that were used involved written work and the use of testing machines for assessing human performance. Their work was closely linked to that of Galton. Eames and Eames (1973) shows several pictures of the laboratory work involved, one of which is given as Fig. 6.38.

Ross (1941) includes a chapter on the historical development of measurement in education. It is clear from his account that tests were devised which attempted to measure complex situations of interconnected and apparently unravellable variables giving the results as simplistic singular quantities such as, pass or fail in the two-state measurement case or an IQ number in the extended scale situation. He covers such measurement parameters as examinations, tests, character, personality and intelligence.

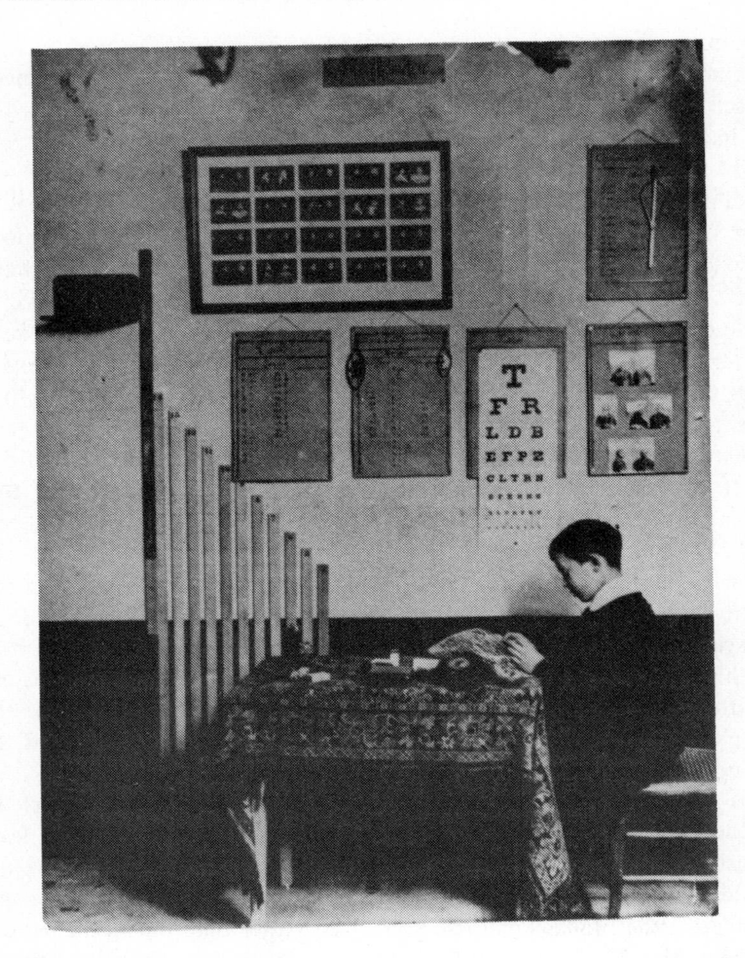

Fig. 6.38 *Measurement in education by Binet c1908*
[Eames and Eames (1973), p.45. Courtesy Les Demoiselles Binet, Paris]

Ross's chapter on the characteristics of a satisfactory measuring instrument is not about the development of hardware but is concerned with the structuring level of a measurement. (By dictionary definition, an instrument is 'a thing with or through which something is done or effected'). Undoubtedly workers in the empirical sciences would be pleased to be able to use instrumentation in the same hardware forms as can the sister exact sciences but, as yet, the stages of development existing, in face of the peculiar types of problems, have severely limited the extent to which this can be done.

Another example of the mode of operation of measurement in the empirical sciences can be seen from the reported work of the US government on assessment of the proficiency of tradesmen. Chapman (1921) is the final report of a committee set up to establish an instrument through which the army could select trades personnel.

It does not directly debate the issue of the philosophy concerned as does Ross (1941) but exemplifies the practice used through the procedures suggested.

In 1941 there was published an account of the experience gained with the use of one of the tests used in the counselling process. That work led to more extensive book, Darley and Hagenah (1955). It is concerned with the test known as the 'strong vocational interest blank'. It is but one more example of the struggle that the inexact sciences are engaged upon owing to the highly complex interactive, interconnected and inaccessible nature of the situation that they need to quantify. A more recent further example is Bunker (1964) on the measurement of office work. The range of this kind of account is great, each group reporting their particular experience with yet another application of the same fundamental problem.

Many apparently new fields have emerged in this century that are concerned with metrology of the less exact sciences. Such endeavours as biometrics, econometrics, anthropometrics are similar in that they are based upon the application of statistical procedures to differing applications.

This proliferation of seemingly new, unique, endeavours has not helped the emergence of a single basic field of measurement science. In the recent times there has emerged clear evidence, such as overlapping citation and interactive groups, that these formerly isolated endeavours are coming together as they each begin to recognise the commonality of their procedures.

If the experience of the measurement of such variables as temperature is any indication, see Ellis (1966) for a translation of a most enlightening 1896 discussion, it is to be expected that the empirical sciences will gradually and steadily find ways to utilise hardware forms of measuring instrumentation, providing more rigidity of attribute codification and which yield more highly filtered data for conversion into knowledge.

Bibliography

Dates given are editions consulted. Earlier editions are noted to provide a guide to date when the work was first state-of-the-art.
Most of the books are not in print; difficulty will be experienced in obtaining many of the works. National libraries and National interlibrary loan services should generally be able to supply copies.

ABC (1977): 'Museums and galleries in Great Britain and Ireland', ABC Travel Guides Ltd, Dunstable, England (Published July of each year)

ABRAMSON, A. (1955): 'Electronic motion pictures: a history of the television camera', Facsimile reprint, (Arno Press, New York, *c*1974)

ADDYMAN, F.T. (1904): 'Agricultural analysis' (Longmans Green, London)

Admiralty Compass Dept. (1941): 'BR.9 Manual of the Admiralty gyro compass (Sperry type) 1941'. Admiralty Compass Department CD 210/41, June 1941, HMSO, London (Replaces issue of 1931)

AFRICA, T.W. (1968): 'Science and the state in Greece and Rome' (Wiley, London)

Air Ministry (1939): 'Manual of air navigation–Vol. I'. Air Pub. 1234, HMSO, London

AIRY, W. (1896): 'Autobiography of Sir George Biddell Airy' (Cambridge University Press)

AKED, C.K. (1971): 'Electricity, magnetism and clocks', *Antiquarian Horology,* Dec., pp.398-415

ALLEN, H.S. (1925): 'Photo-electricity' (Longmans Green, London)

ALLEN, H.S., and MAXWELL, R.S. (1944): 'A textbook of heat' (Macmillan, London) Part I, 1944; Part 2, 1945, 1939 (1st edn.)

ALLSOP, F.C. (1902): 'Practical electric bell fitting' (E. & F. N. Spon, London) and (Spon & Chamberlain, New York)

ALTHIN, T.K. (1948): 'C.E. Johansson, 1846-1943: the master of measurement' (Johansson, Stockholm)

ANDERSON, D.L. (1964): 'The discovery of the electron' (Van Nostrand, Reinhold, New York)

ANDRADE, E.N. da C. (1960): 'A brief history of the Royal Society', Royal Society, London

ANGWIN, A.S. (1951): 'Telecommunications'. Proceedings of the Engineering Conference, Institutions of Civil, of Mechanical, and of Electrical Engineers, London, pp. 352-361

APPLETON, G.V. (1932): 'Thermionic vacuum tubes' (Methuen, London)

ARCHER, G.L. (1938): 'History of radio to 1926' (American Historical Society, New York)

ARDLEY, W.G. (1953): 'The history of flow measurement by differential pressure. IV – the development of venturi recorders from 1894-1911', *Instrument Engineer,* **1**, pp. 83-86

ARMITAGE, A. (1962): 'William Herschel' (Thomas Nelson, London)

ARMITAGE, A. (1966*a*): 'Edmond Halley' (Nelson, London)

ARMITAGE, A. (1966*b*): 'John Kepler' (Faber & Faber, London)

ARMYTAGE, W.H.G. (1957): 'Sir Richard Gregory – his life and work' (Macmillan, London)

Arno Press (1974): 'Eyewitness of early American telegraphy'; Three first-person documents of 1845, 1914, reprinted in facsimile, (Arno Press, New York)

ARTZT, M. (1945): 'Survey of d-c amplifiers', *Electronics*, **18**, **8**, pp.112-118
ASTBURY, N.F. (1952): 'Industrial magnetic testing'. Institute of Physics, London
AUSTIN, G.L. (1883): 'Water analysis: a handbook for water drinkers' (Lee & Shepard, Boston)
AVEBURY, (1903): 'Essays & addresses 1900-1903' (Macmillan, London)
BACHMAN, W.S., BAUER, B.B., and GOLDMARK, P.C. (1962): 'Disk recording and reproduction', *Proc. IRE*, **50**, pp. 738-744
BAILEY, A.E. (1977): 'The measures of man', *Proc IEE*, **124**, (1), pp.83-88
BAILEY, C. (1923): 'The legacy of Rome' (Clarendon Press, Oxford)
BAKER, E.C. (1976): 'Sir William Preece, F.R.S. – Victorian engineer extraordinary' (Hutchinson, London)
BAKER, J.F. (1951): 'The development and trend of University education in Engineering'. Proceedings of the Joint Engineering Conference, Institutions of Civil, and of Mechanical, and of Electrical Engineers, London, pp. 261-264
BAKER, W.J. (1970): 'A history of the Marconi company' (Methuen, London)
BALDWIN, F.G.C. (1925): 'The history of the telephone in the United Kingdom' (Chapman & Hall, London)
BARBER, T.W. (1906): 'The engineer's sketch book of mechanical movements, devices, appliances, contrivances and details, (Spon, London)
BARLOW, P., and HERSCHEL, J.F.W. (1848): 'The encyclopaedia of mechanical philosophy' (John Joseph Griffin & Co., London) (Similar work published 1854, by Richard Griffin, Glasgow)
BARNES, B.R., and SILVERMAN, S. (1934): 'Brownian motion as a natural limit to all measuring processes', *Rev. Mod. Phys*. **6**, pp. 162-192
BARRY, P.D. (1969): 'George Boole – a miscellany' (Cork University Press, Dublin)
BATCHER, R.R. (1962)' 'The influence of product design on radio progress', *Proc. IRE*, **50**, pp. 1289-1305
BASTEL, M.C. (1962)' 'Film recording and reproduction', *ibid*., **50**, pp. 745-751
BATSON, R.G., and HYDE, J.H. (1931*a*): 'Mechanical testing – Vol. I: testing of materials of construction' (Chapman & Hall, London)
BATSON, R.G., and HYDE, J.H. (1931*b*): 'Mechanical testing – Vol. 2, testing of prime movers, machines, structures and engineering apparatus' (Chapman & Hall, London)
BATTEN, M. (1968): 'Discovery by chance – science and the unexpected' (Funk & Wagnalls, New York)
BAUER, B.B. (1962): 'A century of microphones', *Proc. IRE*, **50**, pp. 719-729
BBC (1958): 'The history of science' (Melbourne University Press, Victoria) Australia, 1951 (1st edn., London)
BECKMANN, J. (1846)' 'A history of inventions, discoveries and origins', (Henry G. Bohn, London, 2 Vols 4th edn.)
BEGGS, J.S. (1955): 'Mechanism' (McGraw-Hill, New York)
BELEVITCH, V. (1962): 'Summary of the history of circuit theory', *Proc. IRE*, **50**, pp. 848-855
BELL, A.E. (1947): 'Christian Huygens and the development of science in the seventeenth century' (Edward Arnold, London)
BELL, A.G. (1908): 'The Bell telephone: the deposition of Alexander Graham Bell in the suit by the United States to annul the Bell patents'. Facsimile reprint, (Arno Press, New York, *c*1974)
BELL, G. (1978): 'Computer generations', *Electron. News*, April, pp. 36-37
Bell Telephone (1957): 'Overall view of exhibit on the telephone in America' (at Smithsonian Institution), Bell Telephone System and Independent Telephone Companies; copies available National Museum of History and Technology, Smithsonian Institution, Washington, DC
BENCE JONES, H. (1896): 'Life and letters of Faraday' (Longmans Green, London, 2 Vols.)
BENDICK, J. (1947): 'How much and how many: the story of weights and measures' (McGraw-Hill, New York)
BENEDICT, R.R. (1951): 'Introduction to industrial electronics' (Prentice-Hall, New York)

BENNETT, A.R., and LAWS, W.H. (1895, 1910): 'The telephone systems of the continent of Europe' and 'The development of the telephone in Europe'. Facsimile editions bound as one volume, (Arno Press, New York, 1974)

BENNETT, J.A. (1964): 'Resistive thin films and thin film resistors history, science and technology', *Electron. Components & App.* ,**5**, pp. 737-748

BENNETT, S. (1979): 'History of control engineering' (Peter Peregrinus, London)

BENTON, W.A. (1941): 'The early history of the spring balance' *Trans. Newcomen Soc. Study Hist. Eng. & Technol.*, **22**, pp. 65-78, plus plates VII, VIII, IX and X.

BERANEK, L.L. (1962): 'Electroacoustic measuring equipment and techniques', *Proc. IRE*,**50** pp. 762-768

BERKELEY, E.C. (1949): 'Giant brains or machines that think' (Wiley, New York)

BERRIMAN, A.E. (1953): 'Historical metrology. A new analysis of the archaeological and historical evidence relating to weights and measures' (Dent, London)

BERRY, A.J. (1960): 'Henry Cavendish – his life and scientific work' (Hutchinson, London)

BHOLANATH, L. (1970): 'Maharajah Jaisingh's astronomical observatory, Delhi'. Pt. Shib Lal, Jantar Mantar, Delhi, India (Date approximate)

BILLMEIR, J.A. (1953): 'Scientific instruments (13-19th Century)'. Catalogue of Collection of J.A. Billmeir, as exhibited at Museum of History of Science, Oxford, 1955

BINGLEY, F.J. (1962): 'A half-century of television reception', *Proc. IRE*, **50**, pp. 799-805

BISWAS, A.K. (1970): 'History of hydrology' (North Holland, Amsterdam)

BLACK, H.S. (1934): 'Stabilized feed-back amplifiers', *Bell Sys. Tech. J.*, **13**, pp. 1-18. See also 'Stabilized feed-back amplifiers' *Electr. Eng.*, **53**, p. 114, 1934

BLACK, R.M. (1972): 'Electric cables in Victorian times' (HMSO, London)

BLAKE, G.C. (1928): 'History of radio telegraphy and telephony'. Facsimile of 1928 edition, (Arno Press, New York)

BONNARD, A. (1961): 'Greek civilisation – Vol. 3: from Euripides to Alexandria' (George Allen & Unwin, London)

BORING, E.G. (1929): 'A history of experimental psychology' (Century, New York)

BORTH, P.E.A. L.A. (1895): 'Gesammelte Werke', translated from German, (Macmillan, London, 3 Vols.)

BOS, H.J.M. (1968): 'Mechanical instruments in the Utrecht University Museum' (Utrecht University Museum, Utrecht)

BOSMAN, D. (1978): 'Systematic design of instrument systems', *J. Phys. E.*, **11**, pp. 97-105

BOTSFORD, G.W., and SIHLER, E.G. (1915): 'Hellenic civilization' (Columbia University Press, New York)

BOTTOM, V.E. (1962): 'Piezo-electric effect and applications in electrical communication', *Proc. IRE*, **50**, pp.929-931

BOWERS, B. (1969): 'R.E.B. Crompton – an account of his electrical work' (HMSO, London)

BOWERS, B. (1970): 'X-rays' (HMSO, London)

BOWERS, B. (1975): 'Sir Charles Wheatstone FRS, 1802-1875' (HMSO, London)

BRADBURY, S. (1967): 'The evolution of the microscope' (Pergamon, Oxford)

BRADDICK, H.J.J. (1966): 'The physics of experimental method' (Chapman & Hall, London), 1954 (1st edn.)

BRAKEL, J.V. (1977): 'Meten in de empirische wetenschappen'. University of Utrecht (Vakgroep Onderzoek van de Grondslagen van de Natuurkunde), June

BRAUER, E. (1909): 'The construction of the balance'. Incorporated Society of Inspectors of Weights and Measures, London. (Translated by H.C. Walters.)

BRAUN, E., and MACDONALD, S. (1978): 'Revolution in miniature: the history and impact of semiconductor electronics' (Cambridge University Press)

BRAY, F. (1930): 'Light' (Edward Arnold, London) 1927 (1st edn.)

BREASTED, J.H. (1944): 'Ancient times – a history of the early world' (Ginn & Co., Boston), 1916 (1st edn.)

BREWSTER, D. (1813). 'Treatise on new philosophical instruments' (John Murray, Edinburgh)

BRIGGS, L.J. (1949): 'NBS War Research – the National Bureau of Standards in World War II', National Bureau of Standards, Washington, DC

BRIGHT, C. (1898): 'Submarine telegraphs: their history, construction and working'. Facsimile reprint of 1898 edition, (Arno Press, New York, *c*1974)

British Association (1931): 'London and the advancement of science', British Association, London

BRITTAIN, J.E. (1977): 'Turning points in American Electrical Engineering' (IEEE Press, New York) (Selected historic papers)

BRODETSKY, S. (1927): 'Sir Isaac Newton – a brief account of his life and work' (Methuen, London)

BRODRICK, B.S.J. (1964): 'Galileo – the man, his work, his misfortunes' (Chapman & Hall, London)

BROTHERS, J.T. (1962): 'Historical development of components parts field', *Proc. IRE,* **50**, pp. 912-919

BROTHERTON, M. (1972): 'The magic crystal: how the transistor revolutionized electronics'. Bell Laboratories, PE-111

BROUGHAM, L. (1855): 'Lives of philosophers' (Richard Griffin, Glasgow)

BROWN, A.E., and JEFFCOTT, H.A. (1970): 'Absolutely mad inventions' (Dover, New York), 1932 (1st edn.)

BRUCE, R.V. (1973): 'Alexander Graham Bell and the conquest of solitude' (Little Brown, Boston)

BUNKER, L.H. (1964): 'Measuring office work' (Isaac Pitman, Lond)

BURGESS, G.K., and LE CHATELIER, H. (1912): 'The measurement of high temperatures' (John Wiley, New York)

BURSTALL, A.F. (1970): 'A history of mechanical engineering' (Faber & Faber, London) 1963 (1st edn.)

BURTT, P. (1926): 'Control on the railways' (George Allen & Unwin, London)

BURY, R.G. (1952): 'Plato (with an English translation) IX, Laws', (Heinemann, London) Vol.1, 1926 (1st edn.)

BUSHNELL, O.J., and TURNBULL, A.G. (1920): 'Meter testing and electrical measurements'. (American Technical Soc., Chicago.)

BUTLER, R.R. (1947): 'Scientific discovery' (English Universities Press, London)

BUTTERFIELD, A.D. (1906): 'A history of the determination of the figure of the earth from arc measurements' (Davis Press, Worcester, Mass.)

BYRN, E.W. (1900): 'The progress of invention in the nineteenth century' (Munn, New York)

CADELL, C.S., and O'KANE, B.J. (1951): 'Forty years progress in air radio'. Proceedings of the Joint Conference, Institutions of Civil, of Mechanical and of Electrical Engineers, London, pp. 245-261

CAJORI, F. (1962): 'A history of physics' (Dover, New York), 1929 (1st edn.)

CALVERT, H.R. (1971): 'Scientific trade cards' (HMSO, London)

Cambridge Instrument Co. (1945): '50 years of scientific instrument manufacture', *Engineering,* **159**, pp. 361-363; pp. 401-403; pp. 461-463; pp. 501-502

Cambridge Instrument Company (1955): '75 years'. Internal publication, Cambridge Instrument Co., Cambridge, England

CAMPBELL, L., and GARNETT, W.C. (1884): 'The life of James Clerk-Maxwell' (Macmillan, London)

CAMPBELL, N.R. and RITCHIE, D. (1929): 'Photo-electric cells' (Isaac Pitman, London)

CANDLER, C. (1951): 'Modern interferometers' (Hilger, London)

CARDWELL, D.S.L. (1972): 'The organisation of science in England' (Heineman, London), 1957 (1st edn.)

CARHART, H.S., and PATTERSON, G.W. (1897): 'Electrical Measurements' (Allyn & Bacon, Boston)

CARNEAL, G. (1930): 'A conqueror of space: an authorized biography of the life and work of Lee De Forest' (Constable, London)
CAROE, G.M. (1978): William Henry Bragg' (Cambridge University Press)
CARPENTER, R.C. (1892): 'A text book of experimental engineering for engineers and students in engineering laboratories' (Wiley, New York)
CARR, F.G.G. (1967): 'The pictorial history of Maritime Greenwich'. (Pitkin Pictorials Ltd., London)
CARTER, E.F. (1974): 'Dictionary of inventions and discoveries' (Frederick Muller, London) 1966 (1st edn.)
CARTER, J., and MUIR, P.H. (1967): 'Printing and the mind of man' (Cassell, London)
CASSON, H.N. (1929): 'Kelvin – his amazing life and worldwide influence', (Efficiency Mag. London)
CELENT, C.M. (1965): 'The art of xerography', *Electron. World,* **74**, pp.25-28; 72; 74
CHAFFEE, E.L. (1933): 'Theory of thermionic vacuum tubes' (McGraw-Hill, New York)
CHALDECOTT, J.A. (1954): 'Handbook of the collection relating to heat and cold: Part II' (HMSO, London)
CHALDECOTT, J.A. (1976): 'Temperature, measurement & control, Part II'. (Science Museum, London) 1955 (1st edn.)
CHALMERS, B. (1939): 'The physical examination of metals – vol. I, optical methods' (Edward Arnold, London)
CHALMERS, B., and QUARRELL, A.G. (1941): 'The physical examination of metals – Vol. 2, electrical methods' (Edward Arnold, London)
CHAPMAN, J.C. (1921): 'Trade tests, the scientific measurement of trade proficiency' (George G. Harrap, London)
CHAPMAN, S. (1936): 'The Earth's magnetism' (Methuen, London)
CHESTNUT, H. (1962): 'Automatic control and electronics', *Proc. IRE,* **50**, pp.787-792
CHEW, V.K. (1967): 'Talking machines 1877-1914 – some aspects of the early history of the gramophone' (HMSO, London)
CHEW, V.K. (1968): 'Physics for princes' (HMSO, London)
CHIPMAN, R.A. (1964): 'The earliest electromagnetic instruments'. Paper 38, pp. 21-36, from United States National Museum Bulletin 240: Smithsonian Institution, Washington, DC
CHROUST, G. (1974): 'Bibliography for the history wall of data processing'. (IBM, Vienna)
CHURCH, W.C. (1890): 'The life of John Ericsson' (Sampson Low, Marston, Searle and Rivington, London, 2 vols.)
CLARKE, A.R. (1866): 'Comparisons of the standards of length of England, France, Belgium, Prussia, Russia, India, Australia' (HMSO, London)
CLARK-KENNEDY, A.E. (1929): 'Stephen Hales, D.D., F.R.S. – an eighteenth century biography' (Cambridge University Press)
CLARK, R.W. (1977): 'Edison, the man who made the future' (G.P.Putnam's Sons, New York)
CLAY, R.S., and COURT, T.H. (1932): 'The history of the microscope' (Charles Griffin, London)
COCHRANE, R.C. (1966): 'Measures for progress – a history of the National Bureau of Standards'. National Bureau of Standards, (Washington DC)
COCK, C.M., DYER, H.H., and DELL, R. (1951): 'Electric traction and signalling'. Proceedings of the Conference Institutions of Civil, of Mechanical and of Electrical Engineers, London, pp. 78-91
COHEN, I.B. (1975): 'Benjamin Franklin – scientist and statesman' (Charles Scribner's Sons, New York)
COHEN, M.R., and DRABKIN, I.E., (1948): 'A source book in Greek science' (McGraw-Hill, New York)
COOGAN, C.K. (1966): 'Diffraction gratings', *Electron. Aust.*, **28**, p. 4, pp. 8-11, pp. 13, 14
COUGHANOWR, D.R., and KOPPEL, L.B. (1965): 'Process systems analysis and control' (McGraw-Hill, New York)
COULSON, T. (1950): 'Joseph Henry – his life and work' (Princeton University Press, Princeton)

CRANK, J. (1947): 'The differential analyser' (Longmans Green, London)
CROMBIE, A.C., (1969): 'Augustine to Galileo: 1 – science in the Middle Ages'. (Penguin Books, Harmondsworth, England) 1952 (1st edn.)
CROMPTON, R.E. (1928): 'R.E.Crompton' (Constable, London)
CROWTHER, J.A. (1944): 'Handbook of industrial radiology' (Edward Arnold, London)
CROWTHER, J.G. (1960): 'Founders of British Science' (Cresset Press, London)
CSIR (1947): 'A text book of radar'. Radio-physics Laboratory, Council for Scientific and Industrial Research, Australia. (Angus and Robertson, Sydney)
CUMBERBATCH, E.P. (1921): 'Essentials of medical electricity' (Henry Kimpton, London)
DALBY, W.E. (1904): 'Charlottenburg', the Berlin Technical High School, *Technics*, **1**, pp. 9-15 and pp. 143-150
Danish Journal (1977): 'Hans Christian Orsted'. Collected papers on 200th anniversary of his birth', *Danish J.*, Copenhagen
DARK, E.P. (1934): 'Diathermy in general practice', (Angus & Robertson, Sydney)
DARLEY, J.G., and HAGENAH, T. (1955): 'Vocational interest measurement – theory and practise' (University of Minnesota Press, Minneapolis)
DARLING, C.R. (1911): 'Pyrometry' (E. & F. N. Spon, London)
DARNELL, P.S. (1958): 'History, present status and future developments of electronic components', *IRE Trans.*, **CP-5**, 3, pp. 124-129
DAVIES, D.A., and STIFF, G.M. (1969): 'Diffraction grating ruling in Australia', *Appl. Opt.*, **8**, pp. 1379-1384
DAVIES, I.L. (1976): 'Electronics and the aeroplane', *Proc. IEE*, **123**, (1), pp. 13-15
DAVIS, A.B. (1979): 'An introduction to the history of medicine and its technology' for Greenwood Press, Connecticut. (In preparation, Nov. 1978)
DAVIS, A.B., and MERZBACH, U.C. (1972): 'Graphic recording in psychology: the kymograph' (American Psychological Association, Washington)
DAVIS, A.B., and MERZBACH, U.C., (1975): 'A national inventory of historic psychological apparatus', *J. Hist. Behavioural Sci.*, **11**, pp. 284-286
DAY, A.L., and SOSMAN, R.B. (1911): 'High temperature gas thermometry'. Carnegie Institution of Washington, Pub. No. 157, Washington, DC
DE CARLE, D. (1965): 'Horology' (English Universities Press, London)
DE JUHASZ, K.J. (1934): 'The engine indicator – its design, theory and special applications' (Instruments Publishing Co., New York)
DELLOW, E.L. (1970): 'Measuring and testing in science and technology' (David & Charles, Newton Abbot, UK)
Department of Commerce (1921): 'War work of the Bureau of Standards'. Misc. Pub. No. 46, (Now NBS) Washington, DC
Department of Commerce (1924): 'Radio instruments and measurements'. Bureau of Standards, Circular 74, (Now NBS), Washington, DC
DESCHANEL, A.P. (1891): 'Elementary treatise on natural Philosophy' (Blackie, London, 12th edn.) (4 parts translated by J.D.Everett)
DE SOLLA PRICE, D. (1959*a*): 'An ancient Greek computer', *Sci.Am.*, **200**, 6, pp. 60-67
DE SOLLA PRICE, D. (1959*b*): 'On the origin of clockwork, perpetual motion devices and the compass'. Paper 6, pp. 81-112, United States National Museum Bulletin 218, Smithsonian Institution, Washington, DC
DE SOLLA PRICE, D. (1974): 'Gears from the Greeks: the Antikythera mechanism – a calendar computer from *c*80BC', *Trans. Am. Philos. Soc.*, **64**, 7, pp. 3-70. Also casebound from Science History Publications, New York
Deutsches Museum (1971): 'Deutsches Museum of masterpieces of natural science and technology – illustrated guide' (Deutsches Museum, Munich)
DE VRIES, L. (1971): 'Victorian Inventions' (American Heritage Press, New York)
DIBNER, B. (1954): 'Ten founding fathers of the electrical science' (Burndy Library, Norwalk, Conn.) (Ampere, Faraday, Franklin, Gauss, Geuricke, Gilbert, Henry, Maxwell, Ohm, Volta)

DIBNER, B. (1957): 'Early electrical machines' (Burndy Library, Norwalk, Conn.)
DIBNER, B. (1964): 'Allessandro Volta and the electric battery' (Franklin Watts, New York)
DIBNER, B. (1971): 'Luigi Galvani' (Burndy Library, Norwalk, Conn.)
DICKINSON, C.J. (1950): 'Electrophysiological technique' (Electronic Engineering (Publishing House), London)
DICKINSON, H.W., and VOWLES, H.P. (1949): 'James Watt and the Industrial Revolution' (Longman Green, London), 1943 (1st edn.)
DICKSON, W.K.L., and DICKSON, A. (1894): 'The life and inventions of Thomas Alva Edison' (Chatto & Windus, London)
DISNEY, A.N. (1928): 'Origin and development of the microscope'. Catalogues of Collections of the Royal Microscopial Society, Publishers of same, London.
DOBELL, C. (1932): 'Antony van Leeuwenhoek and his "Little animals" ' (John Bale & Danielsson, London)
DONALD, T.H. (1925): 'Our manufacturing industries' (Isaac Pitman, London)
DONNELLY, M.C. (1973): 'A short history of observatories' (Astronomical) (University of Oregon Books, Eugene)
DONOVAN, F.R. (1963): 'The many worlds of Benjamin Franklin' (American Heritage Pub. Co., New York)
DOOLEY, J.C., and BARLOW, B.C., (1976): 'Gravimetry in Australia, 1819-1976:, *BMR J. Aust. Geol. & Geophys.*, **1**, pp. 261-272
DOUGLAS, J.C. (1877): 'A manual of telegraph construction' (Charles Griffin, London)
DOWSE, T.S. (1889): 'Lectures on massage and electricity' (John Wright, Bristol) or (Adams & Co., London)
DRAKE, J.A. (1879): 'The practical mechanic: comprising a clear exposition of the principles and practice of mechanism, with their application to the industrial arts' (J.W. Lukenbach, Philadelphia)
DRAPER, A.L., and LOCKWOOD, M. (1939): 'The story of astronomy' (Dial, New York)
DRESNER, S. (1971): 'Units of measurement; an encyclopaedic dictionary of units both scientific and popular and the quantities they measure' (Harvey Miller & Medcalf, Aylesbury)
DREYER, J.L.E. (1890): 'Tycho Brahe – a picture of scientific life and work in the sixteenth century' (Adam & Charles Black, Edinburgh)
DRYSDALE, C.V., and JOLLEY, A.C. (1924): 'Electrical measuring instruments' Parts 1, 2. (Ernest Benn, London)
DUBBEY, J.M. (1978): 'The mathematical work of Charles Babbage' (Cambridge University Press)
DUMMER, G.W.A. (1977): 'Electronic inventions 1745-1976' (Pergamon Press, Oxford)
DU MONCEL, T.A.L. (1879): 'The telephone microphone and the phonograph'. Facsimile reprint, (Arno Press, New York, *c*1974)
DUNLAP, O.E. (1937): 'Marconi, the man and his wireless' (Macmillan, London)
DUNLAP, O.E., (1944): 'Radio's 100 men of science' (Harper & Bros., New York)
DUNNINGTON, G.W. (1955): 'Carl Friedrich Gauss – titan of science' (Exposition Press, New York)
DUNSHEATH, P. (1962): 'A history of electrical engineering' (Faber and Faber, London, 2 Vols.)
DURANT, W. (1944): 'Caesar and Christ' (Simon & Schuster, New York)
DYER, F.L., and MARTIN, T.C. (1910): 'Edison his life and inventions' (Harper & Brothers, New York)
DYSON, J. (1970): 'Interferometry as a measuring tool' (Machinery Publishing Co., Sussex)
EAMES, C., and EAMES, R. (1973): 'A computer perspective' (Harvard University Press)
ECKHARDT, G.H. (1936): 'Electronic television'. Facsimile reprint, (Arno Press, New York)
EDGCUMBE, K. (1908): 'Industrial electrical measuring instruments' (Archibald Constable, London)
Edison Electric Institute (1940): 'Electrical metermens' handbook'. Meters and Services Committee, Edison Electric Institute. Pub. G-8, New York, 1923 (4th edn.)

Electronics (1946): 'ENIAC: War Department unveils 18,000 tube robot calculator', *Electronics,* **19**, p.308

ELLIS, B. (1966): 'Basic concepts of measurement' (Cambridge University Press)

ELLIS, K. (1973): 'Man and measurement' (Priory Press, London)

Engineer (1949): 'The development of the magnetic tape recorder', *Engineer,* **184**, p.313

'ESPINASSE, M. (1956): 'Robert Hooke' (Heineman, London)

ETI (1975): 'Early radio patents' *Electron. Today.,* June, pp.26-30 Sydney Edition

EVANS, K.M. (1966): '1891-1966, 75 years Brown Boveri' (Brown Boveri, Baden, Switzerland) (Also in French and German)

EVE, A.S. (1939): 'Rutherford – being the life and letters of the Right Honourable Lord Rutherford' (Cambridge University Press)

EVE, A.S., and CREASEY, C.H. (1945): 'Life and work of John Tyndall' (Macmillan, London)

EVERETT, J.D. (1876): 'Units and physical constants' (Macmillan, London)

EVERITT, C.W.F. (1974): 'James Clerk-Maxwell – physicist and natural philosopher: (Charles Scribner's Sons, New York)

EVERITT, W.L. (1976): 'Telecommunications – the resource not depleted by use – a historical and philosophical resume', *Proc. IEEE,* **64**, pp. 1292-1299

EVERSON, G. (1949): 'The story of television: the life of Philo T. Farnsworth'. Facsimile reprint. (Arno Press, New York, *c*1974)

EWING, A. (1933): 'An engineers outlook' (Methuen, London)

FAGEN, M.D. (1975): 'A history of engineering and science in the Bell System – the early years 1875-1975' (Bell Telephone Labs. Inc., Murray Hill, New Jersey)

FAHIE, J.J. (1884): 'A history of electric telegraphy to the year 1837: Chiefly compiled from original sources and hitherto unpublished documents'. Facsimile report, (Arno Press, New York, *c*1974)

FAHIE, J.J. (1903): 'Galileo – his life and work' (John Murray, London)

FARRINGTON, B. (1936): 'Science in antiquity: (Thornton Butterworth, London)

FARRINGTON, B. (1949): 'Francis Bacon: philosopher of industrial science' (Schuman, New York)

FEATHER, N. (1940) 'Lord Rutherford' (Priory Press, London)

FEINSTEIN, A.R. (1971): 'On exorcising the ghost of Gauss and the curse of Kelvin', *Clinical Pharmacology &Therapeutics*, **12**, pp. 1003-1016

FEIRER, J.L. (1977): 'SI metric handbook' (Wiley, London)

FELGENTRAEGER, W. (1932): 'Fein Waagen Wagungen und Gewichte' (Springer, Berlin)

FERGUSON, E.S. (1962): 'Kinematics of mechanisms from the time of watt'. Paper 27, 185-230, from United States National Museum Bulletin 228, Smithsonian Institution, Washington, DC

FERGUSON, E.S. (1968): 'Biliography of the history of technology' (MIT Press)

FERGUSON, E.S. (1977): 'The mind's eye: nonverbal thought in technology'. *Science,* **197**, 4306, pp. 827-36

FERRANTI, G.Z., DE., and INCE, R. (1956): 'The life and letters of Sebastian Ziani de Ferranti' (Williams & Norgate, London) 1934 (1st edn.)

FESSENDEN, H.M. (1940): 'Fessenden; builder of tomorrow'. Facsimile reprint, (Arno Press, *c*1974)

FINK, D.G. (1938): 'Engineering electronics' (McGraw-Hill, London)

FINK, D.G. (1976): 'Perspectives on television: the role played by the two NTSC's in preparing television service for the American public'. *Proc. IEEE,* **64**, pp. 1322-1331

FINKELSTEIN, L. (1963): 'The principles of measurement', *Trans. Soc. Instrum. Tech.,* **15**, pp. 181-189

FINKELSTEIN, L. (1975): 'Fundamental concepts of measurement: definition and scales', *Meas. & Control,* 8, pp. 105-111

FINKELSTEIN, L. (1976): 'Preliminary study of the economic significance of industrial measurement and instrumentation technology for the United Kingdom'. Internal report DSS/LF/110, City University, London

FINN, B.S. (1963): 'Developments in thermoelectricity, 1850-1920'. Ph.D. dissertation, University of Wisconsin, University Microfilm, Ann Arbor, 63-3923

FINN, B.S. (1966): 'Alexander Graham Bell's experiments with the variable resistance transmitter', *Smithsonian J. Hist.,* **1**, 4, pp.1-16

FINN, B.S. (1971): 'Output of eighteenth century electrostatic machines', *Brit. J. Hist. Sci.*, **5**, **19**, pp.289-291

FINN, B.S. (1973): 'Submarine telegraphy – the grand Victorian technology' (HMSO, London)

FINN, B.S. (1976*a*): 'Franklin as electrician', *Proc. IEEE,* **64**, pp. 1270-1273

FINN, B.S. (1976*b*): 'Growing pains at the cross roads of the world: a submarine cable station in the 1870's', *ibin.*, **64**, pp. 1287-1292

FIRTH, I. (1969): 'N-rays – ghost of scandal past', *New Sci.,* **44**, pp. 642-643

FLEMING, A.P.M. (1951): 'The practical training of mechanical and electrical engineers'. Proceedings of the Joint Engineering Conference, Institutions of Civil, of Mechanical and of Electrical Engineers, London, pp. 286-298

FLEMING, J.A. (1919): 'The thermionic valve and its developments in radiotelegraphy and radiotelephony' (Wireless Press, London)

FLOOD, J.E. (1976): 'Alexander Graham Bell and the invention of the telephone', *Proc. IEE,* **123**, pp. 1387-1388. Full paper M94 available, IEE Library, London

FOLEY, M. (1953): 'De Viribus Electricitatis in Motu Musculari Commentarius' by Luigi Galvani, 1971. Translated by M. Foley, Burndy Library Pub. No. 10, Norwalk, Conn.

FORBES, R.J. (1969): 'Martinus van Marum – life and work' (H.D.Tjeenk Willink Zoon, Haarlem,) 6 Vols. by other editors and authors published subsequently.

FORREST, D.W. (1974): 'Francis Galton – the life and work of Victorian genius' (Paul Elek, London)

FOURNIER D'ALBE, E.E. (1923): 'The life of Sir William Crookes' (Fisher Unwin, London)

FRANKEL, T. (1977): 'Timekeepers: from sundials to atomic clocks'. National Museum of History and Technology (Smithsonian Institution Press, Washington, DC)

FRANKLIN, W.T. (1818): 'Memoirs of the life and writings of Benjamin Franklin' (Henry Colburn, London, 6 Vols.)

FRANKLIN, W.T. (1850): 'The autobiography of Benjamin Franklin' (H.G.Bohn, London)

FRASER, P.M. (1971): 'Eratosthenes of Cyrene' from *Proc. British Acad.*, LVI, Oxford University Press, London

FRAZIER, A.H. (1964): 'Daniel Farrand Henry's cup type "Telegraphic" current meter', *Technology & Culture*, V, 4, pp. 541-565

FRAZIER, A.H. (1967): 'William Gunn Price and the Price current meters'. Contributions to the Museum of History and Technology; Paper 70, pp. 39-68, Smithsonian Press, Washington, DC

FREUNDLICH, M.M. (1963): 'Origin of the electron microscope', *Science,* **142**, pp. 185-188

FREY, H. (1872): 'The microscope and microscopial technology' (William Wood, New York)

FRISCH, O. (1977): 'The first sub-atomic particle', *New Sci.,* **77**, pp. 408-410

FRISINGER, H.H. (1977): 'The history of meteorology: to 1800' (American Meteorological Society, Boston)

FRISON, K.E. (1965): 'Catalogue of Henri van Heurk collection of microscopes' Koninklijke Maatschappij voor Dierkunde van Antwerpen. (Date approximate)

FRITH, H., and RAWSON, W.S. (1896): 'Coil and current or the triumphs of electricity' (Ward, Lock & Co., London)

GABOR, D. (1972): 'Holography, 1948-1971', *Science,* 177, pp. 299-313

GADE, J.A. (1947): 'The life and times of Tycho Brahe' (Princeton University Press)

GARBEDIAN, H.G. (1940): 'Thomas Alva Edison – builder of civilization' (Hamilton, London)

GARDINER, K.R., and D.L., (1965): 'Andre-Marie Ampere and his English aquaintances', *Brit. J. Hist. Sci.,* **2**, p.235

GARTMANN, H. (1960): 'Science as history' (Hodder & Stoughton, London) (Translated from the German by A.G. Readett)

GEDDES, K. (1974): 'Guglielmo Marconi, 1874-1937' (HMSO, London)
General Electric (1976): 'The General Electric story'. Vol. 1 – The Edison era, 1876-1892: Vol. 2 The Steinmetz era, 1892-1923, (Photo histories). Schenectady Elfun Society Territorial Council, New York
GERNSHEIM, H. and A. (1969): 'The history of photography 1685-1914' (McGraw-Hill, New York
GEYMONAT, L. (1965): 'Galileo Galilei – a biography and inquiry into his philosophy of science' (McGraw-Hill, New York)
GIBBS, S. L. (1976): 'Greek and Roman Sundials' (Yale University Press)
GIBSON, C.R. (1915): 'Electricity of today' (Seeley, Service & Co. London)
GILBERT, K.R. (1971): 'Textile machinery' (HMSO, London)
GILLMOR, C.S. (1971): 'Coulomb and the evolution of physics and engineering in eighteenth century France' (Princeton University Press)
GINZTON, E.L. (1944): 'D-C amplifier design technique', *Electronics,* **17**, 3, pp.98-102
GLASSÈR, O. (1933): 'Wilhelm Conrad Rontgen' (John Bale, Sons & Danielsson, London) also Charles C. Thomas, Springfield, USA (1934)
GLASSER, O., QUIMBY, E.H., TAYLOR, L.S., and WEATHERWAX, J.L. (1947): 'Physical foundations of radiology' (Hoeber, New York) 1944 (1st edn.)
GLAZEBROOK, R.T. (1902): 'Heat – an elementary text-book, theoretical and practical' (Cambridge University Press)
GLAZEBROOK, R.T. (1922): 'A dictionary of applied physics' (Macmillan, London) (Vol. 1, 2 – 1922; 3, 4, 5 – 1923)
GLEGG, G.L. (1969): 'The design of design' (Cambridge University Press)
GODFRAY, H. (1880): 'A treatise on astronomy' (Macmillan, London)
GOLDBERG, E.A. (1950); 'Stabilization of d-c amplifiers', *RCA Rev.*, **11**, pp. 296-300
GOLDING, H. (1932): 'The wonder book of electricity' (Ward, Lock & Co., London)
GOLDING, E.W. (1940): 'Electrical measurements and measuring instruments'. (Pitman, London)
GOLDSTINE, H.H. (1972): 'The computer from Pascal to von Neumann' (Princeton University Press)
GOODMAN, N.G. (1956): 'The ingenious Dr.Franklin – selected scientific letters' (University of Pennsylvania Press)
GORDON, J.E.H. (1880): 'A physical treatise on electricity and magnetism' (Sampson Low, Marston, Searle and Rivingham, London, 2 Vols.)
GORDON, M.M. (1869): 'The home life of Sir David Brewster' (Edmonston & Douglas, Edinburgh)
GRANT, E. (1978): 'Physical science in the Middle Ages' (Cambridge University Press)
GRANT, K. (1952): 'The life and work of Sir William Bragg' (University of Queensland Press, Brisbane, Australia)
GRAY, A. (1893): 'Absolute measurements in electricity and magnetism' (Macmillan, London) 1884 (1st edn.)
GREEN, G., and LLOYD, J.T. (1970): 'Kelvin's instruments and the Kelvin Museum' (University of Glasgow)
GREENWOOD, I.A., HOLDAM, J.V., and MACRAE, D. (1948): 'Electronic instruments' (McGraw-Hill)
GREIG, J. (1970): 'John Hopkinson – electrical engineer' (HMSO, London)
GRIFFITHS, E.A. (1920): 'Engineering instruments and meters' (G.Routledge, London)
GRIFFITHS, E. (1947): 'Methods of measuring temperature' (Charles Griffin, London) 1925 (1st edn.)
GRIFFITHS, R. (1943): 'Thermostats and temperature-regulating instruments' (Charles Griffin, London)
GRIMAL, P. (1963): 'The civilization of Rome' (George Allen & Unwin, London) (Translated from the French edn. of 1960)
Grolier Society (1946): 'The book of popular science' (Grolier Society Inc., New York, 12 Vols.), 1924 (1st edn.)

GUNTHER, R.T. (1935): 'Handbook of the Museum of the History of Science in the Old Ashmolean Building, Oxford' (Oxford University Press)

GUNTHER, R.T. (1939): 'The old Ashmolean and its historic scientific collections' Private printing, available Reading University Library, England

HACKMANN, W.D. (1973): 'Alexander Bain's short history of the electric clock' (1852) Occasional Paper No. 3. (Turner and Devereux, London)

HALDANE, J.S., and GRAHAM, J.I. (1935): 'Methods of air analysis' (Charles Griffin, London)

HALL, C. (1926): 'Triumphs of invention' (Blackie, London)

HALLOCK, W., and WADE, H.T. (1906): 'Outlines of the evolution of weights and measures and the metric system' (Macmillan, New York)

HALLOWS, R.W. (1946): 'Radar – radiolocation simply explained' (Chapman & Hall, London)

Halstead Press (1974): 'Wireless telegraphy' (Halstead Press, New York)

HAMARNEH, S. (1964): 'History of division of medical sciences'. Paper 43, pp. 269-300 from United States National Museum Bulletin 240; Smithsonian Institution, Washington, DC

HANCOCK, H.E., (1950): 'Wireless at sea: the first fifty years'. Facsimile reprint, (Arno Press, *c*1974)

HANDSCOMBE, E. (1970): 'Electrical measuring instruments' (Wykeham Publications, London)

HANSCOM, C.D. (1961): 'Dates in American telephone technology' Bell Telephone Lab. Inc., USA

HANSON, N.R. (1958): 'Patterns of discovery: an inquiry into the conceptual foundations of science' (Cambridge University Press)

HARKNESS, W. (1890): 'The progress of science as exemplified in the art of weighing and measuring', from Smithsonian Report for 1888 pp. 597-633, Government Printing Office, Washington, DC

HARMAN, G. (1973): 'Human memory and knowledge' (Greenwood Press, London)

HARRE, R. (1969): 'Some nineteenth century British scientists' (Pergamon Press)

HARRIS, L.E. (1961): 'The two Netherlanders – Humphrey Bradley and Cornelius Drebbel'. (W. Heffer, Cambridge)

HARROW, L., and WILSON, P.L. (1976): 'Science and technology in Islam' (Crescent Moon Press, London)

HART, I.B. (1924): 'Makers of science – mathematics, physics, astronomy' (Oxford University Press)

HART, I.B., (1961): 'The world of Leonardo da Vinci – man of science, engineer and dreamer of flight' (Macdonald, London)

HARTLEY, H. (1960): 'The Royal Society: Its origins and founders' (Royal Society, London)

HARTREE, D.R. (1946): 'The ENIAC, an electronic calculating machine', *Nature,* 157, 3990, p.527

HASLEGRAVE, H.L. (1951): 'The contribution of the British Technical Colleges to engineering education'. Proceedings of the Joint Engineering Conference, Institutions of Civil, of Mechanical, and of Electrical Engineers, London, pp. 265-278

HAWKINS, L.A. (1950): 'The story of General Electric Research'. General Electric Co. adapted from 'Adventure into the unknown' by L.A.Hawkins, William Morrow, New York

HAWKS, E. (1927): 'Pioneers of wireless'. Facsimile reprint, (Arno Press, New York, *c*1974)

HAWKS, E. (1929): 'The book of electrical wonders' (Harrap, London)

HAY, A. (1907): 'Alternating currents – their theory, generation and transformation' (Harper & Bros., London) 1905 (1st edn.)

HEILBRON, J.L. (1974): 'H.G.J.Moseley – the life and letters of an English Physicist 1887-1915' (University of California Press, Berkeley)

HEISING, R.A. (1962): 'Modulation methods' *Proc. IRE,* **50**, pp. 896-901

HERBERT-GUSTAR, L., and NOTT, P.A. (1974): 'Earthquake Milne and the Isle of Wight' (Vectis Biographies, Isle of Wight)

HERBERT, T.E. (1923): 'Telephony – an elementary exposition of the telephone system of the British Post Office' (Isaac Pitman, London)

HERING, D.W. (1924): 'Foibles and fallacies of science: an account of celebrated scientific

vagaries' (Van Nostrand, New York)
HERIVEL, J. (1975): 'Joseph Fourier – the man and the physicist' (Clarendon Press, Oxford)
HEROLD, E.W. (1962): 'The impact of receiving tubes on broadcast and t.v. receivers', *Proc.IRE*, **50**, pp. 805-809
HEROLD, E.W. (1976): 'A history of colour television displays', *Proc. IEEE*, **64**, pp. 1331-1338
HERSCHEL, J.F.W. (1851): 'Manual of scientific inquiry' (John Murray, London)
HETT, W.S. (1955): 'Aristotle, minor works' (William Heinemann, London)
HEWITT, C.W. (1934): 'An elementary course of air navigation' (John Hamilton, London)
Hewlett-Packard (1977): 'Hewlett Packard: a company history'. Pub. Rel. Dept., Palo Alto, USA
HIBBERT, W. (1910): 'Popular electricity' (Cassell, London)
HIGGINS, T.J. (1961): 'A biographical bibliography of electrical engineers and electrophysicists Pt. 1', *Technology & Culture*, **2**, pp. 28-32, pp. 146-165
HILL, J.A. (1932): 'Letters from Sir Oliver Lodge' (Cassell, London)
HILLIARD, J.K. (1962): 'The history of stereophonic sound reproduction' *Proc. IRE.*, **50**, pp. 776-780
HILLIER, M. (1976): 'Automata and mechanical toys – an illustrated history' (Jupiter Books, London)
HINDLE, B. (1966): 'Technology in early America' (University of North Carolina Press)
HMSO (1938): 'Admiralty navigational manual – Vol. 1' (HMSO, London)
HMSO (1957): 'Books on engineering – a subject catalogue of books in the Science Library' (HMSO, London)
HMSO (1970): 'Science Museum publications' (HMSO, London)
HMSO (1972): 'Bases of measurement'. National Physical Laboratory, Ref. No. 56-6946 5/72 (HMSO, London)
HMSO (1977): 'The wireless show, 130 classic radio receivers – 1920s to 1950s' (HMSO, London)
HOBART, H.M. (1911): 'A dictionary of electrical engineering' (Gresham, London)
HOGBEN, L. (1938): 'Science for the citizen' (Allen & Unwin, London)
HOGG, J. (1867): 'The microscope – its history, construction and application' (George Routledge & Sons, London) 1854 (1st edn.)
HOLBROOK, M. (1972): 'Scientific instruments of the 17th and 18th centuries and their makers' (Batsford, London) (Translation of work in French by M. Daumas, 1950)
HOLDEN, E.S. (1881): 'Sir William Herschel – his life and works' (W.H.Allen, London)
HOLMAN, S.W. (1894): 'Discussion on the precision of measurements' (Wiley)
HOLTON, G. (1978): 'The scientific imagination' (Cambridge University Press)
HOPKINSON, B. (1901): 'Original papers' (Cambridge University Press, 2 Vols.)
HORNER, J.G. (1905): 'The encyclopaedia of practical engineering' (Virtue and Co., London, 10 Vols.)
HOSKIN, M.A. (1963): 'William Herschel – and the construction of the heavens' (Oldburne, London)
HOUNSHELL, D.A. (1973): 'Manuscripts in U.S. Depositories relating to the history of electrical science and technology'. Div. Electricity and Nuclear Energy, Smithsonian Institution, Washington, DC
HOUNSHELL, D.A. (1976): 'Bell and Gray: contrasts in style, politics and etiquette, *Proc. IEEE*, **64**, pp. 1305-1314
HOUSTON, E.J. (1892): 'A dictionary of electrical words' (W.J.Johnston, New York) 1889 (1st edn.)
HOUSTON, E.J. (1893): 'Electrical measurements and other advanced primes of electricity' (W.J.Johnston, New York)
HOWARTH, O.J.R. (1931): 'The British Association for the Advancement of Science: a retrospect 1831-1931'. (British Association, London)
HOWSE, D. and HUTCHINSON, B. (1969): 'The clocks and watches of Captain James Cook'. Collection of reprints from 'Antiquarian horology' Antiquarian Horological Society, London (5 papers with index)

HOWSE, D. (1973): 'National Maritime Museum – guide to the Old Royal Observatory, Greenwich, England
HOWSE, D. (1975): 'Greenwich Observatory – the buildings and instruments' (Taylor & Francis, London)
HUDSON, H.K. (1931): 'Radio charts the air-course'. *Radio News & Short Wave Radio (New York)*, February, pp. 708-709, p.746
HUDSON, K., and NICHOLLS, A. (1974): 'The directory of museums' (Macmillan, London)
HUGHES, T.P. (1971): 'Elmer Sperry – inventor and engineer' (Johns Hopkins Press, London)
HUGILL, A.L. (1978): 'Gravimeter design'. Internal Report DSS/AH 171, September, Department of Systems Science, City University, London
HUNT, F.V. (1954): 'Electroacoustics' (John Wiley, New York)
HUNT, I., and DRAPER, W.W. (1964): 'Lightning in his hand – the life story of Nikola Tesla' (Sage Books, Denver)
HUNTOON, R.D. (1967): 'Concept of a national measurement system', *Science,* **158**, pp. 67-71
HUXLEY, T.H. (1882): 'Science and culture and other essays' (Macmillan, London)
IEE (1955): 'Thermionic valves 1904-1954'. IEE, London (Series of papers).
IEEE (1976): 'Special issue – two centuries in retrospect', *Proc IEEE,* **64**, pp.1267-1428
ILES, G. (1912): 'Leading American inventors' (Holt, New York) (Includes Morse and Ericsson)
Instrument Engineer (1954): 'Letter from America', *Instrument Eng.,* **1**, p. 114
International Hydrographic Bureau (1926): 'Tide predicting machines'. Special Publication No. 13, July.
IRE (1962): 'Communications and electronics 1912-1962, *Proc. IRE,* **50**, 28 Sections of papers in an Anniversary Issue, May, pp. 658-1447
IRWIN, J.T. (1925): 'Oscillographs' (Isaac Pitman, London)
ISING, G. (1926): 'A natural limit for the sensibility of galvanometers', *Philos. Mag.,* **1**, pp. 827-834
JACKSON, E, (1975): 'Technician and craft education – a simple survey of the present state', *Meas. & Control.,* **8**, pp. 84-86
JACOBS, D.H. (1943): 'Fundamentals of optical engineering' (McGraw-Hill, New York)
JACOT DE BOINOD, B.L., and COLLIER, D.M.B. (1935): 'Marconi, master of space' (Hutchinson London)
JAFFE, B. (1961): 'Michelson and the speed of light' (Heinemann, London)
JAHANS, G.A. (1931): 'Paper testing and chemistry for printers' (Isaac Pitman, London)
JAMES-KING, W. (1962*a*): 'The development of electrical technology in the 19th century: 1 – The electrochemical cell and the electromagnet'. Paper 28, pp. 231-271 from United States National Museum Bulletin 228: Smithsonian Institution, Washington, DC
JAMES-KING, W. (1962*b*): 'The development of electrical technology in the 19th century: 3 – The early arc light and generator'. Paper 30, pp. 333-407, from United States National Museum Bulletin 228: Smithsonian Institution, Washington, DC
JAMES-KING, W. (1962*c*): 'The development of electrical technology in the 19th century: 2 – The telegraph and telephone'. Paper 29, pp. 273-332. From United States National Museum Bulletin 228, Smithsonian Institution, Washington, DC
JAMMES, A. (1973): 'Williams H. Fox Talbot – inventor of the negative positive process' (Collier Books, New York)
JEWKES, J., SAWERS, D., and STILLERMAN, R. (1958): 'The sources of invention' (Macmillan, London)
JOHNSON, J.B. (1928): 'Thermal agitation of electricity in conductors'. *Phys. Rev.,* **32**, Series 2, pp. 97-109
JOHNSON, J.H., and RANDELL, W.L., (1946): 'Colonel R.E.B. Crompton', (Longmans Green, London) 1945 (1st edn.)
JOLLY, W.P. (1972): 'Marconi' (Constable, London)
JONES, B. (1870): 'Life and letters of Faraday' (Longmans Green, London, 2 Vols.)
JONES, F.A. (1907): 'Thomas Alva Edison' (Hodder & Stoughton, London)

JONES, J.C., and THORNLEY, D.G. (1963): 'Design methods' (Pergamon Press, London)

JONES, R.V. and MCCOMBIE, C.W. (1952): 'Brownian fluctuations in galvanometers and galvanometer amplifiers', *Philos. Trans. R. Soc. (London)*, A, 881, 224, pp. 205-230

JONES, R.V. (1967*a*): 'Instruments and advancement of learning'. *Trans. Soc. Instrum. Tech.*, **19**, pp. 3-11

JONES, R.V. (1967*b*): 'The measurement and control of small displacements', *Phys. Bull.*, **18**, pp. 325-336

JONES, R.V. (1968): 'More and more about less and less', *Proc. R. Inst.*, **43**, 202, pp. 323-345

JONES, R.V. (1970): 'The pursuit of measurement', *Proc. IEE*, **117**, (6) pp. 1185-1191

JONES, R.V. (1971*a*): 'Domesday book of British Science', *New Sci.*, **49**, pp. 481-483

JONES, R.V. (1971*b*): 'Plenum in vacuo or vacuum and the advancement of man' *Chemistry & Industry*, pp. 436-442

JONES, R.V. (1972): 'Some turning-points in infra-red history', *Radio & Electron. Eng.*, **42**, pp. 117-126

JONES, R.V. (1973): 'The R.W. Paul instrument fund', *J. Phys. E.*, **6**, pp. 944-947

JONES, R.V. (1975): 'Some contributions to a discussion held by the Royal Society on the effects of the two world wars on the organization and development of science in the United Kingdom, 1974', *Proc R. Soc. Lond.* A, 342, pp. 441-445, pp. 481-490, pp. 549-554, pp. 575-579, pp. 581-586. (Eight other papers were presented)

JOSEPHSON, M. (1961): 'Edison' (Eyre & Spottiswoode, London)

JOWETT, B. (1953): 'The dialogues of Plato' (Clarendon Press, Oxford)

KARP, W.(1965): 'The Smithsonian Institution'. (Smithsonian Institution with Editors of American Heritage Magazine, Washington, DC)

KELLY, M. (1953): 'The first five years of the transistor', *Bell Tel. Mag.*, Summer

KEMP, P. (1950): 'Electrical engineering – theory and practice' (Pitman, London, 3 Vols.)

KEMPE, H.R. (1892): 'A handbook of electrical testing' (E. & F. N. Spon, London)

KENT, W.G. (1895): 'Catalogue of apparatus for the measurement of water' (George Kent, London)

KENT, W.G. (1912): 'An appreciation of two great workers in hydraulics (G.B. Venturi and C.Herschel)'. Privately printed. (Blades East and Blades, London)

KENT, W.G. (1930): 'Meters and gauges for waterworks, sewage plants, irrigation schemes'. Catalogue, (George Kent Ltd., London)

KENT, W.G. (1950): 'Measurement and automatic control in steam power plants' Pub. No. 913/950

KEYNES, G, (1960): 'A bibliography of Dr.Robert Hooke' (Clarendon Press, Oxford)

KIMBALL, D.S. (1949): 'The book of popular science' (The Grolier Society Inc., New York) 1924 (1st edn.)

KING, A.G. (1925): 'Kelvin the man' (Hodder & Stoughton, London)

KING, E.E. (1921): 'Railway signalling' (McGraw-Hill, New York)

KING, H.C. (1955): 'History of the telescope' (Griffin, London)

KING, R. (1973): 'Michael Faraday of the Royal Institution'. Royal Institution, London

KIRKMAN, M.M. (1904): 'The science of railways' 20 Vols. Vol. IV contains signalling, XX telegraphs and telephones (World Railway Publishing Co., New York)

KISCH, B. (1965): 'Scales and weights: an historical outline' (Yale University Press)

KITCHINER, W. (1824): 'The economy of the eyes: precepts for the improvement and preservation of the sight . . . ' (Hurst Robinson, London)

KLEIN, H.A. (1975): 'The world of measurements' (Allen & Unwin, London)

KLEMM, F. (1959): 'A history of western technology' (George Allen & Unwin, London) (Translated from German by F.Klemm, 1954 (1st edn.)

KLEMPERER, O. (1939): 'Electron optics' (Cambridge University Press)

KLOEFFLER, R.G. (1949): 'Industrial electronics and control' (Wiley, New York)

KNIGHT, D. (1975): 'Sources for the History of Science, 1660-1914' (Cambridge University Press)

KNOTT, C.G. (1911): 'Life and scientific work of Peter Guthrie Tait' (Cambridge University Press)

KNOWLES-MIDDLETON, W.E., and SPILHAUS, A.F. (1953): 'Meteorological instruments' (University of Toronto Press) also Oxford University Press 1941 (1st edn.)

KNOWLES-MIDDLETON, W.E. (1964): 'The history of the barometer' (Johns Hopkin Press, Baltimore)

KNOWLES-MIDDLETON, W.E. (1966): 'History of the thermometer and its uses in meteorology' (Johns Hopkins Press, Baltimore)

KNOWLES-MIDDLETON, W.E. (1969*a*): 'Catalog of meteorological instruments in the Museum of History and Technology'. (Smithsonian Institution Press, Washington, DC)

KNOWLES-MIDDLETON, W.E. (1969*b*): 'Invention of the meteorological instruments' (Johns Hopkins Press, Baltimore)

KNOX, G. D. (1917): 'All about electricity' (Cassell, London) 1914 (1st edn.)

KOENIGSBERGER, L. (1906): 'Hermann von Helmholtz' (Oxford University Press)

KOLLER, L.R. (1937): 'The physics of electron tubes' (McGraw-Hill, New York)

KOLLRAUSCH, F. (1883): 'An introduction to physical measurements' (J. & A. Churchill, London)

KRANTZ, D.H., LUCE, R.D., SUPPES, P., and TVERSKY, A. (1971): 'Foundations of measurement – Vol. 1 (Academic Press, New York) (Vol. 2 to be published)

KURZ, O. (1975): 'European clocks and watches in the Near East'. Studies of the Warburg Institute, Vol. 34. Warburg Institute, London and E.J.Brill, Leiden.

KYLSTRA, P.H. (1977*a*): 'The use of the early telephone in phonetic research', *Phonographic Bull.,* December

KYLSTRA, P.H. (1977*b*): '1877-1977 the centenary of the phonograph', *ibid.,* April

LANCHESTER, G.H. (1974): 'Yellow Emperor's south-pointing chariot' The China Society, Victoria Street, London

LANGFORD, J.J. (1966): 'Galileo, science and the Church' (Desclee Co., New York)

LANGFORD-SMITH, F. (1955): 'Radiotron designers handbook'. (Wireless Press, for Amalgamated Wireless Valve Co. Pty. Ltd. Sydney) 1934 (1st edn.)

LANPHIER, R.C. (1925): 'Electric meters-history and progress'. Sangamo Electric Co., Springfield, Illinois

LARDNER, D. (1862): 'The Museum of science and art' (Walton & Maberley, London) (12 vols. in 6 cases appeared over several years)

LARMOR, J. (1937): 'Origins of Clerk-Maxwell's electric ideas as described in familiar letters to William Thomson' (Cambridge University Press)

LARSEN, E. (1960): 'Ideas and invention' (Spring Books, London)

LAVEN, W.J., and CITTERT EYMERS, J.G. Van (1967): 'Electrostatic Instruments in the Utrecht University Museum' (Utrecht University Museum, Utrecht)

LAYTON, E.T. (1976): 'Scientists and engineers: the evolution of the IRE', *Proc. IEEE,* **64**, pp. 1390-1392

LE CHATELIER, H., and BOUDOUARD, O.L. (1904): 'High temperature measurements' (Wiley, New York) 1900 (1st edn.) (Translated by G.K.Burgess)

LEE, G. (1947): 'Oliver Heaviside' (Longmans Green, London)

LEECH, D.J. (1972): 'Management of engineering design' (John Wiley, London)

LENZEN, V.F., and MULTHAUF, R.P. (1965): 'Development of gravity pendulums in the 19th century'. Paper 44, pp.301-348, from United States National Museum Bulletin 240: Smithsonian Institution, Washington, DC

LESSING, L. (1969): 'Man of high fidelity: Edwin Howard Armstrong' (Bantam Books, London) 1956 (1st edn.)

LEWIS JONES, H. (1900): 'Medical electricity'. H.K.Lewis, London

LEY, W. (1967): 'Otto Hahn – a scientific autobiography' (Macgibbon & Kee, London)

LILLEY, S. (1951): 'The development of scientific instruments in the seventeenth century' *in* 'The history of science' (Cohen, London)

Linderman Library (1973): 'Nicolaus Copernicus 1473-1973: his 'revolutions' and his revolution: catalogue of an exhibition of manuscripts and books'. Linderman Library, Lehigh University,

Bethlehem, Pennsylvania

LINKLATER, E. (1972): 'The voyage of the Challenger' (John Murray, London)

LOCKYER, T.M. and LOCKYER, W.L. (1928): 'Life and work of Sir Norman Lockyer' (Macmillan, London)

LODGE, O.J. (1892): 'Modern views of electricity' (Macmillan, London)

LODGE, O.J. (1893): 'Pioneers of science' (Macmillan, London)

LODGE, O.J. (1900): 'Signalling through space without wires: being a description of the work of Hertz and his successors'. Facsimile reprint, (Arno Press, New York, *c*1974)

LODGE, O.J. (1910): 'Electrons, or the nature and properties of negative electricity' (Bell & Sons, London)

LODGE, O.J. (1931): 'Past years' (Hodder & Stoughton, London); Scribner, New York, 1932

LORENZEN, E. (1966): 'Technological studies in ancient metrology'. (Nyt Nordisk Forlag, Arnold Busck, Copenhagen)

LOUIS, R.V. (1974): 'Biography index' (H.W.Wilson Co., New York) (1st edn. 1970, quarterly cumulative index, 9 vols. to 1974)

MABEE, C. (1943): 'The American Leonardo: a life of Samuel F.B. Morse' (Knopf, New York)

MABIE, H.H., and OCVIRK, F.W. (1958): 'Mechanisms and dynamics of machinery' (John Wiley, New York) 1957 (1st edn.)

MABON, P.C. (1975): 'Mission communications – the story of Bell Laboratories' (Bell Laboratories Inc., Murray Hill, New Jersey)

MACCURDY (1938): 'The notebooks of Leonardo da Vinci' (Reynal & Hitchcock, New York)

MACDOUGAL, D.T. (1908): 'Textbook of plant physiology' (Longmans, London) 1901 (1st edn.)

MACGREGOR-MORRIS, J.T., and HENLEY, J.A. (1936): 'Cathode ray oscillography' (Chapman & Hall, London)

MACKENZIE, C.D. (1928): 'Alexander Graham Bell; the man who contracted space' (Houghton Mifflin, New York)

MACLAURIN, W.R. (1949): 'Invention and innovation in the radio industry' (Macmillan, New York)

MAGILL, E.M. (1916): 'Notes on Galvanism and Faradism', (H.K.Lewis, London)

MAGNUS, R. (1949): 'Goethe as a scientist' (Henry Schuman, New York)

MALMGREN, E. (1975): 'Telemuseum – en korthistorik', *Daedalus*, pp. 37-64

MANUEL, F.E. (1968): 'A portrait of Isaac Newton' (Belknap Press, Harvard University Press, Cambridge, Mass.)

MARCONI, D.P. (1962): 'My father, Marconi' (McGraw-Hill)

MARGENAU, H., BERGAMINI, D. and EDITORS of LIFE (1966): 'The scientist', (Time-Life International, Netherlands)

MARLAND, E.A. (1964): 'Early electrical communication' (Abelard-Schuman, London)

MARMERY, J.V. (1895): 'Progress of science' (Chapman & Hall, London)

MARSHALL, R.C. (1978): 'Ideas to order' *Electron. & Power,* **24**, pp. 54-57

MARSTEN, J. (1962): 'Resistors – a survey of the evolution of the field'. *Proc. IRE,* **50**, pp. 920-924

MARTIN, L.C. (1924): 'Optical measuring instruments – their construction, theory and use' (Blackie & Son, Glasgow)

MARTIN, T. (1961): 'The Royal Institution' (Royal Institution, London)

MARTIN, T.C. (1894): 'The inventions, researches and writings of Nikola Tesla'. The Electrical Engineer, New York

MASON, H.L. (1977): 'Catalog of artifacts on display in the NBS Museum' Reprint NBSIR 76-1125, US Department of Commerce

MATSCHOSS, C. (1925): 'Das Deutsche Museum' (VDI Verlag, Berlin)

MATTHEWS, W. (1968): 'British autobiographies' (Archon Books, California)

MAY, C.P. (1964): 'James Clerk-Maxwell and electromagnetism' (Chatto & Windus, London)

MAYR, O. (1970): 'The origins of feedback control' (MIT Press, USA) (Translated from German language version by same author)

MAYR, O. (1971a): 'Feedback mechanisms in the historical collections of the National Museum of History and Technology'. (Smithsonian Institution Press, Washington DC)
MAYR, O. (1971*b*): 'Victorian physicists and speed regulation: an encounter between science and technology', *Notes & Records, R. Soc., London,* **26**, pp. 205-228
MAYR, O. (1976): 'Philosophers and machines'. (Science History Publications, New York)
MAZZOTTO, D. (1906): 'Wireless telegraphy and telephony' (Whittaker, London)
MCARTHUR, E'D' (1936): 'Electronics and electron tubes' (Wiley, New York)
MCCOMBIE, C.W. (1953): 'Fluctuation theory in physical measurements' in Reports on Progress in Physics'. Vol. XVI, The Physical Society, London
MCCONNELL, A. (1977): 'First steps in seismology', *Geophys. Mag.,* XLIX, 5, pp. 292-294
McGraw-Hill (1973): 'Encyclopaedia of world biography' (McGraw-Hill, New York)
MCGUCKEN, W. (1960): 'Nineteenth century spectroscopy – development of the understanding of spectra, 1802-1897' (Johns Hopkins Press, Baltimore)
MCKENDRICK, J.G. (1894): 'Life in motion – or nerve and muscle' (Adam & Charles Black, London)
MCKENZIE (1944): 'Magnetism and electricity' (Cambridge University Press) 1938 (1st edn.)
MCKIE, D. (1972): 'The scientific periodical from 1665 to 1798' *in* 'Natural Philosophy through the eighteenth century'. A.Ferguson, (Taylor & Francis, London) 1948 (1st edn.)
MCLELLAN, J.A., and DEWEY, J. (1907): 'The psychology of number' (Appleton and Co., New York) (1st edn. 1895)
MCMAHON, A.M. (1976): 'Corporate technology: The social origins of the American Institute of Electrical Engineers', *Proc. IEEE,* **64**, pp. 1383-1390
MCMULLIN, E. (1967): 'Galileo – man of science' (Basic Books, London)
MCNICOL, D. (1913): 'American telegraph practice' (McGraw-Hill)
MCNICOL, D. (1925): 'Printing telegraph systems' (International Textbook Co., Scranton, Pennsylvania)
MCNICOL, D. (1946): 'Radio's conquest of space: the experimental rise in radio communication'. Facsimile reprint, (Arno Press, New York, *c*1974.)
MEDLOCK, R., (1952): 'Oxygen analysis, *Instrum. Eng.,* **1**, pp.3-10
MEETHAM, A.R. (1952): 'Atmosphere pollution – its origins and prevention' (Pergamon Press, London)
MESCH, F. (1976): 'The contribution of systems theory and control engineering to measurement science'. Survey Lecture SL5, VIIth IMEKO Congress, London
MICHEL, H. (1966*a*): 'Scientific instruments in art and history' (Viking Press, New York)
MICHEL, H. (1966*b*): 'Instruments des sciences'. (Rhode-Saint-Genese, Belgium)
MICHELSON, A.A. (1928): 'Studies in optics' (University of Chicago Press)
MILLER, L.B. ; VOSS, F., and HUSSEY, J.M. (1972): 'The Lazzaroni – science and scientists in mid-nineteenth century America'. (Smithsonian Institution Press, Washington, DC)
MILLIKAN, R.A. (1951); 'The autobiography of Robert A. Millikan' (Macdonald, London)
MILNER, P., and PENGILLEY, C.J., (1972): 'Towards technology as a science'. Studies in technology and science No. 2, Internal publication, Department of Mechanical Engineering, University of Melbourne, Australia. (Additional notes were prepared for lectures in 1974).
MINCK, J. (1977): 'The national measurement system – a hidden giant' *NCSL Newsletter,* **17**, pp. 3-5
MITCHELL, F.H. (1951): 'Fundamentals of electronics' (Addison-Wesley)
M'KENDRICK, J.G. (1899): 'Hermann Ludwig Ferdinand van Helmholtz'. (T. Fisher Unwin, London)
MOLDENKE, H.N., and MOLDENKE, A.L. (1951): 'The mysterious silphium', *New York Botanical Garden J.,* **1**, pp. 140-142
MOLELLA, A.P. (1976): 'The electric motor, the telegraph, and Joseph Henry's theory of technological progress', *Proc. IEEE,* **64**, pp. 1273-1278
MOODY, E.A., and CLAGETT, M. (1952): 'The medieval science of weights' (University of Wisconsin Press, Madison)

MORE, L.T. (1934): 'Isaac Newton: a biography; 1642-1727' (Scribner, New York)
MORECROFT, J.H. (1936): 'Electron tubes and their applications' (John Wiley, New York), 1933 (1st edn.)
MORGAN, B. (1953): 'Men and discoveries'. (Scientific Book Club, London)
MORGAN, R.W. (1969): 'Sir Francis Galton (1822-1910)' *in* 'Some nineteenth century British scientists', by R. Harre (Pergamon Press, London)
MORTLOCK, J.R., and DAVIES, M.W.H. (1952): 'Power system analysis' (Chapman & Hall, London)
MORSE, E.L. (1914): 'Samuel F.B. Morse – his letters and journals' (Houghton Mifflin Co., New York, 2 Vols.)
MOTTELAY, P.F. (1922): 'Biographical history of electricity and magnetism' (Charles Griffin & Co., London) Arno Press Reprint *c*1974
MULLINEUX-WALMSLEY, R. (1910): 'Electricity in the service of man'. (Cassell and Co., London) Vol. 1, A Vol. 2 does not seem to have been published. See also earlier version, Urbanitzky (1886)
MULTHAUF, R.P. (1961*a*): 'A catalogue of instruments and models in the possession of the American Philosophical Society'. (American Philosophical Society, Philadelphia)
MULTHAUF, R.P. (1961*b*): 'The introduction of self-registering meteorological instruments'. Paper 23, pp. 95-116 from United States National Museum Bulletin 228; Smithsonian Instiution, Washington, DC
MULTHAUF, R.P. (1962): 'Holcomb, Fitz and Peate: three 19th century American telescope makers'. Paper 26, pp. 155-184, from United States Museum Bulletin 228, Smithsonian Institution, Washington, DC
MURRAY, F.J. (1948): 'The theory of mathematical machines' (Kings Crown Press, New York), 1947 (1st edn.)
MYERS, C.S. (1928): 'A text book of experimental psychology, with laboratory exercises, Part 1. Text book' (Cambridge University Press)
NASR, S.H. (1976): 'Islamic science – an illustrated study' (World of Islam Festival Pub. Co., London)
NEAL, H.E. (1958): 'The telescope' (Julian Messner, New York)
NEEDHAM, J. (1954): 'Science and civilisation in China'. (Cambridge University Press, Cambridge) Vol. I. Introductory Orientations; Vol. II, History of Scientific Thought; Volume III Mathematics and the Sciences of the Heavens and the Earth; Volume IV, Physics and Physical Technology – Part 1 Physics, Part 2 Mechanical Engineering, Part 3 Civil Engineering and Nautics; Volume V, Chemistry and Chemical Technology Part 2 Spagyrical Discovery and Invention: Magisteries of Gold and Immortality; Part 3 Historical Survey, from Cinnabar Elixirs to Synthetic Insulin, Part 4 Apparatus, Theory and Comparative Macrobiotics. Further volumes in preparation after 1978. (An abridged version was begun by C.A.Ronan in 1978).
NEEDHAM, J. (1959): 'Science and civilisation in China – Vol. III' (Cambridge University Press) Sections 19-25
NEEDHAM, J. (1962): Section 26: 'Science and civilisation in China – Vol. IV' (Cambridge University Press) Part 1. 'Physics'
NEEDHAM, J. (1965): 'Science and civilisation in China – Vol. IV, Pt.2, (Cambridge University Press) (Section 27) Mechanical Engineering
NEEDHAM, J. (1970): 'Clerks and craftsmen in China and the West' (Cambridge University Press)
NELSON, (1881): 'Triumphs of invention and discovery in art and science' (T.Nelson, London)
NEUBURGER, A. (1930): 'The technical arts and sciences of the Ancients' (Methuen, London)
NEUGEBAUER, O. (1975): 'A history of ancient mathematical astronomy – Vol. 1.', (Spring-Verlag, Berlin)
NEVILLE, R.H.C. (1885): 'On private installations of electric lighting'. *Proc. Inst. Mech. Eng.*, pp. 376-412, plus plates 50-54 incl.
New Caxton (1969): 'The New Caxton Encyclopaedia' (Caxton Pub. Co., London, 20 Vols.)

NEWTON, R.R. (1977): 'The crime of Claudius Ptolemy' (Johns Hopkins University Press, Baltimore)

NICHOLS, E.L. (1894): 'The galvanometer – a series of lectures' (McIlroy and Emmet, New York)

NITSKE, W.R. (1971): 'The life of Wilhelm Conrad Rontgen – discoverer of the X-ray' (University of Arizona Press, Tucson)

NORBERG, A.L. (1976): 'The origins of the electronics industry on the Pacific coast', *Proc. IEEE,* **64**, pp. 1314-1322

NORTH, J. (1969): 'Sir Norman Lockyer (1836-1920)' *in* Harre, E.: 'Some nineteenth century British scientists' (Pergamon Press, London)

OATLEY, C.W. (1932): 'Wireless receivers' (Methuen, London)

OBERG, E., and JONES, F.D. (1917): 'Machinery's encyclopaedia' (Industrial Press, New York)

O'DEA, W.T. (1939): 'The automatic telephone'. HMSO, London (for Science Museum, London)

Odhams Press (1948): 'How and why it works' (Odhams, London)

OLDHAM, F. (1954): 'Thomas Young – Natural philosopher 1773-1829' (Cambridge University Press)

OLIVER, B.M. (1962): 'Digital display of measurements in instrumentation; *Proc. IRE,* **50**, pp. 1170-1172

OLSON, H.F. (1943): 'Dynamic analogies' (Van Nostrand, London)

OLSON, H.F. (1962): 'Loud speakers', *Proc. IRE,* **50**, pp. 730-737

O'NEILL, H. (1934): 'The hardness of metals and its measurement' (Chapman & Hall, London)

O'NEILL, J.J. (1944): 'Prodigal genius: the life of Nicola Tesla' (Washburn, New York)

ORE, O. (1953): 'Cardano – the gambling scholar' (Princeton University Press)

OSTROFF, E. (1977): 'Photography'. National Museum of History and Technology, Smithsonian Institution Press, Washington, DC

OWEN, D. (1937): 'Alternating current measurements' (Methuen, London)

PAGE, A.W. (1941): 'The Bell Telephone System' (Harper, New York)

PAGE, C.H., and VIGOUREUX, P. (1975): 'The International Bureau of Weights and Measures 1875-1975'. English translation of BIPM Centennial Volume; NBS Special Pub. 420, US Dept. of Commerce, Washington, DC

PAGE, R.M. (1962): 'The origin of radar' (Anchor Book, New York)

PANOFSKY, W., and KLEMM, F. (1970): 'Deutsches Museum, Munich' (Peter-Winkler-Verlag, Munich)

PAPANEK, V. (1975): 'Design for the real world: making to measure' (Thames & Hudson, London)

PARR, G.D.A. (1903): 'Electrical engineering measuring instruments' (Blackie, London)

PASSER, H.C. (1953): 'The electrical manufacturers 1875-1900' (Harvard University Press)

PEARSALL, R. (1974): 'Collecting and restoring scientific instruments' (David & Charles, Newton Abbot, UK)

PEARSON, K. (1914): 'The life, letters and labours of Francis Galton' (Cambridge University Press 4 Vols.)

PENICK, D.B. (1935): 'Direct-current amplifier circuits for use with the electrometer tube; *Rev. Sci. Instrum.,* **6**, pp. 115-120

PEPPER, J.H. (1874): 'Cyclopaedic science simplified' (Frederick Warne, London)

PERCY LUND HUMPHRIES (1924): 'Kelvin centenary oration and addresses commemorative'. (Percy Lund Humphries, London.)

PERRY, C.C., and LISSNER, H.R. (1955): 'The strain gage primer' (McGraw-Hill, New York)

PERRY, J. (1955): 'The story of standards' (Funk & Wagnalls, New York)

PETRIE, W.T. (1934): 'Measures and weights' (Methuen, London)

PFANZAGL, J. (1968): 'Theory of measurement' (Physica Verlag, Wurzburg)

PHILIPS (1957): '300 years of microscopy'. *Professional Profile,* No. 3. pp.22-29. N.V.Philips' Gloeilampenfabrieken, Eindhoven.

Physical Society (1954): 'Rutherford – by those who knew him'. Collection of five lectures, Physical Society, London

PIPPING, G. (1977): 'The Chamber of Physics' (Almqvist & Wiksell, Stockholm)

Pitkin Pictorials (1974): 'The National Maritime Museum'. (Pitkin Pictorials Ltd., London)

PLEDGE, H.T. (1966): 'Science since 1500'. (HMSO, London) 1939 (1st edn.)

PODOLSKY, L. (1962): 'Capacitors', *Proc. IRE*, **50**, pp. 924-928

POHL, R.W. (1930): 'Physical principles of electricity and magnetism'. Translated by W.M. Deans, (Blackie and Sons, London)

POLE, W. (1888): 'The life of Sir William Siemens' (John Murray, London, 3 Vols.)

Post and Telegraphs (1959): 'Post und. Telegraphen Museum – Jubilaumsfuhrer 1889-1959'. Museum of Post and Telegraphs, Vienna

POST, R.C. (1976*a*): '1876-A centennial exhibition'. National Museum of History and Technology, Smithsonian Institution, Washington, DC

POST, R.C. (1976*b*): 'Stray sparks from the induction coil: the Volta Prize and the Page Patent', *Proc. IEEE*, **64**, pp. 1279-1286

POYNTING, J.H. (1894): 'The mean density of the earth' (C. Griffin and Co., London)

PREECE, W.H., and MAIER, J. (1889): 'The telephone' (Whittaker, London)

PRIME, S.I. (1974): 'The life of Samuel F.B. Morse' (Arno Press, New York) Facsimile of 1875 ed. by D. Appleton, New York.

PRIOR, F.J. (1919): 'Operation of trains and station work and telegraph'. F.J.Drake, Chicago 1907 (1st edn.)

PROUT, H.G. (1922): 'The life of George Westinghouse' (Benn Bros., London)

PUGH, J. (1975): 'Calculating machines and instruments'. (Science Museum, London) 1st edn. by D. Baxandall, 1926.

QUILL, H. (1966): 'John Harrison – the man who found longitude' (John Baker, London)

RANDELL, B. (1973): 'The origins of digital computers' (Springer Verlag, Berlin)

RATCLIFFE, J.A. (1929): 'Physical principles of wireless' (Methuen, London)

RATHBURN, J.B. (1929): 'Aeroplane construction, operation and maintenance' (John R. Stanton Co., Chicago)

RAY, W., and M. (1974): 'The art of invention: patent models and their makers' (Pyne Press, Princeton)

Rayleigh, Lord (1943): 'The life of Sir J.J.Thomson' (Cambridge University Press)

READE, L. (1963): 'Marconi and discovery of wireless' (Faber & Faber, London)

REED, L.C. (1903): 'American meter practice' (McGraw-Hill, New York)

REID, J.D. (1879): 'The telegraph in America: its founders, promoters and noted men'. Facsimile reprint, (Arno Press, New York, *c* 1974)

REINGOLD, N. (1966): 'Science in nineteenth century America', (Macmillan, London)

REISER, S.J. (1978): 'Medicine and the reign of technology' (Cambridge University Press)

RETI, L., and DIBNER, B. (1969): 'Leonardo da Vinci – technologist' (Burndy Library, Conn.)

REYNOLDS, O. (1892): 'Memoir of James Prescott Joule'. Manchester Literary and Philosophical Society, Manchester

REYNOLDS, T.S., and BERNSTEIN, T. (1976): 'The damnable alternating current'. *Proc. IEEE*, **64**, pp. 1339-1343

RHODES, F.L. (1929): 'Beginnings of telephony'. Facsimile reprint, (Arno Press, New York, *c*1974)

RICHESON, A.W. (1966): 'English land measuring to 1800' (MIT Press, Cambridge, Mass.)

RICHTER, J.P. (1939): 'The literary works of Leonardo da Vinci compiled and edited from the original manuscripts' (Oxford University Press)

RIDGEWAY, W. (1897): 'The origin of metallic currency and weight standards'. (Cambridge University Press)

RIENITS, R. and T. (1968): 'The voyages of Captain Cook' (Paul Hamlyn, London)

RISDON, P.J. (1924): 'Wireless' (Ward, Lock and Co., London)

RISDON, P.J. (1935): 'Television really explained' (Foulsham's Wireless Guides, London)

ROBINSON, H.W., and ADAMS, W. (1968): 'The Diary of Robert Hooke – for period 1672-1680' (Wykeham Publications, London)

ROBINSON, J. (1822): 'A system of mechanical philosophy'. (John Murray, London, 4 Vols.)

RODERICK, G.W., and STEPHENS, M.D. (1973): 'Scientific and technical education in nineteenth century England: a symposium' (Barnes & Noble, New York)

ROHR, L.O.M. Von (1899): 'Theorie und geschichte des photographischen objectivs'. (Theory and History of the photographic lens) (Springer, Berlin)

ROLT, R.H. (1929): 'Gauges and fine measurements' (Macmillan, London, 2 Vols.)

RONAN, C.A. (1970): 'Edmond Halley – genius in eclipse' (Macdonald, London)

RONAN, C.A. (1975): 'Greenwich Observatory – 300 years of astronomy'. (Times Newspapers Ltd., London)

ROOSEBOOM, M. (1956): 'Microscopium'. Communication No. 95, National Museum for History of Science, Leyden, Netherlands

ROSS, C.C. (1941): 'Measurement in today's schools' (Prentice-Hall, New York)

ROWLAND, J. (1957): 'Ernest Rutherford: atom pioneer'. (Philosophical, New York)

ROWLAND, J. (1963): 'The radar man – the story of Sir Robert Watson-Watt' (Lutterworth Press, London)

Royal Air Force (1934): 'Flying training manual, part 1 flying instruction'. Air Pub. 129. (HMSO, London)

Royal Society (1924): 'Phases of modern science' (A. & F. Denny, London)

RUITER, J.H., and MURPHY, R.G. (1961): 'Basic industrial electronic controls' (Holt, Rinehart and Winston, New York)

RUSH, P., and O'KEEFE, J. (1964): 'Weights and measures'. (Roy, New York)

RUSSELL, B. (1931): 'The scientific outlook' (W.W.Norton and Co., New York)

SANDERS, L. (1947): 'A short history of weighing'. W. & T. Avery Limited, Soho Foundry, Birmingham, England. (Revised and republished in 1960)

SAS, R.K., and PIDDUCK, F.B. (1947): 'The metre-kilogram-second system of electrical units' (Methuen, London)

SAWYER, R.A. (1945): 'Experimental spectroscopy' (Chapman & Hall, London)

SCHAEDEL, C. (1973): 'Men and machines of the Australian Flying Corps 1914-1919'. Kookaburra Tech. Pubs., Dandenong, Australia

SCHARFFENBERG, H. and WENDEL, R. (1976): 'A brief historical review of microfilming', *Jena Rev.,* **1**, p.4

SCHIAVO, G.E. (1958): 'Antonio Meucci – inventor of the telephone'. (Vigo Press, New York)

SCHIERBEEK, A. (1959): 'Measuring the invisible world – the life and works of Antoni van Leeuwenhoek F.R.S.' (Abelard-Schuman, London)

SCHLESINGER, K., and RAMBERG, E.G., (1962): 'Beam-deflection and photo devices', *Proc. IRE,* **50**, pp. 991-1005

SCHOFIELD, R.E. (1966): 'A scientific autobiography of Joseph Priestley (1733-1804)' (MIT Press, Cambridge, Mass.)

Scientific American (1955): 'Automatic control' (G.Bell & Sons, London)

SCOTT, J.F. (1952): 'The scientific work of Rene Descartes 1596-1650' (Taylor & Francis, London)

SCOTT, N.R. (1960): 'Analog and digital computer technology' (McGraw-Hill, New York)

SCROGGIE, M.G. (1945): 'Radio laboratory handbook'. Office of Wireless World, (Iliffe and Sons, London)

SCROGGIE, M.G. (1952): 'Television' (Blackie & Son, London) 1935 (1st edn.)

SEED, J.R. (1973): 'National measurement system study'. *NSCL Newsletter,* **13**, pp. 19-21

SERRELL, R., ASTRAHAN, M.M., PATTERSON, G.W., and PYNE, I.B., (1962): 'The evolution of computing machines and systems', *Proc. IRE,* **50**, *pp. 1039-1058*

SHANNON, C.E. and WEAVER, W. (1949): 'The mathematical theory of communication' (University of Illinois Press, USA)

SHERCLIFF, J.A. (1962): 'The theory of electromagnetic flow-measurement' (Cambridge University Press)

SHERWOOD TAYLOR, F. (1936): 'The world of science' (Heinemann, London)

SHERWOOD TAYLOR, F. (1945): 'Science, past and present' (Willam Heineman, London)

SHERWOOD TAYLOR, F. (1950): 'Medieval scientific instruments', *Discovery,* Vol. XI, pp. 282-287

SHERWOOD TAYLOR, F. (1972): 'The teaching of the physical sciences at the end of the eighteenth century' *in* FERGUSON, A. 'Natural philosophy through the eighteenth century', (Taylor & Francis, London) 1948 (1st edn.)

SHIERS, G. (1972): 'Bibliography of the history of electronics' (Scarecrow Press, Metuchen, N.J.)

SHIERS, G. (1977): 'Historical notes on television before 1900' *SMPTE J.*, **86**, pp. 129-137

SIDGWICK, J.B. (1953): 'William Herschel – explorer of the heavens'. (Faber & Faber, London)

SIME, J. (1900): 'William Herschel and his work' (Scribner, New York)

SIMER, S.B. (1936): 'Portable hearing aid', *Radio News & Short Wave Radio, (New York),* August, **92**, pp. 106-107

SINCLAIR, D.B. (1962): 'The measuring devices of electronics', *Proc. IRE,* **50**, pp. 1164-1172

SINGER, C., HOLMYARD, E.J., and HALL, A.R. (1954): 'A history of technology'. (Oxford Univ. Press, Oxford) Vol. 1. – 'From early times to fall of Ancient Empires'. (Later vols. may also be of interest.)

SINGER, C., HOLMYARD, E.J., HALL, A.R., and WILLIAMS, T.I. (1956): 'A history of technology'. Vol. II. – 'The Mediterranean civilizations and the Middle Ages, *c*700 BC to *c*AD1500' (Oxford University Press, Oxford.)

SIVOWITCH, E.N. (1970): 'A technological survey of broadcasting's "Pre-history", 1876-1920', *J. Broadcasting,* XV, pp. 1-20

SKINNER, F.G. (1967): 'Weights and measures: their ancient origins and their development in Great Britain up to AD1855' (HMSO, London)

SLOANE, T. O'CONOR, (1894): 'The standard electrical dictionary' (Crosby Lockwood, London)

SMILES, S. (1879): 'Industrial biography: iron workers and tool makers' (Murray, London)

SMITH, D.J. (1940): 'Newnes complete engineer' (George Newnes, London, 4 Vols.) Plus data sheet set.

SMITH, J.W., and BURR, W.H. (1930): 'Memoir of Clemens Hershel', *Am. Soc. Civil Engrs.,* LVI, pt.2

Smithsonian (1968): 'Exhibits in the Museum of History and Technology'. Smithsonian Publication 4720, (Smithsonian Institution Press, Washington, DC)

SMITH-ROSE, R.L. (1948): 'James Clerk-Maxwell, F.R.S. 1831-1879' (Longmans Green, London)

SOKAL, M.M., DAVIS, A.B., and MERZBACH, U.C., (1976): 'Laboratory instruments in the history of psychology'. *J. History of Behavioural Sci.,* **12**, pp. 59-64

SONNENBERG, G.J. (1951): 'Radar and electronic navigation' (George Newnes, London)

SOOTIN, H. (1954): 'Michael Faraday' (Blackie, London)

SPENCER-JONES, H. (1972): 'Astronomy through the eighteenth century', in 'Natural philosophy through the eighteenth century'. A.Ferguson, (Taylor & Francis, London) (1st edn. 1948).

SPERRY, H.M. (1913): 'Electric interlocking handbook'. (General Railway Signal Co., Rochester)

SQUIER, G.O. (1933): 'Telling the world' (Williams & Wilkins, Baltimore)

STANLEY, W.F. (1901): 'Surveying and levelling instruments' (E & F Spon, London) (1st edn. 1890)

STEIN, P.K. (1964): 'Measurement engineering', Stein Engng. Services, Phoenix, USA

STEIN, P. (1967): 'Thirty years of bonded resistance strain gauges, 1938-1968. From magic to science through discovery of uncontrolled variables'. Pub. No. 11, Measurement Engineering Laboratory, Arizona State University, USA

STEIN, P.K. (1970*a*): 'The role of latent information in information processing in measuring systems', *Shock & Vibration Bull.,* **41**, pp. 81-107

STEIN, P.K. (1970*b*): 'Sensors as information processors', *Res. Dev.,* June, pp. 33-40

STILL, A. (1944): 'Soul of Amber: the background of electrical science' (Murray Hill Books, New York)

STILL, E.W., (1957): 'Into thin air'. Normalair Ltd. Yoevil, England

STINE, W.M. (1900): 'Photometrical measurements' (Macmillan, New York)

St James' and St. Martin's (1976): 'The blue book'. (St James' Press, London and St. Martin's Press, New York)

STOCK, J.T. (1969): 'Development of the chemical balance'. (HMSO, London)

Stockholm Museum (1976): 'Telecommunications Museum', *Tele,* **2**, pp. 9-16

STONER, J. (1962): 'The book of my life'. Translation of 'De Vita Propia Liber', by Jerome Cardan. (Dover Publications, London)

STRANDH, S. and GOTTLIEB, J. (1975): 'Telegrafen, telefonen'. Tekniska Museet, Stockholm.

STRONG, J. (1942): 'Modern physical laboratory practice' (Blackie, London) 1938 (1st edn.)

STRUTT, R.J. (1968): 'Life of John William Strutt – third Baron Rayliegh' (University of Wisconsin, Press)

STURGEON, W. (1837): 'Annals of electricity, magnetism and chemistry – Vol. 1' p.1837

SUMMER, W. (1957): 'Photosensitors – a treatise on photo-electric devices and their application to industry' (Chapman & Hall, London)

SUSSKIND, C. (1976): 'American contributions to electronics: coming of age and some more', *Proc. IEEE,* **64**, pp. 1300-1305

SUTTON, H. (1890): 'Tele-photography' *The Telegraphic J. & Electr. Rev.,* **27**, pp. 549-551

SVOBODA, A. (1948): 'Computing mechanisms and linkages' (McGraw-Hill, New York)

SWAN, K.R. (1948): 'Sir Joseph Swan' (Longmans Green, London)

SWAN, M.E., and K.R. (1929): 'Sir Joseph Wilson Swan F.R.S.' (Ernest Benn, London)

SWEENEY, R.J. (1953): 'Measurement technique in mechanical engineering'. (Wiley, New York)

SWINYARD, W.O. (1962): 'The development of the art of radio receiving from the 1920s to the present', *Proc. IRE,* **50**, pp. 793-798

SYDENHAM, P.H. (1974): 'Maxtrix TV – a solid-state picture transmission system', *Electron. Today Int.* (Sydney edition). October, pp. 24-30

SYDENHAM, P.H. (1975): 'Radio – the true pioneers', *ibid.*, Part 1, March, pp. 22-27; Part 2, April, pp.16-21

SYDENHAM, P. H. (1976): 'The instrument science discipline and its curricula', *J. Phys. E.,* **9**, p.230

SYDENHAM, P.H. (1977): 'M & C History series – I Source guide to measurement technology material, *Meas. & Control,* **10**, pp. 257-265

SYDENHAM, P.H. (1978*a*): 'M & C History series – 2 Source guide to history of measurement technology material'. *ibid.*, **11**, pp. 103-116

SYDENHAM, P.H. (1978*b*): 'Early geophysical practice – the BMR Instrument collection', *BMR J. Austral. Geol. & Geophys.,* **2**, pp. 241-248

TAKAHASI, H. (1952): 'Generalized theory of thermal fluctuations', *J. Phys. Soc. Jpn.,* **7**, pp. 439-446

TAYLOR, F. SHERWOOD (1955): 'An illustrated history of science' (Heineman, London) see also Sherwood Taylor

TERMAN, F.E. (1976): 'A brief history of electrical engineering education', *Proc. IEEE,* **64**, pp. 1399-1407

TERRILL, H.M. and ULREY, C.T., (1930): 'X-ray technology – the production, measurement and applications of X-rays' (Chapman & Hall, London)

Tesla Museum (1956): 'Nikola Tesla – lectures, patents, articles'. Nikola Tesla Museum, Beogard, Yugoslavia

THODAY, A.G. (1971): 'Astronomy. 2: astronomical telescopes' (HMSO, London)

THOMPSON, J.S., and THOMPSON H.G. (1920): 'Silvanus Phillips Thompson – his life and letters' (T. Fisher Unwin, London)

THOMPSON, S.P. (1883): 'Philipp Reis – inventor of the telephone'. Facsimile reprint, (Arno Press, New York, *c*1974)

THOMPSON, S.P. (1898): 'Michael Faraday – his life and work' (Cassell & Co., London)

THOMPSON, S.P. (1910): 'Life of William Thomson, Baron Kelvin of Largs', (MacMillan, London 2 Vols.)

THOMSON, G.P. (1955): 'The foreseeable future' (Cambridge University Press)

THOMSON, G.P. (1964): 'J.J. Thomson – and the Cavendish Laboratory in his day' (Nelson, London)

THOMSON, W., and TAIT, P.G. (1883): 'Treatise on natural philosophy' (Cambridge University Press) (Many versions were printed beginning 1863; Thomson also listed as Lord Kelvin)

THORPE, J.F.T. (1937): 'Thorpe's dictionary of applied chemistry' (Longmans Green, London)

THORPE, T.E. (1906): 'Joseph Priestley' (J.M.Dent, London)

THRELFALL, R. (1898): 'On laboratory arts' (MacMillan, London)

THRELFALL, R., and POLLOCK, J.A. (1899): 'On the quartz thread gravity balance', *Philos. Trans. A.*, **193**, pp. 215-258

TILTMAN, R.F. (1933): 'Baird of television: the life story of John Logie Baird'. Facsimile reprint, (Arno Press, New York)

TIMBS, J. (1860): 'Stories of inventors and discoverers in science and the useful arts' (Kent, London)

TIMOSHENKO, S. (1953): 'History of strength of materials' (McGraw-Hill, New York)

TORD HALL, (1970): 'Carl Friedrich Gauss' (MIT Press, Cambridge, Mass.)

TORREY, H.C., and WHITMER, C.A. (1948): 'Crystal rectifiers' (McGraw-Hill, New York)

TRASK, M. (1971): 'The story of cybernetics'. (Studio Vista, London)

TREWMAN, H.F. (1949): 'Electronics in the factory' (Isaac Pitman, London)

TUCKER, D.G. (1970): 'The invention of frequency modulation in 1902:, *Radio & Electron. Eng.*, **40**, pp. 33-37

TUNZELMANN, G.W. DE (1890): 'Electricity in modern life' (Walter Scott, London)

TUPHOLME, C.H.S. (1942): 'Modern engineering' (Faber & Faber, London)

TURNBULL, H.W. (1959): 'The correspondence of Isaac Newton, – vol. 1', Others followed (Cambridge University Press)

TURNER, G.L'E. (1973): 'Van Marum's scientific instruments in Teyler's Museum'. (Noordhoff Int. Pub., Leyden)

TYNDALL, J. (1863): 'Heat – a mode of motion' (Longmans Green, London) Later editions published

TYNDALL, J. (1870): 'Faraday as a discoverer' (Longmans Green, London) (Reprinted T.Y.Crowell, New York, 1961)

TYNE, G.F.J. (1977): 'Saga of the vacuum tube' (H.W.Sams, Indianapolis)

Union Switch (1928): 'Electro-pneumatic interlocking'. Union Switch and Signal Co., Swissvale, Pa.

University of Utrecht (1977): 'NG200 Natuurkundig Gezelschap te Utrecht 1777-1977'. University of Utrecht, the Netherlands

University of Utrecht (undated): 'Electrical instruments of the nineteenth century in the University Museum'. University of Utrecht Museum, The Netherlands

UNWIN, D.L. (1974): 'The start of the scientific instruments explosion'. Internal Publication, Cambridge Instrument Co., Cambridge, England

URBANITZKY, A.R. VON (1886): 'Electricity in the service of man', English language version edited by R. Wormell, Cassell, London. Original version of later, revised, MullineuxWalmsley (1910)

US Army (1967): 'The Billings microscope collection of the Medical Museum of Armed Forces Institute of Pathology'. The American Registry of Pathology, Washington, DC

VALLEY, G.E., and WALLMAN, H. (1948): 'Vacuum tube amplifiers' (McGraw-Hill, New York)

VAN HEURCK, H.I. (1893): 'The microscope – its construction and management' (Crosby, Lockwood & Sons, London)

VERMAN, L.C. (1973): 'Standardisation: a new discipline' (Archon Books, Hamden)

Vienna Technical Museum (1974): 'Technisches Museum fur Industrie und Gewerbe in Wein', Published by that Museum

VOGEL, W.P. Jr. (1949): 'Precision, people and progress' (Leeds & Northrup Co., Philadelphia)

VOLTA, A. (1800): 'On the electricity excited by the mere contact of conducting substances of different kinds', *Philos. Mag. London*, **7**, pp. 289-311

VON WALTERHAUSEN, W.S. (1856): 'Gauss – a memorial'. S. Hirzel, Leipzig. Translated in 1966 by H.W. Gauss, Colorado, USA
VON WEIHER, S. (1975): 'Werner von Siemens – a life in the service of science, technology and industry', Musterschmidt, Gottingen, Zurich
VON WEIHER, S., and GEOTZELER, H. (1977): 'The Siemens company – its historical role in the progress of electrical engineering'. Siemens Aktiengesellschaft, Berlin
WALKER, E. (1866): 'Terrestrial and cosmical magnetism'. (Adams prize essay for 1865) (Deighton, Bell & Co., Cambridge)
WALKER, G.W. (1913): 'Modern seismology' (Longmans Green, London)
WALKER, J. (1887): 'The theory and use of a physical balance' (Clarendon Press, Oxford)
WALKER, R.C., and LANCE, T.M.C. (1933): 'Photoelectric cell applications' (Isaac Pitman, London)
WARD, F.A.B. (1966): 'Descriptive catalogue of the collection illustrating time measurement'. (HMSO, London)
WARD, F.A.B. (1970): 'Time measurement' (Science Museum, London) 1936 (1st edn.)
WARD, F.A.B. (1973): 'Clocks and watches 1: weight-driven clocks'. (HMSO, London)
WARD LOCK (1955): 'The wonder book of aircraft' (Ward, Lock & Co., London)
WARNER, D.J. (1968): 'Alvan Clark and Sons – Artists in Optics', (Smithsonian Institution Press, Washington, D.C.) (US National Museum Bull. 274)
WARTNABY, J. (1957): 'Seismology', (HMSO, London)
WARTNABY, J. (1968): 'Surveying: instruments and methods'. (HMSO, London)
WASSERMAN, P. (1973): 'Museum media'. (Gale Research Co., Detroit)
WATSON, C. (1969): 'William Thomson, Lord Kelvin (1824-1907)', *in* HARRE, R. 'Some nineteenth century British scientists'. (Pergamon Press, London)
WATSON, R.D. and J.M. (1977): 'The great Melbourne Telescope'. *Aust. Physicist,* **14**, pp. 182-185
WATSON, T.A. (1913): 'The birth and babyhood of the telephone'. American Telephone and Telegraph Co. Convention of Telephone Pioneers of America, Chicago.
WATSON-WATT, R.A. (1959): 'The pulse of radar; the autobiography of Sir Robert Watson Watt' (Dial, New York)
WAX, N. (1954): 'Selected papers on noise and stochastic processes', (Dover, New York)
WEBER, R.L. (1939): 'Manual of heat and temperature measurement' (Ann Arbor, Michigan)
WEBSTER, (1970): 'Webster's biographical dictionary' (G. & C. Merriam Co., Springfield, Mass.)
WEIHE, V.I. (1962): 'Fifty years in aeronautical navigational electronics', *Proc. IRE,* **50**, pp. 658-663
WELLS, D.A. (1856): 'Science popularly explained: the principles of natural and physical science' (W. Kent, London)
WESTAWAY, F.W. (1937): 'Scientific method: its philosophical basis and its modes of application' (Hillman-Curl, New York)
WESTFALL, R.S. (1978): 'The construction of modern science' (Cambridge University Press)
Westinghouse (1920): 'Handbook of Westinghouse watthour meters'. Westinghouse Electric and Manufacturing Co., Pub. 5150-B., East Pittsburgh, Pa.
Westinghouse (1921): 'Handbook of Westinghouse indicating instruments'. Westinghouse Electric and Manufacturing Co., Pub. 5245., Newark, N.J.
WHIPPLE, R.S. (1931): 'A brief history of the London makers of scientific instruments', *in* 'London and the advancement of science'. (British Association, London)
WHIPPLE, R.S. (1972): 'Scientific instruments in the eighteenth century', *in* 'Natural philosophy through the eighteenth century' (A.Ferguson, Taylor & Francis, London) 1948 (1st edn.)
WHITE, F.A. (1961): 'American industrial research laboratories'. Public Affairs Press, Washington, DC (or Ph.D. Dissertation, Univ. Microfilm, Ann Arbor 59-3296)
WHITE, F.S. (1827): 'A history of inventions and discoveries: Alphabetically arranged' (C. & J. Rivington, London)
WHITE, W.C. (1962): 'Early history of industrial electronics', *Proc. IRE,* **50**, pp. 1129-1135
WHITEHEAD, E.S. (1939): 'A short account of the life and work of John Joseph Fahie' (Hodder

& Stoughton, London)
WHITEHEAD, S. (1951): 'British achievements in electrical measuring instruments'. Proceedings of the Joint Conference, Institutions of Civil, of Mechanical, and of Electrical Engineers, London, pp. 473-490
WHITEHEAD, T.N. (1954): 'The design and use of instruments and accurate mechanism – underlying principles' (Dover, New York) 1933 (1st edn.)
WHYTE, A.G. (1930): 'Forty years of electrical progress – the story of G.E.C.' (Ernest Benn, London)
WILE, F.W. (1926): 'Emile Berliner: maker of the microphone'. Facsimile reprint, (Arno Press, New York, *c*1974)
WILLIAMS, A.J. (1973): 'Bits of recorder history'. *J. Dynamic Systems, Measurement & Control, Trans. ASME,* 95, series G, 1, pp.6-16
WILLIAMS, T.I. (1969): 'A biographical dictionary of scientists' (Adam & Charles Black, London) (also Wiley). (Halsted Press, New York, 1974, 2nd edn.)
WILLIAMS, W.E. (1930): 'Applications of interferometry' (Methuen, London)
WILLOUGHBY, S. (1891): 'The rise and extension of submarine telegraphy'. Facsimile reprint, (Arno Press, New York, *c*1974)
WILSON, G. (1851): 'The life of the Hon. Henry Cavendish'. Cavendish Society, London.
WOLF, A. (1935): 'A history of science, technology and philosophy in the sixteenth and seventeenth centuries' (Allen & Unwin, London) (Revised 1950 by same publisher)
WOLF, A. (1938): 'A history of science, technology and philosophy in the eighteenth century'. (Allen & Unwin, London) (Revised by D. McKie, 1952, same publisher)
WOOD, W.P., and CORK, J.M. (1941): 'Pyrometry' (McGraw-Hill, New York)
WOODBURY, D.O. (1944): 'Beloved scientist: Elihu Thompson, a guiding spirit of the elctrical age' (McGraw-Hill, New York)
WOODBURY, D.O. (1949): 'A measure for greatness' (McGraw-Hill, New York)
WOODWARD, C.D. (1972): 'BSI: the story of standards'. British Standards Institution, London
WOOLHOUSE, W.S.B. (1856): 'The measures, weights and moneys of all nations: and an analysis of the Christian, Hebrew and Mahometan Calendars' (John Weale, London)
WYNTER, H., and TURNER, A. (1975): 'Scientific instruments'. (Studio Vista, London) also Scribner, New York, 1976
YOST, E. (1962): 'Modern Americans in science and technology' (Dodd, Mead & Co., New York) 1941 (1st edn.)
YOUNG, A.P. (1948): 'Lord Kelvin' (Longmans Green, London)
ZEEMAN, P. (1913): 'Researches in magneto-optics – with special reference to the magnetic resolution of spectrum lines' (Macmillan, London)
ZUPKO, R.E. (1968): 'A dictionary of English weights and measures from Anglo Saxon Times to the nineteenth century' (University Press, Wisconsin)
ZWORYKIN, V.K., and WILSON, E.D., (1934): 'Photocells and their applications' (Wiley, New York)

Appendix 1

Biographies relevant to instrument history

AIRY, George Biddell (1801-1892) ... Airy (1896)
AMPERE, Andre Marie (1775-1836) ... Gardiner and Gardiner (1965)
ANGUS, Donald J. (1887-1966) ... Hounshell (1973)
ARMSTRONG, Edwin Howard (1890-1954) ... Lessing (1969)

BABBAGE, Charles (1791-1871) ... Dubbey (1978)
BAIRD, John Logie (1888-1946) ... Tiltman (1933)
BELL, Alexander Graham (1847-1922) ... Mackenzie (1928), Bruce (1973)
BERLINER, Emile (1851-1929) ... Wile (1926)
BLACK, Joseph (1728-1799) ... Brougham (1855)
BOOLE, George (1815-1864) ... Barry (1969)
BOYLE, Robert (1627-1691) ... Crowther (1960)
BRAGG, William Henry (1862-1942) ... Grant (1952), Caroe (1978)
BRAHE, Tycho (1596-1601) ... Dreyer (1890), Gade (1947)
BREWSTER, David (1781-1868) ... Gordon (1869)
BRUSH, Charles Francis (1849-1929) ... Hounshell (1973)

CARDANO, Gerolamo (1501-1576) ... Ore (1953), Stoner (1962)
CAVENDISH, Henry (1731-1810) ... Wilson (1851), Brougham (1855), Berry (1960)
CLARK, Alvan (1804-1887) ... Warner (1968)
CLERK-MAXWELL, James (1831-1879) ... Campbell and Garnett (1884), Smith-Rose (1948), May (1964), Everitt (1974)
COPERNICUS, Nicolaus (1473-1973) ... Linderman Library (1973)
COULOMB, Charles Augustin (1736-1806) ... Gillmor (1971)
CROMPTON, Rookes Evelyn Bell (1845-1940) ... Crompton (1928), Johnson and Randell (1946), Bowers (1969)
CROOKES, William (1832-1919) ... Fournier (1923)

DE FOREST, Lee (1876-1961) . . . . Carneal (1930)
DESCARTES, Rene (1596-1650) . . . . Scott (1952)
DREBBEL, Cornelius (1572-1633) . . . . Harris (1961)

EDISON, Thomas Alva (1847-1931) . . . . Dickson and Dickson (1894), Jones (1907), Dyer and Martin (1910), Garbedian (1940), Josephson (1961), Clark (1977)
ERATOSTHENES of Cyrene (*c*273-*c*192 BC) . . . . Fraser (1971)
ERICSSON, John (1803-1889) . . . . Church (1890), Iles (1912)

FAHIE, John Joseph (1846-1934) . . . . Whitehead (1939)
FARADAY, Michael (1791-1867) . . . . Bence-Jones (1869), Tyndall (1870), Thompson (1898), Sootin (1954), King (1973)
FARNSWORTH, Philo T. (1907-1971) . . . . Everson (1949)
FERRANTI, Sebastian Ziani de (1864-1930) . . . . De Ferranti and Ince (1956)
FESSENDEN, Reginald Aubrey (1866-1932) . . . . Fessenden (1940)
FOURIER, Jean Baptiste Joseph (1768-1830) . . . . Herivel (1975)
FRANKLIN, Benjamin (1706-1790) . . . . Franklin (1818), Franklin (1850), Goodman (1956), Donovan (1963), Cohen (1975)

GALILEI, Galileo (1564-1642) . . . . Fahie (1903), Brodrick (1964), Geymonat (1965), Langford (1966), McMullin (1967)
GALTON, Francis (1822-1910) . . . . Pearson (1914), Morgan (1969), Forrest (1974)
GALVANI, Luigi (1737-1798) . . . . Dibner (1971)
GAUSS, Carl Friedrich (1777-1855) . . . . Von Walterhausen (1856), Dunnington (1955), Tord Hall (1970)
GOETHE, Johann Wolfgang von (1749-1832) . . . . Magnus (1949)
GREGORY, Richard Arman (1864-1952) . . . . Armytage (1957)

HAHN, Otto (1879-1968) . . . . Ley (1967)
HALES, Stephen (1677-1760) . . . . Clark-Kennedy (1929)
HALLEY, Edmond (1656-1742) . . . . Armitage (1966*a*), Ronan (1970)
HARRISON, John (1693-1776) . . . . Quill (1966)
HEAVISIDE, Oliver (1850-1925) . . . . Lee (1947)
HELMHOLTZ, Hermann Ludwig Ferdinand von (1821-1894) . . . McKendrick (1899), Koenigsberger (1906)
HENRY, Joseph (1797-1878) . . . . Coulson (1950)
HERSCHEL, Clemens (1842-1930) . . . . Kent (1912), Smith and Burr (1930)
HERSCHEL, William (1738-1822) . . . . Holden (1881), Sime (1900), Sidgwick (1953), Armitage (1962), Hoskin (1963)
HERTZ, Heinrich Rudolf (1857-1894) . . . . Borth (1895)
HOOKE, Robert (1635-1703) . . . . 'Espinasse (1956), Crowther (1960), Keynes (1960), Robinson and Adams (1968)

HOPKINSON, John (1849-1898) . . . . Hopkinson (1901), Greig (1970)
HUYGENS, Christian (1629-1695) . . . . Bell (1950)

JOHANSSON, C.E. (1864-1943) . . . . Althin (1948)
JOULE, James Prescott (1818-1889) . . . . Reynolds (1892)

KEPLER, Johannes (1571-1630) . . . . Armitage (1966*b*)
LEEUWENHOEK, Antoni van (1632-1723) . . . . . . Dobell (1932), Schierbeek (1959)
LEONARDO DA VINCI, (1452-1519) . . . . MacCurdy (1938), Hart (1961), Reti and Dibner (1969)
LOCKYER, Norman (1836-1920) . . . . . Lockyer and Lockyer (1928), North (1969)
LODGE, Oliver Joseph (1851-1940) . . . . Lodge (1931), Hill (1932)

MARCONI, Guglielmo (1874-1937) . . . . . . Jacot and Collier (1935), Dunlap (1937) Marconi (1962), Reade (1963), Jolly (1972) Geddes (1974)
MARUM, Martinus van (1750-1837) . . . . Forbes (1969)
MEUCCI, Antonio (1808-1889) . . . . Schiavo (1958)
MICHELSON, Albert Abraham (1852-1931) . . . . Jaffe (1961), Reingold (1966)
MILLIKAN, Robert Andrews (1868-1953) . . . . Millikan (1951)
MILNE, John (1850-1913) . . . . Herbert-Gustar and Nott (1974)
MORSE, Samuel Finley Breese (1791-1872) . . . . Iles (1912), Morse (1914), Maybee (1943), Prime (1974)
MOSELEY, Henry Gwyn Jeffreys (1887-1915) . . . . Heilbron (1974)

NEWTON, Isaac (1642-1727) . . . . Brodetsky (1927), More (1934), Turnbull (1959), Crowther (1960), Manuel (1968)

OERSTED, Hans Christian (1777-1851) . . . . Danish Journal (1977)

PREECE, William Henry (1834-1913) . . . . Baker (1976)
PRICE, William Gunn (1853-1928) . . . . Frazier (1967)
PRIESTLEY, Joseph (1733-1804) . . . . Thorpe (1906), Schofield (1966)
PTOLEMY (PTOLEMAEUS), Claudius (138-180 AD) . . . . Newton (1977)

RAYLIEGH, Lord (1842-1919) . . . . Strutt (1968)
REIS, Philipp (1834-1874) . . . . Thompson (1883)
RONTGEN, Wilhelm Conrad (1845-1923) . . . . Glasser (1933), Nitscke (1971)
RUTHERFORD, Ernest (1871-1937) . . . . Eve (1939), Feather (1940), Physical Society (1954), Rowland (1957)

SIEMENS, Charles William (1823-1883) . . . . Pole (188)
SIEMENS, Ernst Werner von (1816-1892) . . . . Von Weiher (1975)
SPERRY, Elmer (1860-1930) . . . . Hughes (1971)
SWAN, Joseph Wilson (1828-1914) . . . . Swan (1929), Swan (1948)

TAIT, Peter Guthrie (1831-1901) . . . . Knott (1911)

TALBOT, William Henry Fox (1800-1877) . . . . Jammes (1973)

TESLA, Nicola (1856-1943) . . . . Martin (1894), O'Neill (1944), Tesla Museum (1956), Hunt and Draper (1964)

THOMPSON, Silvanus Phillips (1851-1916) . . . . . . Thompson and Thompson (1920)

THOMPSON, Elihu (1853-1937) . . . . Woodbury (1944)

THOMPSON, Joseph John (1856-1940) . . . . Rayliegh (1943), Thomson (1964)

THOMSON, William (1824-1907) . . . . Thompson (1910), Percy Lund Humphries (1924), King (1925), Casson (1929), Swan (1948), Young (1948), Watson (1969)

TYNDALL, John (1820-1893) . . . . Eve and Creasey (1945)

VENTURI, Giovanni Battista (1746-1822) . . . . Kent (1912)

VOLTA, Allessandro (1745-1827) . . . . Dibner (1964)

WATSON-WATT, Robert A. (1892-) . . . . Watson-Watt (1959), Page (1962), Rowland (1963)

WATT, James (1736-1819) . . . . Dickinson and Vowles (1949)

WESTINGHOUSE, George (1846-1914) . . . . Prout (1922)

WESTON, Edward (1850-1936) . . . . Woodbury (1949)

WHEATSTONE, Charles (1802-1875) . . . . Bowers (1975)

WREN, Christopher (1632-1723) . . . . Crowther (1960)

YOUNG, Thomas (1773-1829) . . . . Oldham (1954)

ELECTRICAL . . . . Mottelay (1922), Butler (1947), Morgan (1953), Dibner (1954), Higgins (1961), Williams (1969), Hounshell (1973, Mabon (1975)

ELECTRONICS . . . . Shiers (1972), Mabon (1975)

ENGINEERS in 1850-1930 period . . . . Ewing (1933)

GENERAL BIOGRAPHICAL DICTIONARIES . . . . Matthews (1968), Webster (1970), McGraw-Hill (1973), Louis (1974), St. James and St. Martins (1976)

GEOPHYSICISTS of 19th Century, USA (Lazzaroni) . . . . Miller *et al* (1972)

SCIENTISTS . . . . Williams (1969)

SCIENTISTS of USA in 1800-1950 period . . . . Yost (1962), Mabon (1975)

TECHNOLOGISTS of USA in period 1800-1950 . . . . . . Yost (1962), Mabon (1975)

ROYAL INSTITUTION – Scientists and Engineers . . . . Martin (1961)

ROYAL SOCIETY . . . . Andrade (1960), Hartley (1960)

TELEGRAPH in America . . . . Reid (1879)

WIRELESS PIONEERS . . . . Lodge (1900), Hawks (1927), Dunlap (1944)

Appendix 2

Collections containing instruments

An instrument is the realisation of a physical principle, captured in isolation, applied to some task of measurement. Its existence portrays the art of the possible in the presence of available materials, skills and knowledge of fundamental principles then existing. Like many of man's constructions these, too, record valuable information about the past.

This guide provides an introduction to the availability of instrument artefacts, where they are and how to contact the Institution concerned. Throughout the text reference has been made to some of these sources. Presented here are brief, personal accounts of some of the collections, most of which have been visited in the course of this study. As instruments have traditionally been seen more as part of some form of situation, rather than as artefacts in their own right, they are often to be found scattered over a very wide range of Institutions. Where possible, for not all museum staff see that publications of their holdings are necessary, reference is given to booklets and more substantial guides that can be purchased or, if out of print, at least consulted.

Highlights of some museums are given as an indication of the inventory held. No satisfactory form of documentation exists that enables a search to be made to locate specific items. It is a matter of laboriously contacting likely places by correspondence, purchasing photographs, colour projection slides and similar material. This service is generally available from the larger museums for reasonable charges.

Rarely does a collection feature 'sensing' as the prime theme. Usually, it is necessary to conduct a total inspection of available exhibits on display to establish where a sensor was involved. For example, to study material on temperature sensing in the collections of the Science Museum, London, it was necessary to investigate a range of exhibits – from fire-fighting (bimetallic alarm sensing) through meteorology (radiosonde packages) to specific displays, such as that on heat. To complicate matters, the current museum-display trend is toward 'in-everyday-context' displays, based on lifestyle reconstructions on the historical development of a major social facility (broadcasting, for example). Given scarce and limited display space, which is the usual case in museums, this makes for fewer exhibitions of collections of sameness.

Established museums invariably possess vastly greater stocks of artefacts than can be exhibited. Serious, in-depth study will usually require help by the curators concerned (who are, from experience, keen to be of assistance). They can establish a wider degree of availability from stock book entries and catalogues held by museums.

This list is an updated version of one previously published, Sydenham (1977). That account has been extended to include several museums not visited. Entries are arranged according to geographical region. The importance of making prior arrangements for visits concerned with specific artefacts or people cannot be over-stressed. Some museums are only opened by appointment. The opening times given were correct at the time of writing, but, of course, may subsequently have been altered.

Continental (European) collections

Deutsches Museum
8 Munchen. 26 Postfach, West Germany
(Open daily 0900-1700, except certain religious holidays)

In 1880 Oskar von Miller, inspired by the collections of the Science Museum, South Kensington, London, and the Conservatoire des Arts et Metiers, Paris, began a campaign to build a superb science and technology museum in Germany. In 1903 he took positive action and it was opened in 1925. During 1944-1945 its fabric was virtually totally destroyed by war, but by 1970, 90% of it was restored to its original state.

In layout and style it resembles the organisation of the Science Museum, London – many working exhibits of machines, demonstrations, dioramas, collections of likeness.

Being a general science and technology museum, it has many measuring devices and other instruments spread throughout the collection. Notable artefacts and displays include Hulsmeyer's 1904 'Radar' ideas; ultrasonic depth sounders of 1937 vintage; some of Ohm's original apparatus; many early galvanometers (Nadel, 1830, for example) and other early electrostatic and electromagnetic detectors. Equipment used by Ampere to investigate the relationship between electricity and magnetism, is also available.

Other original apparatus include the Hertz electromagnetic wave experimental equipment and Fedderson's photographic method used to demonstrate that the current in a spark-gap oscillates when fed from an inductor-capacitor source.

In the field of telecommunications the museum possesses an original 1790 speaking machine of Wolfgang van Kempelen and von Sommering's 1810 electrochemical telegraph (Fig. 4.16). A small history-of-telephony section includes a model of Reis's 1861 apparatus and an original of Bell's 1876 device. Notable early tape and wire recorders and television devices are also on display.

The Chemistry display includes the original apparatus (Fig. 2.34), used to demonstrate nuclear fission (Otto Hahn and Fritz Strassmann, 1938). This most sobering display of quite rudimentary electronics gives a vivid idea of what can be achieved with little equipment. Overall, this collection is outstanding, containing so many original works of relevance.

The museum had three levels of guide available – a pamphlet in German, a small-size but extensive guide book, Deutsches Museum (1971) (four languages available), and a hard-cover descriptive work, Panofsky and Klem (1970). An extensive guide was also published in 1925, Matschoss (1925), but this is not in print today.

This museum has a large research staff. Serious enquiries should be addressed to the curators for further help.

Ability to read German would be a distinct advantage during a visit, for notices are only given in that language. An extensive range of postcards and slides are available for purchase.

Museum of Astronomy of the Observatory of Paris
61 Avenue de l'Observatoire
75014, Paris, France

This was established in 1667 and today contains astronomical instruments of the 16th to 19th centuries.

Conservatoire National des Arts et Metiers, Musee National Des Techniques
292, Rue Saint-Martin
75141 Paris. Cedex 03., France.
(Open daily 1200-1745, Sun. 1000-1730)

Here are held many important technological artefacts of French origin, for example, Pascal's calculator and the laboratory of Lavoisier. Catalogues are available on calculating instruments (A), mechanics (B), physical apparatus (GA), geodesy and photogrammetry (H), weights and measures (K), automata (Z), the laboratory of Lavoisier and the history of the Academie des Sciences.

Technisches Museum fur Industrie und Gewerke
(Technical Museum for Industry, Crafts and Trades, Vienna)
Mariahilfer Strasse 212, Vienna XIV, Austria.
(Open daily 0900-1300/1600)

Incorporated are the Museum of Posts and Telegraphs and Museum of Austrian Railways. Its origins go back to several early 19th-century collections which were brought together in 1913.

This museum is by no means as extensive as the Deutsches Museum but does contain much important historic equipment on three floors devoted to scientific and technological equipment.

Display is very much as static collections of sameness. Modern trends have not caught up. Many areas are as they were put down many years ago. This can be a virtue for research.

Recently devised is a display of the historic development of indicating instruments. This is important because it includes many makes seldom seen in the Western museums (Tarsa, Schuckert, Tauber, Egger, Bruchner and Nadir).

Of particular note is the 1729 Braun calculating machine of superb mechanical manufacture. The Schreibautomat, by Knaus (1760) is a magnificent automaton to be seen here.

This museum features many optical lens developments, the first calculated 4-element Petval lens, 1870, being notable.

The original aim of this museum was to foster public awareness of Austrian 'firsts' in technology. It, therefore, contains many original artefacts related to indigenous development.

The collection of telegraphic and telephonic apparatus is most extensive. The Posts and Telegraphs exhibitions cover the whole gamut of the theme with about one-sixth being devoted to documented hardware – the rest relates to postage, transport and more. In the Railway Museum (Signalling Section) there is a very large, but poorly arranged, collection of hardware relating to early telegraphy and signalling.

An outstanding exhibit in this museum is a recently completed history wall of data processing. Inspired by the original IBM version (see USA), Prof. H. Zemanek of IBM, Vienna, assembled a history of data processing composed of photographs, old documents and artefacts. A pictorial guide to this was not available, the documentation at the time of a visit being a bibliography, Chroust (1974).

A small guide book is available in the German language, Vienna Technical Museum (1974). A 10-page English language supplement is added giving maps and a key to some 150 important technological artefacts. Posts and Telegraphs Museum description was available, Posts and Telegraphs (1959). There is no bookshop or stall, and no cards or slides were offered for sale.

Museum voor de Geschiedenis van de Wetenschappen
(History of Science Museum)
Kon Meerstraat, Ghent, Belgium.
(Opening hours restricted to two afternoons per week)

This collection contains a small number of well documented and well presented electrostatic instrument devices – generators, meters and discharge arrangements. The museum was instituted by the Ghent University and illustrates the work of many pioneer scientists and inventors (to quote a tourist brochure).

Teyler's Museum
Damstraat 21
Haarlem, The Netherlands
(Open daily)

Around 1790, M. Van Marum put together an unusually extensive collection of physical apparatus and instruments. Today this is housed on display in Haarlem. The artefacts on display appear to be in two sections, those acquired under Van Marum's direction and an equally large collection of apparatus postdating his period.

Those of Van Marum have been extensively catalogued, Turner (1973). This catalogue is profusely illustrated and contains a long bibliography on scientific instruments of the 18th century.

The second part of the collection contains apparatus of the electromagnetic era which began around 1800. The collection is extensive but is, apparently, not all catalogued. Arrangement is in classical glass cases classified by sameness of application. A brief pamphlet on the museum is available in the Dutch language.

University of Utrecht Museum
8 Trans
Utrecht, The Netherlands
(Open weekdays 1000-1700, Saturdays 1400-1700)

In 1919 Dr. van Cittert rediscovered in an attic a large collection of physical instruments from the Physical Society and the Academy. Cataloguing began in 1923 and today a large part of the collection is catalogued – electrical instruments, Laven and van Cittert-Eymers (1967), University of Utrecht (undated); mechanical instruments, Bos (1968). A descriptive account of the history of the University display also exists, University of Utrecht (1977).

This collection includes the original metric standards introduced in Holland, the surviving original Leeuwenhoek microscope and Huygen lenses, one of which was that used to discover Saturn. Utrecht University has many important associations with history of science.

Institute for Theoretical Geodesy
University of Bonn
17 Nuss-Allee, Bonn 53
West Germany
(Visit by appointment)

A large collection of geodetic and survey instruments is held. They form part of teaching and have been assembled as now-obsolete apparatus purchased when new as teaching aids. They are not catalogued. It contains instruments from earliest to current times.

Technological University of Delft
Delft, The Netherlands
(Visit by appointment in both cases)

Department of Electrical Engineering
The basement of the building is used as a store and display for a collection of thermionic devices. Some tens of thousands are displayed. Another section displays thermionic laboratory equipment purchased or made in the University dating from the early 19th century. Higher-level student project apparatus of considerable extent – such as full working electrical mechanical telephone exchange, are also stored there. These date from 1900 onward. No catalogues are published but an inventory of valves and measuring instruments exists.

Department of Applied Physics
A small display of 19th and early 20th century scientific instruments is displayed in a foyer.

Continental (Eastern Europe) collections

Narondi Technicke Muzeum
(Narondi Technical Museum)
Prague, Czechoslovakia
(Open each day, exact times unknown)

This museum is recently built and provides a show-piece for people of Prague. Its displays are not extensive, the museum being at an early growth-stage.

A fine display on photography is available. The collections are small and have very little that is not found in western museums. It is striking that the artefacts are mainly of British and US origin – Harrison's clock, Bell and Hughes' telephones. Kelvin instruments, Davy lamps.

There were no (in 1974) guide books, cards or slides available and there was no help available for visitors not speaking the local language.

Scandanavia

Tekniska Museet Med Telemuseum
(National Museum of Science and Technology)
N. Djurgarden, Museivagen 7
Stockholm, Sweden
(Open daily 1000-1600, Sat. and Sun. 1200-1600)

A section of this museum is for teletechnique – the use of telegraphy and telephony. It is especially well stocked and displayed. The collection was begun in 1853. In 1937 an inadequate museum was opened in Central Stockholm. This was moved to its current site in 1975. A description of the current display in the English language is available, Stockholm Museum (1976). More details of the museum's history were published by Malmgren (1975). Considerable account and contemporary etchings of early Swedish telegraphy and telephone equipment is given in a facsimile reprint from the Telemuseum, Strandh and Gottlieb (1975).

Royal Swedish Academy of Sciences
Stockholm, Sweden.

A collection, not exhibited, that has grown since it was begun around 1898. Pipping (1977) describes the collection.

United Kingdom (Provincial)

Aberdeen University – Department of Natural Philosophy
(Visit by appointment)

This museum was created in 1973 by Prof. R.V. Jones using a large room to display artefacts connected historically with the University (Telescopes predominate). It also serves to teach students through physical implementation of early discovered physical principles (mainly by Clerk-Maxwell).

Avery Historical Museum
W. & T. Avery Pty. Ltd., Smethwick,
Warley, Warwickshire
(Visit by appointment)

During a long involvement in weighing-device manufacture, W. & T. Avery have assembled a most comprehensive and extensive collection of artefacts related specifically to mechanical weighing. It does not include devices having output transduced into non-visual control signal forms. Its value to this survey is the embodied wealth of ideas in mechanical design. The collection is housed in the Soho Foundry (of Boulton) building.

No guide, as such, has been published but Sanders (1947) used illustrations of artefacts from the collection.

Birmingham Museum of Science and Industry
Newhall Street, Birmingham
Warwickshire
(Open 1000-1700 most days; religious holidays – closed)

This museum serves a large city and is a development of the nearby Museum and Art Gallery. It was formed in 1950 to preserve examples of engines and machine tools which would be representative of good engineering practice. A modern-science section is included.

From the sensing technology point of view there is little to see. The science section has on display eight turn-of-the-century common indicating instruments – Kelvin ampere balance (1890), electrostatic voltmeter (1899), various hot-wire meters and ampere meters of Ayrton.

No guide book or catalogue was available, the best being a one-sheet pamphlet guide listing notable artefacts. Additional information is available on some items through the Museum's Museograph Series.

The Cambridge Instrument Co. Ltd.
Melbourn, Royston
Hertfordshire
(Visit by appointment with Assistant Company Secretary)

This company has a long association (it was founded in 1880) with scientific measuring instruments, and illustrated historical accounts were compiled on its 50th and 75th anniversaries, Cambridge Instrument Company (1945), Cambridge Instrument Company (1955). The most recent historical report is a very brief study of the company, written in 1974, Unwin (1974).

The company maintains a collection of many of its early products (in storage). An internal catalogue is available that lists each item, its serial number, year of manufacture and the subsidiary that made it. The majority of the archives are now held in the Whipple Museum, Cambridge.

Cambridge University – Whipple Museum of History of Science
Free School Lane, Cambridge
(Visit by appointment)

This collection comprises historic scientific instruments relating to the 16-19th centuries. It was not visited, being closed for alterations at the time.

Glasgow University – Department of Physics
(Visit by appointment)

This University has important connections with Lord Kelvin. Many of Kelvin's commerical products (*c*1880-1910) are displayed in cabinets of a lecture room. A specific museum has been created, Green and Lloyd (1970).

Oxford University – Museum of History of Science
Old Ashmolean Building, Broad Street, Oxford
(Weekdays 1030-1300, 1430-1600)

This collection has its beginnings around 1680 when the building was erected as one of the first museums. It houses several collections of scientific instruments, created by individuals.

Modern (post 1900) sensors are little represented, the bulk of the measuring equipment being mechanical and classical-optical designs. Important electrical pieces include Lindeman's electrometer (1924), thermopiles, induction coils, Wollaston's thimble voltaic cell, early X-ray tubes, frictional electricity sets and early indicating instruments.

No handbook is available but one was produced in the past, Gunther (1935). This 160-page guide describes many of the important pieces. A later work by Gunther (1939) adds further detail by referring to sections of the various Annual Reports published by the museum up to 1939.

Many other tertiary institutions possess collections but they are seldom cared for adequately, there usually being no continuing institutional mechanism to ensure continuity when the devotee is no longer available to take an interest.

United Kingdom, London

London possesses a wide range of museums (120 are listed!) BAC (1977). Of these, instrument technology is to be seen in about four places.

National Maritime Museum
Romney Road, Greenwich, SE10
(Open 1000-1800, Sundays 1430-1800. Closes 1700 in Winter)

A variety of measuring instruments used in navigation and astronomy is to be seen in this museum. It is located at the above address and in the nearby Old Royal Observatory in Greenwich Park. The collection relates mainly to pre-19th century history. Guides published include a short-form version of the Observatory by Howse (1973) and a larger book Ronan (1975). Pictorial guides of the National Maritime Museum are available, Carr (1967), Pitkin (1974).

Science Museum
Exhibition Road, South Kensington, SW7
(Open 1000-1800, except Sundays 1430-1800)

This museum rates in the three largest world history of science and technology institutions.

Five large galleries house collections arranged in many ways – on themes, on sameness, on requirement and on physical variable. Galleries most useful to this study were fire-fighting appliances, meteorology, time, map-making and surveying, astronomy, electric power, weighing and measuring, photography, atomic physics, navigation, magnetism and electricity, X-rays, heat, optics, geophysics, telecommunications, telegraphy and radio.

A list, compiled by the writer, of 'modern' sensors to be seen contained hundreds of artefacts spread across the above galleries. What is on show is but part of the total available. The exhibitions of the development of radio, telegraphy, telephony and television use innumerable artefacts to illustrate changes in time.

Originals include a 1922 VTVM, Wheatstone's equipment, Marconi apparatus, Fleming valves, Froud ship-roll recorder of 1872, Braun 1916 gyrocompass, Baird's TV of 1925, Tyndall's heat apparatus, Callendar's recorders (1899), Boys' radiometers (1886) and Babbage's calculating engines.

The Science Museum publishes many books, HMSO (1970). They range from tutorial works for schools and light interest to serious studies. Four inexpensive short guides were available but no extensive guide work exists. There are, however, some specialist handbooks on measurement, trade cards, chemical balances, X-rays and heat. A wide range of coloured postcards includes several instruments. Copies of photographs are obtainable, selection being made from a large catalogue held in the Enquiry Office.

Royal Institution – Faraday Museum
21 Albermarle Street, London W1
(Visit by appointment)

The Royal Institution began as a venture to teach use of scientific ways in everyday life. Faraday began his career there around 1804 and served the RI until his death in 1867. One of the rooms he worked in – his magnet laboratory – has been restored as a museum. It contains his workbooks, experimental apparatus and related equipment. A guide book has been published, King (1973).

IBA Television Gallery
70 Brompton Road, London SW3
(Visit by appointment)

This permanent exhibition covers the entire field of historical and modern television. The displays include a few models and originals of early video-sensing methods.

National Physical Laboratory Museum
Teddington, Middlesex
(Visit by appointment)

In 1978 the finishing touches were made to enable the NPL to declare its museum open. It contains apparatus from past NPL work.

United States of America

Hindle (1966) is a survey guide to technology in early America. It contains a short *Directory of artefact collections* by Lucius Ellsworth. Using this Direcory, collections clearly relating to sensing technology interests are to be found in the Henry Ford Museum, Smithsonian Institution, Franklin Institute, Museum of New York, Western Union Telegraph Museum, Bell Telephone Laboratories, US Army Signal Corps Museum, Chicago Museum and the IBM Exhibition Center in New York.

Wasserman (1973) is a valuable guide to media materials available from US museums. Hudson and Nicholls (1974) is a primary source of 26 000 museums (world wide) through which the enthusiastic reader can seek more detail. A listing concerned specifically with public and private depositories associated with electrical technology and science has been compiled by Belfield (1977). It also includes Canadian material.

Smithsonian Institution – Museum of History and Technology
Washington, DC
(Open 1000-1750 daily)

In 1835 an Englishman, James Smithson, left a bequest to the United States' people that was to be used for '... an establishment for the increase and diffusion of knowledge among men'. The final outcome was this Institution, Karp (1965), which operates several extensive museum collections as well as effective research groups. The National Museum of History and Technology, one of the Museums, was relocated in its present modern building in 1964.

This is the third of the three best science and technological museums in the world.

The display naturally has a good proportion of early US designs, such as the Morse telegraph apparatus but as there was relatively low indigenous activity in the 19th century, the bulk of the collections of early artefacts is of European origin. There is a strong emphasis on 19th and early 20th-century artefacts of measurement, presumably because of greater availability.

Being little more than a decade old, all displays were current and stimulating. Another most significant feature of this Institution is the markedly great research effort taking place behind the public facade. In sensing technology, many highly trained research workers have published museums' reports on both specific and general history of a variety of measurement devices. Air-conditioned, highly-organised storage systems house the 90% of the collection not on display.

Many relevant and well produced catalogues have been published, examples being on meteorological instruments (130 pp., about 200 photographs), and feedback mechanisms (similar). These provide source material for very serious study.

A comprehensive, colour-illustrated guide book (128 pp.) is available, Smithsonian (1968). The Museum houses a large bookshop specialising in material relevant to the themes contained in the Museum.

Another museum of the Smithsonian Institution complex that is of partial relevance is the National Air and Space Museum. While concentrating on space and aircraft, it does serve as a minor source of flight instrumentation.

The Air-traffic Control Exhibition shows the advances in hardware technique made in each decade from the 1920s to the 1960s. No documentation was available on this.

IBM Exhibit Center
Madison Avenue and 57th Street, New York City.
(Open during business hours)

This Center displays changing exhibitions featuring a wide range of scientific themes. Highly relevant to a sensing technology interest is *A computer perspective.* This history wall, using six, room-high panels, shows the development that led to the modern computer. The display consists of densely-packed photographs with the occasional presence of an actual artefact. It begins in Babbage's time (early 19th century) with descriptions of calculating devices. Instruments are intertwined to show their relevance to the modern calculator. The value of the exhibit is the wealth of fact and the great source of ideas involved. The advertising consultants who produced the IBM display prepared an illustrated guide, Eames and Eames (1973), which presents the hundreds of photographs and captions of the display along with their sources of reference.

National Bureau of Standards Museum
Office of Information Activities
National Bureau of Standards
Gaithersburg, Maryland
(Visit by appointment)

This museum was dedicated in 1966 being 'designed to preserve and display apparatus and other memorabilia illustrative of the past scientific work of NBS . . .'

Over 500 items are held, many relating to measurement. These are gradually being catalogued, see Mason (1977), with the eventual creation of an 'Archive of the NBS museum'.

Australia

Bureau of Mineral Resources, Geology and Geophysics
Department of National Development
P.O. Box 378, Canberra City
A.C.T. 2601
(Visit by appointment)

A collection of early 20th century geophysical apparatus is held in store (at a considerable distance from the above situation). It has been catalogued, Sydenham (1978*b*).

Index